Abrupt Climatic Change

Evidence and Implications

NATO ASI Series

Advanced Science Institutes Series

A Series presenting the results of activities sponsored by the NATO Science Committee, which aims at the dissemination of advanced scientific and technological knowledge, with a view to strengthening links between scientific communities.

The series is published by an international board of publishers in conjunction with the NATO Scientific Affairs Division

A Life Sciences B Physics	Plenum Publishing Corporation London and New York
C Mathematical and Physical Sciences	D. Reidel Publishing Company Dordrecht, Boston, Lancaster and Tokyo
D Behavioural and Social Sciences E Applied Sciences	Martinus Nijhoff Publishers Dordrecht, Boston and Lancaster
F Computer and Systems Sciences G Ecological Sciences H Cell Biology	Springer-Verlag Berlin, Heidelberg, New York, London, Paris, and Tokyo

Series C: Mathematical and Physical Sciences Vol. 216

Abrupt Climatic Change

Evidence and Implications

edited by

W. H. Berger

Scripps Institution of Oceanography,
La Jolla, California, U.S.A.

and

L. D. Labeyrie

Centre de Faibles Radioactivités,
Gif-sur-Yvette, France

D. Reidel Publishing Company

Dordrecht / Boston / Lancaster / Tokyo

Published in cooperation with NATO Scientific Affairs Division

Proceedings of the NATO Advanced Research Workshop on
Abrupt Climatic Change - Evidence and Implications
St. Hugues de Biviers, France
16-22 October 1985

Library of Congress Cataloging in Publication Data

NATO Advanced Research Workshop on Abrupt Climatic Change (1985: St. Hugues de Biviers, France)
Abrupt climatic change.

(NATO ASI series. Series C, Mathematical and physical sciences; vol. 216)
"Proceedings of the NATO Advanced Research Workshop on Abrupt Climatic Change, St. Hugues de Biviers, France, October 1985"–T.p. verso.
Bibliography: p.
Includes index.
1. Climatic changes–Congresses. 2. Paleoclimatology–Congresses. 3. Glacial epoch–Congresses. I. Berger, Wolfgang H. II. Labeyrie, L. D. (Laurent D.), 1946– . III. Title. IV. Series: NATO ASI series. Series C, Mathematical and physical sciences; no. 216.
QC981.8.C5N38 1985 551.6 87–20780
ISBN 90–277–2604–3

Published by D. Reidel Publishing Company
P.O. Box 17, 3300 AA Dordrecht, Holland

Sold and distributed in the U.S.A. and Canada
by Kluwer Academic Publishers,
101 Philip Drive, Assinippi Park, Norwell, MA 02061, U.S.A.

In all other countries, sold and distributed
by Kluwer Academic Publishers Group,
P.O. Box 322, 3300 AH Dordrecht, Holland

D. Reidel Publishing Company is a member of the Kluwer Academic Publishers Group

Printed in The Netherlands

Mer-de-Glace, in a mid-19th century etching as published in *The Forms of Water*, by John Tyndall, 1889, compared with present-day condition (photo J.F.B. Mitchell, Oct 20, 1985). Note decrease of ice volume within the last century.

TABLE OF CONTENTS

PREFACE

Our motivation for calling a conference on climatic change was to stimulate interdisciplinary exchange between researchers in various fields concerned with climatic research, which include meteorology, oceanography, limnology, palynology, glaciology, dendrochronology, and climate modeling. The philosophy behind this attempt at cross-fertilization is much the same as that of previous conferences on climatic change in the NATO-series ("Climatic variations and variability, facts and theories", Berger, 1981; "Milankovitch and climate", Berger et al., 1984).

The past is the key to the range of future possibilities. It is for this reason that we stressed history and case studies in convening the present symposium on abrupt climatic change, without however forgetting that modeling, conceptual and mathematical, ultimately provides the understanding necessary for prediction. We attempted to strike a balance between these complementary aspects of current climatological research, aiming at a symposium situated in scope about halfway between "Climate in Earth history" (Berger and Crowell, 1982) and "Climate processes and climate sensitivity" (Hansen and Takahashi, 1984). We concentrated, therefore, on the last 20,000 years, where the time-scales are potentially good enough to document rapid change.

We should perhaps explain what phenomenon the phrase "abrupt climatic change" is supposed to describe. An objective definition is possible through mathematical statistics (Goossens and Berger, this volume). Less technically, "abrupt change" denotes a rupture in the established range of experience, that is, a surprisingly fast transition from one state to another. Experience, of course, is a defined quantity, which has an arbitrary start or an arbitrary window length along a series of climatic states, which may be described in terms of means, variability, trends, or even the frequency of unusual events such as El Niños. "Surprise" likewise is an arbitrary quantity, and depends on where one draws the line between the expected and the improbable. Thus, "abrupt change" is whatever one chooses to make it: the important criterion is the usefulness of the concept in communication and in focussing research.

Our interest in abrupt change derives from the expectation of finding hitherto hidden but potentially important mechanisms within the climate machine, performing at the extremes of the range. Such mechanisms may be expected to derive from two kinds of processes: strong positive feedback (e.g., albedo, CO_2) and transient geochemical reservoirs (e.g., ice, fresh-water bodies, labile carbon stores). The hope of finding the one dominant factor in producing a given abrupt change may be misleading, of course. As John Imbrie (this volume) reminds us in discussing the case of glacial-interglacial transitions, it may be exactly the arithmetic addition of a plethora of different factors which produces the rapid change. If so, such a period of abrupt change is no more instructive (or even less so) than the longer series within which it resides.

As it turned out, the last glacial-interglacial transition proved to be the major topic of the conference, with about one half of the contributions addressing the evidence for rapid climatic fluctuations in this interval, and the question regarding underlying mechanisms. Other topics were: the last millenium, the Holocene, the Pleistocene, and climate modeling. A summary of workshop findings, and the abstracts of the conference on abrupt climatic change were published shortly after the conference (Berger and Labeyrie, 1986).

The conference, and its theme of "abrupt climatic change" was conceived by Andre Berger, who also offered much appreciated advice in setting goals and priorities. Other colleagues who offered advice on topics and potential speakers were T. Barnett, W. Broecker, T. Crowley, W. Dansgaard, J.-C Duplessy, H. Flohn, J. Imbrie, J. Mangerud, S. Schneider, and E. Vincent. Their help was crucial in the initial planning stage. D. Raynaud took charge of the local logistics: his efforts played a leading role in getting the conference going, and making the meeting a most pleasant experience.

We met in St. Hugues de Biviers, near Grenoble, in a former monastery situated among vinyards at the foot of the immense north wall of the ice-carved valley of the Isere, in October, 1985. Our special thanks go to L. Lorius, director, and the members of the Laboratoire de Glaciologie et de Géophysique de l'Environnement. We are grateful to J. Roquemora, M. Poinot, F. Rapetto, and A.M. de Bois Remond, for the assistance they rendered during the conference. Camera-ready copy was produced at S.I.O., with the able assistance of Susan Burke, Malana Tabak and Heike Schmidt. Funding was by NATO and by the U.S. National Science Foundation (ATM 85-04150).

The Editors

References

Berger, A. (ed), 1981. *Climatic variations and variability: facts and theories*, Reidel, Dordrecht, 795pp.

Berger, A., et al. (eds) 1984. *Milankovitch and climate*, 2 vols., Reidel, Dordrecht, vol. 1, 510pp.; vol. 2, 895pp.

Berger, W.H., Crowell, J.C. (eds), 1982. *Climate in Earth history. Studies in Geophysics*, National Academy Press, Washington, D.C., 198pp.

Berger, W.H. and Labeyrie, L.D. (eds), 1985. *The book of abstracts and reports from the conference on abrupt climatic change*, SIO Reference 86-8, University of California, Scripps Institution of Oceanography, La Jolla, 274pp.

Goossens, C. and Berger, A., 1987. How to recognize an abrupt climatic change?, this volume.

Hansen, J.E. and Takahashi, T. (eds), 1984. *Climate processes and climate sensitivity*, Geophys. Monogr. 29, Amer. Geophys. Union, 368pp.

Imbrie, J., 1987. Abrupt terminations of Late Pleistocene ice ages: A simple Milankovitch explanation, this volume.

Dryas octopetala

LIST OF AUTHORS AND PARTICIPANTS

M. Arnold, Centre des Faibles Radioactivités, Laboratoire mixte, CNRS-CEA, P.O. Box 1, F.91190 Gif sur Yvette, France

E. Bard, Centre des Faibles Radioactivités, Laboratoire mixte, CNRS-CEA, P.O. Box 1, F.91190 Gif sur Yvette, France

J.L. de Beaulieu, Laboratoire de Botanique historique et Palynologie, UA CNRS n° 1152, Faculté des Sciences et Technique St. Jérôme, 13397 Marseille Cedex 13, France

André Berger, Institut d'Astronomie et de Géophysique G. Lemaitre, Université Catholique de Louvain, 2 Chemin du Cyclotron, 1348 Louvain-la-Neuve, Belgium

W.H. Berger, Scripps Institution of Oceanography, University of California, San Diego, La Jolla, CA 92093, U.S.A.

Wolfgang Bludau, Botanisches Institut, Universität Stuttgart-Hohenheim, Garbenstr. 30, D-7000 Stuttgart 70, F.R.G.

R. Bonneville, Laboratoire de Géologie du Quaternaire, Case 907, C.N.R.S., Faculté des Sciences de Luminy, 13288 Marseille Cedex 9, France

Raymond S. Bradley, Geology & Geography Department, University of Massachusetts, Amherst, MA 01003-0026, U.S.A.

S.K. Burke, Scripps Institution of Oceanography, University of California, San Diego, La Jolla, CA 92093, U.S.A.

Paul A. Colinvaux, Departments of Zoology and Anthropology and Byrd Polar Research Center, The Ohio State University, Columbus, OH 43210, USA

Edward R. Cook, Tree-Ring Laboratory, Lamont-Doherty Geological Observatory of Columbia University, Palisades, NY 10964, U.S.A.

G.R. Coope, Department of Geological Sciences, University of Birmingham, PO Box 363, Birmingham B15 2TT, England

Richard G. Craig, Department of Geology, Kent State University, Kent, OH 44242, U.S.A.

H.N. Dalfes, TÜBITAK/TBAE, Yer Bilimleri Bölümü, P.O. Box 74, 41410 Gebze, Kocaeli, Turkey

W. Dansgaard, Geophysical Institute, Haraldsgade 6, DK-2200 Copenhagen N, Denmark

Robert J. Delmas, Laboratoire de Glaciologie et Géophysique de l'Environnement B.P. 96 38402, St. Martin D'Heres Cedex, France

Anicet Dirk Dubois, University of Antwerp (RUCA), Department of Botany, Groenenborgerlaan 171, B, 2020 Antwerp, Belgium

J.C. Duplessy, Centre des Faibles Radioactivités, Laboratoire mixte CNRS-CEA, P.O. Box 1, F.91190 Gif sur Yvette, France

J. Duprat, Département de Géologie et Océanographie, Université de Bordeaux, 1 Avenue des Facultés, F.33405 Talence, France

G.P. Eberli, Department of Geology, University of Florida, Gainesville, Florida 32611, U.S.A.

Hermann Flohn, University of Bonn, F.R.G.

B. Frenzel, Botanisches Institut, Universität Stuttgart-Hohenheim, Garbenstr. 30, D-7000 Stuttgart 70, F.R.G.

C. Genthon, Israel, Dept. Physico-Chimie, Cen-Saclay BP n° 2, 91190 Gif sur Yvette, France

Glenn A. Goodfriend, Isotope Department, The Weizmann Institute of Science, 76100 Rehovot, Israel

C. Goossens, Institut d'Astronomie et de Géophysique G. Lemaître , Université Catholique de Louvain, 2 Chemin du Cyclotron, 1348 Louvain-la-Neuve, Belgium

J. Guiot, Laboratoire de Botanique historique et Palynologie, UA CNRS n° 1152, Faculté des Sciences et Techniques St. Jérôme, 13397 Marseille Cedex 13, France

Jane Hsuing, Department of Earth, Atmospheric and Planetary Sciences, Room 54-1522, Massachusetts Institute of Technology, Cambridge, MA 02139, USA

John Imbrie, Department of Geological Sciences, Brown University, Providence, RI 02912-1846, U.S.A.

Gordon C. Jacoby, Tree-Ring Laboratory, Lamont-Doherty Geological Observatory of Columbia University, Palisades, NY 10964, U.S.A.

Eystein Jansen, Department of Geology Sect. B, University of Bergen, Allégaten 41, N-5007 Bergen, Norway

P.D. Jones, Climatic Research Unit, School of Environmental Sciences, University of East Anglia, Norwich, NR4 7TJ, U.K.

S. Joussaume, Laboratoire de Météorologie Dynamique, 24. rue Lhomond, 75231 Paris Cedex 05, France 2

Jean Jouzel, Laboratoire de Géochimie Isotopique/LODYC, (UA CNRS 1206) CEA/IRDI/DESICP, D.P.C., B.P. 2, 91191 Gif sur Yvette Cedex, France

Lloyd Keigwin, Woods Hole Oceanographic Institution, Woods Hole, MA 02543, U.S.A.

L.D. Labeyrie, Centre des Faibles Radioactivités, Laboratoire Mixte C.N.R.S.-C.E.A., 91190 Gif-sur-Yvette, France

M. Labracherie, Université Bordeaux, Lab. Oceanologie et Sédimentologie, 1 Av. des Facultés, 33405 Talence, France

Michel Legrand, Laboratoire de Glaciologie et Géophysique de l'Environnement B.P. 96 38402, St. Martin D'Heres Cedex, France

Kam-biu Liu, Department of Geography & Anthropology, Louisiana State University, Baton Rouge, LA 70803, U.S.A.

Claude Lorius, Laboratoire de Glaciologie et Géophysique, de l'Environnement B.P. 96 38402, St. Martin D'Heres Cedex, France

Mordeckai Magaritz, Isotope Research Department, The Weizmann Institute of Science, 76100 Rehovot, Israel

Jan Mangerud, University of Bergen, Department of Geology, Sec. B, Allegt 41, N-5007 Bergen, Norway

Peter Mayes, Tree-Ring Laboratory, Lamont-Doherty Geological Observatory of Columbia University, Palisades, NY 10964, U.S.A.

Andrew McIntyre, Lamont-Doherty Geological Observatory of Columbia University, Palisades, NY 10964, U.S.A.

J.A. McKenzie, Department of Geology, University of Florida, Gainesville, FL 32611, U.S.A.

Liliane Merlivat, Laboratoire de Géochimie Isotopique/LODYC, (UA CNRS 1206) CEA/IRDI/DESICP, D.P.C, B.P. 2, 91191 Gif Sur Yvette Cedex, France

J.F.B. Mitchell, Meteorological Office, Bracknell, Berkshire, RG12 2SZ, England

Ellen Mosley-Thompson, Institute of Polar Studies, The Ohio State University, Columbus, OH 43210, U.S.A.

J. Moyes, Département de Géologie et Océanographie, Université de Bordeaux, 1 Avenue des Facultés, F. 33405 Talence, France

Reginald E. Newell, Department of Earth, Atmospheric and Planetary Sciences, Room 54-1522, Massachusetts Institute of Technology, Cambridge, MA 02139, U.S.A.

Gerald R. North, Department of Meteorology, Texas A & M University, College Station, TX 77843, U.S.A.

Jonathan T. Overpeck, Lamont-Doherty Geological Observatory, Palisades, N.Y. 10964, U.S.A.

Jürgen Pätzold, Geologisch-Palaontologisch Institut und Museum, Christian-Albrechts-Universität, Olshaussenstrasse 40/60, D-2300 Kiel, West Germany

A. Perrott, School of Geography, Mansfield Road, Oxford, OX1 3TB, England

Dorothy M. Peteet, NASA/Goddard Space Flight Center, Goddard Institute for Space Studies, 2880 Broadway, New York, NY 10025, U.S.A.

Jean-Robert Petit, Laboratoire de Glaciologie et Géophysique de l'Environnement B.P. 96 38402, St. Martin D'Heres Cedex, France

A. Pons, Laboratoire de Botanique historique et Palynologie, UA CNRS n° 1152, Faculté des Sciences et Technique St. Jérôme, 13397 Marseille Cedex 13, France

S.C.B. Raper, Climatic Research Unit, School of Environmental Sciences, University of East Anglia, Norwich NR4 7TJ, U.K.

D. Raynaud, Laboratoire de Glaciologie et Géophysique de l'Environnement, 2 Rue Molieres BP n°. 96 , 38402 St. Martin-D'Heres Cedex, France

M. Reille, Laboratoire de Botanique historique et Palynologie, UA CNRS n° 1152, Faculté des Sciences et Technique St. Jérôme, 13397 Marseille Cedex 13, France

Pierre Rognon, Department of Earth Sciences, Universite Pierre et Marie Curie, 4 Place Jussieu, 75252 Paris Cedex 05, France

Michael Sarnthein, Geological-Palaeontological Institute, Christian Albrecht University, Olshausenstr. 40, 2300 Kiel, F.R.G.

Stephen H. Schneider, National Center for Atmospheric Research, P.O. Box 3000, Boulder, CO 80307, U.S.A.

C. Schurrmans, K.N.M.I. Postbus 201, 3730-AR, de Bilt, Holland

Guy Seret, U.C.L., Institute de Géologie, Laboratoire Paléontologie et Paléogéographie, Place L. Pasteur, 3, B-1238 Louvain-la-Neuve, Belgium

F. Serre-Bachet, Laboratoire de Botanique historique et Palynologie, UA 1152 du CNRS, Faculté des Sciences et Techniques, St. Jérôme, 13397 Marseille Cedex 13, France

F.A. Street-Perrott, Tropical Palaeoenvironments Res. Group, School of Geography, Mansfield Road, Oxford, OX1 3TB, England

L. Tessier, Laboratoire Botanique historique, Rue H. Poincarré, Unité Associée au C.N.R.S., n° 1152, Faculté des Sciences et Techniques St. Jérôme, 13397 Marseille Cedex 13, France

Jörn Thiede, Geological-Palaeontological Institute, Christian Albrecht University, Olshausenstr. 40, 2300 Kiel, F.R.G.

Hans Thierstein, Geological Institute ETH-Zentrum, CH - 8092 Zurich, Switzerland

Lonnie G. Thompson, Institute of Polar Studies, The Ohio State University, Columbus, OH 43210, U.S.A.

Jean-Louis Turon, Université Bordeaux, Département de Géologie et Océanographie, 1 Av. des Facultés, 33405 Talence, France

Colette Vergnaud-Grazzini, Laboratoire de Géologie Dynamique, Tour 24, Université Pierre et Marie Curie, 4 Place Jussieu, 75230 Paris Cedex 05, France

Edith Vincent, Département de Géologie Dynamique, Université Pierre et Marie Curie, 4 Place Jussieu, 75005 Paris, France

T.M.L. Wigley, Climatic Research Unit, School of Environmental Sciences, University of East Anglia, Norwich NR4 7TJ, U.K.

Kyaw Winn, Geological-Palaeontological Institute, Christian Albrecht University, Olshausenstr. 40, 2300 Kiel, F.R.G.

R. Zahn, Geological-Palaeontological Institute, Christian Albrecht University, Olshausenstr. 40, 2300 Kiel, F.R.G.

Part I.

Introduction

ABRUPT CLIMATIC CHANGE - AN INTRODUCTION

W.H. Berger[1] and L.D. Labeyrie[2]

[1]Scripps Institution of Oceanography
University of California, San Diego
La Jolla California 92093

[2]Centre des Faibles Radioactivites
Laboratoire Mixte C.N.R.S.-C.E.A
91190 Gif sur Yvette - France

IMPACT OF CLIMATIC CHANGE

History is rich in examples of the impact of climatic change on the well-being of human populations, and the various possible connections between climatic variability, food shortages, epidemic diseases, and large-scale migrations have generated much discussion in recent years (Ladurie, 1971; Lamb, 1977, 1982; Schneider, 1976; Rotberg and Rabb, 1981). Famine due to bad harvests resulting from bad weather is a phenomenon as old as agriculture itself. However, with the rapid growth of populations since the industrial revolution, and the evolution of a world food and energy market, famines now affect more people and more nations than ever before. Thus, the vicissitudes of climate have begun to loom large in world affairs.

A substantial proportion of all people who lived during the last 1,000 years are alive today. Whereas for tens of thousands of years the human population levied but a minor tax on the ecologic support systems of the planet, we now find a totally different situation. We find the greater part of suitable land areas in economic use, with prairies turned to wheat fields, wetlands to rice paddies, and natural forests replaced by managed timber or retreating before the demands for pasture and croplands. We find water resources in full use for agriculture, industry, hydropower, and removal of wastes. We find the ocean fished beyond maximum yield, with mariculture increasing as hunting success declines.

In this situation of intense human management and taxation of ecosystems, the ecologic and economic impacts of climatic change are bound to be greatly amplified. On the most elementary level, a population living at subsistence, and barred from seeking greener pasture by its abundant neighbours, will experience starvation as a result of even a relatively minor decrease of food production. In a tight world market, a serious decrease of harvests in industrialized countries will

W. H. Berger and L. D. Labeyrie (eds.), Abrupt Climatic Change, 3–22.

result in higher food prices for all, with potentially devastating effects on food-importing countries with weak economies.

Similarly, any other climate-generated strain on resources (energy, fertilizer) is likely to be passed on to the economically weak, as the strong bid up the price. In addition, and ironically, industrial countries using a combination of yield maximization through chemical fertilization and incentives for non-production and other price-support devices, are likely to suffer surplus problems whenever yields increase, with disruptive effects on national economies and international relations. Thus, any major climatic change is likely to result in economic (and political) stress, within nations and in the world market.

THE CLIMATE PREDICAMENT

The climate predicament is a clash of parallel trends with opposite effects: the increasing efforts at maximization of yields from ecosystems, for human consumption, and the increasing potential for disruptive climatic change, generated as a side-effect of the maximization effort.

Concurrently with the evolution of increased global vulnerability to climatic change, the increasing demand for energy has introduced a new human-geological factor with unprecedented potential for fostering rapid climatic change. We refer, of course, to the alteration of the chemistry of the atmosphere, and especially to the addition of carbon dioxide, which has already led to an increase in the atmosphere by more than 20 percent of pre-industrial levels. The ramifications of the carbon dioxide problem have been extensively discussed in the "state-of-the-art" report sponsored by the U.S. Department of Energy (MacCracken and Luther, 1986a, b; Strain and Cure, 1986; Trabalka, 1986; White, 1986). The addition of this greenhouse gas results in global warming: a doubling of atmospheric CO_2 is thought to produce a temperature increase of between 2 to 3°C. Thus, an addition of 20% might be expected to deliver an increase of 0.5°C. Indeed, there has been an overall increase in temperature of this magnitude in the northern hemisphere, during this century. It is not clear, to be sure, whether this warming is due to natural or man-made causes.

Besides carbon dioxide, there are trace gases to be considered, both man-made (chlorofluorocarbon, nitrous oxide) and natural (methane), the latter being increased by excess release of natural gases from human disturbance of environmental conditions. The content of these gases in the atmosphere apparently is rising rapidly, and their contribution to the greenhouse effect may rival that of the industrial CO_2 (Ramanathan et al., 1985). In addition, human activities can influence climate through the atmospheric content of suspended particles (aerosols), through stress on the vegetation from acid deposition (and the attendant changes in albedo), and through large-scale changes in the water budget and soil moisture (irrigation, desertification) (Flohn, 1985). Much of the resulting climate modification is regional in scope, but may affect the larger picture as well.

Anticipated effects of severe global climate modification go well beyond the impact on agriculture, energy use, water use and fishing activities. Perhaps the most discouraging of the predicted consequences of substantial warming is the rise of sealevel expected from the melting of glaciers and polar ice. Since the mid-1930's sealevel seems to have been rising at a rate near 2mm/yr, partly due to thermal expansion of the ocean's water. This rate could be increased to some 10mm/yr upon doubling of CO_2 (or equivalent addition of other trace gases) (Meier et al., 1985). People concentrate along the coast, and they farm the fertile coastal lowlands. Such a rise in sealevel, combined with the usual (or intensified) storm action, would soon result in disaster, along many coastal regions.

ABRUPT CHANGE AND THE ELEMENT OF SURPRISE

Can one, in principle, predict abrupt climatic change from long-term trends and a knowledge of how the climate system works? Probably not. The system is very complicated indeed (Fig. 1). The physics involved (planetary and surface albedo, heat transport, cloud formation and dissipation, evaporation and precipitation, etc.) are reasonably well understood, but because of the large number of feedback loops small errors in a deterministic model system are likely to grow in unpredictable ways. In fact, it has been suggested that a significant portion of climate variability is probabilistic in nature (Mitchell, 1976). To keep models manageable, approximations and statistical relationships based upon modern climate characteristics are abundantly introduced. This is why it is necessary to test the performance of climate models against historical records in the broadest sense, within different climatic contexts.

Of course, even greatly improved models will never be entirely reliable in predicting the future. This is true not only because climate feedbacks from global ocean circulation and from continental ice dynamics are extremely complicated, but also because we know next to nothing on the changes of solar activity, or on the factors controlling volcanism, parameters with major implications for climatic change (e.g., Eddy, 1977; Pollack et al., 1976).

An extremely demanding task is to define the role of the ocean in a period of rapid climatic change. On the one hand, the ocean can moderate change, through its large heat capacity. On the other, it can accelerate climatic trends, through a decrease in the rate of exchange between ocean and atmosphere. Consider, for example, the possibility that warming reduces the rate of formation of deep and bottom waters in high latitudes. This would cut down on the uptake of trace gases from the atmosphere by the deep sea. Thus, greenhouse gases might keep building up in the atmosphere even though input from human activity were held steady or declined, through voluntary or involuntary restrictions.

Similarly, a reduction of deep-water formation in the North Atlantic, which is associated with cross-equatorial transport of tropical water masses from south to north, could substantially change the heat budget of the entire North Atlantic region. As Stommel (1980)

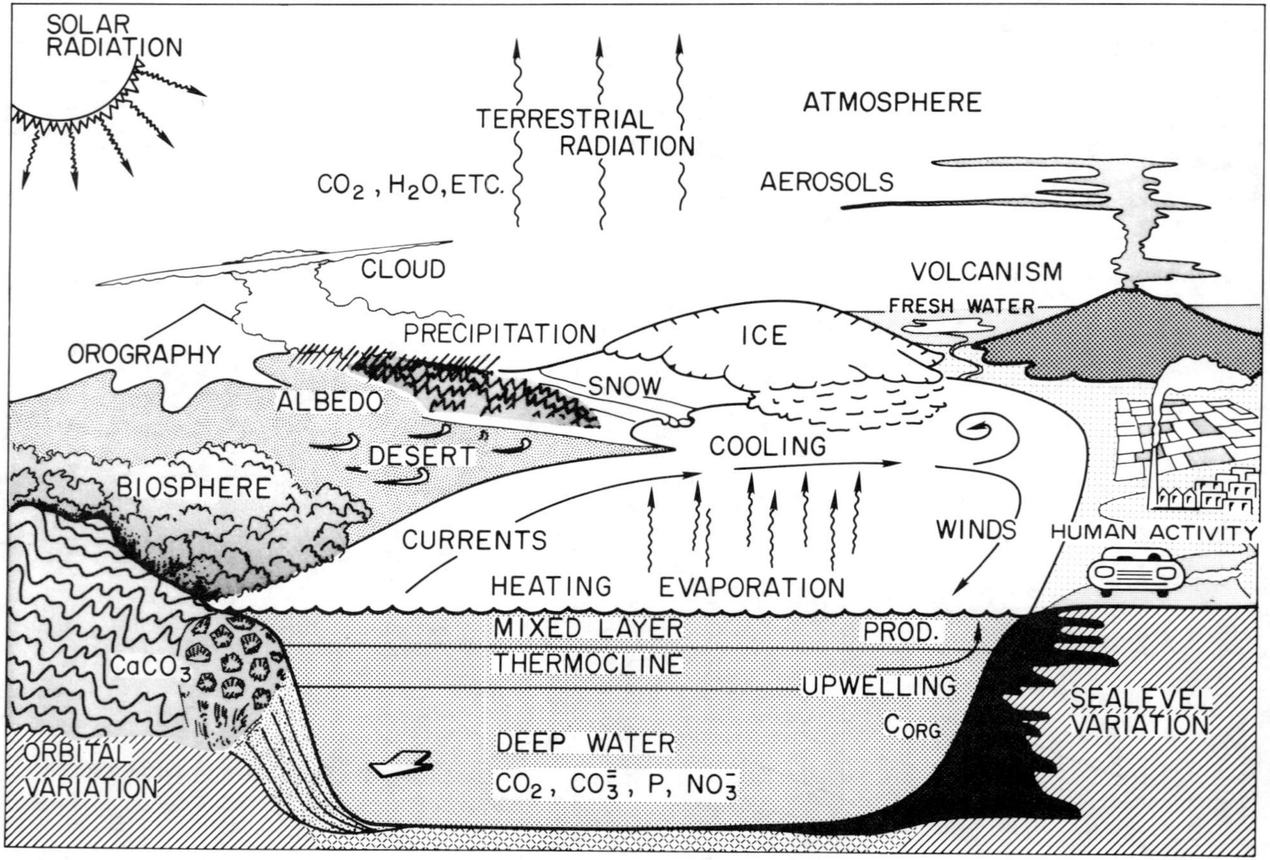

Figure 1. Major components of the climate system. Main sources of instability: ice mass, volcanism, variable deep-water production, upwelling, and human activities. Positive feedback: albedo, CO_2 (and trace gases).

pointed out, the magnitude of cross-latitudinal heat transport by ocean currents is poorly known: what we do know is that there are strong asymmetries in the global system, asymmetries which are intimately tied to present climate patterns.

Surprises, then, may be lurking within these poorly understood oceanic heat transport processes, whose modification might greatly accelerate climatic change once the process is started. Other potential sources of surprise are sea-ice formation and the response of coastal upwelling to warming: both might develop strong positive feedback, through albedo and carbon dioxide changes. On land, for example, permafrost melting may bring unexpected consequences, through release of methane (Bell, 1982). Surging glaciers could change the rules in the response of ice and sealevel to warming (Barth and Titus, 1984). These are examples of "transient reservoirs" (Berger, 1982a) which can produce disturbance of unexpected magnitude. In addition, changes in ice-loading could conceivably trigger volcanic activity of unexpected intensity.

Another aspect of the lack of understanding of climate dynamics is the question of detection of abrupt climatic change. As far as we know, such change may be occurring right now. New records of summer heat and winter cold have been common throughout the last decade, in areas where longterm records have been kept. However, since the realm of natural fluctuations has not been clearly exceeded by our evolving climate patterns we cannot be sure that this trend will continue into the future. The signal generated by human climate modification may still just be within the background noise of climate variation.

The effects of surprise within climate dynamics can only be studied in history, where improbable events become reality through the sheer passage of time.

THE LAST MILLENIUM

The last 150 years, with their instrumental records of climatic change, provide the best chance for acquiring accurate and geographically extensive experience against which the abruptness of change can be gauged. A rapid warming between 1910 and 1940 is evident in such records (Wigley and Jones, 1981). Jones et al. (this volume) contemplate the possibility that this warming (by 0.5°C globally) is near the maximum rate for global-mean change, independent of the strength of forcing, because of the thermal inertia of the ocean. A similarly rapid increase has now been experienced in the years between 1970 and 1985 (0.3°C), and this rise, according to modeling by Wigley et al. (1986a) is more or less as expected from the effects of added greenhouse gases (Fig. 2).

According to these studies, then, it appears that abrupt climatic change is with us already. We cannot be certain that this is so because, lacking instrumental records on a global scale, we do not know whether similar warming periods were commonplace in past centuries. To get perspective on this question we must turn to proxy-records, that is, to the clues provided by tree-rings, pollen, corals, lake and ocean

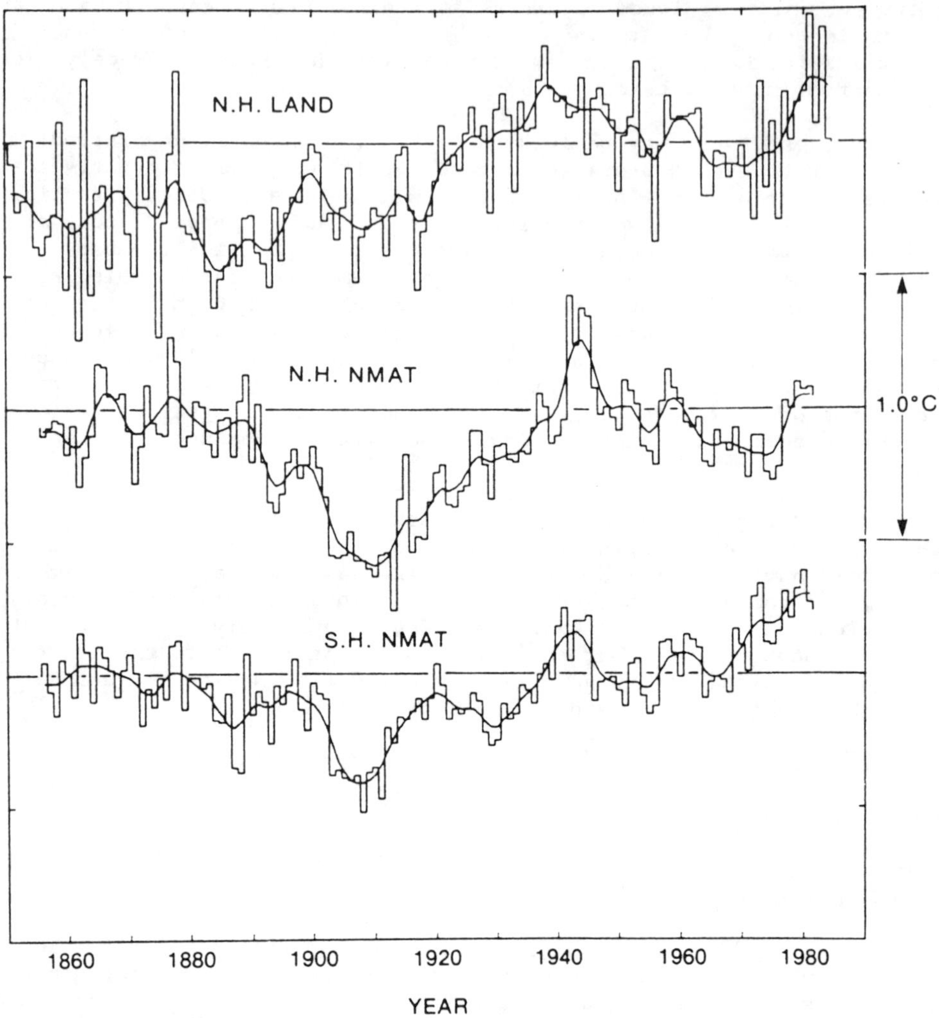

Figure 2. Surface air temperature fluctuations in the northern and southern hemisphere. NMAT: night-time marine air temperatures. From Wigley et al. (1986b).

sediments, and ice strata, for example (Sawyer, 1966; Fritts, 1971; Oeschger et al., 1980; Harding, 1982; Hughes et al., 1982; Mörner and Karlén, 1984; Webb, 1985). In regions of human settlement, recorded history can yield valuable information (Lamb, 1972, 1977, 1982). Ship-logs provide scattered data at sea (Fieux, 1978; Barnett, 1984).

For all kinds of proxies, calibration with the actual measures of climate (temperature, precipitation, seasonality, winds, etc.) is necessary and is a difficult and time-consuming undertaking (e.g., Imbrie and Kipp, 1971; Webb and Bryson, 1972; Atkinson et al., 1987). Usually, the response characteristics of a proxy are complex, and several proxy-records must be analyzed simultaneously to obtain the climate descriptors.

Only one period was identified by the working group on the last millenium (Bradley et al., 1986), as having witnessed abrupt climatic change in many areas of the world: the period centered on A.D. 1820 ± 10. This period contains the infamous "year without a summer" (1816), embedded within a series of extremely cool summers, from 1812 to 1817. It was marked by a series of major volcanic events including the enormous Tambora eruption in 1815. The years following this eruption are remarkable for calamity: widespread famine, typhus and plague in Europe and the eastern Mediterranean, and cholera spreading globally from Bengal. One of the greatest floods of the Nile is recorded for 1817/1818. The 1820's in contrast were distinguished by "genial warm springs and autumns", and 1826 produced the warmest summer in 300 years of temperature observation in England (Lamb, 1982, p. 241).

Other abrupt changes seen in the various proxy series appear to have been of regional significance. However, no prejudice against further study is intended by this finding. Firstly, coverage is spotty, and no effort was made to comprehensively review all the evidence. Secondly, abruptness is to some degree in the eyes of the beholder. The profound changes, in the North Atlantic region, from the warm climate in early medieval times, to a climate rich in cold and wet summers, and in devastating winter storms (Fig. 3), surely warrant a careful search for possible causes. The drastic increase in variability of the weather in the "Little Ice Age" (as opposed to simple cooling) is of special interest in this context.

The $\delta^{18}O$-record in Greenland ice (Dansgaard et al., 1971) suggests that the last few centuries did contain the coldest interval in the North Atlantic region, within the last 8000 years. This point must be kept in mind when discussing the warming trend since the industrial revolution. Volcanic activity may have been very important in producing a series of cold extremes, within the "Little Ice Age" (Bradley, this volume). With respect to the timing of volcanism and its effects, the ice core record has proved to be especially valuable (Hammer et al., 1980).

THE HOLOCENE

Moving back in time and digging deeper into the middens of climate history, the record becomes ever more beset by problems of dating,

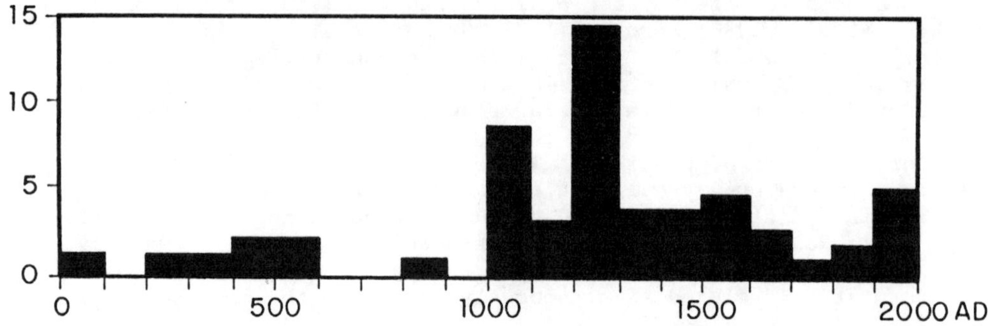

Figure 3. Reports of severe storm floods on the coasts of the North Sea and English Channel, according to Lamb (1982). 20th century value is 4/3 times the 1900 to 1975 total.

geographic coverage becomes less satisfactory, and the continuity of climate proxy series becomes more difficult to establish.

Considering the Holocene in its entirety, there is evidence for rapid climatic change in almost all regions, for the various indicators studied (Street-Perrott et al., 1986). A systematic search for those events which are widely correlated and represent large-scale, short-term climate shifts should prove extremely useful. It was noted that the complexity of Holocene climate is not done justice by such terms as "Hypsithermal" and "Little Ice Age", which tend to obscure the great range of Holocene climatic states.

There is, nevertheless, a consensus that the early Holocene, on the whole, was different from the late Holocene. It may have been up to 1°C warmer, with quite distinct precipitation patterns (Fig. 4). It has been suggested that this kind of pattern may come to establish itself again as the planet warms, in the near future (Kellogg and Schware, 1981). However, it is by no means established beyond doubt that the global climate was indeed warmer during the "Altithermal" (Webb and Wigley, 1986).

In the present context the question is of interest just how quickly one pattern can change into the other, which requires intercontinental correlation on a rather fine scale (that is, decades). Such correlation is difficult to accomplish, to say the least.

It is remarkable how little is known about the causes of climatic change in the Holocene, this most familiar of all geologic periods. Solar variability, volcanicity, tropospheric aerosols, and carbon dioxide, among other factors, have been invoked to produce change, but no clear lines of evidence have been developed. More detailed time-series of well-documented proxy records are needed, especially in sensitive regions. Paleohydrological indicators (δD, $\delta^{18}O$, lake sediments, glaciers) appear to be especially well-suited for reconstruction. The record in ice, because of continuity and high resolution of multiple

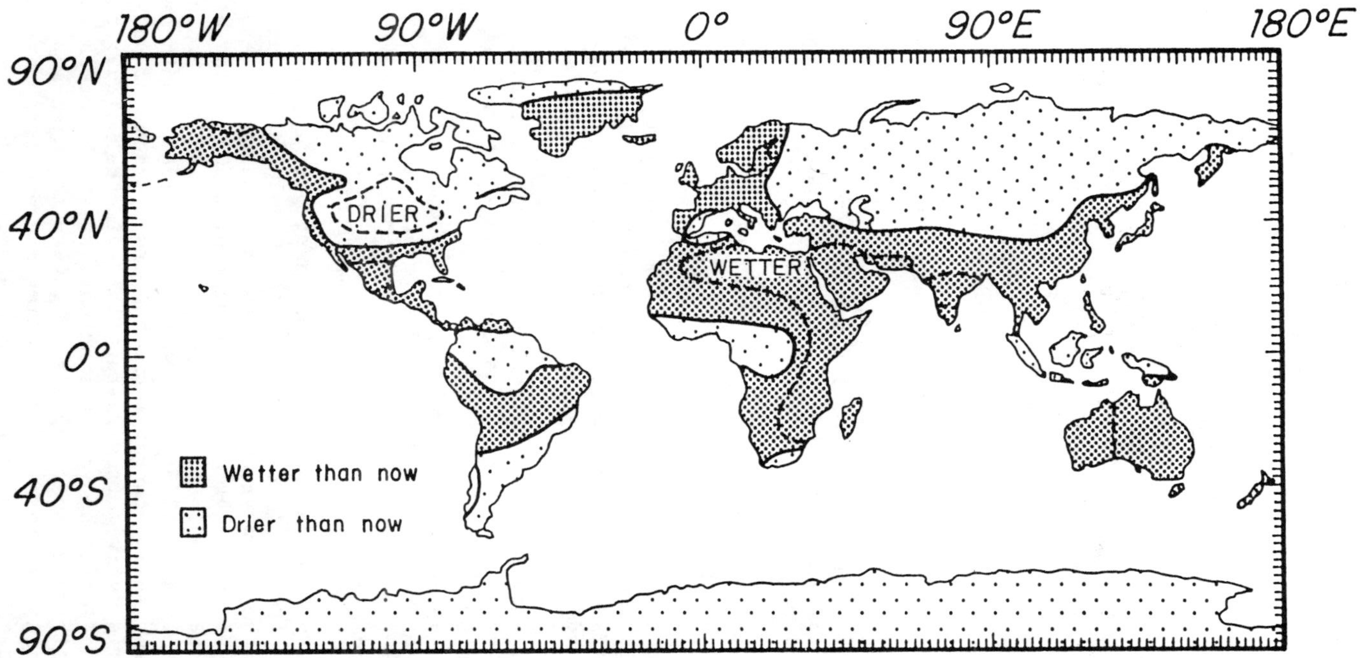

Figure 4. Anticipated soil moisture patterns on a warmer Earth. Scenario based on paleoclimatic recostructions of the Altithermal Period (8000 to 4500 years ago), comparisons of warm and cold years in recent times, in the northern hemisphere, and a climate model experiment. From Kellogg and Schware (1981).

climate-related signals, holds great promise for elucidating Holocene climate dynamics (Oeschger, 1980; Langway et al., 1985; Dansgaard, this volume). A glaring shortage exists for data from the oceans, and more profiles are needed from cores with a high sedimentation rate.

There is, it appears, a possiblility that the ^{14}C-record in tree-rings contains information about ocean-atmosphere mixing patterns. Frenzel (1975) pointed out some time ago that phases of changing climate coincide with steps in the ^{14}C record indicating increased $^{14}C/^{12}C$ ratios followed by decreased $^{14}C/^{12}C$ ratios. This holds true, according to Frenzel, for major changes at about 4200 yr bp, 3600 yr bp, and 2500 yr bp, as well as possibly for others. One conceivable way to produce such a correlation, using the ^{14}C-model of Keir (1983), would be an association of a change in the rate of deep-water production with climatic change. For instance, if the production of North Atlantic deep water were decreased, the $^{14}C/^{12}C$ ratio in the atmosphere would increase, and cross-equatorial heat flow into the North Atlantic would diminish. Upon resumption of the original rate of deep water formation, the $^{14}C/^{12}C$ ratio in the atmosphere would drop (from dilution with old CO_2 from the deep ocean), and cross-equatorial heatflow would come back to its former level. Other scenarios are possible, of course, including changes in solar activity.

THE GLACIAL-HOLOCENE TRANSITION

The record of the Glacial-Holocene transition emerged as a centerpiece of attention in the conference, partly because of the interests of the conveners, but also because climatic changes in this period show maximal abruptness and amplitude (e.g. Flohn, 1986). Among the pressing questions which were identified by the working groups (Peteet et al., 1986; Dalfes et al., 1986) are the following: (a) quantification of the history of ice retreat and associated sealevel change, (b) history of growth and retreat of small glaciers, (c) information about the changing radiocarbon age of deep water, (d) recovery of seasonality data, (e) improved coverage of rainforest history and desert behavior, (f) climate response functions for ecological systems, and (g) increased quantity and quality of ice-core information.

A few remarks are in order about some of these points. The rather confusing array of different time-scales for the Glacial-Holocene transition, as derived from the oxygen isotope record in deep-sea carbonates (see Ruddiman and Duplessy, 1985) is now being replaced by a set of time-scales which show much closer agreement (Berger, 1982b; Sarnthein et al., 1982; Berger et al., 1985; Mix and Ruddiman, 1985; Duplessy et al., 1986). These new scales converge on the Norden scheme of Mangerud et al. (1974) which is derived from terrestrial evidence. Thus, the steplike nature of deglaciation, and the abruptness of the steps which has long been recognized on land, appears to govern the deep-sea record as well. Another point, which has far reaching consequences for the understanding of the deglaciation period is the approximately 2 kyr lead of the Southern Ocean warming (1600-1700 yrs bp) (Labeyrie et al., 1986).

The possibility of obtaining separate radiocarbon ages for surface-water and deep water in the ocean is an entirely new development, based on accelerator dating which uses samples in the milligram range (Andree et al., 1984; Duplessy et al., 1986). The difference in radiocarbon ages of benthic and planktonic species in deep-sea carbonates can yield valuable clues to changes in deep circulation: a slowing of deep mixing should increase the difference. The global heat budget could be readily modified by such changes in mixing, which have been suggested by various authors in connection with on-top input of (low density) meltwater to the ocean. Available accelerator dates (Andree et al., 1986) have been interpreted to support the notion of substantially altered circulation pattern during glacial and deglacial time (Berger, in press).

The new method is especially interesting in the context of the change in atmospheric carbon dioxide which occurred across the Glacial-Holocene transition (Berner et al., 1980; Delmas et al., 1980). A change in the partitioning of carbon dioxide between ocean and atmosphere is the favored way of changing atmospheric CO_2 on short notice (see articles in Sundquist and Broecker, 1985). Through the greenhouse mechanisms and associated feedbacks, such CO_2 changes could contribute substantially to the rate of climatic change within the Glacial-Holocene transition.

THE YOUNGER DRYAS AND BEYOND

The period of the Younger Dryas (named after the remains of the arctic flower *Dryas octopetala*, in periglacial deposits near Alleröd, Denmark) is of particular interest when discussing large-scale abrupt climatic change. The Younger Dryas occurs in the middle of the Glacial-Holocene transition, between 11,000 and 10,000 years ago, and represents a brief episode of a return to almost fully glacial conditions. As Mangerud (this volume) points out, deglaciation did not just pause, it reversed its course, and glaciers grew and expanded, at least in northern Europe. The beginning of the Younger Dryas was extremely abrupt, as was its cessation (e.g. Lowe et al., 1980) (Fig. 5).

As yet there is no satisfactory explanation for this reversal of the general glacial-to-postglacial warming trend. One is tempted to invoke the analogy of chaotic behavior of a system with strong positive feedback, midway between stable extremes. Seemingly small disturbances could produce large effects under such conditions. Point disturbances could come, for example, from volcanism or from the sudden introduction of fresh water through draining of large dammed lakes (e.g., the Baltic Ice Sea). Sea-ice formation, presumably, could be the major amplifying mechanism, translating such disturbances into large albedo changes.

The important role of the North Atlantic region as a climate amplifier (Ruddiman and McIntyre, 1981; Ruddiman and McIntyre, 1984) becomes especially apparent in the context of abrupt climatic change during the G/H transition. The heat transport to the northern North Atlantic is a crucial factor in the climate dynamics of the entire region, and most likely the northern hemisphere as a whole.

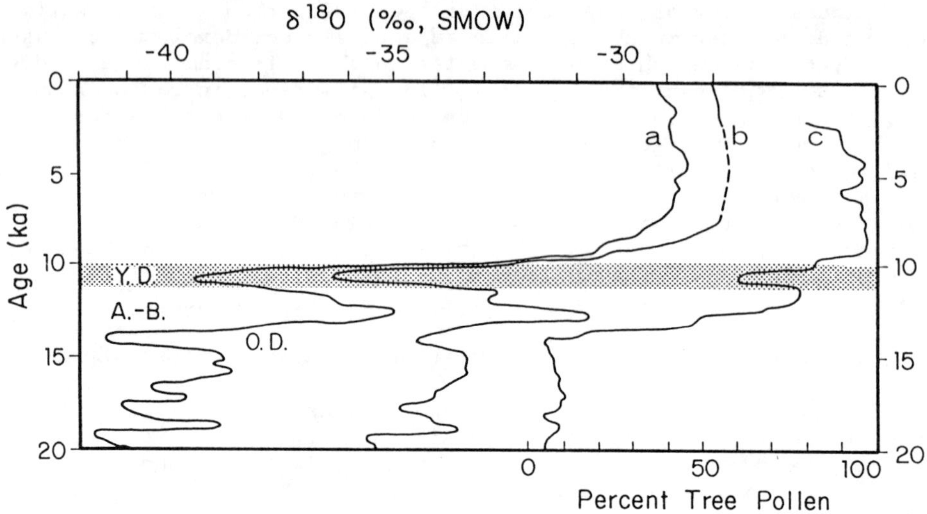

Figure 5. The Alleröd/Younger Dryas Oscillation in the North Atlantic realm. a and b, ice core records, according to Dansgaard et al. (1984). a, Camp Century,; b, Dye 3. Note that the lines cross. c, pollen record of Woillard and Mook (1982), Grand Pile, SE France; percent tree pollen of total (tree + grass). Age scales slightly adjusted to agree with the Norden time scale of Mangerud et al. (1974). O.D., Older Dryas cold period; A.-B., Alleröd-Bölling warm period; Y.D. (shaded zone) Younger Dryas cold episode.

In the typical Holocene configuration, warm waters from the Gulf Stream extension flow northward into the Norwegian Sea, and (relatively warm) salty North Atlantic deep water forms here. During glacial configurations surface flow is west to east toward the Portuguese coast, with most of the flow returning clockwise to the Caribbean (CLIMAP, 1976). Heat transport to high latitudes is cut off, and much less NADW forms than now, involving less saline (and colder) water with a higher CO_2 content (Duplessy et al., 1980; Labeyrie and Duplessy, 1985).

It stands to reason that the reduction or shutdown of NADW during deglaciation (Olausson, 1965) should have had severe climatic consequences, because of the link to northward heat-transport (Weyl, 1968). Such shutdown has been invoked to explain the origin of the Younger Dryas (Berger and Killingley, 1982; Berger, 1985; Broecker et al., 1985), but the dynamics are as yet entirely unclear. It is encouraging to see a vigorous effort at modeling the effects of a change in rate of NADW formation, on the downwind climate in northern Europe (Rind et al., 1986; Jones et al., this volume; Schneider et al., this volume).

Beyond the Glacial-Holocene transition there are other periods of rapid climatic change which are worth studying. To do this properly, it will be necessary, among other things, to extract that portion of climate variability which is not linearly related to changes in insolation patterns, and to analyze the residual. One must beware, of course, of accepting Milankovitch forcing as a description of physical climate dynamics. Instead, the system's responses to forcing can be used to extract constraints which are necessary to model such dynamics.

ON MODELS

The mathematical representation of the climate machine is a most complex and intricate task (Hansen and Takahashi, 1984; Gates, 1986), and the necessary simplifications degrade the explanatory and predictive power of climate models considerably. Reasons include unrealistically small geographic coverage, coarse grid scale, and the need to tune to present climatic conditions. Nevertheless, since the present and imminent climatic change from greenhouse gases may be quite unprecedented in history, modeling is essential. One reason why this will require great effort is that climate can only be modeled correctly by including ocean-atmosphere interaction, that is, ocean circulation (Fig. 6).

Climate is global in nature, and since much heat in the North Atlantic comes from the South Atlantic (Stommel, 1980) and perhaps even from the Indian Ocean (Gordon, 1985), a self-contained model of the North Atlantic just will not accurately represent the range of possible variations in this region. The need for large-scale geographic data coverage, and for seasonal effects, is obvious. Such data are likely to have the greatest impact on models (Joussaume, this volume; North et al., this volume).

One important way in which models can be used directly is to calculate distributions of climatic tracers, such as isotopic species or dust in ice (Joussaume and Jouzel, this volume). Extension of this methodology to other proxy records should prove extremely useful. It forces a thorough investigation of the meaning of a climate proxy.

Models are, despite drawbacks, the basis for logical and systematic discussion of the physical principles underlying climatic change. Thus, the interaction between the data-oriented and the model-oriented students of climate is absolutely vital. As "educational machines" (a phrase coined by S. Schneider) models serve their purpose by keeping in focus what counts most in doing paleoclimatic research: understanding the dynamics of the changing climate system. Without such understanding there can be no extrapolation into the future.

To illustrate this last point, consider the extrapolation of climate evolution over the next several thousand years, which is based on the astronomical theory of climate variation (Berger, 1981; Imbrie and Imbrie, 1984). This theory, which invokes the "Milankovitch mechanism", says that we are well on the way to next ice-age. If this is so, and if the usual projections of CO_2-input and CO_2-titration by deep-sea carbonates (e.g., Bacastow and Keeling, 1973) are correct, then the Milankovitch forcing-mechanism and the climatic reality on the globe will be increasingly out of phase, for the next two to three thousand years. When the time comes to correct this lag, a most unusually abrupt climatic change might be expected to occur. It is unlikely, of course,

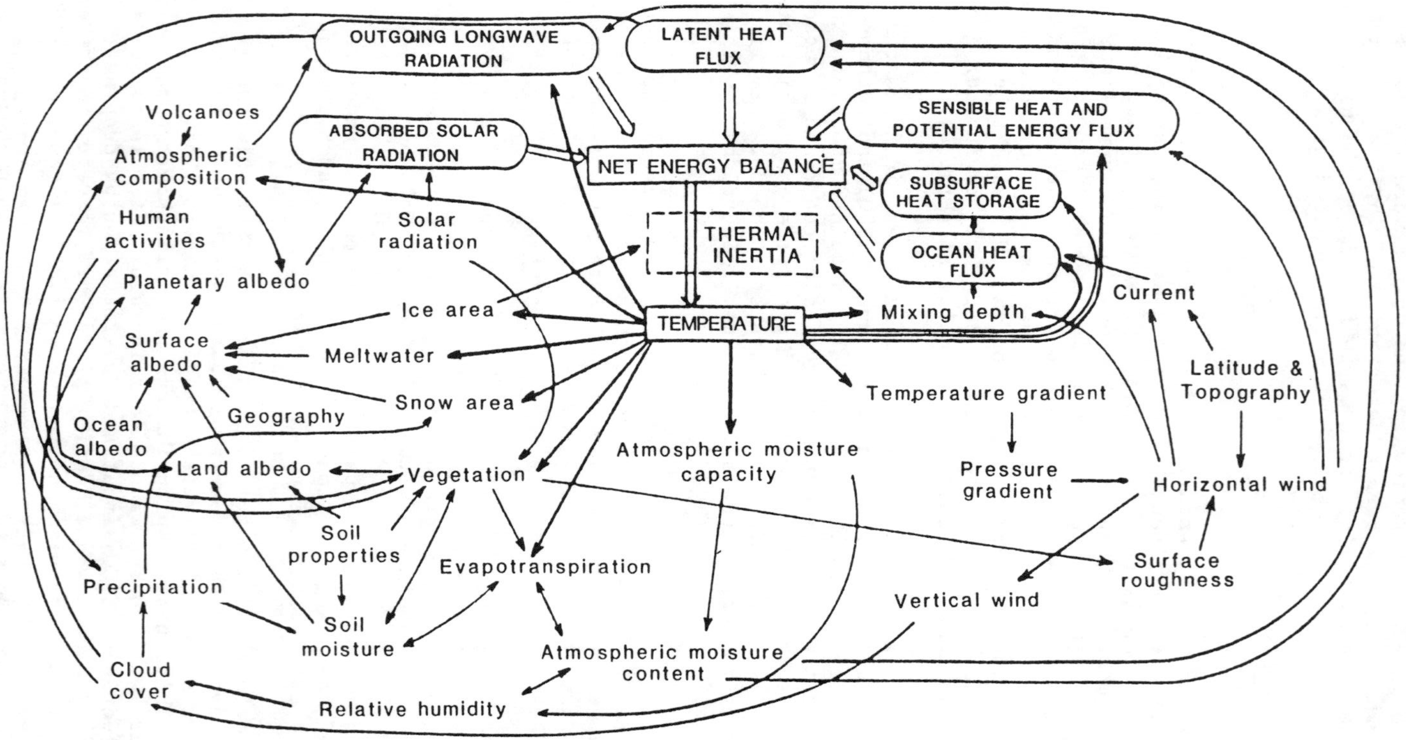

Figure 6. Flow diagram for climate modeling, showing feedback loops. From Robock (1985).

that such a prediction would influence decisions of policy-makers in our own time.

REFERENCES

Andree, M., Beer, J., Oeschger, H., Broecker, W., Mix, A., Ragano, N., O'Hara, P., Bonani, G., Hofmann, H.J., Morenzoni, E., Nessi, M., Suter, M. and Wölfli, W., 1984. ^{14}C measurements on foraminifera of deep sea Core V28-238 and their preliminiary interpretation, *Nuclear Instruments and Methods in Physics Research*, B5: 340-345.

Andree, M., Oeschger, H., Broecker, W.S., Beavan, N., Klas, M., Mix, A., Bonani, G., Hofmann, H.J., Suter, M., Wölfli, W. and Peng, T.H., 1986. Limits on the ventilation rate for the deep ocean over the last 12000 years, *Climate Dynamics*, 1: 53-62.

Atkinson, T.C., Briffa, K.R. and Coope, G.R., 1987. Seasonal temperatures in Britain during the past 22,000 years, reconstructed using beetle remains, *Nature*, 325: 587-592.

Bacastow, R. and Keeling, C.D., 1973. Atmospheric carbon dioxide and radiocarbon in the natural carbon cycle: II. Changes from A.D. 1700 to 2070 as deduced from a geochemical model. In: Woodwell, G.M., and Pecan, E.V. (eds.) *Carbon and the Biosphere*, U.S. Atomic Energy Commission, Springfield, Va., p. 86-136.

Barnett, T.P., 1984. Longterm trends in surface temperature over the oceans, *Monthly Weath. Rev.*, 112: 303-312.

Barth, M.C. and Titus, J.G. (eds.) 1984. *Greenhouse Effect and Sea Level Rise*, Van Nostrand Reinhold, New York, 325pp.

Bell, P.R., 1982. Methane hydrate and the carbon dioxide question. In: Clark, W.C. (ed) *Carbon Dioxide Review 1982*, Clarendon Press, Oxford, p. 401-406.

Berger, A. (ed), 1981. *Climatic Variations and Variability: Facts and Theories*, Reidel, Dordrecht, 795pp.

Berger, W.H., 1982a. Climate steps in ocean history - lessons from the Pleistocene. In: Berger, W.H., Crowell, J.C. (eds) 1982. *Climate in Earth History*. *Studies in Geophysics*, National Academy Press, Washington, D.C., 198pp.

Berger, W.H., 1982b. On the definition of the Pleistocene-Holocene boundary in deep-sea sediments, *Sveriges geologiska undersökning, Avh. Uppsat.*, 76 (7), ser.C (794): 270-280.

Berger, W.H., 1985. Carbon dioxide increase and climate prediction: clues from deep-sea carbonates, *Episodes*, 8: 163-168.

Berger, W.H., in press. Counterpoint deep water production during the last 12,000 years: evidence from foraminiferal radiocarbon, *Marine Geology*.

Berger, W.H. and Killingley, J.S., 1982. The Worthington Effect and the origin of the Younger Dryas, *J. Mar. Res.*, 40: 27-38.

Berger, W.H., Killingley, J.S. and Vincent, E., 1985. Timing of deglaciation from an oxygen isotope curve for Atlantic deep-sea sediments, *Nature*, 314: 156-158.

Berner, W., Oeschger, H. and Stauffer, B., 1980. Information on the CO_2 cycle from ice core studies, *Radiocarbon*, 22: 227-235.

Bradley, R.S., 1987. The explosive volcanic eruption record in Northern Hemisphere temperature records, this volume.

Broecker, W.S., 1982. Ocean chemistry during glacial time, *Geochim. Cosmochim. Acta*, 46: 1689-1705.
Broecker, W.S., Peteet, D.M. and Rind, D., 1985. Does the ocean-atmosphere system have more than one stable mode of operation?, *Nature*, 315: 21-26.
CLIMAP Project Members, 1976. The surface of the ice-age Earth, *Science*, 191: 1131-1137.
Dalfes, H.N., 1987. Detection of abrupt climatic changes in deep-sea sediment cores, this volume.
Dalfes, H.N.., Jansen, E., et al., 1986. Late Glacial to Interglacial Transition: the oceanic record. In: W.H. Berger and L.D. Laberyie (eds.) *The Book of Abstracts and Reports from the Conference On : Abrupt Climatic Change*, SIO Reference Series, 86-6, p. 38-41.
Dansgaard, W., Johnsen, S.J., Clausen, H.B. and Langway, C.C., 1971. Climatic record revealed by the Camp Century ice core. In: Turekian, K.K. (ed.) *The Late Cenozoic Glacial Ages*, Yale University Press, New Haven, p.37-56.
Dansgaard, W., Johnsen, S.J., Clausen, H.B., Dahl-Jensen, D., Gundestrup, N. and Hammer, C.U., 1984. In: J.E. Hansen and T. Takahashi (eds.) *Climate Processes and Climate Sensitivity. Geopysical Monograph* 29, American Geophysical Union, Washington DC, p.288-298.
Dansgaard, W., 1987. Ice core evidence of abrupt climatic change, this volume.
Delmas, R.J., Ascencio, J.-M. and Legrand, M., 1980. Polar ice evidence that atmospheric CO_2 20,000 years BP was 50% of present, *Nature*, 284: 155-157.
Duplessy, J.-C., Moyes, J., and Pujol, C., 1980. Deep water formation in the North Atlantic Ocean during the last ice age, *Nature*, 286: 479-482.
Duplessy, J.-C., Arnold, M., Maurice, P., Bard, E., Duprat, J., and Moyes, J., 1986. Direct dating of the oxygen-isotope record of the last deglaciation by ^{14}C accelerator mass spectrometry, *Nature*, 320: 350-352.
Eddy, J.A., 1977. Climate and the changing sun, *Climatic Change*, 1: 173-190.
Fieux, M.M., 1978. Evolution a long terme des temperatures de surface de la mer. In: *Evolution of planetary atmospheres and climatology of the earth*, C.N.E.S.
Ed., Paris, p. 465-471.
Flohn, H., 1985. *Das Problem der Klimaänderungen in Vergangenheit und Zukunft*, Wissenschaftliche Buchgesellschaft, Darmstadt, 228pp.
Flohn, H., 1986. Singular events and catastrophes now and in climatic history, *Naturwissenschaften*, 73: 136-149.
Frenzel, B., 1975. The distribution pattern of Holocene climatic change in the northern hemisphere. In: Lamb, H. (convener) *Proceedings of the WMO/IAMAP symposium on long-term climatic fluctuations*, World Meteorological Organization, Geneva, p. 105-118.
Fritts, H.C., 1971. Dendroclimatology and Dendroecology. *Quaternary Research*, 1: 419-449.
Gates, W.L., 1986. Modeling as a means of studying the climate system. In: MacCracken and Luther, (eds.), 1986, *Projecting the Climatic*

Effects of Increasing Carbon Dioxide, U.S. Dept. of Energy, Washington D.C., p.57-79.
Gordon, A., 1985. Indian-Atlantic transfer of thermocline water at the Agulhas retroflection, *Science*, 227: 1030-1033.
Hammer, C.U., Clausen, H.B. and Dansgaard, W., 1980. Greenland ice sheet evidence of post-glacial volcanism and its climatic impact, *Nature*, 288: 230-235.
Hansen, J.E. and Takahashi, T. (eds), 1984. *Climate processes and climate sensitivity*, Geophys. Monogr. 29, Amer. Geophys. Union, 368pp.
Harding, A. (ed.), 1982. *Climatic Change in Later Prehistory*, Edinburgh University Press, Edinburgh, 210pp.
Hughes, M.K., Kelly, P.M., Pilcher, J.R. and Lamarche, V.C. (eds.), 1982. *Climate from Tree Rings*, Cambridge University Press, Cambridge, 223pp.
Imbrie, J. and Kipp, N.G., 1971. A new micropaleontological method for quantitative paleoclimatology: application to a Late Pleistocene Caribbean core. In: K.K. Turekian (ed), *The Late Cenozoic Glacial Ages*, Yale Univ. Press, New Haven, 71-181.
Imbrie, J. and Imbrie, J.Z., 1984. Modeling the climatic response to orbital variations, *Science*, 270: 943-953.
Jones, P.D., Wigley, T.M.L. and Raper, S.C.B., 1986. The rapidity of CO_2-induced climatic change: a comparison of observations and model results. In: W.H. Berger and L.D. Labeyrie (eds.) *The Book of Abstracts and Reports from the Conference on: Abrupt Climatic Change*. SIO Reference Series, 86-6: 140-144.
Jones, P.D., Wigley, T.M.L. and Raper, S.C.B., 1987. The rapidity of CO_2-induced climatic change: observations, model results and palaeoclimatic implications, this volume.
Keir, R.S., 1983. Reduction of thermohaline circulation during deglaciation: the effect on atmospheric radiocarbon and CO_2, *Earth Planet. Sci. Lett.*, 64: 445-456.
Kellogg, W.W. and Schware, R., 1981. *Climate Change and Society*. Westview Press, Boulder, Co., 178pp.
Labeyrie, L.D. and Duplessy, J.C., 1985. Changes in the oceanic $^{13}C/^{12}C$ ratio during the last 140,000 years: high latitude surface water records, *Palaeogeogr., Palaeoclimatol., Palaeoecol.*, 50: 217-240.
Labeyrie, L.D., Pichon, J.J., Labracherie, M., Ippolito, P., Duprat, J. and Duplessy, J.C., 1986. Melting history of Antarctica during the past 60,000 years, *Nature*, 322: 701-706.
Ladurie, E.L., 1971. *Times of feast, times of famine: a history of climate since the year 1000*, Doubleday, Garden City, New York.
Lamb, H.H., 1972. *Climate: present, past and future. Volume 1: Fundamentals and climate now*, Methuen, London, 613pp.
Lamb, H.H., 1977. *Climate: present, past and future. Volume 2: Climatic history and the future*, Methuen, London, 835pp.
Lamb, H.H., 1982. *Climate, history, and the modern world*, Methuen, London, 387pp.
Langway, C.C., Oeschger, H. and Dansgaard (eds), 1985. *Greenland ice core: geophysics, geochemistry, and the environment*, Amer. Geophys. Union Geophys. Monogr. 33, 118pp.
Lowe, J.J., Gray, J.M. and Robinson, J.E. (eds.), 1980. *Studies in the Lateglacial of North-West Europe*. Pergamon Press, Oxford.

MacCracken, M.C. and Luther, F.M. (eds) 1986a. *Projecting the climatic effects of increasing carbon dioxide*. U.S. Dept. Energy, U.S. Govt. Printing Office, Washington D.C., 381pp.
MacCracken, M.C. and Luther, F.M. (eds) 1986b. *Detecting the climatic effects of increasing carbon dioxide*. U.S. Dept. Energy, U.S. Govt. Printing Office, Washington D.C., 198pp.
Mangerud, J., Andersen, S.T., Berglund, B.E. and Donner, J.J., 1974. Quaternary stratigraphy of Norden, a proposal for terminology and classification, *Boreas*, 3: 109-128.
Meier, M.F., et al., 1985. *Glaciers, ice sheets, and sea level: effect of a CO_2- induced climatic change*. Report of a Workshop, Seattle, 1984, National Technical Information Service, U.S. Dept. Commerce, Springfield, Va., 330pp.
Mitchell, J.M., 1976. An overview of climatic variability and its causal mechanism, *Quaternary Research*, 6: 481-493.
Mix, A.C. and Ruddiman, W.F., 1985. Structure and timing of the last deglaciation: oxygen-isotope evidence, *Quaternary Science Reviews*, 4: 59-108.
Moerner, N.-A. and Karlen, W. (eds.) 1984. *Climatic Changes on a Yearly to Millenial Basis*, Reidel, Dordrecht, 667pp.
Oeschger, H., 1980. In der Natur gespeicherte Geschichte von Umweltsvorgängen. In: H.Oeschger, B.Messerli, M.Svilar (eds) *Das Klima, Analysen und Modelle Geschichte und Zukunft*, Springer-Verlag, Berlin, p.209-236.
Oeschger H., Messerli, B., Svilar, M. (eds), 1980. *Das Klima, Analysen und Modelle, Geschichte und Zukunft*, Springer-Verlag, Berlin, 296pp.
Olausson, E., 1965. Evidence of climatic changes in North Atlantic deep-sea cores, with remarks on isotopic paleotemperature analysis. *Progress in Oceanography*, 3: 221-252.
Peteet, D., Raynaud, D., et al., 1986. Land and ice records of abrupt climatic changes between the last glacial maximum and the Holocene. In: W.H. Berger and L.D. Labeyrie (eds.) *The Book of Abstracts and Reports from the Conference on: Abrupt Climatic Change*, SIO Reference Series, 86-6: 25-37.
Pollack, J.B., Toon, O.B., Sagan, C., Summers, A., Baldwin, B. and van Camp, W., 1976. Volcanic explosions and climatic change: a theoretical assessment, *J. Geopys. Res.*, 81: 1071-1083.
Ramanathan, V., Cicerone, R.J., Singh, H.B. and Kiehl, J.T., 1985. Trace gas trends and their potential role in climatic change, *J. Geophys. Res.*, 90 (D3): 5547-5566.
Rind, D., Peteet, D., Broecker, W., McIntyre, A. and Ruddiman, W., 1986. The impact of cold North Atlantic sea surface temperatures on climate: implications for the Younger Dryas cooling (11-10 k), *Climate Dynamics*, 1: 3-33.
Robock, A., 1985. An updated climate feedback diagram. *Bull. Amer. Meteorol. Soc.*, 66: 786-787.
Rotberg, R.I. and Rabb, T.K. (eds), 1981. *Climate and History*, Princeton University Press, Princeton, 280pp.
Ruddiman, W.F. and Duplessy, J.-C., 1985. Conference on the last deglaciation: timing and mechanism, *Quat. Res.*, 23: 1-17.
Ruddiman, W.F. and McIntyre, A., 1981. Oceanic mechanisms for amplification of the 23,000-year ice-volume cycle, *Science*, 212: 617-627.

Ruddiman, W.F. and McIntyre, A., 1984. An evaluation of ocean-climate theories on the North Atlantic. In: A.L. Berger et al. (eds.) *Milankovitch and Climate*, Part 2, D. Reidel, Dordrecht, p. 671-686.
Ruddiman, W.F., Molfino, B., Esmay, A. and Pokras, E., 1980. Evidence bearing on the mechanism of rapid deglaciation, *Climatic Change*, 3: 65-87.
Sarnthein, M., Erlenkeuser, H. and Zahn, R., 1982. Termination I: the response of continental climate in the subtropics as recorded in deep-sea sediments, *Bulletin de l'Institut Geologique du Bassin d'Aquitaine*, 31: 393-407.
Sawyer, J.S. (ed.) 1966. *World Climate from 8000 to 0 B.C.* Royal Meteorological Society, London, 229pp.
Schneider, S.H., 1976. *The Genesis Strategy*, Plenum Press, New York, 419pp.
Schneider, S.H., Peteet, D.M. and North, G.R., 1987. A climate model intercomparison for the Younger Dryas and its implications for paleoclimatic data collection, this volume.
Stommel, H., 1980. Asymmetry of interoceanic fresh-water and heat fluxes, *Proc Natl. Acad. Sci.* USA, Geophysics, 77 (5): 2377-2381.
Strain, B.R. and Cure, J.D. (eds), 1986. *Direct effects of increasing carbon dioxide on vegetation*, U.S. Dept. Energy, U.S. Govt. Printing Office, Washington D.C., 286pp.
Street-Perrott, F.A., et al 1986. Abrupt climatic changes during the Holocene. In: W.H. Berger and L.D. Labeyrie (eds.), *The Book of Abstracts and Reports from the Conference on: Abrupt Climatic Change*, SIO Reference 86-6, p. 15-24.
Sundquist, E.T. and Broecker, W.S. (eds.), 1985. *The Carbon Cycle and Atmospheric* CO_2*: Natural Variations Archean to Present*, Geophysical Monograph, 32, American Geophysical Union, Washington DC, 627pp.
Trabalka, J.R. (ed), 1986. *Atmospheric carbon dioxide and the global carbon cycle*, U.S. Dept. Energy, U.S. Govt. Printing Office, Washington D.C., 316pp.
Webb, T., 1985. Holocene palynology and climate. In: A.D. Hecht (ed.) *Paleoclimate Analysis and Modeling*, Wiley, New York, p. 163-195.
Webb, T. and Bryson, R.A., 1972. Late and post-glacial climatic change in the northern midwest U.S.A.: quantitative estimates derived from fossil pollen spectra by multivariate statistical analysis, *Quaternary Res.*, 2: 70-115.
Webb, T. and Wigley, T.M.L., 1986. What past climates can indicate about a warmer world. In: M.C.MacCracken and F.M.Luther (eds.) *Projecting the Climatic Effects of Increasing Carbon Dioxide*, U.S. Dpt. of Energy, Washington, DC, p. 237-257.
Weyl, P.K., 1968. The role of the oceans in climate change: a theory of the ice ages, *Met. Monogr.* 8: 37-62.
White, M.R. (ed), 1986. *Characterization of Information Requirements for Studies of* CO_2 *Effects: Water Resources, Agriculture, Fisheries, Forests and Human Health*, U.S. Dept. Energy, U.S. Govt. Printing Office, Washington D.C., 235pp.
Wigley, T.M.L. and Jones, P.D., 1981. Detecting CO_2-induced climatic change, *Nature*, 292: 205-208.

Wigley, T.M.L., Ingram, M.J. and Farmer, G. (eds), 1981. *Climate and History: Studies in Past Climates and their Impact on Man*, Cambridge University Press, Cambridge, England.

Wigley, T.M.L., Jones, P.D. and Kelly, P.M., 1986a. Empirical climate studies. Warm world scenarios and the detection of climatic change induced by radiatively active gases. In: B. Bolin et al. (eds) *The Greenhouse Effect, Climatic Change, and Ecosystems*, International Council of Scientific Unions, SCOPE Series 29, Wiley, New York.

Wigley, T.M.L., Angell, J.K. and Jones, P.D., 1986b. Analysis of the temperature record. In: M.C. MacCracken and F.M. Luther (eds.) *Detecting the Climatic Effects of Increasing Carbon Dioxide*, U.S. Gvt. Printing Office, Washington D.C., p. 55-90.

Woillard, G. and Mook, W.G., 1982. Carbon-14 dates at Grande Pile: correlation of land and sea chronologies, *Science*, 215: 159-161.

AIR-SEA INTERACTION PROCESSES AS MODELS FOR ABRUPT CLIMATIC CHANGES

Hermann Flohn

University of Bonn
F.R.G.

ABSTRACT. The existence of various kinds of abrupt changes suggests a chaotic almost-intransitive behavior of the climatic system (E. Lorenz) with its high number of degrees of freedom and its non-linearities.

Taking the parallel fluctuations of ambient temperature and carbon dioxide in Greenland ice-cores as an example, the rapid changes of CO_2, water vapour and mid-tropospheric temperatures observed during the last 20-30 years in relation to El Niño-Southern Oscillation events in the equatorial region suggest a mechanism for a rational interpretation of such fluctuations under natural conditions.

INTRODUCTION

During recent years, the occurrence of abrupt climatic changes during the earth's history has aroused much interest (Flohn, 1979; Kukla, 1980; Ghazi, 1983; Mörner and Karlén, 1984). Such cases - at a time-scale comparable to human life, and thus much smaller than variations of the earth's long governed Aristotelian natural philosophy: *Natura non facit saltum*. But they seem to fit into recent ideas on instability of deterministic systems as in theoretical physics. It was first E. Lorenz (1963, 1970) who suggested the possibility of instabilities of the climatic system and described its behavior as "almost-intransitive". Even in the interdiurnal variability of the mid-latitude troposphere, two equilibrium stages alternate, one of them is known as the "blocking event" (C.E. Leith, 1984). The same seems to be true, on a much larger scale, for the transitions between glacial and interglacial climates, in spite of the fact that the continental ice-sheets in the northern hemisphere need about 10^4a (years) to build or to decay. Recent investigations have shown the occurrence of catastrophes in the behavior of radically simplified deterministic climate models (Fraedrich, 1979) and the existence of strange attractors in observed series of different time-scales: Nicolis and Nicolis (1984) for palaeoclimatic deep-sea cores spanning 10^6a, and Hense (1987) for recent 100a climatic series within the "Southern Oscillation". Such a behavior has been predicted by the theory of nonlinear dynamics with more than two degrees of freedom (Schuster, 1984). It is also endorsed by the occurrence of several recent cases of climatic jumps and chaotic time-series at a regional scale (Flohn, 1986a).

W. H. Berger and L. D. Labeyrie (eds.), Abrupt Climatic Change, 23–30.

In this paper a classification of observed abrupt changes is proposed first. Then ENSO (El Niño-Southern Oscillation) events, with accompanying gas exchanges between air and sea, are discussed as a conceptual model for abrupt changes.

DEFINITION AND TYPES OF ABRUPT CHANGES

At a recent EEC workshop (Ghazi, 1983) the following working definition of an abrupt change has been proposed:

An abrupt climatic change has a time-scale of between 50 and 200 years, while the temperature difference is of the order of about half the difference between glacial and interglacial, i.e. 2-3K.

Rognon (1983) has described such events as climatic crises including sudden changes in the precipitation regime. This is certainly justified; at least the onset of the moist period in the arid and semiarid tropics around 12,000 BP (Rognon, 1983; Heine, 1983; Emiliani, 1975) is an example which had a catastrophic character in some areas. Such cases suggest the need for an extension of the above definition.

One can distinguish two families of abrupt climatic changes - one affecting primarily the thermal situation, the other the moisture balance - and in each of them two classes: positive and negative changes. The latter is not obvious; Broecker and van Donk (1970) described, in a discussion of ocean bottom cores with a time resolution of a few ka, abrupt terminations of glacial periods, but no abrupt beginnings. Evidence of catastrophic rains ("deluges") has been found at two dates: around 12 ka and perhaps at about 5.5 ka, but no example of a sudden onset of a drought period at the ka-scale has been described up to now. Naturally, both families can also occur in a combined form.

Examples of abrupt coolings are indeed frequent - one of the best documented cases is the sudden recurrence of a near-glacial climate in the European area at the beginning of the Younger Dryas epoch (about 10.8 ka ago). But the warmings at the beginning of the Bölling-Alleröd warm phase (about 3 ka ago) and at the transition Younger Dryas- Preboreal (about 10.3 ka ago) are of similar intensity and time-scale (Duplessy et al., 1981). Coope (this volume) has described their abruptness using the passive transport of climate-sensitive beetles as an indicator.

The best example of a catastrophic rainfall peak is the 12 ka-event causing a sudden glaciation at Mexican volcanic peaks (Heine, 1983, 1984). Its duration has been estimated to be less than 200 years.

For the last glaciation, between 40 and 30 ka ago, Oeschger and Stauffer (1983) and Dansgaard et al. (1984) demonstrated that a group of abrupt, simultaneous changes of ^{18}O (indicating ambient temperatures) and CO_2 in a Greenland ice core, occur between nearly constant values of both parameters at irregular time intervals of 1-3 ka in a time- scale of the order 100a. Similar changes have been reported from the transition between Late Glacial and Holocene (Oeschger et al., 1984). While the CO_2 changes are of the order 70-80ppm, the $\delta^{18}O$ changes (of the order $5^o/oo$) have been converted to a surface temperature difference of 8°C (Weber and Flohn, 1984). This figure should be revised to a more representative cloud temperature change of 5°C. Parallel trends of temperature in Greenland and Switzerland between 14 and 10 ka had been demonstrated by Oeschger et al. (1983, 1984).

ENSO EVENTS AS MODELS AND TRIGGERS FOR ABRUPT CLIMATIC CHANGES?

During recent years, investigations of El Niño-Southern Oscillation (ENSO) events (e.g. Rasmussen and Carpenter, 1982; Arkin, 1983; Wright, 1985) have shown that these events - with a duration of 1-2 years - are not limited to the Indo-Pacific area, but may trigger nearly simultaneous, and in many areas of the globe, other climatic anomalies (teleconnections). This is frequently the case with severe droughts in northeastern Brazil and in parts of Indonesia and Australia. Since the climatic system atmosphere-ocean- biosphere has very many degrees of freedom, other events may occur also at different times. Weak correlations do exist between ENSO events and drought occurrences in Africa and even with anomalies in high latitudes (Reiter, 1978; van Loon and Shea, 1985). This is especially true for strong events as in 1902, 1918, 1940/1, 1972 and for the most extreme case of 1982/3.

A negative correlation between these ENSO events and the intensity of the Indian summer monsoon rains (preceding the ENSO peak by some months) is common, but absent in some years (Rasmussen and Carpenter, 1983). Other anomalies notably above North America and the Atlantic are less frequent; clear and unambiguous correlations with the Arctic are hitherto unknown.

Some of the most important ENSO anomalies are:

a) Expansion of very warm water (28-29°C) from the western Pacific along the equator towards the east. It is driven by westerly surface winds and is correlated with high convective activity and rainfall at the central and eastern Pacific, releasing latent heat thus warming the troposphere.

b) Suppression of upwelling of cool ocean waters along the equator in the eastern and central Pacific and along the west coast of South America.

c) Large-scale atmospheric subsidence and drought in the Indonesian area (Flohn, 1986b), due to low-level divergence and high-level convergence (near the tropopause); intensification of both Pacific Hadley cells with increased subsidence in the subtropical latitudes and high intensity of the subtropical jet-stream in the upper troposphere at both hemispheres.

In the context of palaeoclimatic events one should discuss the possibility of natural changes of the composition of the atmosphere, effective at a time-scale of 10-100 years, notably the role of greenhouse gases. Most important in this respect are H_2O and CO_2, which are responsible for about 65 and 25 percent of the total greenhouse effect (Dickinson, 1982).

While condensation at the sea surface is (in the tropics) a rare event, evaporation increases strongly from upwelling episodes (with cold water, stable stratification and weaker winds) to warm water episodes (with unstable stratification and consequently stronger surface winds). Henning and Flohn (1980) and others have shown that in typical upwelling episodes evaporation drops to values near 1mm/d, in contrast to (normal) values of 4-5mm/d above warm water. Under ENSO conditions water vapour is convectively transported into higher layers and (as a positive feedback mechanism) enhances high convection and rainfall.

Following Bacastow (1976), many authors have found a distinct positive correlation between sea surface temperature (SST) in the upwelling regions of the Pacific, and the detrended and deseasonalized course of the CO_2 content of the atmosphere. A rational interpretation of this relation is still controversial (cf. Gammon et al., 1985), since extratropical ocean sinks and biological processes (on land and in the photic layer of the ocean, see also Weber and Flohn, 1984) may contribute in a complex way.

ROLE OF THE CO_2 AND H_2O FOR ABRUPT CLIMATIC CHANGE

The essential point is that in the atmosphere *both* main greenhouse gases, H_2O and CO_2, increase during ENSO events and decrease during cold (upwelling) episodes. In the case of CO_2 this effect is only small when compared with the large-scale circulation of CO_2 in a balanced ocean-atmosphere budget, i.e. with CO_2 flux from ocean to air in low latitudes (with CO_2-supersaturated water) and from air to ocean in high latitudes (with CO_2-undersaturated water). Due to the long residence time of CO_2 in the atmosphere (about 4 years), the distribution of CO_2 over the globe is rather uniform (cf. Gammon et al., 1985, Fig. 3.13 and 3.20).

H_2O, however, stays only 9 days in the atmosphere before raining out. In the tropics the speed of the vertically integrated transport vector is only of the order of 250km/d (estimated from Peixoto and Oort, 1983) with a component towards the Intertropical Convergence Zone, so that little additional water vapour in the tropics can escape into higher latitudes. Recent studies have indicated (Weber et al.,1986) that in coincidence with a general rise of tropospheric temperatures on the order of 0.04°C/a the water vapour content of the air above the equatorial Pacific above 700hPa (about 3km) has also increased by 0.6-1% per year during the last 20 years, while the annual increase of CO_2 is now of the order 1.6ppm or about 0.5%. These trends - especially the trends of temperature and H_2O observed since 1950 or 1960 - may have contributed to the extraordinary intensity of the recent 1982/3 ENSO event. It should be remembered, however, that in contrast to H_2O, the present CO_2 increase is not a natural, but a man-made phenomenon, and that H_2O operates, in the tropics, not only as a greenhouse gas, but even more effectively through phase changes and release of latent heat.

Weber and Flohn (1984) have confirmed that the global CO_2 growth rate varies strongly between cold and warm episodes of SST in the equatorial Pacific. The differences between these episodes are equivalent to an increase of 1.14-1.36 ppm/a, i.e. nearly the same as the anthropogenic increase. This evidence suggests that under natural conditions - in the tropics - both H_2O and CO_2 are subject to (parallel) changes on the order of 0.5-1% per year. Assuming that the frequency and intensity of ENSO-type events can vary with time - this should be carefully checked, e.g. based on coral growth rates and similar proxy data - this suggests the occurrence of natural changes in the content of both greenhouse gases on the order of 20-30% at a time-scale on the order of 100 years or so as observed in the Greenland ice (Dansgaard et al., 1984; Oeschger et al., 1984).

Referring to CO_2 changes alone, Flohn (1984) had proposed an effective feedback mechanism for climatic change. But visualizing a simultaneous effect of CO_2 and H_2O the situation is more complex; CO_2 stays much longer, spreads globally and has its radiative effect mainly in high latitudes while H_2O remains in the tropical belt (at first in the Pacific alone) and operates there also via phase changes. In this way increasing CO_2 leads to decreasing equator-pole thermal difference; H_2O tends to increase it. This difference controls the intensity of both the tropical (Hadley) cell and the extratropical (Ferrel) cell of the atmospheric circulation. Such a joint operation does not allow a simple estimate of the impact of the increase of both gases on circulation and climate without using complex, interactive general circulation models. External triggers of such dramatic changes in the composition of the atmosphere have been proposed earlier (Flohn, 1984): a cluster (or longer absence) of heavy volcanic eruptions of the explosive type, or a surge of heavy shelf-ice from the Antarctic or the Arctic (Denton and Hughes, 1981).

CONCLUSIONS

The simultaneous occurrence of parallel changes of ambient temperature and CO_2 content in Greenland ice cores, at a time-scale of 5°C and 70ppm per 100 years, has something in common with actual climatic anomalies known as ENSO events. Both are accompanied by parallel changes of CO_2 - which spreads rapidly over the whole globe - and (probably) H_2O, which remains in the tropics, causing heavy rainfall and mid-tropospheric warming. The joint impact of both gases on the atmospheric circulation and on climatic patterns is ambiguous. Most climate models describe CO_2- induced warming including a feedback effect of H_2O as evaporated from slowly warming ocean surfaces. During ENSO cases, however, the sudden shift between "cold" and "warm" events is much faster and more effective than sea surface warming by the greenhouse effect alone. Under palaeoclimatic conditions, it is conceivable that at irregular time intervals in the order of 2000 years short-lived periods with increased frequency (and intensity) of ENSO events produce (or remove), during 100 years or so, as much CO_2 as is needed for such abrupt changes as described, and simultaneously cause an increase (or drop) of water vapour above tropical latitudes of some 10% or more.

Up to now, CO_2 variations of this time-scale have not been found in Antarctica - but investigations with a higher time-resolution are now under way. Nevertheless, the conditions for climatic change and for the inclusion of ambient air into the ice are different in both cases. Greenland is situated at the advancing part of an orographically forced, quasi-stationary upper trough centered around Long. 80-90°W with strong, relatively moist upper winds from the SW; Antarctica forms the center of a permanent cold low in the upper troposphere, with large distances from the periphery into its central part and with a more or less autonomous desert climate. Therefore, with small accumulation as in interior Antarctica (<5cm water equivalent per year), the inclusion of air bubbles must be much slower that at Greenland. This may create long time lags between $\delta^{18}O$ (in snow) and CO_2 signals or even suppress short-living CO_2 peaks.

No unambiguous interpretation can be given for the 1-3 ka interval between these abrupt changes.

REFERENCES

Arkin, P.A. et al., 1983. El Niño - Southern Oscillation Event Quick Look Atlas, NOAA National Meteorological Center, Climate Analysis Center, Washington, D.C.

Bacastow, R., 1976. Modulation of atmospheric carbon dioxide by the Southern Oscillation, *Nature*, 261: 116-118.

Broecker, W.S. and van Donk, J., 1970. Insolation changes, ice volumes and the ^{18}O record in deep-sea cores, *Review of Geophysics and Space Physics*, 8: 169-198.

Coope, G.R., (this volume). Fossil beetle assemblages as evidence for sudden and intense climatic changes in the British Isles during the last 45,000 years.

Dansgaard, W. et al., 1984. North-Atlantic climatic oscillations revealed by deep Greenland ice cores, Geophysical Monograph, 29, Maurice Ewing Vol. 5: 283-298.

Denton, G.H. and Hughes, T., 1981. *The Last Great Ice Shields*, J. Wiley, New York, XVIII + 484 pp.

Dickinson, R.E., 1982. Modeling climate changes due to carbon dioxide increases. In: W.C. Clark (ed.), *Carbon Dioxide Review*, Clarendon Press Oxford, 101-142.

Duplessy, J.C. et al., 1981. Deglacial warming of the northeastern Atlantic Ocean: correlation with the palaeoclimatic evolution of the European continent, *Palaeogeogr. Palaeoclimat. Palaeoecol.*, 35: 121-144.

Emiliani, C. et al., 1975. Paleoclimatic analysis of the Late Quaternary cores from the northeastern Gulf of Mexico, *Science*, 189: 1083-1088.

Flohn, H., 1979. On time-scales and causes of abrupt palaeoclimatic events, *Quaternary Res.*, 12: 135-149.

Flohn, H., 1984. A possible mechanism of abrupt climatic change. In: Mörner and Karlén (eds.), *Climatic Changes on a Yearly to Millennial Basis*, D. Reidel Pub. Co., Dordrecht, 521-531.

Flohn, H., 1986a. Singular events and catastrophes now and in climatic history, *Naturwissenschaften*, 73: 136-149.

Flohn, H., 1986b. Indonesian droughts and their teleconnections, *Berliner Geographische Studien*, 20: 251-265.
Fraedrich, K., 1979. Catastrophes and resilience of a zero-dimensional climate system with ice-albedo and greenhouse feedback, *Quart. Journ. Royal Met. Soc.*, 105: 147-167.
Gammon, R.H., Sundquist, E.T. and Fraser, P.J., 1985. History of carbon dioxide in the atmosphere. In: J.R. Trabalka (ed.), *Atmospheric Carbon Dioxide and the Global Carbon Cycles*, U.S. Dept. Energy, DOE/ER-0239, XXIII + 315 pp. (p. 25-62)
Heine, K., 1983. Ein außergewöhnlicher Gletschervorstoß in Mexiko vor 12 000 Jahren, *Catena*, 10: 1-25.
Heine, K., 1984. The classical Late Weichselian climatic fluctuations in Mexico. In: Mörner, N.A. and Karlén, W. (eds.), *Climatic Changes on a Yearly to Millennial Basis*, D. Reidel Publ. Co., Dordrecht, p. 95-115.
Henning, D. and Flohn, H., 1980. Some aspects of evaporation and sensible heat flux in the tropical Atlantic, *Contrib. Atmos. Physics*, 53: 430-441.
Hense, A., 1987. On the possible existence of a strange attractor for the Southern Oscillation, *Contrib. Atmos. Physics*, 60: 34-47.
Kukla, G., 1980. End of the last glaciation: a predictive model for the future?, *Palaeoecology of Africa*, 12: 395-408.
Leith, C.E., 1984. Global climate research. In: Houghton, J.T. (ed.), *The Global Climate*, Cambridge Univ. Press, p. 13-24.
Lorenz, E.N., 1963. Deterministic nonperiodic flow, *Journ. Atmos. Sci.*, 20: 131-141.
Lorenz, E.N., 1970. Climatic change as a mathematical problem, Journ. Appl. Meteor., 9: 325-329.
Moerner, N.-A., and Karlen, W. (eds.), 1984. *Climatic Changes on a Yearly to Millenial Basis*, Reidel Publ. Co., Dordrecht, 667pp.
Nicolis, C. and G., 1984. Is there a climatic attractor?, *Nature*, 311: 529-532.
Oeschger, H., Beer, J., Siegenthaler, U., Stauffer, B., Dansgaard, W. and Langway, C.C., 1983. Late glacial climate history from ice cores. In: Ghazi, A. (ed.), *Paleoclimatic Research and Models*, D. Reidel Publ. Co., Dordrecht, p. 95-107.
Oeschger, H. and Stauffer, B., 1983 (in print). Review of the history of the atmospheric CO_2 recorded in ice cores, Paper 6th ORNL Life Sciences Symposium on the Global Carbon Cycle, Knoxville, Tenn., Nov. 1983.
Oeschger, H. et al., 1984. Late-glacial climate history from ice cores, Geophysical Monograph 29, Maurice Ewing Vol. 5: 299-306.
Peixoto, J.P. and Oort, A.H., 1983. The atmospheric branch of the hydrological cycle and climate. In: A. Street-Perrott et al. (eds.), *Variations in the Global Water Budget*, D. Reidel Pub. Co., Dordrecht, 5-65.
Rasmusson, E.M. and Carpenter, T.H., 1982, Variations in tropical sea surface temperature and surface wind fields associated with the Southern Oscillation/ El Niño, *Monthly Weather Rev.*, 110: 354-384.
Rasmusson, E.M. and Carpenter, T.H., 1983. The relationship between eastern equatorial Pacific sea surface temperatures and rainfall over India and Sri Lanka, *Monthly Weather Rev.*, 111: 517-528.

Reiter, E.R., 1978. Long-term wind variability in the tropical Pacific, its possible causes and effects, *Monthly Weather Rev.*, 106: 324-330.
Rognon, P., 1983. Les crises climatiques de courte durée (quelques année à quelque siècles) et leur enregistrement dans la sédimentation continentale. In: Ghazi, A. (Ed.), *Paleoclimatic Research and Models*, D. Reidel Publ. Comp., Dordrecht, p. 114-123.
Schuster, H.G., 1984. *Deterministic Chaos - An Introduction*, Weinheim.
van Loon, H. and Shea, D.J., 1985. The Southern Oscillation, Part IV: the precursors south of 15°S to the extremes of the oscillation, *Monthly Weather Rev.*, 113: 2063-2074.
Weber, K.-H. and Flohn, H., 1984. Oceanic upwelling and air-sea exchange of carbon dioxide and water vapour as a key for large-scale climatic change?, *Bonner Meteorologische Abhandlungen*, 31: 73-98.
Weber, K.-H. et al., 1986. Variation des Wasserdampfes in der unteren Troposphäre zwischen 20°N and 20°S während der letzten 20 Jahre, *Ann. Meteor. Soc.*, 23: 5-6.
Wright, P.B., 1985. The Southern Oscillation: An ocean-atmosphere feedback system?, *Bull. Amer. Meteor. Soc.*, 66: 398-412.

HOW TO RECOGNIZE AN ABRUPT CLIMATIC CHANGE?

C. Goossens and A. Berger

Institut d'Astronomie et
de Géophysique G. Lemaitre
Université Catholique de Louvain
2 Chemin du Cyclotron
1348 Louvain-la-Neuve
Belgium

ABSTRACT. Climatic changes have been detected by means of an appropriate statistical test, the Mann-Kendall rank statistic. Most of the instrumental temperature series exhibit a general warming which support the reality of a 20th century northern hemisphere temperature trend noted by many authors. It clearly appears that this long term warming over which higher frequencies variations are superimposed started generally around 1900 A.D.±25 years and is still presently observed. This continuous warming, corresponding to the low frequency part of the 20th century records, might thus be associated with the observed atmospheric CO_2 increase during this period. Proxy climatic data over more than 700 kyr show mostly a great number of oscillations marking the transition from full glacial to full interglacial conditions. For the last 150 kyr, results of the test confirm the temperature maximum around 120 kyr BP (before present) followed by a decrease until a minimum around 20 kyr BP and a climatic optimum some 7 kyr ago.

DEFINITION OF CLIMATIC CHANGES

The Earth's climate has always been changing and the magnitude of these changes has varied from place to place and from time to time. The facts are, of course, that climate does fluctuate quite noticeably from year to year and that there are gradual long-term changes in climate which make one decade, one century or one millennium different from the one before.

Such yearly fluctuations and long-term changes result from natural external processes or from internal influences on the complex system that determines the Earth's climate, a system which includes the atmosphere, the oceans, the land and ice masses and the biosphere.

Before discussing any property of the climate system or how a climatic change can arise, it is useful to define the terms used in this

W. H. Berger and L. D. Labeyrie (eds.), Abrupt Climatic Change, 31–45.

context. Let us begin with a general definition of climate and weather (Lamb, 1972, chap. I and Berger, 1981):

Climate may be understood as "the sum total of the weather experienced at a place in the course of the years and over the years. It comprises not only those conditions that can obviously be described as normal but also the extremes and all the variations". *Weather* in turn is taken to mean "the totality of atmospheric conditions at any particular place and time".

The term "climatic change" has been used extensively in recent years, by scientists as well as journalists and others outside the scientific community, without being defined in any precise way. In fact, one finds in literature reference to climatic change, climatic fluctuation, climatic oscillation, climatic trend, climatic vacillation and climatic variation.

In order to avoid confusion, it is convenient to define each term as a particular form of a general climatic change, following the definitions proposed by Mitchell (1966) and used, among others, by the United States Committee for the Garp (1975), Berger (1981), WCRP (1981), Newell (1981), Kates et al. (1983) and Hare (1983).

- the expression *climatic change* is the completely general term that encompasses all forms of climatic inconstancy, regardless of their statistical nature or physical causes (Mitchell, 1966). In fact, climatic change occurs on a variety of time scales. On a long time-scale, for example, there are the glacial periods, when the volume of water locked up as ice tripled and temperature dropped by 10°C or more in middle latitudes. Another type of change on a shorter scale occurred in Europe and perhaps elsewhere from a warmer-than-present period in the thirteenth century to a colder-than-present period centered in the seventeenth century and often called the Little Ice Age. The amplitude of the change in the mean air temperature was probably only 1° to 2°C. On an even shorter time scale there are temporary coolings comprising a year or two, which amount up to 1°-2°C. Some of these events, like that of 1815, have had catastrophic effects on agriculture (Newell, 1981).

- a *climatic trend* is a climatic change characterized by a smooth, monotonic increase or decrease *of average value* in the period of record (Mitchell, 1966). This trend is not restricted to a linear change with time but characterized by only one minimum and one maximum at the end points of the record.

- a *climatic discontinuity* is an abrupt and permanent change during the period of record from one average value to another. It is what we are going to call an abrupt climatic change.

- a *climatic fluctuation* is any form of systematic change, whether regular or irregular, characterized by at least two maxima (or minima) and one minimum (or maximum) included in the record.

- a *climatic variation* is a fluctuation whose characteristic time scale is sufficiently long to result in appreciable differences between successive means (or normal) taken over a specific interval of time, usually a number of decades.

- a *climatic oscillation* is a fluctuation in which the variable tends to move gradually and smoothly between successive maxima and minima.

- a *climatic vacillation* is a fluctuation in which the climatic variable tends to dwell alternately around two (or more) average values, and to drift from one average value to the other at regular or irregular intervals.

- a *climatic periodicity* is an oscillation or a vacillation in which the successive maxima and minima occur at approximately equal intervals of time.

- a *climatic variability* is best thought of as the manner of variation of the climatic parameters within the typical averaging period. Suitable measures of variability include such parameters as the standard deviations of continuous elements like temperature and pressure (Kates et al., 1983; Hare, 1983).

In these definitions, we have not given any precise time intervals to distinguish between climatic variations, fluctuations and oscillations since the distinctions are, like the definition of climate itself, a matter of viewpoint and application. General discussions of climate, therefore, usually require a specification of the application time scale: inter-annual, decadal and so on.

To sum up, we have determined a relationship between the terms of the above glossary which may be visualized by reference to the following diagram:

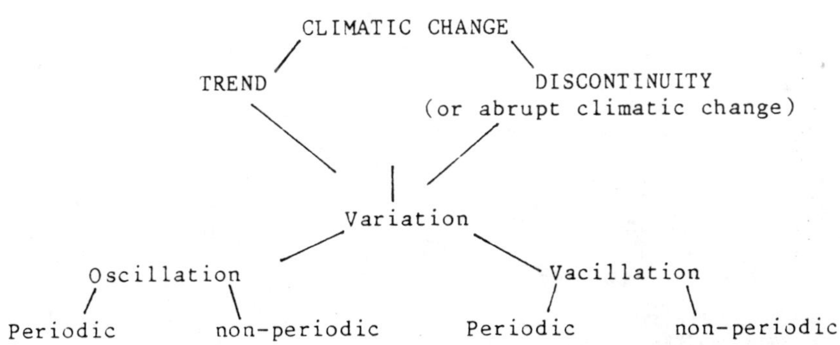

PROXY AND INSTRUMENTAL DATA

To study the gradual changes in climate over millennium, century or decade, proxy data (from deep-sea cores, for example) and instrumental ata have been used.

Deep-sea sediment cores contain the fossils of species which lived at various depths of the ocean and can yield a measure of the temperature of the environment in which they lived. Especially useful in this regard are the benthic and planktonic foraminifera, which inhabit the sea floor and surface ocean layers respectively. Twelve of the most accurate existing oxygen isotope records have been selected from the paleoclimatic data bank (PACDATA, Berger and Pestiaux, 1982). They have been classified in three groups (Berger and Pestiaux, 1984) covering respectively more than 700 kyr for group L3, 400 kyr (only the last oxygen isotope stages) for group L2 and less than 125 kyr (the last Interglacial-Glacial cycle) for group L1. The chronologies of these records have been built either by absolute dating or/and by isotopic stratigraphy without making any a priori assumption of a relationship between a master curve and the analyzed isotopic record. The location, name and source of the cores used are given in Table 1.

Table 1. Location, name and reference of the oxygen isotope record (from PACDATA, Berger and Pestiaux, 1982)

NAME	LATITUDE	LONGITUDE	DEPTH (M)	PLANCTONIC (P) OR BENTHIC (B)	REFERENCE
L1	130 KYR				
MD76131	15 32N	72 34E	1230	P	D
				B	
MD76135	12 26N	50 31E	1895	P	D
				B	
MD73025	43 49N	51 16E	3284	P	D
				B	
L2	400 KYR				
RC11120	43 31S	79 52E	3913	P	S
V1928	10 22S	84 39W	2720	B	S
V1929	03 35S	83 13W	3157	B	S
V19240	30 35S	13 17E	3013	P	S
L3	730 KYR				
MD73004	04 58S	61 40E	3440	P	D
V28238	01 01N	160 29E	3120	P	S
V28239	03 15N	159 11E	3490	P	S
502B	11 30N	79 23W	3051	P	
P64089	14 55N	69 05W	4098	P	
M13519	05 40N	19 51W	2862	P	

D = DUPLESSY
S = SHACKLETON

Instrumental records of past climate date back from only two or three centuries. The annual temperature anomalies for the Northern Hemisphere (Jones, 1985a), relative to the 1951-70 reference period, the air surface temperature of the Arctic region (65-85°N) (Jones, 1985b), the average temperature series of Gruza and Ran'kova (1979) for selected latitudes (20°N, 45°N, 55°N, 65°N, 80°N and 30-85°N) (Table 2), the reconstructed temperature series for Central England (Manley, 1974), Paris (Detwiller, 1981) and De Bilt (Labrijn, 1945, Schuurmans, 1981) as well as most of the instrumental Western European series (Table 3) have been analysed. The basic data set is constructed from the monthly station data, originally published by the Smithsonian Institution in the various volumes of the World Weather Record up to 1960, digitized by the National Centre for Atmospheric Research (Jenne, 1975) and updated by NOAA in Monthly Climatic data for the World (MCDW).

ABRUPT CLIMATIC CHANGE

The analysis of abrupt climatic changes has been made using the test of Mann-Kendall (Mann, 1945; Sneyers, 1975; Goossens, 1983). This nonparametric test is based on the fact that, under the hypothesis of a stable climate, the succession of climatological values must be independent and the probability distribution must remain always the same (simple random series). The efficiency of this test has been demonstrated using a Monte Carlo technique (Goossens and Berger, 1985). The null hypothesis of no change in a time series (x_i, $1 \le i \le N$) is accepted if the probability $\alpha_1 = P(|u|<|u(d_N)|)$, associated with the standardized value of the test is greater than a given significance level α_0 (in practice, α_0 equal to 0.05) :

$$u(d_N) = (d_N - E(d_N))/\sqrt{\text{var } d_N} \quad (1)$$

In this formula, $d_N = \Sigma_i\ m_i$, m_i is the number of terms x_j preceeding each term x_i in the chronology and smaller then x_i:

$$m_i = \{x_j\ :\ j<i \text{ and } x_j<x_i\} \quad (2)$$

Its distribution law under the null hypothesis, is asymptotically normal with the following mean and variance:

$$E(d_N) = N(N+1)/4$$
$$\text{var } d_N = N(N-1)(2N+5)/72. \quad (3)$$

If the null hypothesis has been rejected, i.e. that a change does exist somewhere in the time series, the sign of $u(d_N)$ will in addition allow us to conclude that this change was towards an increase of the variable for $u(d_N)>0$ or towards a decrease for $u(d_N)<0$ (refer to Sneyers, 1975 and Goossens and Berger, 1986 for demonstration and examples). Using such a simple and objective method, the beginning of an abrupt climatic change may then be detected by a graphical representation of the sequential version of the statistic of the test $u(d_i)$, $1 \le i \le N$. The graphical representation of this ensemble of all

Table 2. Sources of the average Northern Hemisphere temperature series.

AVERAGE NORTHERN HEMISPHERE TEMPERATURES SERIES		
N.H. Temperatures Anomalies	1881-1981	Jones et al. (1982)
N.H. Temperatures Anomalies	1851-1984	Jones (1985a)
Arctic Temperatures	1851-1984	Jones (1985b)
N.H. Temperatures (30-85 N)	1891-1976	Gruza and Ran'kova (1979)
N.H. Temperatures (80 N)	1891-1976	"
N.H. Temperatures (65 N)	1891-1976	"
N.H. Temperatures (55 N)	1891-1976	"
N.H. Temperatures (45 N)	1891-1976	"
N.H. Temperatures (20 N)	1891-1976	"

Table 3. A list of most of the European stations where instrumental temperature series are available over more than 100 years.

COUNTRY	STATION	PERIODS OF OBSERVATIONS
Austria	Wien	1755-1981
Belgium	Uccle	1833-1969
France	Paris	1874-1960
	Nantes	1881-1981
	Marseille	1871-1981
Germany	Berlin	1769-1970
	Frankfurt	1833-1970
	Hohenpeissenberg	1781-1970
Greece	Athens	1859-1977
Italy	Milan	1866-1978
Nederlands	De Bilt	1849-1980
Portugal	Lisboa	1864-1977
Spain	Madrid	1860-1979
	Palma	1866-1979
United	Aberdeen	1871-1981
Kingdom	Edinburgh	1764-1960
	Greenwich	1841-1960
Ireland	Valentia	1867-1979
Switzerland	Bale	1755-1970
	Zurich	1864-1981
	Santis	1883-1981
	Genève	1768-1970

Reference : World Weather Records
Monthly climatic data for the world

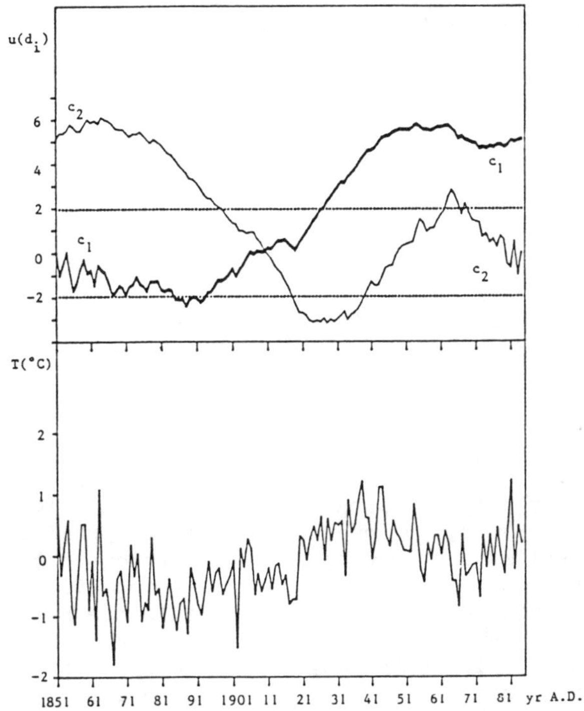

Figure 1. Lower part: Surface air temperature for Arctic region (65-85°N) (1851-1984) (Jones, 1985b). Upper part: Values based on the sequential version of the Mann-Kendall test. Dashed lines indicated the value of the 0.05 significance levels. The clear intersection of curves c_1 and c_2 between the 5% levels of significance localizes the abrupt change towards a warming around 1910.

$u(d_i)$, $1 \le i \le N$ along the time axis is denoted c_1. The same principle may also be applied to the retrograde series. In this case, the curve is labelled by c_2.

As shown in Goossens and Berger (1985), among the numerous tests generally used to detect a climatic change (low pass-filter, Cramer's test, rank Spearman statistic ...), this Mann-Kendall rank statistic seems the most appropriate, its sequential version allowing the detection of the approximate beginning of an abrupt climatic change. Applying the meaning of $u(d_N)$ and of its associated 5% significance levels to all $u(d_i)$ $1 \le i \le N$, which generate both the normal and the retrograde time series used for drawing the curves c_1 and c_2, we can conclude that the intersection of curves c_1 and c_2 localizes the timing of the change, provided this intersection is situated between the critical values of the 5% level of significance (Goossens and Berger, 1985). Moreover, the existence of a warming (or a cooling) trend may also be put forward if the intersection of the two curves c_1 and c_2 is localized above the critical value of the 5% level of significance. In this case, the maximum (or the minimum) of curve c_1 corresponds to the beginning of the trend.

ABRUPT CHANGES IN INSTRUMENTAL TEMPERATURE RECORDS

We have applied the test of Mann-Kendall to the instrumental records in order to try to detect abrupt climatic changes, if any (Figs 1 to 4). Most of the temperature series present a warming which supports the reality of the 20th century Northern Hemisphere warming noted by many authors (Mitchell, 1963; Kukla et al., 1977; Wigley, 1980). From the results obtained (for an extensive description of the results refer to Goossens and Berger, 1986), we may conclude that for the high latitudes (65°–85°N), the intersection of c_1 and c_2 localizes the abrupt change towards a warming around 1910 (Fig. 1). Superimposed on this change we observed a long-term oscillation which exhibits a warming up to the 1950's, followed by a cooling up to around 1970. Since then, the values increase again. For stations situated between 50°N and 60°N, the main change is observed around 1900 (Fig. 2). Stations situated between 40°N and 50°N exhibit the main change to a warmer climate later on, around 1925 (Fig. 3). For all these latitudes, we are still experiencing this long-term warm climate now. In fact, this change towards a warming is observed everywhere except in low latitudes where the long-term climate is steady and only short-term fluctuations are present (Fig. 4).

Although it is not the purpose of this paper to discuss the cause(s) of the detected warming and of the early 20th century continuity break, the following remarks must be made. First, it is true that pre-1900 data were obtained by sometimes questionable measuring and averaging techniques affecting the homogeneity of the records. It is also known that the cities are experiencing a urban warming which progressively grows with time (Kukla et al., 1986). Although Jones et al. (1986) tried to remove this urban warming from the northern hemisphere temperature series, we must continue to be fairly cautious about the true hemispheric temperature trends. However, the signal found for the yearly mean (Goossens and Berger, 1986) is mainly concentrated in the intermediate seasons (spring and fall) and not in winter, as it must have been found if entirely related to the heat island effect. Second, if the hemisphere and the globe were warming (what we found quite likely for the 1900-1985 interval), the question is how much of that warming was due to some long term natural variability and how much to the CO_2 (Lough et al., 1983). Although a full range of possible causes was not investigated as done by Gilliland (1982) who considered volcanic, solar and CO_2 forcing and by Hansen et al. (1981), the transient response to the CO_2 increase over the last 100 years obtained by Tricot and Berger (1986) with a simple time dependent atmosphere-ocean coupled model is coherent with the amplitude and the time scale of the warming described in this paper. In other words, without claiming necessarily that there is a one-to-one cause to effect

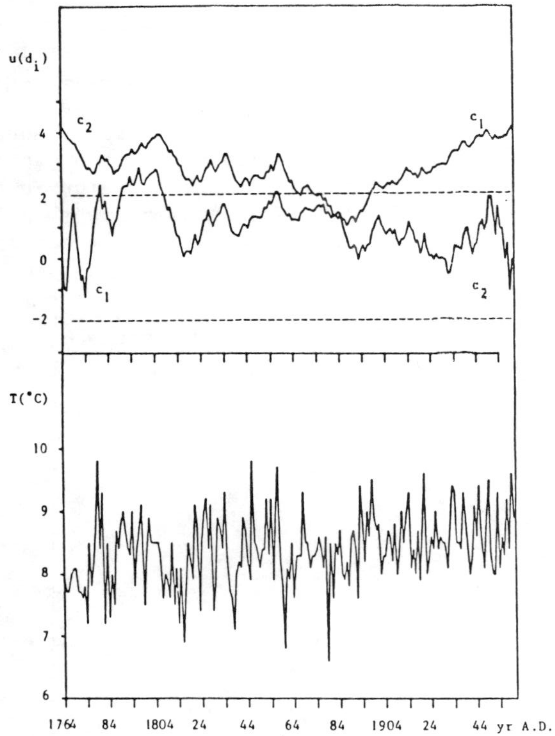

Figure 2. Lower part: Mean annual air temperature for Edinburgh (55°54N-3°12W) (1764-1960). Upper part : Values based on the sequential version of the Mann-Kendall test. Dashed lines indicated the value of the 0.05 significance levels. The intersection of curves c_1 and c_2 between the 5% levels of significance localizes the abrupt change towards a warming around 1884.

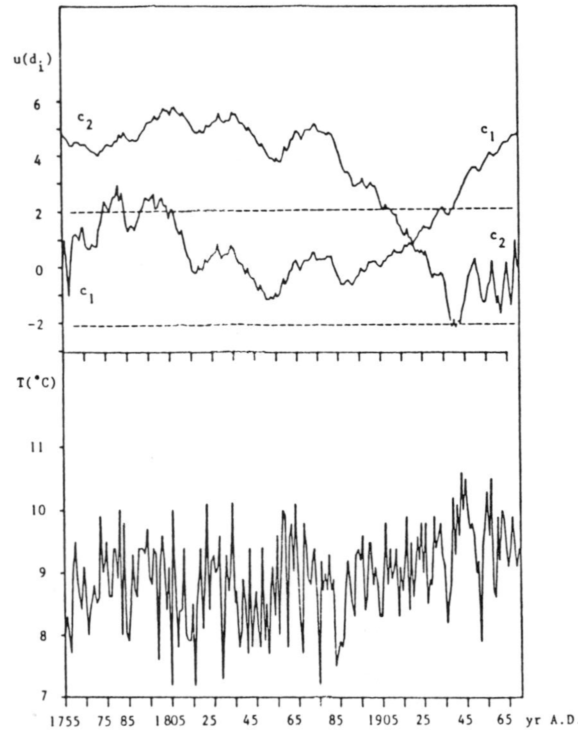

Figure 3. Lower part: Mean annual air temperature for Bâle (47°33N - 7°35W) (1755-1970). Upper part: Values based on the sequential version of the Mann-Kendall test. Dashed lines indicated the value of the 0.05 significance levels. The clear intersection of curves c_1 and c_2 between the 5% levels of significance localizes the abrupt change towards a warming around 1920.

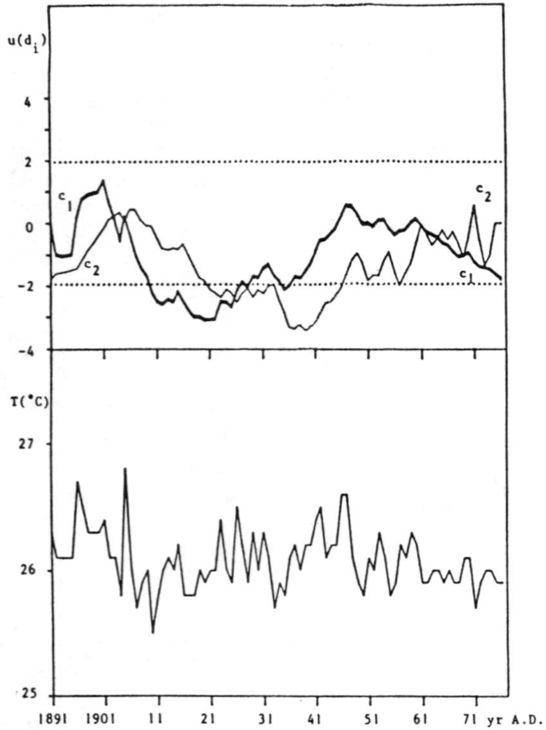

Figure 4. Lower part: Low-latitudes (20°N) air temperature (1891-1976) Gruza and Ran'kova, 1979). Upper part: Values based on the sequential version of the Mann-Kendall test. Dashed lines indicated the value of the 0.05 significance levels. The large number of intersections between curves c_1 and c_2 prevents the singling out of any change towards a warming or cooling.

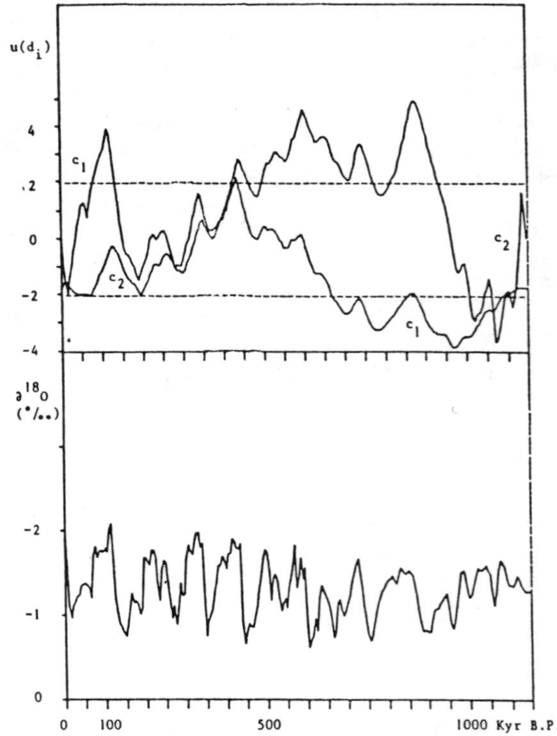

Figure 5. Lower part: Basic chronology of the isotope record V28-238 (L3 group) (Shackleton and Opdyke, 1976). Upper part: Values based on the sequential version of the Mann-Kendall test. Dashed lines indicated the value of the 0.05 significance levels. The test does not clearly single out any of the climatic changes as an abrupt change.

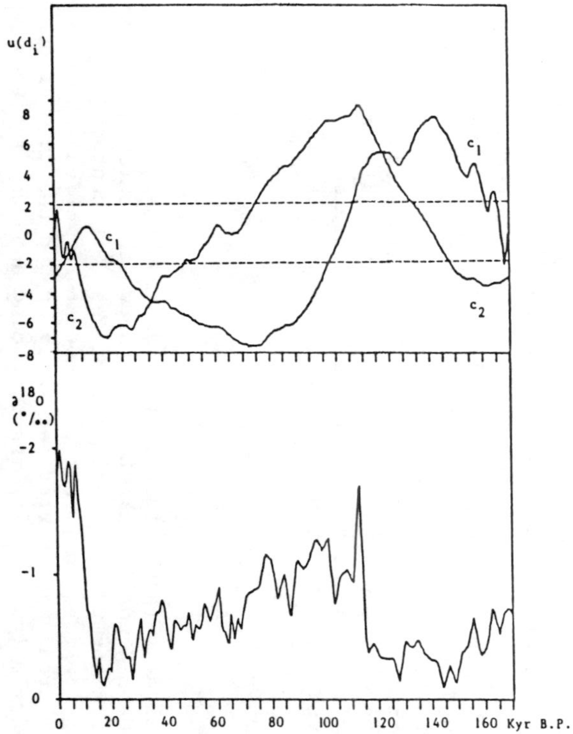

Figure 6. Lower part: Basic chronology of the isotope record MD73004 (L1 group) (Duplessy, unpublished). Upper part: Values based on the sequential version of the Mann-Kendall test, Dashed lines indicated the value of the 0.05 significance levels. The test does not clearly single out any of the climatic changes as an abrupt change.

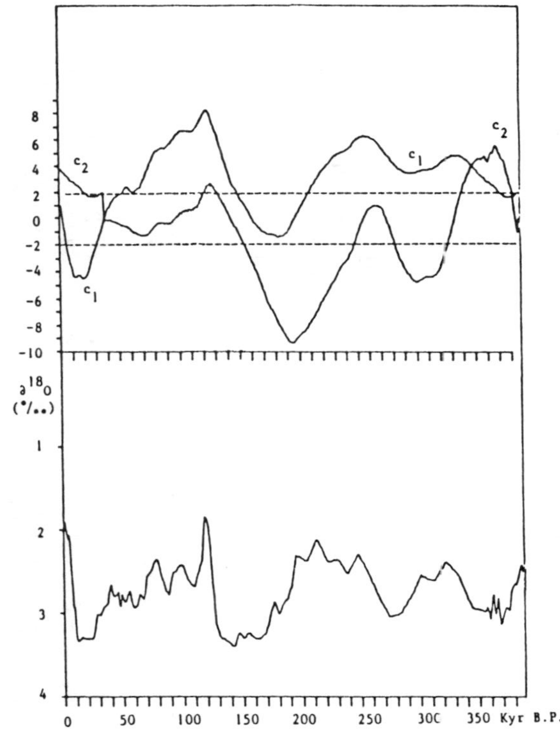

Figure 7. Lower part: Basic chronology of the isotope record RC11120 (L2 group) (Shackleton, 1977). Upper part: Values based on the sequential version of the Mann-Kendall test. Dashed lines indicated the value of the 0.05 significance levels. The test does not clearly single out any of the climatic changes as an abrupt change.

relationship, it is at least worthwhile to note that the CO_2 induced warming of the 20th century is not inconsistent with the warming observed through a careful statistical analysis of the records over the same period.

ABRUPT CHANGES IN CLIMATE PROXY DATA

The oxygen isotope records show a great number of oscillations between glacials and interglacials. The interpretation of the c_1 and c_2 curves is very difficult because no clear crossing is situated within the significance levels. The records display actually a saw-tooth shape (Goossens and Berger, 1985) where only maxima and minima can usually be localized. Consequently, contrary to the result obtained from the 20 century records, our test does not clearly single out any of the climatic changes as an "abrupt change". Results only confirm the main known climatic features of the Quaternary.

Deep-sea oxygen isotope records in category L3 (Fig. 5, core V28-238) show a great number of these oscillations which mark the alternation of glacial and interglacial periods during the Pleistocene: the Earth's climate was characterized in both hemispheres by periods of rising and falling temperature. It must however be underlined that contrary to the cores younger than 500 kyr BP, core V28-238 and others belonging to the L3 group do not present a similar behavior. This could be linked, for example, to a variable sedimentation rate during the last several hundred thousand years, to the resolving power of each indicator or to their geographical localization.

For deep-sea cores of L1 and L2 categories (Fig. 6 for L1 group and Fig. 7 for L2 group), the results of the test confirm the maximum around 120 kyr BP (the Eemian interglacial), followed by a decrease until a minimum around 20 kyr BP (time of the last extensive continental glaciation during the Pleistocene). After this period, the climate warmed again up to the climatic optimum ≈7 kyr BP), this increase marking the beginning of the present interglacial. It is also observed that the transition from glacial to interglacial conditions is definitely more abrupt than the transition between interglacial to glacial.

CONCLUSIONS

For instrumental data (approximately the last 200 years), the application of the Mann-Kendall rank statistic shows the existence of an abrupt climatic change towards the beginning of the 20th century. Considering only the low frequency part of temperature series, it seems clear that a warming trend has started generally around the turn of our century (1900±20 years), its exact timing depending upon the latitude of the region analysed. Such a long term warm climate is still experienced now. It is only the high frequency climatic fluctuations superimposed on it which display a maximum in the 1940-1950's and two minima, one centered around the end of the 19th century and the other around 1970. This long term warming might be related to (or at least is not in opposition with) the calculated CO_2 warming which is based on the continuous 20th century atmospheric CO_2 increase.

For paleoclimatic records, the results of the test show a great number of oscillations between glacials and interglacials. In such a case, the detection of an abrupt climatic change requires the splitting of the initial records into sub-periods of approximately 150 kyr before the application of the test.

The next step of this analysis, in order to bridge the gap between the modern instrumental records and less accurate, more general paleoclimatic record spanning many glacial periods, should be the application of this test to the high resolution tree-ring chronologies.

ACKNOWLEDGEMENTS

Thanks are due to Dr. G. Mélard (Institut de Statistique, Université Libre de Bruxelles, Belgique), Dr. J.M. Mitchell, Jr. (National Oceanic and Atmospheric Administration, Silver Spring), to Dr. A. Pittock (CSIRO, Division of Atmospheric Physics, Australia), and to Dr. R. Sneyers (Institut Royal Météorologique, Bruxelles, Belgique) for their helpful suggestions, to the Climatic Research Unit of East Anglia for providing the European data, to Prof. C. Schuurmans (University of Utrecht and KMI, De Bilt) for providing the data of De Bilt and to Dr. Detwiller (Direction de la Météorologie, Paris) for providing the data of Paris.

REFERENCES

Berger, A. (Ed.), 1981. *Climatic Variations and Variability: Facts and Theories*, D. Reidel Publ. Company, Dordrecht, Holland.

Berger, A. and Pestiaux, P., 1982. An introduction to the IAG Pacdata bank, *Contribution n°28*, Institut d'Astronomie et de Géophysique G. Lemaitre, Université Catholique de Louvain.

Berger, A. and Pestiaux, P., 1984. Astronomical frequencies in Paleoclimatic data. In: *Proceedings of Beijing Climate Symposium*, October 1984.

Detwiller, J., 1981. Les températures annuelles à Paris durant les 300 dernières années, *La Météorologie*, VI, 25: 103-110.

Gilliland, R.L., 1982. Solar, volcanic, and CO_2 forcing of recent climatic changes, *Climatic Change*, 4: 111-132.

Goossens, C., 1983. Etude de l'homogenéité et de la stabilité des séries climatiques, Scientific Report 1983/1, Institut d'Astronomie et de Géophysique, G. Lemaitre, Université Catholique de Louvain.

Goossens, C. and Berger, A., 1986. Annual and seasonal climatic variations over the Northern Hemisphere and Europe during the last century, *Annalae Geophysicae*, 4(B4): 385-399.

Goossens, Chr. and Berger, A., 1985. How to recognize an abrupt climatic change? *Scientific Report 1985/5*, Institut d'Astronomie et de Géophysique G. Lemaitre, Université Catholique de Louvain.

Gruza, G.V. and Ran'kova, E. Ya., 1979. Data on the structure and variability of climate-air temperature sea level. Northern Hemisphere, *Central Institute for the Collection of Hydrometeorological Information - World Data Centre*, Obninsk, USSR, 204pp.,(in Russian).

Hansen, J., Johnson, D., Lacis, A., Lebedeff, S., Lee, P., Rind, D. and Russell, G., 1981. Climate impact of increasing atmosphere carbon dioxide, *Science*, 213: 957-966.

Hare, K., 1983. Climate and desertification : a revised analysis, *World Climate Programme*, 44, WMO, Geneva.

Jenne, R., 1975. Data sets for meteorological research, *NCAR-TN/IA-III*. National Centre for Atmospheric Research, Boulder, 194pp.

Jones, P.D., 1985a. Northern Hemisphere temperatures 1851-1984, *Clim. Mon.*, 14(1): 14-21.

Jones, P.D., 1985b. Arctic temperatures 1851-1984, *Clim. Mon.*, 14(2): 43-50.

Jones, P.D., Raper, S.C.B., Bradley, R.S., Diaz, H.F., Kelly, P.M. and Wigley, T.M.L., 1986. Northern Hemisphere surface air temperature variations: 1851-1984, *Journal of Climate and Applied Meteorology*, 25: 161-179.

Kates, R.W. et al., 1983. Improving the science of climatic impact assessment, *SCOPE* Report, J. Wiley & Sons Chicester.

Kukla, G.J., Angell, J.K., Korshover, J., Dronia, H., Hoshiai, M., Namias, J., Rodewald, M., Yamamoto, L. and Iwashima, I., 1977. New data in climatic trends, *Nature*, 270: 573-580.

Kukla, G.J., Gavin J. and Karl, T.R., 1986. Urban warming, *Journal of Climate and Applied Meteorology*, 25: 1265-1270.

Labrijn, A., 1945. The climate of the Netherlands during the past 2 1/2 centuries, *Med. Verh. Kon. Ned. met. Inst.*, 49: 102.

Lamb, H., 1972. *Climate: Present, Past and Future*. Vol. 1, Methuen and Co., London.

Lough, J.M., Wigley, T.M.L. and Palutikoff, J.P., 1983. Climate and climate impact scenarios for Europe in a warmer world, *J. of Climate and Appl. Mett*, 22(10): 1673-1684.

Manley, G., 1974. Central England temperature: monthly means 1659-1973, *Quart. J. Meteor. Soc.*, 100: 387-405.

Mann, H.B., 1945. Non-parametric test against trend, *Econometrika*, 13: 245-259.

Mitchell, J.M., 1963. On the world-wide pattern of secular temperature change, *Proc. UNESCO-WMO Symposium of Climatic Changes of Climate*, Unesco Arid Zone Research Series. 20, Unesco, Paris, 161-181.

Mitchell, J.M., 1966. Climatic Change, *Technical Note* 79, WMO, Geneva.

Newell, R.E., 1981. Climate and the ocean. In: *Climate, Past and Present*, Skinner B.J. (Ed.), Williams Kaufmann, Inc., California.

Schuurmans, C., 1981. Central England and De Bilt series. In: *Climatic Variations and Variability: Facts and Theories*, A. Berger (Ed.), D. Reidel Publ. Company, Dordrecht, Holland.

Shackleton, N.J., 1977. The isotope stratigraphic record of late Pleistocene, *Philos. Trans. R. Soc. London*, B.280: 169-182.

Shackleton, N.J. and Opdyke, N.D., 1976. Oxygen isotope and paleomagnetic stratigraphic of equatorial Pacific core V28-238: oxygen isotope temperature and ice volumes on a 10^5 year and 10^6 year scale, *Quaternary Research*, 3: 39-55.

Sneyers, R., 1975. Sur l'analyse statistique des séries d'observations, *Technical Note* 143, WMO, Geneva.

Tricot, Ch. and Berger, A., 1986. Modelling the equilibrium and transient responses of temperature to past and future trace gases concentrations, *Climate Dynamics*.

United States Committee for the Global Atmospheric Research Program, 1975. *Understanding Climatic Change*, A program for Action, National Academy of Sciences, Washington D.C.

WCRP, 1981. Preliminary plan for the world climate research programme, *World Climate Research Programme*, WMO, Geneva.

Wigley, T.M.L., Jones, P.D. and Kelly, P.M., 1980. Scenario for a warm, high CO_2 world, *Nature*, 283: 17-21.

THE RAPIDITY OF CO_2-INDUCED CLIMATIC CHANGE: OBSERVATIONS, MODEL RESULTS AND PALAEOCLIMATIC IMPLICATIONS

P.D. Jones, T.M.L. Wigley and S.C.B. Raper

Climatic Research Unit
School of Environmental Sciences
University of East Anglia
Norwich NR4 7TJ, U.K.

ABSTRACT. The rates of global and regional temperature changes due to increasing greenhouse gas concentrations are estimated using a multibox, upwelling-diffusion climate model. The model is also used to estimate how these changes might be perturbed by changes in deep water formation rate. The results are interpreted in a palaeoclimatic context and it is shown that the magnitude and rapidity of observed late-glacial changes in climate are compatible with large shifts in the rate of formation of North Atlantic Deep Water.

INTRODUCTION

The rapidity of climatic change is controlled, not only by externally imposed forcing, but also by internal factors within the climate system. The most important of these is oceanic thermal inertia, which considerably dampens the response of the climate system to external forcing. In this paper we use the example of the changes in climate due to projected future changes in the atmospheric concentrations of the "greenhouse" gases (carbon dioxide and other trace gases) as a guide to the extremes of rapidity of climatic change that might occur in other circumstances. Quantitative estimates of the rates of change of temperature in response to greenhouse gas forcing are made with an upwelling-diffusion climate model which accounts for the damping effect of oceanic thermal inertia on temperature. This model also allows us to simulate the influence of changes in North Atlantic Deep Water (NADW) formation rate, and we show that such changes can cause changes in temperature that are both rapid and large.

RECENT CHANGES IN GLOBAL MEAN TEMPERATURE

Recently, surface land temperature data have been carefully examined for homogeneity. Data errors arising from many sources have been corrected, and new area-average time series for the land masses of the Northern and Southern Hemispheres have been produced back to the mid nineteenth century (Jones et al., 1986a, b). Marine data have also been

W. H. Berger and L. D. Labeyrie (eds.), Abrupt Climatic Change, 47–55.

subjected to detailed analysis and attempts have been made by Folland et al. (1984) and Jones et al. (1986c) to correct for the large and time-varying non-climatic biases that exist in these data. By combining the land and marine data, area-averages representative of both hemispheres and the globe have been produced (Jones et al., 1986c). These are a considerable improvement over earlier such estimates which almost invariably used only land-based data (e.g. Hansen et al., 1981).

The two homogenized marine data series, however, are not in complete accord, especially prior to about 1900. The marine data of Jones et al. (1986c) are compatible with the land data throughout the whole period of record, 1861-1984, while those of Folland et al. (1984) are noticeably warmer than the land prior to 1900. There is still some dispute over which of these two data series is the better. The issue is discussed in detail by Jones et al. (1986c) whose global mean (land plus marine) series is shown here in Figure 1. It can be seen that the globe was relatively cool for the period 1860-1900, and that there has been a warming of about 0.5°C since the late nineteenth century. The early 1980's has been the warmest 5-year period on record.

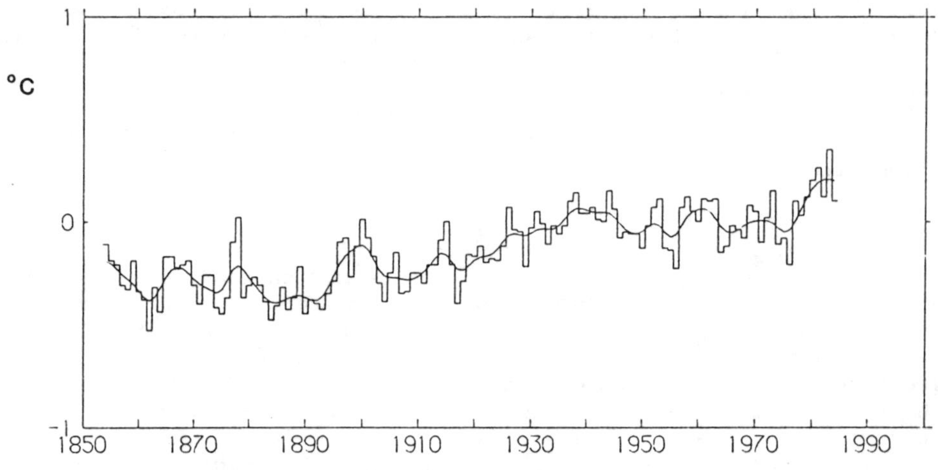

Figure 1. Estimated global mean temperature changes from Jones et al. (1986c). The smooth curve was obtained using a Gaussian filter designed to suppress variations on time scales less than 10 years.

CAUSES OF RECENT CLIMATIC CHANGE

It is now generally thought that the main cause of the large-scale warming trend that has occurred over the last hundred years is the increase in the concentrations of carbon dioxide and other trace gases in the atmosphere, resulting from human activities (MacCracken and Luther, 1985). The shorter time scale fluctuations within this trend may be looked on as natural background noise, caused either by external factors (such as changes in solar irradiance and changes in volcanogenic stratospheric aerosol loading) and/or by internal factors (such as changes in the ocean circulation, changes in surface albedo, changes in cloudiness, etc.).

Model results suggest that the strength of the CO_2 "greenhouse" signal is such that a doubling of atmospheric CO_2 concentration would cause an *equilibrium* global mean surface air temperature increase of 1.5-4.5°C (MacCracken and Luther, 1985; W.M.O., 1986). This is a measure of the sensitivity of the climate system to external forcing. Its uncertainty arises because of our inability to adequately model all of the complex feedback mechanisms within the climate system. These have uncertain magnitude, and even, in the case of clouds, uncertain sign (Schlesinger and Mitchell, 1985).

The equilibrium response, however, does not immediately tell us what the time-dependent response to time-dependent forcing would be. At any given time, there would be a time-dependent lag of the actual temperature change in response to a steadily increasing CO_2 concentration behind the instantaneous equilibrium temperature change because of the damping effect of oceanic thermal inertia. The current magnitude of this lag is uncertain, and it may range from 10 to more than 50 years (Hansen et al., 1984; Wigley and Sclesinger, 1985). Estimates depend on how the ocean is modeled. Because of this transient response uncertainty, it is difficult to determine the climate sensitivity from an analysis of observational data, and quite a wide range of values of the CO_2-doubling temperature change are compatible with an observed warming of 0.5° (Wigley and Schlesinger, 1985).

To estimate future changes in global mean temperature that might occur in response to continually increasing greenhouse gas concentrations, one needs a scenario for the future concentrations of these gases and a suitable climate model. The combined effect of changes in the trace gases methane, nitrous oxide, ozone and chlorofluorocarbons has, in the past, been similar in magnitude to the effect of changes in CO_2, and they will continue to have an important contribution in the future. Their influence can be combined with that of CO_2 to produce an equivalent CO_2 concentration (see Wigley, 1985). In Figure 2 we show the history of past variations of equivalent CO_2 concentrations based on the most recently available data, together with a scenario for future changes based on Wigley (1985) and Dickinson and Cicerone (1986).

This time series can be used as input to a climate model to investigate the rapidity of changes in global mean temperature in response to greenhouse gas forcing. Here we use an upwelling-diffusion model which has four separate "boxes", one each for land and ocean in each hemisphere. The model, which is an extension of the global-mean box-diffusion model of Wigley and Schlesinger (1985), couples, in each hemisphere, an oceanic mixed layer with a diffusive deeper ocean, includes high-latitude down-welling and distributed (compensating) upwelling, and allows for land-ocean and inter-hemispheric mixed layer heat transport.

We consider first the effects of CO_2 and trace gas forcing alone. Figure 3 shows the global-mean warming from 1765 to 2050 for three different climate sensitivities, equilibrium CO_2-doubling temperature

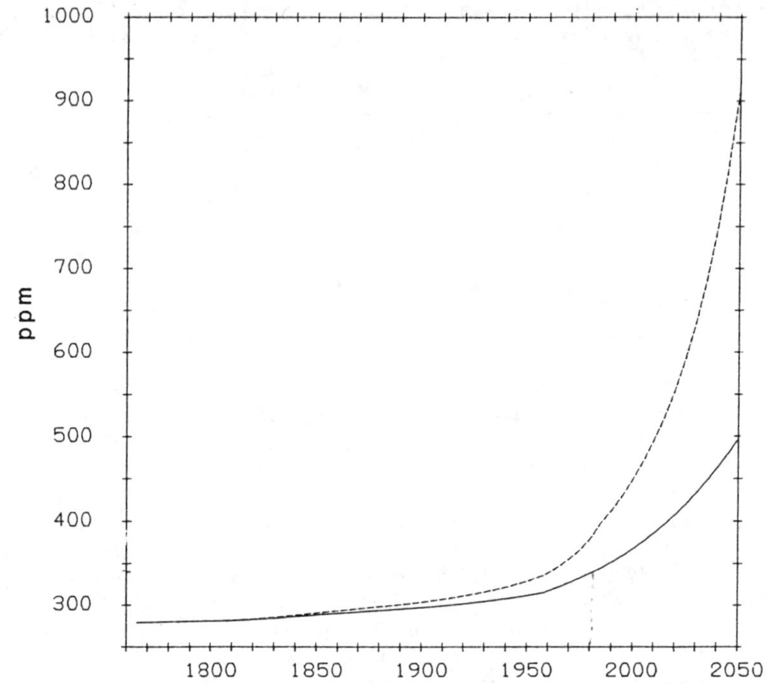

Figure 2. Past and future changes in CO_2 concentration (full curve) and in the equivalent CO_2 concentration (dashed curve). The initial value is 279ppmv in 1765. The difference between the two curves represents the additional radiative forcing due to changes in methane, nitrous oxide and chlorofluorocarbons. Historical data are from Neftel et al. (1985) for CO_2, Stauffer et al. (1985) for methane and Dickinson and Cicerone (1986) for the other trace gases. The CO_2 projection follows Wigley (1985), and the projections for other trace gases are essentially mid-point values from ranges given by Dickinson and Cicerone (1986).

changes of 1.5, 3.0 and 4.5°C. An upwelling rate of 4m/yr and a diffusivity of $2cm^2/sec$ have been assumed. The latter is a high value, chosen deliberately to minimize the rate of response; i.e. to produce conservative estimates of the rapidity of potential future warming.

If Figure 3 is compared with Figure 1 it can be seen that only the lower curves, corresponding to CO_2-doubling temperature changes of 1.5 and 3.0°C, are compatible with the observed century time-scale warming. (For these lower sensitivities, the results are rather insensitive to the choice of the diffusivity parameter). We note, however, that some recent models have given CO_2-doubling temperature changes in the upper end of the accepted 1.5-4.5°C range, around 4°C. These values appear to be too high to be compatible with observations (for details see Wigley and Schlesinger, 1985).

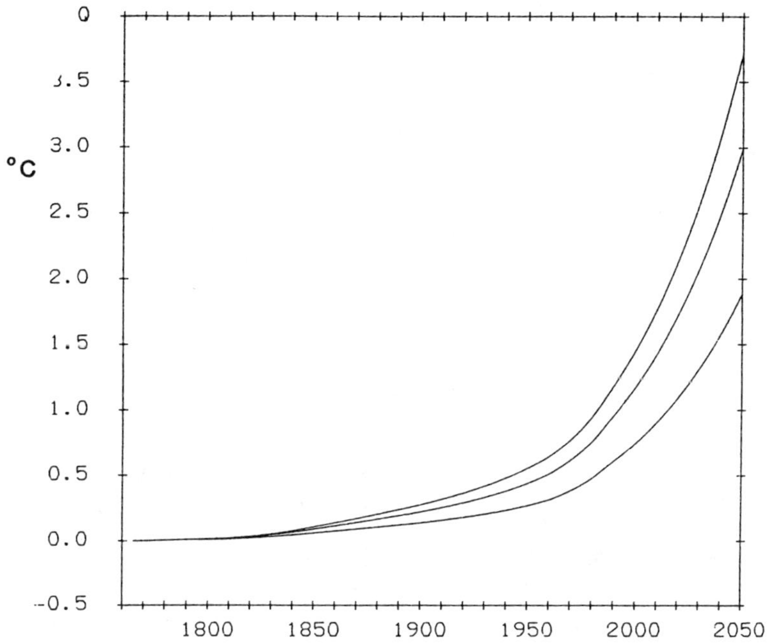

Figure 3. Past and future global mean surface temperature changes due to atmospheric CO_2 and trace gas concentration changes for three values of the CO_2-doubling temperature change, 1.5, 3.0 and 4.5°C. Results obtained are based on the forcing shown in Fig. 2 using a multi-box upwelling-diffusion model with diffusivity of $2 cm^2/sec$ and upwelling rate of 4 m/yr.

The most obvious discrepancy between Figures 1 and 3 is the slight observed global cooling between about 1940 and 1970, which is clearly at odds with greenhouse gas changes as the sole forcing mechanism. This cooling is not globally uniform and is strongest in the Northern Hemisphere (Jones et al., 1986c). Numerous explanations invoking natural causes of climatic change have been proposed for this cooling (reviewed in Wigley et al., 1985, and Wigley et al., 1986). The possibility that has received least attention is that the cooling may reflect a change in the rate of deep water formation, a possibility that has palaeoclimatic significance since it has also been proposed as an important forcing factor in glacial and late-glacial times (see Broecker et al., 1985).

To test the sensitivity of the system to changes in NADW formation rate (w) we "turned off" the production of deep water in the Northern Hemisphere of our model over the period 1950-2000. The response to such a change is quite complicated. Consider first a step function decrease in w. Since the net surface heat balance is not changed, the *equilibrium* surface temperature cannot be changed by changing w (Watts, 1985). However, the equilibrium vertical temperature profile *will* change if w is changed, with warmer intermediate and deep waters for lower w. In order for this to happen, heat must be transferred downward, so the surface temperature of the ocean must initially cool before eventually rewarming (very slowly) to its original value. For a temporary decrease in w (as in the case we have simulated here), the initial cooling response at the surface is quite rapid. Before 50 years have passed, slow warming commences at the surface, and this is then rapidly accelerated when w is returned to its initial value. In our case, all this is superimposed on a general greenhouse-gas induced warming. (For a somewhat different description of the mechanisms by which deep water formation rate changes may affect the climate, see Broecker et al., 1985.)

Reducing Northern Hemisphere downwelling/upwelling to zero is, of course, an extreme perturbation. It is not meant to realistically simulate the magnitude of changes in NADW formation rate that might have occurred this century. However, changes in NADW formation rate of almost this severity may have occurred in late-glacial times.

The modeled Northern Hemisphere mean temperature curve for the combined greenhouse gas and deep water formation rate forcing history is shown in Figure 4. Global mean changes are somewhat smaller than those for the Northern Hemisphere. The model response to deep water formation rate changes can be seen to be both large and rapid.

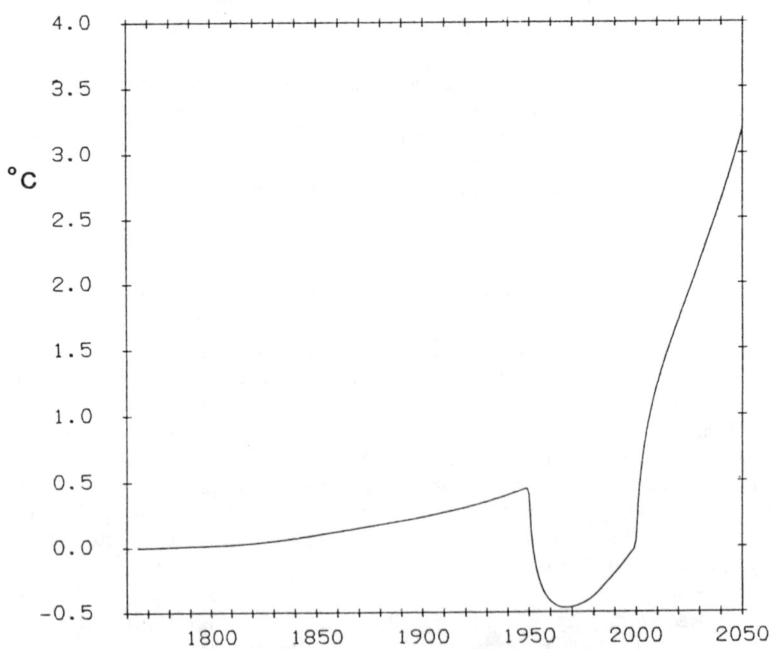

Figure 4. Past and future Northern Hemisphere mean surface temperature changes due to the CO_2 and trace gas forcing shown in Fig. 2, together with a deep water formation scenario in which deep water formation in the Northern Hemisphere is turned off in 1950 and turned back on in 2000. Other conditions as for the 3°C case in Fig. 3.

PALAEOCLIMATIC IMPLICATIONS

These results, both the observations and the model projections, have direct bearing on the interpretation of rapid palaeoclimatic warming or cooling events. Over the period 1880 to 1940, the globe warmed by about 0.5°C, a rate of almost 1°C per century. The simple model projections in Figure 3 suggest that the warming rate of annual mean surface air temperature in the early to mid twentyfirst century may be as much as 3.6°C per century (this value corresponds to a CO_2-doubling temperature change of 3°C). This latter value is in response to a major forcing, albeit anthropogenic, and probably represents an upper bound to the rate of any natural global-mean warming in the absence of major changes in the ocean circulation. The existence of an upper limit is likely, simply because the thermal inertia of the oceans limits the rate of response no matter what the forcing mechanism. Figure 3 shows that the change in the rate of response is much less than the change in the rate of forcing, which is here simulated by changing the climate sensitivity (compare the slopes of the 1.5°C and 4.5°C curves).

Figure 4 shows that much more rapid changes may occur in response to the combined effects of external forcing and the internal forcing of changes in deep water formation rate. Over the period 2000-2050, for a CO_2-doubling temperature change of 3°C, the Northern Hemisphere change due to greenhouse gases alone is 1.9°C (slightly more than for the globe). For the scenario with Northern Hemisphere deep water formation turned off in 1950 and back on in the year 2000 as an additional forcing factor, the change between 2000 and 2050 is 3.1°C. The deep water formation rate change contribution is clearly considerable.

The NADW change scenario assumed here is an extreme one. More gradual changes of smaller magnitude would have smaller effects. On the other hand, even with smaller and more gradual changes, the effects may be larger than the present model indicates because the feedback processes operating in late-glacial times may be inadequately simulated by our model. For example, with ocean temperatures close to freezing point, a small cooling due to a NADW formation rate decline could be considerably amplified by attendant increases in sea-ice extent and snow cover over land.

The model results demonstrate that the response of the climate system to internal, oceanic forcing can be much more rapid than the response to external forcing in the absence of oceanic circulation changes. In most palaeoclimatic situations, the external forcing would be expected to be less severe and less rapid than in the greenhouse gas example considered here, so deep water formation rate changes must be a prime suspect in attempting to explain rapid climate changes in the past.

Regional and smaller scale temperature changes may, of course, be much more rapid than those shown in Figures 3 and 4. General circulation climate models show that the equilibrium high latitude response to a CO_2 doubling in the spring/winter seasons could be up to four times the global mean response. Using the 2000-2050 changes in Figure 3, this would imply a warming of 4x1.9 or ≈8°C over 50 years in response to external forcing for these seasons and for an area the size of, say, the Arctic. This corresponds to a warming rate of 15°C per century.

In the case of internal forcing due to a change in deep water formation rate, however, such a simple extrapolation is not valid. Here, the cooling rapidly reaches a maximum which, in our model results, correspond to a depression of about 1.2°C below the result for no deep water formation rate change (see Figure 4). Thus, the model-based maximum cooling for an area the size of the Arctic and on the seasonal time scale is about 5°C, with the rate of cooling determined by the rate of change of deep water formation rate. Somewhat larger temperature changes might be expected over smaller areas in the vicinity of the forcing, i.e. around the northern North Atlantic. Further amplication might result from a stronger ice-albedo feedback than is assumed in our model.

Thus, the late-glacial temperature changes in England of around 10°C in summer and 20°C in winter over an interval of a few centuries, as inferred from the evidence of coleopteran data (Atkinson et al., 1985; Coope, 1985), should not be viewed with too much surprise, especially if one argues that they were associated with major shifts in the production rate of deep water. The processes that lead to such large deep water formation rate changes, however, clearly require further research. It is likely that the magnitude of these changes results partly from rather special conditions prevailing in the late-glacial. The possibility that similar changes might occur today or in the near future, with similarly large attendant temperature changes, is remote. Nevertheless, the processes that control deep water formation, particularly those that might cause noticeable changes in the formation rate, do have relevance to today, and relatively small changes in deep water formation rate may have appreciable affects on future climate.

REFERENCES

Atkinson, T.C.. Briffa, K.R. and Coope, G.R., 1985. Reconstruction of late-glacial climates from Coleoptera using the mutual climatic range method, Abstracts and reports from the conference on Abrupt Climatic Change, Biviers, France. *Scripps Institution of Oceanography Reference Series 86-8*, 56-59.

Broecker, W.S., Peteet, D.M. and Rind, D., 1985. Does the ocean-atmosphere system have more than one stable mode of operation? *Nature*, 315: 21-26.

Coope, G.R, 1985. Evidence from insect fossils for sudden and intense climatic changes during the last 45,000 years, Abstracts and reports from the conference on Abrupt Climatic Change, Biviers, France, *Scripps Institute of Oceanography Reference Series 86-8*, 90-92.

Dickinson, R.E. and Cicerone, R.J., 1986. Future global warming from atmospheric trace gases, *Nature*, 319: 109-115.

Folland, C.K., Parker, D.E. and Kates, F.E., 1984. Worldwide marine temperature fluctuations, 1856-1981, *Nature*, 310: 670-673.

Hansen, J.E., Johnson, D., Lacis, A., Lebedeff, S., Lee, P., Rind, D. and Russell, G., 1981. Climatic impact of increasing atmospheric carbon dioxide, *Science*, 213: 957-966.

Hansen, J.E., Lacis, A., Rind, D., Russell, G., Stone, P., Fung, I., Ruedy, R. and Lerner, J., 1984. Climate Sensitivity: Analysis of Feedback Mechanisms. In: J.E. Hansen and T. Takahashi (Eds.) *Climate Processes and Climate Sensitivity* (Maurice Ewing Series No. 5), American Geophysical Union, Washington, D.C., 130-163.

Jones, P.D., Raper, S.C.B., Bradley, R.S., Diaz, H.F. Kelly, P.M. and Wigley, T.M.L., 1986a. Northern Hemisphere surface air temperature variations: 1851-1984, *J. Clim. Appl. Met.*, 25: 161-179.

Jones, P.D., Raper, S.C.B. and Wigley, T.M.L., 1986b. Southern Hemisphere surface air temperature variations: 1851-1984, *J. Clim. Appl. Met.*, 25: 1213-1230.

Jones, P.D., Wigley, T.M.L. and Wright, P.B.. 1986c. Global temperature variations, 1861-1984, *Nature*, 322: 430-434.

MacCracken, M.C. and Luther, F.M., Eds., 1985. *Projecting the Climatic Effects of Increasing Carbon Dioxide* (DOE/ER-0237), U.S. Department of Energy, Carbon Dioxide Research Division Washington, D.C.

Neftel, A., Moor, E., Oeschger, H. and Stauffer, B., 1985: Evidence from polar ice cores for the increase in atmospheric CO_2 in the past two centuries, *Nature*, 315: 45-47.

Schlesinger, M. and Mitchell, J.F.B., 1985. Model projections of the equilibrium climatic response to increased carbon dioxide. In: M.C. MacCracken and F.M. Luther (Eds.) *Projecting the Climatic Effects of Increasing Carbon Dioxide* (DOE/ER-0237), U.S. Department of Energy, Carbon Dioxide Research Division, Washington, D.C., 81-147.

Stauffer, B., Fischer, G., Neftel, A. and Oeschger, H., 1985. Increase of atmospheric methane recorded in Antarctic ice core, *Science*, 229: 1386-1388.

Watts, R.G., 1985. Global climate variation due to fluctuations in the rate of deep water formation, *J. Geophys. Res.*, 90: 8067-8070.

Wigley, T.M.L., 1985. Carbon Dioxide, trace gases and global warming, *Climate Monitor*, 13: 133-148.

Wigley, T.M.L., Angell, J.K. and Jones, P.D., 1985. Analysis of the temperature record. In: M.C. MacCracken and F.M. Luther (Eds.), *Detecting the Climatic Effects of Increasing Carbon Dioxide*, (DOE/ER-0235), U.S. Department of Energy, Carbon Dioxide Research Division, Washington, D.C., 55-90.

Wigley, T.M.L., Jones, P.D. and Kelly, P.M., 1986. Empirical climate studies. Warm world scenarios and the detection of climatic change induced by radiatively active gases. In: B. Bolin, B.R. Döös, J. Jäger, and R.A Warrick (Eds.) *The Greenhouse Effect, Climatic Change, and Ecosystems*, SCOPE 29, Wiley, 271-322.

Wigley, T.M.L. and Schlesinger, M.E., 1985. Analytical solution for the effect of increasing CO_2 on global mean temperature, *Nature*, 315: 649-652.

W.M.O. (World Meteorological Organization), 1986. *Report of the International Conference on the Assessment of the Role of Carbon Dioxide and of other Greenhouse Gases in Climate Variations and Associated Impacts*. W.M.O. No. 661, Geneva.

Part II.

The Last Millenium

THE EXPLOSIVE VOLCANIC ERUPTION RECORD IN NORTHERN HEMISPHERE TEMPERATURE RECORDS

Raymond S. Bradley

University of Massachusetts, Amherst

Any attempt to evaluate the effect of explosive volcanic eruptions on climate is made difficult by the inadequacies of both climatic and volcanic data which are available. To obtain the best set of temperature data for the northern hemisphere land areas, a project was recently initiated to compile all available high quality, long-term temperature records and to adjust them for homogeneity where necessary (Bradley et al., 1985; Jones et al., 1985a, b). The data base which was produced represents the optimum large-scale network available with which to evaluate the influence of explosive volcanic eruptions on climate. Data were interpolated to a 5° latitude by 10° longitude grid and used to compute time series of annual and seasonal temperature variations for the northern hemisphere as a whole and for selected zonal bands.

Data on explosive volcanic eruptions are poor and often contradictory. Volcanological criteria used in ranking eruptions may not be the best in terms of potential climatic effects. Using Simkin et al.'s (1981) Volcanic Explosivity Index (V.E.I.), which rates eruption severity on a scale of 0 to 8, it would appear that there were 56 eruptions since 1851 which were equal in magnitude to Mt. Agung (1963) or El Chichon (1982) (V.E.I.= 4). The index does not consider sulphur produced by the eruption and this may be of critical importance in perturbing solar radiation receipts.

Eight major eruptions (V.E.I. >4) have occurred since 1851 and the northern hemisphere temperature record indicates statistically significant, but brief, negative temperature anomalies following some of these. In particular, the eruptions of Askja (Iceland) in 1875, Katmai, (Alaska) in 1912 and Bezimianny (Kamchatka) in 1956 appear to have lowered temperatures. However, the perturbation is principally in fall and summer months (0.4-0.6°C in fall, 0.2-0.4°C in summer). The effect is also different at different latitudes. Mid and high latitude zones are affected by major eruptions at all latitudes. Low latitudes do not always show the influence of high latitude eruptions. There is no consistent influence on Equator-Pole temperature gradients.

W. H. Berger and L. D. Labeyrie (eds.), Abrupt Climatic Change, 59–60.

CONCLUSIONS

The influence of explosive eruptions on temperatures over northern hemisphere land areas is seasonally and latitudinally dependent; the magnitude of any temperature depression is small and short-lived, generally only one or two seasons. Fall and summer seasons are affected most. Temperature depression may be large at high latitudes (0.6°C in summer, >1°C in fall) and this may affect cryospheric and ecological conditions in marginal environments. Overall the explosive volcanic eruption signal is not very strong and often lost in other noise (which may, however, also be related to volcanic eruptions which are poorly known). It is necessary that considerable improvements be made in the volcanic eruption chronology to improve understanding of short-term climatic variability.

REFERENCES

Bradley, R.S., Kelly, P.M., Jones, P.D., Diaz, H.F. and Goodess, C., 1985. *A Climatic Data Bank for the Northern Hemisphere Land Areas 1851-1980*, DOE Technical Report No. TR017, U.S. Dept. Energy Carbon Dioxide Research Division, Washington, D.C. 335 pp.

Jones, P.D., Raper, S.C.B., Bradley, R.S., Diaz, H.F., Kelly, P.M. and Wigley, T.M.L., 1985. Northern Hemisphere Surface Air Temperature Variations 1851-1984, *Journal of Climate and Applied Meteorology* (in press).

Jones, P.D., Raper, S.C.B., Santer, B., Cherry, B.S.G., Goodess, C., Kelly, P.M., Wigley, T.M.L., Bradley, R.S. and Diaz, H.F., 1985. *A Grid Point Surface Air Temperature Data Set for the Northern Hemisphere*, DOE Technical Report No. TR022, U.S. Dept. Energy Carbon Dioxide Research Division, Washington, D.C. 251 pp.

Simkin, T., Siebert, L., McClelland, L., Bridge, D., Newhall, C. and Latter, J.H., 1981. *Smithsonian Institution - Volcanoes of the World: A Regional Directory, Gazetteer and Chronology of Volcanism During the Last 10,000 Years*, Hutchinson Ross Publishing Co., Stroudsburg, Pa., 232 pp.

DECADAL-SCALE PATTERNS OF CLIMATIC CHANGE OVER EASTERN NORTH AMERICA INFERRED FROM TREE RINGS

Edward R. Cook and Peter Mayes

Tree-Ring Laboratory
Lamont-Doherty Geological Observatory
of Columbia University
Palisades, New York 10964

A network of 74 tree-ring chronologies covering much of eastern North America south of the boreal forest has been developed for dendroclimatic studies. The coverage (shown in Figure 1) extends from southern Canada to the southeastern United States and west to the 90th meridian. Ten tree species from the genera *Picea*, *Pinus*, *Tsuga*, *Quercus* and *Juniperus* are represented in the network. The chronologies in the network all begin at or before 1700 and end no earlier than 1972. The purpose of this paper is to describe some preliminary dendroclimatic analyses of the network using principal components analysis (PCA). These analyses reveal tree-ring anomaly patterns across the network between 1810 and 1880 which are both highly persistent and extreme compared to the pre-1810 and post-1880 periods.

The dendroclimatic interpretation of the tree-ring network is complicated by the high diversity of species and sites represented herein. While many of the chronologies are from moderate to well-drained sites where growing season moisture deficiencies are likely to be limiting to growth, several chronologies are from high elevation and high latitude sites where temperature-related variables are more likely to limit tree growth. In addition, there is a geographic bias in species coverage in the network which is inevitably caused by the distributional limits of tree species in the eastern deciduous forest. Thus, oak species predominate in the western half of the network while eastern hemlock is most common in the northeastern quadrant. These complexities make it very difficult to map and interpret tree-ring anomaly patterns across space using the original tree-ring chronologies (e.g. Fritts, 1965) or tree-ring eigenvector patterns (e.g. LaMarche and Fritts, 1971). The dendroclimatic interpretations of the spatial patterns would be inevitably confounded by species and site biases noted above. However, PCA can be used to search for network-scale growth anomalies that are so large as to overwhelm the more geographically limited biases of species and site. This can be done through an examination of the scores of the first eigenvector of tree growth. It is well known (e.g. Richman, 1983) that the loadings of the first eigenvector commonly indicate a pattern of positive covariance between variates across the entire domain. Given this property, the scores of the first eigenvector can be

W. H. Berger and L. D. Labeyrie (eds.), Abrupt Climatic Change, 61–66.

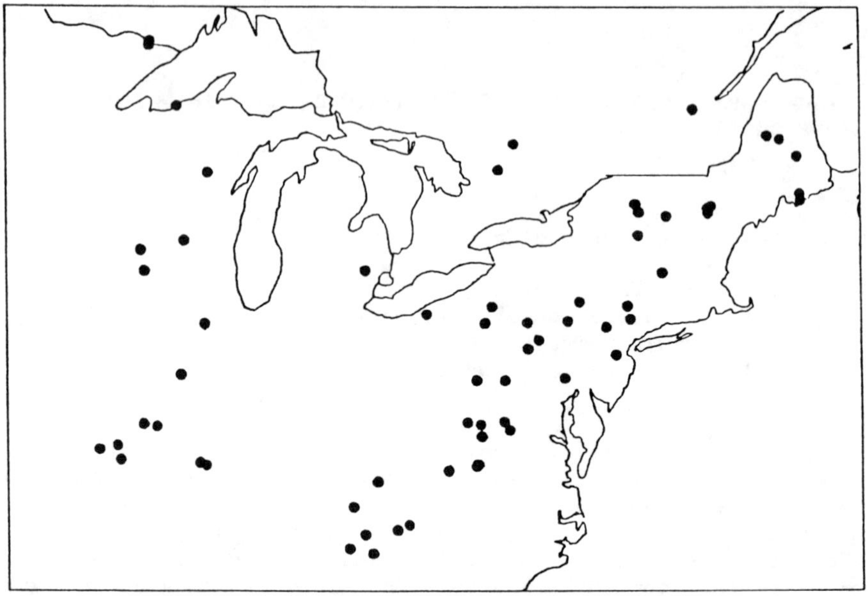

Figure 1. The geographic coverage of the 74 chronology tree-ring network over eastern North America. Some points have more than one chronology. All begin in 1700 or earlier.

used to search for time periods when tree growth was strongly above or below average across the network. Such anomalies should be free of species and geographic biases and, therefore, have some climatic meaning.

The 74 tree-ring chronologies were subjected to PCA as described above. The first eigenvector extracted from the correlation matrix accounted for 15% of the total variance. As anticipated, the loadings of the first eigenvector were all positive across the domain. The corresponding scores for this eigenvector are shown in Figure 2. To emphasize decadal-scale fluctuations in the scores, the series has also been smoothed with a 10-year low-pass filter. Both series show a rather abrupt change of behavior beginning in about 1810 and continuing to about 1880. During this period, highly persistent growth anomalies both above and below the mean are indicated. The 1827-1835 period is the most persistently above average growth period since 1700. For this period, 61 out of 74 chronologies show above average growth. Using a simple sign test, the number of positive anomalies exceed chance alone well in excess of the 99% confidence level. The mean anomaly pooled across all 74 series is +0.119 ±.0159. Since the expected value for treering chronologies is 1.0, the mean anomaly is 7 standard errors above the expected value. There is little doubt that the 1827-1835 anomaly seen in the PCA scores is an exceptional event. Eight of the 13 chronologies showing below average growth for this period are from the southern end

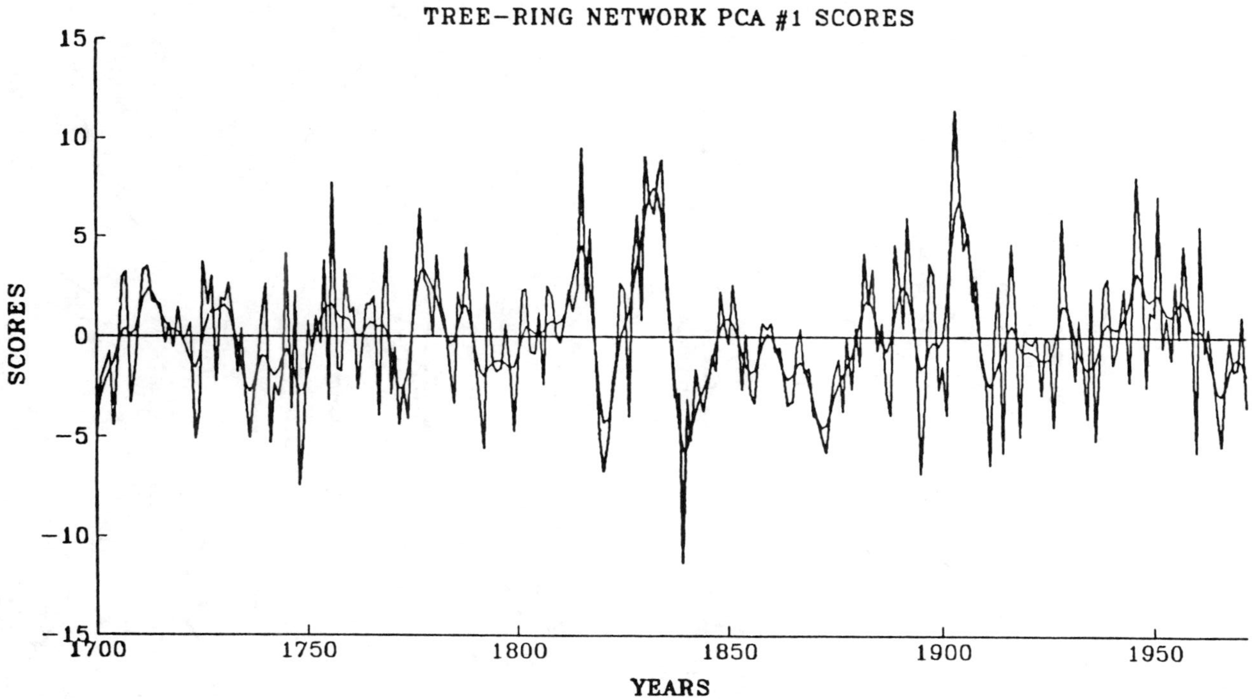

Figure 2. The scores of the first principal component of tree-rings extracted from the correlation matrix of 74 tree-ring chronologies.

of the network. Thus, there appears to be a north-south geographic gradient to the anomaly pattern.

A first-order dendroclimatic interpretation of the 1827-1835 anomaly is possible by using the 1902-1908 positive growth anomaly in the scores as the analogue. Meteorological records are readily available for this more recent anomaly period which ranks second to the 1827-1835 event. Like the earlier anomaly, above average growth is indicated in 65 of the 74 chronologies, and the mean anomaly pooled over all chronologies is +0.117 ±.0113. Thus, this anomaly is equally significant in a broad-scale sense. An examination of July Palmer Drought Severity Indices (PDSI) (Palmer, 1965) over the region for this time period indicates wetter-than-average conditions especially in the northern half of the region. In addition, an analysis of cyclone frequencies (Hayden, 1980) over the area reveal a higher-than-average number of cyclones passing over the northern half of the network during the principal growing season months of April-July. This anomaly map is shown in Figure 3. It is consistent with the July PDSI data and contains a north-south gradient which agrees with the tree-ring anomaly gradient in the 1827-1835 period. Wahl (1968) shows a summer precipitation anomaly map for the eastern United States in the 1830's which is also consistent with the PDSI and cyclone frequency anomalies. That is, he shows wetter-than-average conditions in the Great Lake region and drier-than-

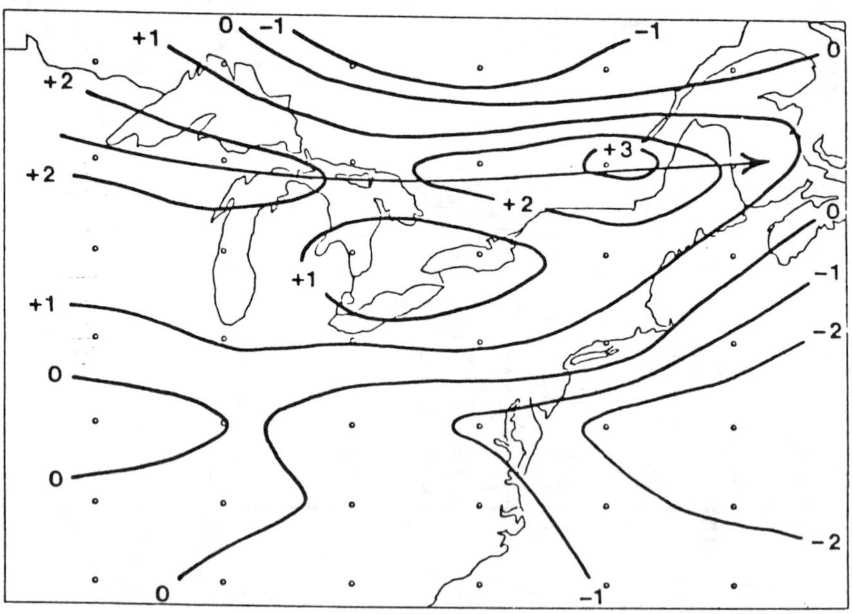

Figure 3. Cyclone frequency anomaly map for the 1902-1908 period of record. The principal axis of cyclone frequency is deliniated by the west-to-east arrow. The open dots are the centers of grid squares through which the cyclones pass.

average conditions in the Middle Atlantic coastal region of the United States. Thus, the meteorological conditions of the 1902-1908 period appear to be a reasonable analogue to the conditions which prevailed in 1827-1835.

The other anomaly in the scores, which is quite remarkable, is a prolonged below-average growth period extending from 1837 to 1879. During this time Period, 60 out of 74 tree-ring chronologies showed below-average growth averaging -0.039 ±.0062. These statistics far exceed the 99% confidence level. The 1837-1879 anomaly contains two shorter-term, more severe below-average growth periods in 1837-1847 and 1861-1879. For the 1837-1847 event, 56 out of 74 chronologies show below-average growth averaging -0.066 ±.0120. Of the 18 Chronologies showing above-average growth, 11 are concentrated in the far southern and southwestern portion of the network. This indicates that the meteorological conditions responsible for poor tree growth were located in the central and northern portions of the network. For the 1861-1879 period, 61 of 74 chronologies show below-average growth averaging -0.051 ± .0086. There is a somewhat weaker indication of above-average growth in the southern-southwestern end of the network. Seven of 13 above-average growth chronologies are concentrated there.

The 1837-1847 anomaly immediately follows the extreme above average anomaly of 1827-1835 and appears to be even more unprecedented in the record. The juxtaposition of these opposite extremes suggests a rapid change in the atmospheric circulation over eastern North America at about 1836 to a new circulation mode which remained stable until about 1880, the widely agreed beginning of the Northern Hemisphere warming trend. There appears to be no modern (i.e. post-1900) analogue for this event in the scores. A similar, persistently negative, tree-ring anomaly occurred at about the same time in trees growing near the northern treeline of eastern Canada (e.g. Cropper and Fritts, 1981; Jacoby, 1983). Trees growing at treeline are frequently limited by cool growing season temperatures. This fact coupled with the coincidence of timing suggests a southward displacement of the summertime position of the polar front in eastern Canada and a deepening of the western Atlantic pressure trough over eastern North America. This situation would allow for the enhanced advection of cool, dry, stable air into the eastern United States and a southward and/or seaward displacement of the principal storm tracks over the region. Such conditions prevailed during the more geographically limited but extremely severe drought over the northeastern United States in the 1960's (Namias, 1966).

One final speculative observation about PCA scores is in order. Very rapid transitions from above-average growth to below-average growth are indicated for the years 1817-1819 and 1835-1839. These abrupt transitions follow the major eruptions of Tambora in the East Indies and Coseguina in Nicaragua (Lamb, 1970). Kelly and Sear (1984) show that large volcanic eruptions can lower Northern Hemisphere temperatures by a statistically significant amount for around 18 months following volcanic eruptions, and generally depress temperatures for two to three years. It is possible, therefore, that some of the fluctuations seen in the PCA scores are in a consequence of the impact of volcanic eruptions on climate.

The dendroclimatic interpretations described above should be regarded as tentative at this time. The next step will be to directly calibrate the tree-ring anomaly patterns against actual meteorological records in the post-1900 period. This will allow for the reconstruction of climate over the network back to 1700 and a re-appraisal of the 1827-1835 and 1837-1879 anomaly periods through an analysis of the tree-ring estimates of climate.

REFERENCES

Cropper, J.P. and Fritts, H.C., 1981. Tree-ring width chronologies from the North American arctic, *Arctic and Alpine Res.*, 13 (3): 245-260.

Fritts, H.C., 1965. Tree-ring evidence for climatic changes in western North America, *Mon. Weather Rev.*, 93 (7): 421-443.

Hayden, B.P., 1980. Atlantic coast extratropical cyclones: characteristic frequency patterns and their secular variations. In: *Proc. Second Conf. on Coastal Meteor.*, 30 Jan.-1 Feb., 1980, Los Angeles, CA, pp. 192-197.

Jacoby, G.C. Jr., 1983. A dendroclimatic study in the forest-tundra ecotone on the east shore of Hudson Bay. In: P. Morisset and S. Payette (eds.), *Tree-Line Ecology*, No. 47, Centre d'etudes nordiques, Université Laval, Quebec, pp. 95-99.

Kelly, P.M. and Sear, C.B., 1984. Climatic impact of explosive volcanic eruptions, *Nature*, 311: 740-743.

LaMarche, V.C. and Fritts, H.C., 1971. Anomaly patterns of climate over the western United States, 1700-1930, derived from principal component analysis of tree-ring data, *Mon. Weather Rev.*, 99 (2): 138-142.

Lamb, H.H., 1970. Volcanic dust in the atmosphere; with a chronology and assessment of its meteorological significance, *Phil. Trans. Roy. Soc.* A, 266 (1178): 425-533.

Namias, J., 1966. Nature and possible causes of the northeastern United States drought during 1962-65, *Mon. Weather Rev.*, 94 (9): 543-554.

Palmer, W.C., 1965. *Meteorlogical Drought*, Weather Bureau Research Paper no. 45, U.S. Dept. of Commerce, Wash. D.C., 58pp.

Richman, M.B., 1983. Rotation of principal components in climatological research. Part 1: Theoretical considerations. Suitable applications and advantages over unrotated solutions. In: *Preprints Eighth Conf. on Probability and Statistics in Atmospheric Sciences*, 16-18 Nov. 1983, Hot Springs, AZ, pp. 59-68.

Wahl, E.W., 1968. A comparison of the climate of the eastern United States during the 1830's with the current normals, *Mon. Weather Rev.*, 96 (2): 73-82.

FACTORS CONTROLLING FREE AIR AND OCEAN TEMPERATURE OF THE LAST 30 YEARS AND EXTRAPOLATION TO THE PAST

Reginald E. Newell and Jane Hsiung

Department of Earth, Atmospheric and Planetary Sciences
Room 54-1522
Massachusetts Institute of Technology
Cambridge, Massachusetts 02139 USA

ABSTRACT. Relationships between global sea surface temperature, tropospheric free air temperature and oceanic surface wind patterns are explored. Data for the past 30 years show that tropical free air temperature changes are jointly related to preceding sea surface temperature changes and to stratospheric aerosol which varies with volcanic activity. Global sea surface temperature patterns show two principal modes: El Niño, which changes from a negative to a positive phase in less than one year; and the cooling trend which began in the middle 1960's and was associated with drought in the Sahel. High latitude oceanic cooling is related to surface wind pattern changes which alter oceanic currents and to changes in energy absorbed by the Indian Ocean which is normally passed on to the Atlantic. The idea of an optimum oceanic energy flow vector is coined, this being the pattern of the curl of the surface wind stress. This pattern is shown to be different for wet and dry years in the Sahel. The pattern would also have changed during the last Ice Age as large-scale waves forced by the fixed mountain and variable ice topography interfered. These changes in the orientation of the oceanic energy flow vector may have accounted for the abrupt environmental changes observed in northwest Europe.

INTRODUCTION

During the past 30 years of reasonably comprehensive climatological records there have been several different types of what might be called abrupt climatic changes. Selected examples are presented here together with comments on the possible physical mechanisms involved and their applicability to changes in the more distant past.

Two main data sets are considered: tropical zonal mean air temperature for the layer 300-700mb and a global set of surface observations from ships.

W. H. Berger and L. D. Labeyrie (eds.), Abrupt Climatic Change, 67–87.

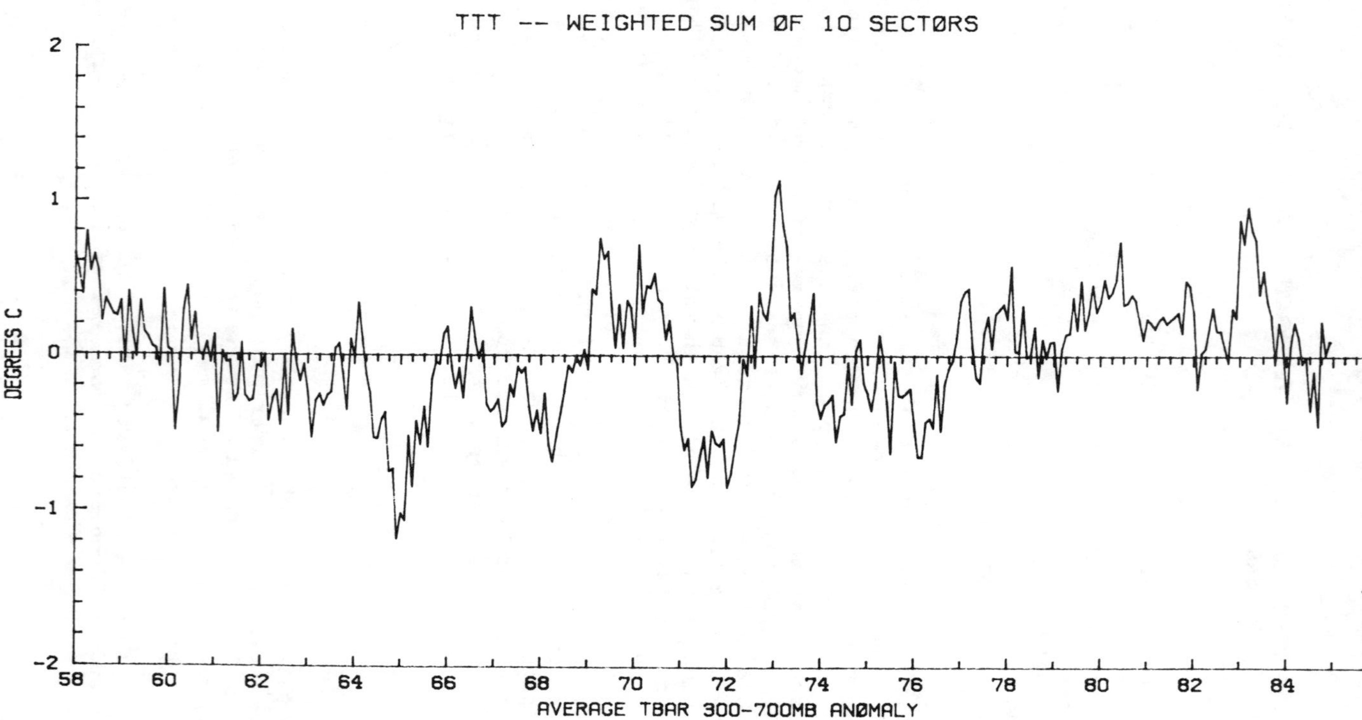

Figure 1. Zonal mean air temperature for the 300-700 mb layer. Monthly anomalies from 1958-1984 means. Area weighted for 10 longitude sectors in the tropics. Ticks represent January of indicated year. Units: degrees Celsius.

ZONAL MEAN AIR TEMPERATURE IN THE TROPICAL TROPOSPHERE

Zonal mean free-air temperature anomalies are obtained from monthly mean values of geopotential height difference between the 700mb and 300mb level for 60 radiosonde stations in the tropics. These include a number of stations near the equator; this region was not sampled by previous studies of tropical free air temperature (e.g., Angell and Korshover, 1975) and its relationship with sea surface temperature (e.g., Newell and Weare, 1976a, b; Navato et al., 1981). The stations have an irregular distribution with longitude thus the tropical zone (23°N-23°S) is divided into ten longitudinal sectors. Long-term monthly mean values of the height difference are obtained for each station and a monthly deviation from the average is found for each station for each month. These are then averaged for each sector and an area weighted average is found for the tropical belt. Geopotential difference anomalies are converted to equivalent temperature change using the equation of state and the hydrostatic equation.

As can be seen from Fig. 1 there are substantial fluctuations on a month-to-month scale with the most abrupt change being an increase of about 2°C during 12 months of 1972. It has been argued that these relatively abrupt changes evident in free-air temperature have two causes: preceding sea surface temperature (SST) changes in the equatorial Pacific and preceding changes in atmospheric transmission due to volcanic aerosol (Newell and Weare, 1976a, b; Newell, 1979; Newell and Hsiung, 1984). It is not considered that the tropical SST changes themselves cause the temperature change, for such a direct effect would be evident only in the boundary layer, but that there is an indirect effect such as modulated latent heat liberation. For example, in an El Niño there is decreased cloud in the western Pacific, an area where additional energy at the surface is used for evaporation, as the water temperature is close to the evaporational limit. Concomitantly cloudiness, moisture, and precipitation increase in the central Pacific and these are associated with atmospheric heating. The increase in latent heat liberation here must have more than offset the decrease in the west to account for the observed net increase in the zonal mean temperature, unless there is an overall decrease in large-scale rising motion and adiabatic cooling. In the volcanic case cooling of the troposphere takes place either because near-infrared radiation is absorbed in the stratosphere, and is therefore depleted in the troposphere where it normally provides heating, or because visible radiation is backscattered by the aerosol so does not reach the ground to contribute to evaporation. If either or both of these ideas are correct, then if one is searching the air temperature for volcanic effects the sea surface influence must first be removed, or vice versa.

The main variability of SST is the El Niño-Southern Oscillation mode (Weare et al., 1976; Hsiung and Newell, 1983). This phenomenon dominates tropical free air temperature changes and, in fact, also shows in middle latitude air temperatures although there is a six-month lag between the SST change and the middle latitude air temperatures (Newell and Weare, 1976b; Navato et al., 1981; Horel and Wallace, 1981; Pan and Oort, 1983). The oceanic effect appears as a positive correlation between sea and air temperature. The volcanic effect is of opposite sign

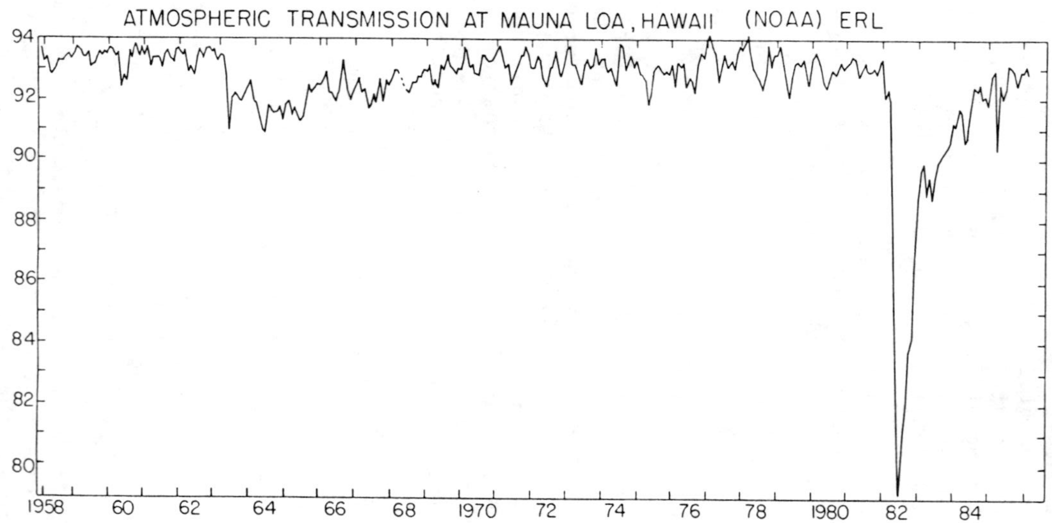

Figure 2. Atmospheric transmission at Mauna Loa, Hawaii. Data provided by Dr. B.G. Mendonca, Geophysical Monitoring for Climatic Change Program, NOAA Environmental Research Labs, Boulder, Colorado. All ticks except the first represent December. Units: percent.

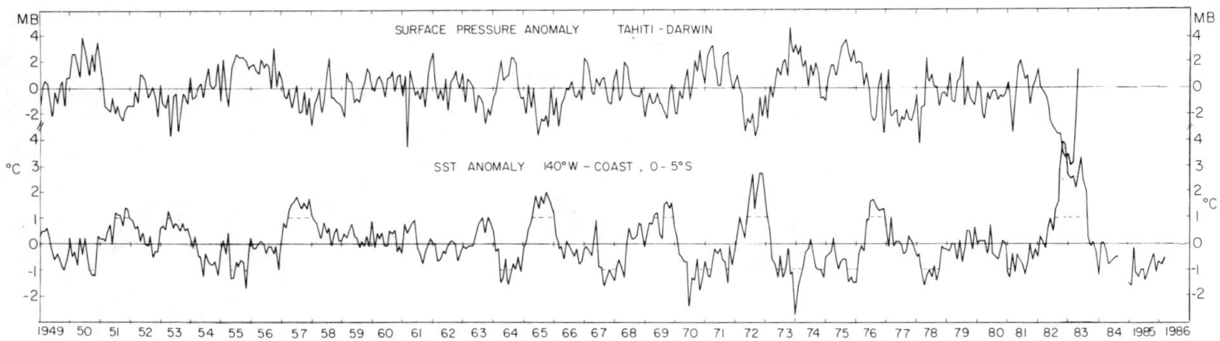

Figure 3. Sea surface temperature anomalies averaged over region 140°W to the coast of South America, 0-5°S. Recent data from Forrest Miller, Interamerican Tropical Tuna Commission, Scripps Institution of Oceanography. Ticks represent January of indicated year. Units: degrees Celsius. Upper curve represents surface pressure anomaly difference between Tahiti and Darwin. Units: millibars.

with greater stratospheric aerosol corresponding to lower tropospheric free air temperatures. A similar study by Parker (1985) has verified the dependence of tropical air temperature on the El Niño mode but did not find a relation with volcanic aerosol. However, comparison of air temperature (Fig. 1) with atmospheric transmission (Fig. 2) and SST (Fig. 3) shows an evident interrelationship. In the studies carried out before the 1982-83 El Niño it was possible to assign 63% of the air temperature variability to the two factors - 47% to SST and 16% to volcanic activity (Newell and Weare, 1976b). Because the extreme 1982-83 El Niño was accompanied by a large volcanic eruption - El Chichon - and these two phenomena give opposite effects on atmospheric temperature there was little sign of the eruption in the air temperature record; as may be seen from Figure 1 air temperature was actually higher than average in 1983.

THE SEA SURFACE TEMPERATURE

A global set of SST measurements covering the 30 year period 1949-79 has been subjected to eigenvector analysis with results described by Hsiung and Newell (1983). The most important mode is that of El Niño (see their Figs. 3 and 4) as was previously found for a set of data for the Pacific Ocean alone (Weare et al., 1976) and for a set of zonal mean SST for the three main oceans (Chiu and Newell, 1983). The first mode is well separated from the others. The time series of the first mode is quite similar to that of the Southern Oscillation index, or zonal mean tropical free air temperature (Fig. 1). Warm periods in the eastern equatorial Pacific occur every 3-6 years. This mode also accounts for the highest fraction of the variance in global surface pressure, tropical rainfall and temperature (Kidson, 1975) as well as the largest fraction of the variance of global 500mb air temperature .

The second eigenvector found by Hsiung and Newell (1983) (their Figs. 5-7) shows an alternation between the hemispheres with the North Atlantic and North Pacific being warm when the southern oceans are cold and vice versa. As is evident from the time series, there is a downward trend starting about 1964, the phase being such that the northern oceans are cooler after this time. This trend is in accord with previous studies of zonal mean SST by Kukla et al. (1977) and Newell and Hsiung (1979). It is also in line with trends in 1000-700mb thickness reported by Boer and Higuchi (1980) and by Douglas et al. (1982).

When the eigenvector analysis is performed for the periods 1949-64 and 1965-79 separately, the patterns resulting for the second eigenvector are somewhat different. In the first period the middle latitude North Pacific and the western middle latitude North Atlantic are out of phase while in the second period they are in phase and both oceans are less broken up in the east-west and north-south direction. In the first period the South Atlantic is out of phase with the western North Atlantic and in the second period with the entire North Atlantic. Evaluation of the autocorrelation function shows that the trend is more important in the second period.

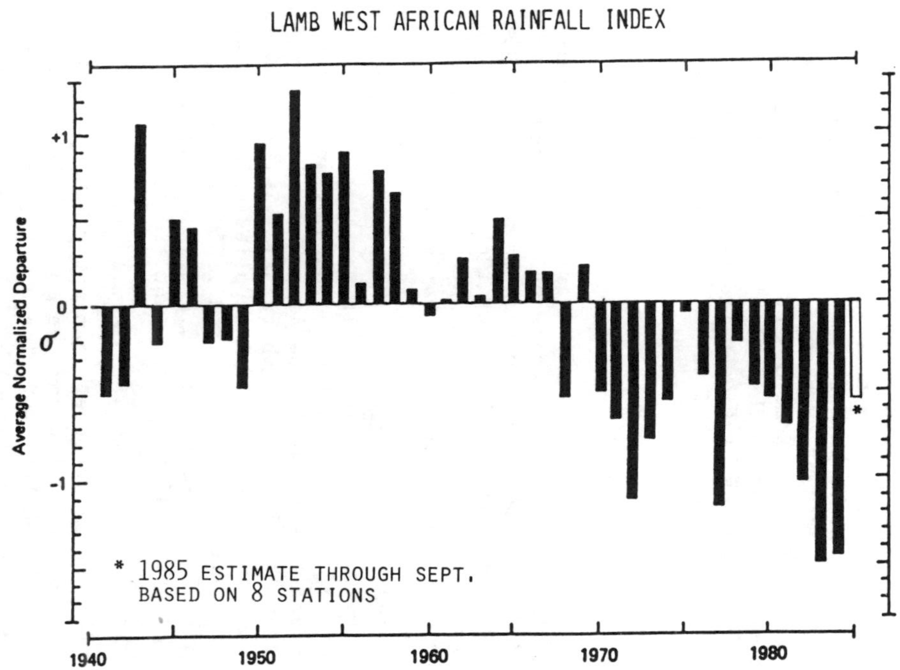

Figure 4. Lamb's West African rainfall index. From Climate Analysis Center, 1985.

At the Biviers meeting we drew attention to the analogy between the time series of the second eigenvector and the time series of rainfall over Africa (see Fig. 4) and suggested that both were manifestations of a global circulation change with drought over the Sahel being associated with colder water in the northern oceans and warmer water in the southern oceans. A similar point of view has been expressed by Folland et al. (1986). We argued that one of the physical mechanisms that may be responsible for this association is a decreased northward flux of energy by the oceans. This flux may be controlled by the wind pattern changes in the tropics and to the north or by changes in the energy received at the surface of the Indian Ocean. Some of the energy entering the Indian Ocean is thought to be passed on to the Atlantic and Pacific. Possible trajectories of the water masses have been discussed by Gordon (1986). Similar wind changes may have occurred in the past and given rise to relatively abrupt climatic changes. The recent wind and SST changes are reviewed below in detail as a prelude to the application of the ideas to the past.

SST anomalies with respect to 1949-79 means have been combined for wet and dry years in the Sahel, these labels being defined as in the work by Newell and Kidson (1984): wet covers the period May 1958 to April 1963, while dry corresponds to December 1969 to December 1973. These periods were initially selected for study because of the availability of wind data over Africa and on the basis of rainfall analysis by Nicholson (1981) and Lamb (1983). Fig. 4 is an updated version of the Lamb index (CAC, 1985). The difference in SST between dry and wet years for August and February is shown in Fig. 5. In August the northern oceans are cooler in dry years while the South Atlantic is warmer than average. There are exceptions in the 45-50°N regions of the Atlantic and various areas of the Pacific where SST is slightly warmer in the dry period. When a two-sided t-test is applied using standard deviation values of 2.5°C in middle latitudes and 1.0°C in the tropics, both based on actual data, values greater than 0.8°C in the northern oceans and 0.5°C in the tropics are slightly different from zero at the 90% confidence level.

Dry Februarys are generally cooler than wet Februarys with some warmer regions in the South Atlantic and the South Pacific, the western tropical Pacific and the North Pacific. These patterns reported at Biviers are quite similar to those found by Folland et al. (1986) from a longer period of data (1901-1985). It is worth noting that the change from wet to dry over Africa or warm to cold SST anomalies in the northern oceans occurs over a period of 5-10 years, considerably longer than the El Niño mode fluctuations. But in paleoclimate studies this sort of mode change could still be classed as abrupt.

WINDS OVER THE OCEAN

There are characteristic changes in the oceanic wind field accompanying the El Niño mode fluctuation, as described for the Pacific from a 100-year data set by Newell et al. (1982). Stronger than average easterlies in the central and west Pacific were found to accompany cold water in the eastern Pacific. We inquired at Biviers - Are there

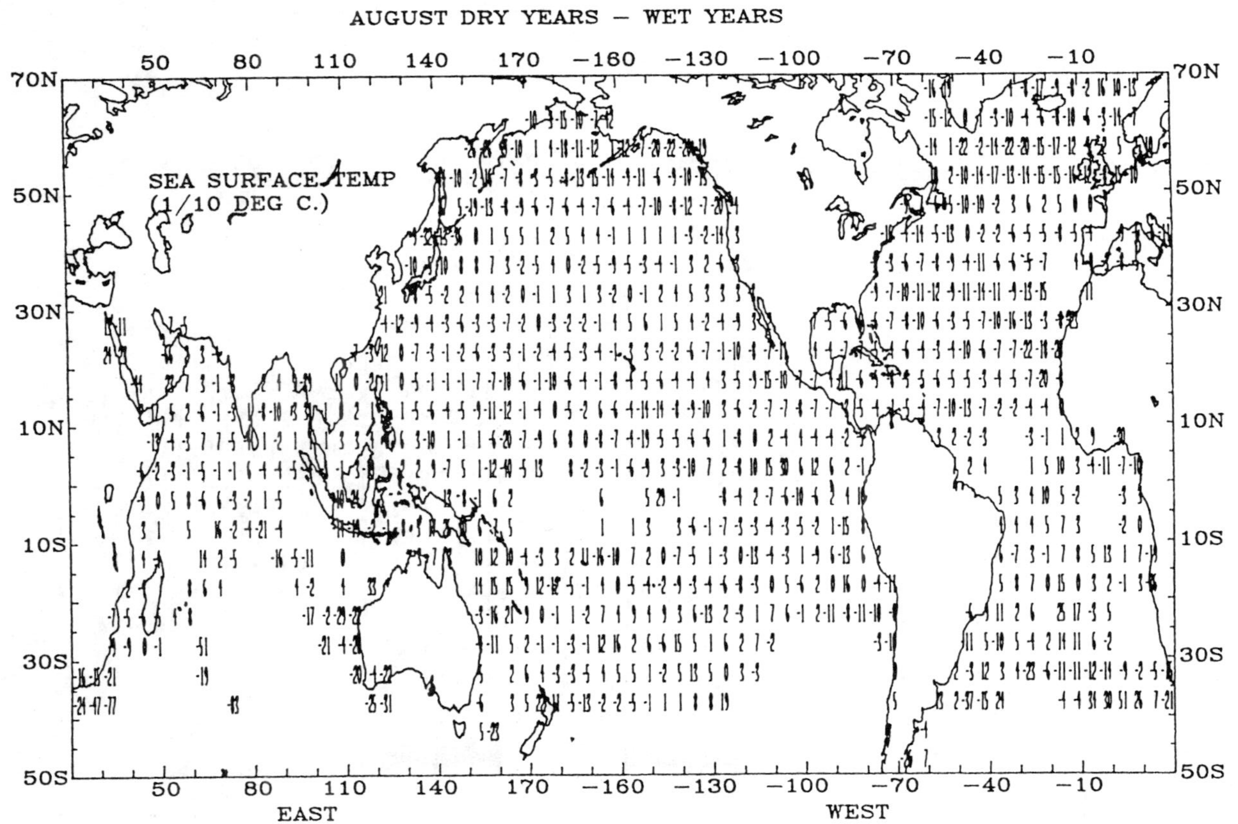

Figure 5a. Sea surface temperature difference for August (1970-1973) - (1958-1963). Units: 1/10 degree Celsius.

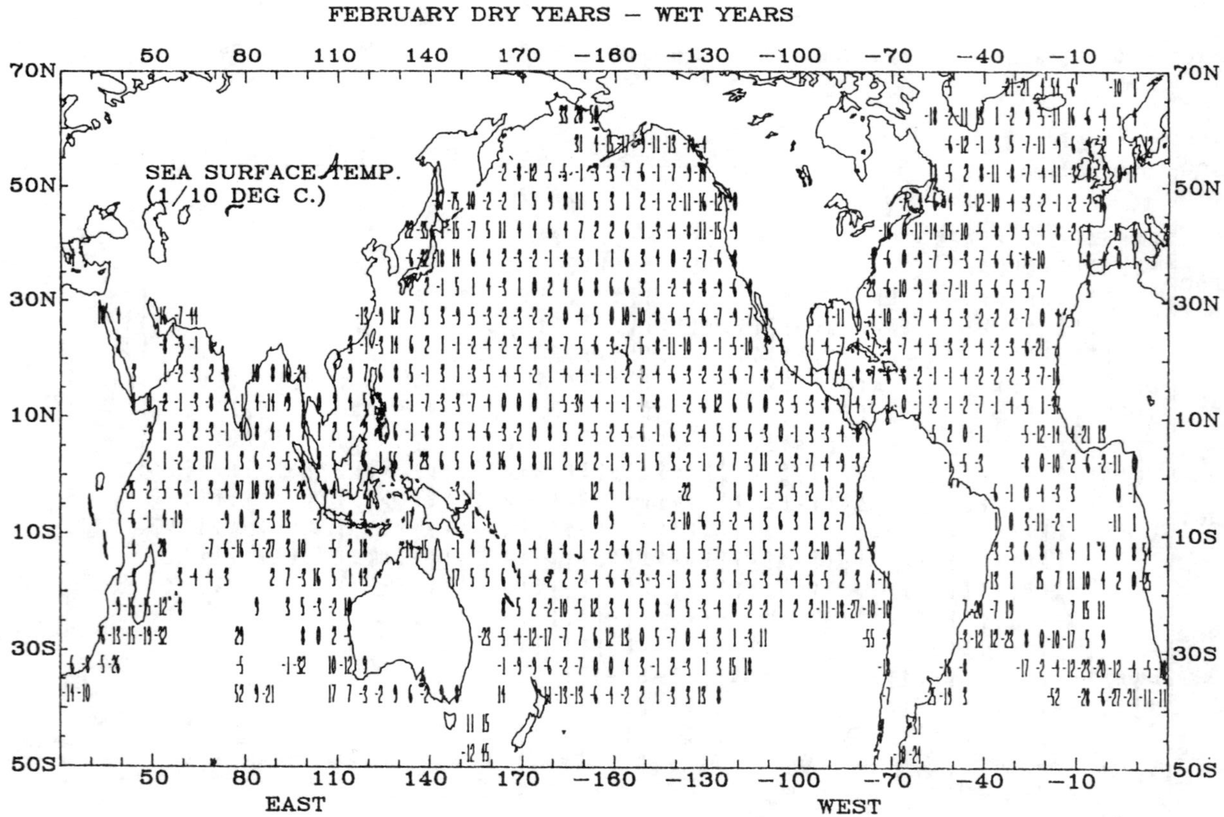

Figure 5b. Sea surface temperature difference for February (1970-1973) - (1958-1963). Units: 1/10 degrees Celsius.

characteristic changes of the oceanic wind field associated with the second most important mode of global SST fluctuation, which might be called the Sahel drought mode, and if so is there a relationship between the wind and SST fluctuations for this mode? To examine this question winds from the same data set of ships observations used for the SST eigenvector analysis were subdivided into two periods, 1949-1964 and 1965-1979. This subdivision is based on the time series of the second eigenvector of global SST; the first and second periods correspond to "wet" and "dry" of the SST section respectively. The two components of wind stress were calculated for each month (from $\tau_x = C_D\rho|V|u$ and $\tau_y = C_D\rho|V|v$ where C_D is a drag coefficient, ρ is density, $|V|$ is the modulus of the vector wind, and u and v are the monthly mean zonal (x) and meridional (y) components of the wind respectively) and monthly averages were evaluated for each of the two periods. The differences in these two means for the months of August are shown in Fig. 6. Over the North Atlantic zonal winds were stronger in the first period. The stronger westerly winds would lead to an increased southward Ekman drift (tending to produce cooling by advection of colder water from the north) as well as increased evaporation (also associated with cooling). But lower temperatures in this region occur in the second period so cannot be due to these effects of the zonal wind. However, the trade winds off northwest Africa were stronger in the second period, this being in the right direction to account for colder temperatures. The Atlantic equatorial easterlies were also stronger in the first period tending to lead to more upwelling and lower temperatures, as observed. In the North Pacific between 65° and 55°N, a latitude belt where there was a significant drop in the zonal wind in the Atlantic in the second period, there was essentially no wind change. However, the Pacific trade winds between 25°N and 10°S were weaker in the second period while the SST in that region was generally colder.

A notable point about the meridional wind stress is that the South-East Trades off Brazil are stronger in the first period. Also the meridional component near 30°S in all oceans is stronger from the south in the second period. The change in the South-East Trades off Brazil observed here is similar to that predicted in the model studies of African moisture flux reported by Folland et al. (1986). This provides evidence that the model circulation and the SST field are mutually compatible. It is worth noting that there is no significant difference in this region in the actual wind stress patterns for February. These patterns are not shown; this is not an important season for Sahelian drought.

It was shown elsewhere that the atmospheric circulation over Africa alters between wet and dry conditions (Kidson, 1977; Newell and Kidson, 1984). The 200mb easterly jet and the southwesterlies of the lower troposphere are weaker in dry conditions, while the 700mb easterly jet over west Africa is stronger. What is not known is to what extent the changing circulation alters the SST patterns.

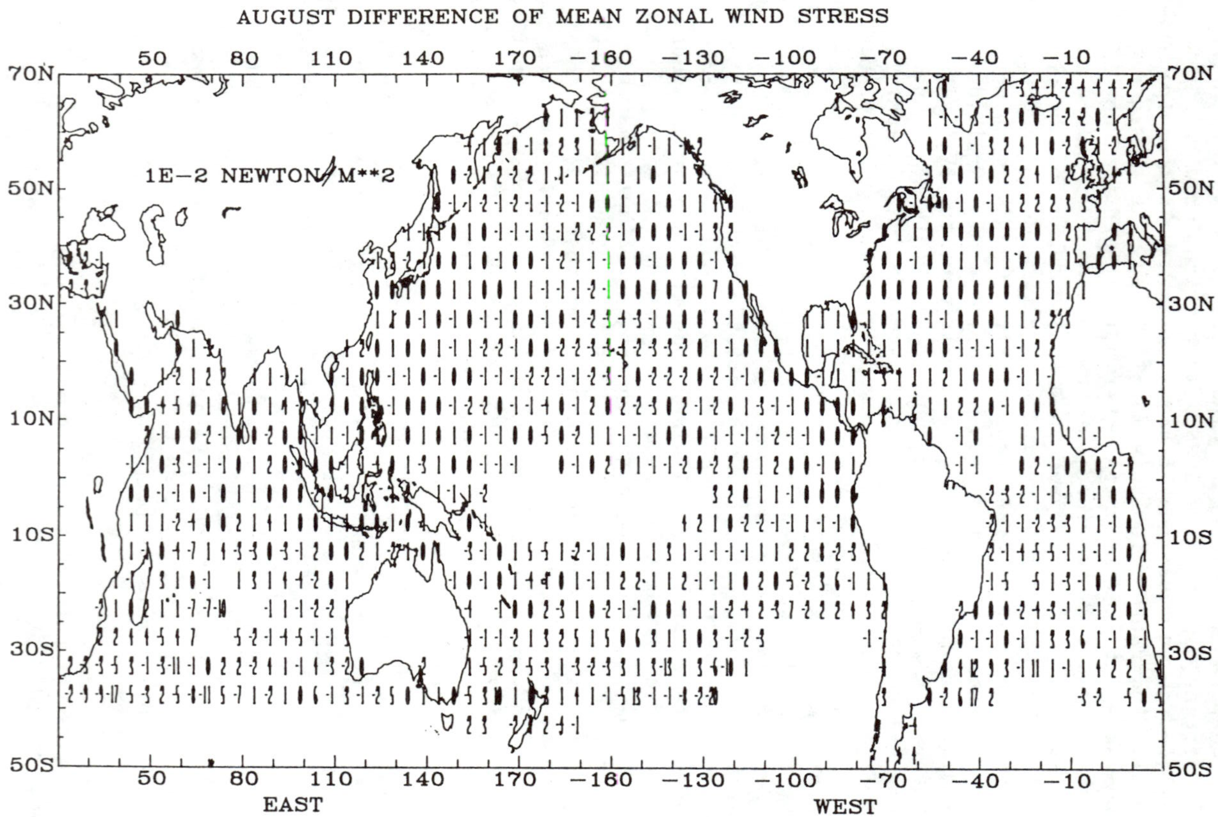

Figure 6a. Difference of August mean zonal wind stress (1949-1964) - (1965-1979). Units: 10^{-2} N/m^2

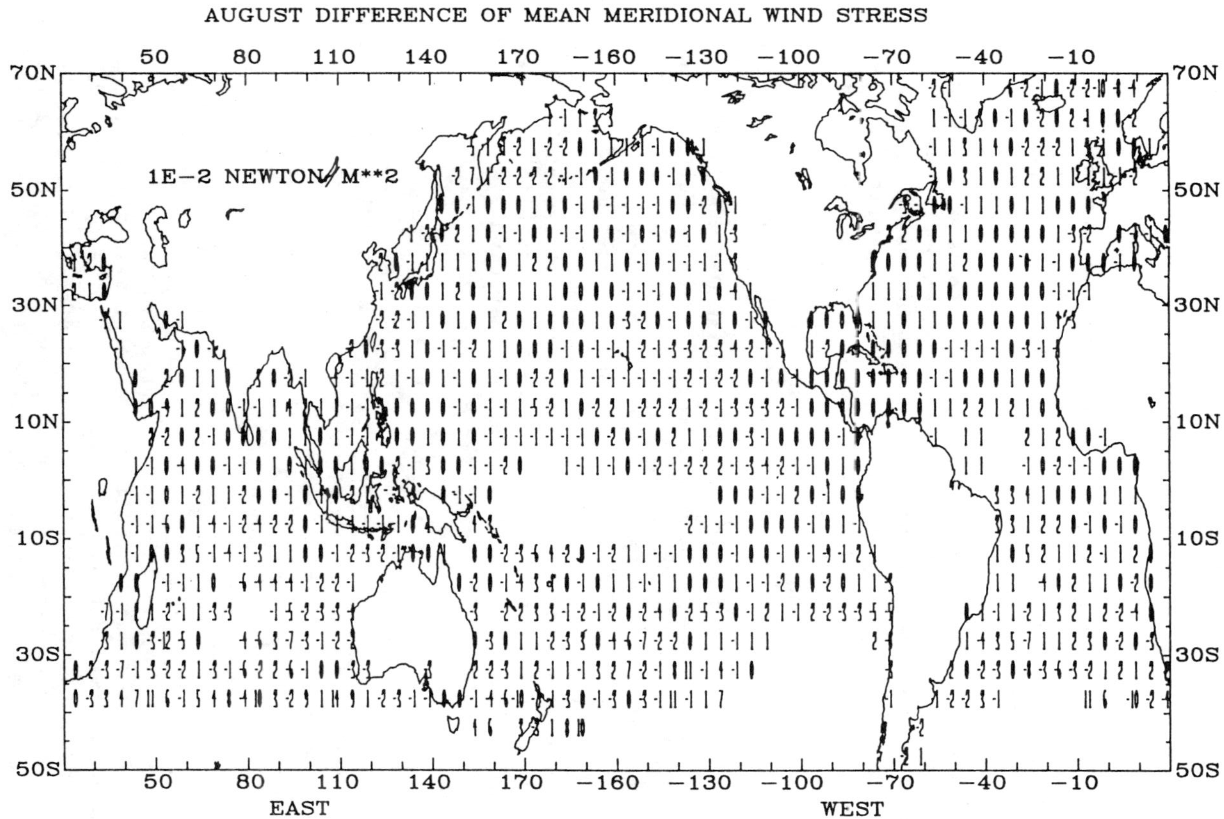

Figure 6b. Difference of August meridional wind stress (1949-1964) - (1965-1979). Units: 10^{-2} N/m^2.

COMMENTS ON THE POSSIBLE MECHANISMS

The North Atlantic and North Pacific SST decreases between the two periods considered cannot be explained directly on the basis of wind stress changes by evaporation or Ekman advection. Another way to cool the high latitude oceans is to supply less energy by advection from lower latitudes. In the case of the Atlantic this energy apparently passes from the Indian Ocean to the South Atlantic and then to the North Atlantic (Gordon, 1986).

Hsiung (1983) proposed that the reason for the SST difference in the North Atlantic between the two periods was that more energy was absorbed by the Indian Ocean in the first period and was therefore available to be passed on to the North Atlantic. She calculated the energy entering the surface by the bulk formulae (see, e.g., Hsiung, 1985) and found that there is less energy entering the Indian Ocean in the second period. The other oceans showed no significant differences. There is thus direct experimental evidence for source modulation.

Modulation of either the source, the Indian Ocean-Atlantic transfer, the South Atlantic-North Atlantic transfer or the North Atlantic transport are all possible mechanisms for producing changes in North Atlantic SST. As far as interhemispheric energy transfer is concerned there is the wind data quoted above which shows weaker South-East Trades near Brazil in the second period and this would be expected to give rise to weaker interhemispheric ocean currents there. It is not known if there are enough local ocean current readings available in this region to check this possibility but there are few measurements which can be used for basin-wide estimates (see Hsiung, 1985).

Finally we examine the possibility that energy transport within the North Atlantic is different between the two periods. We hypothesize that the zero line of the wind stress curl, often associated with the Sverdrup transport, represents an oceanic energy flow vector. The wind stress curl is given by:

$$\frac{\partial \tau_y}{\partial_x} - \frac{\partial \tau_x}{\partial_y} \qquad (1)$$

When the zero line is oriented east-west or north-south there is no energy flow within the ocean by the interchange of warm water from low latitudes with cold water from high latitudes. At some intermediate orientation we hypothesize that the oceanic energy flow is optimum; deviations in either direction from this optimum then, by definition, result in smaller energy transports. The wind stress curl has been evaluated for the wet and dry years used for the SST anomaly evaluation with results shown in Fig. 7. Between 30° and 50°N the line points more towards the north in the Atlantic but more towards the east in the Pacific in the dry years. Both configurations may be far enough from the optimum to diminish the energy flux.

Although precise data are not available it is possible to estimate the energy fluxes involved in the cycle for the Atlantic. If the

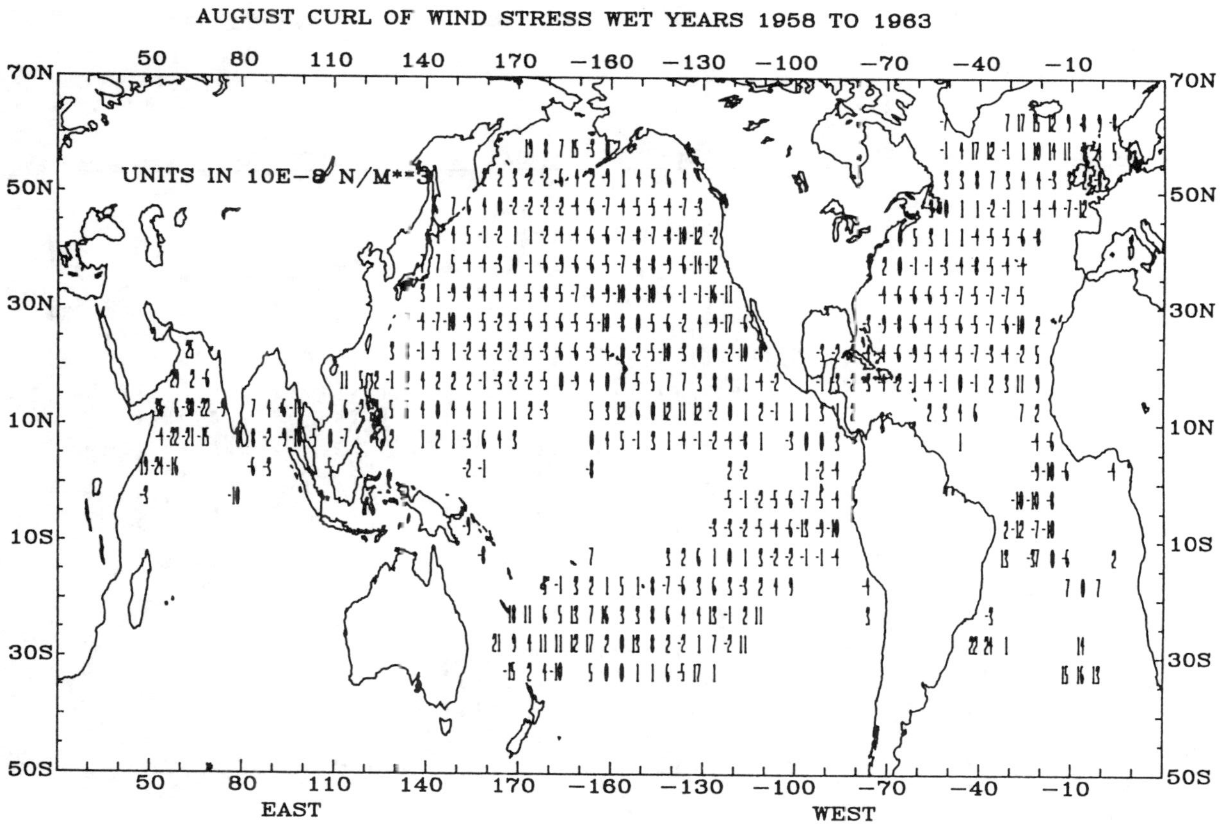

Figure 7a. Curl of August wind stress for wet years (1958-1963). Units: 10^{-8} N/m^3.

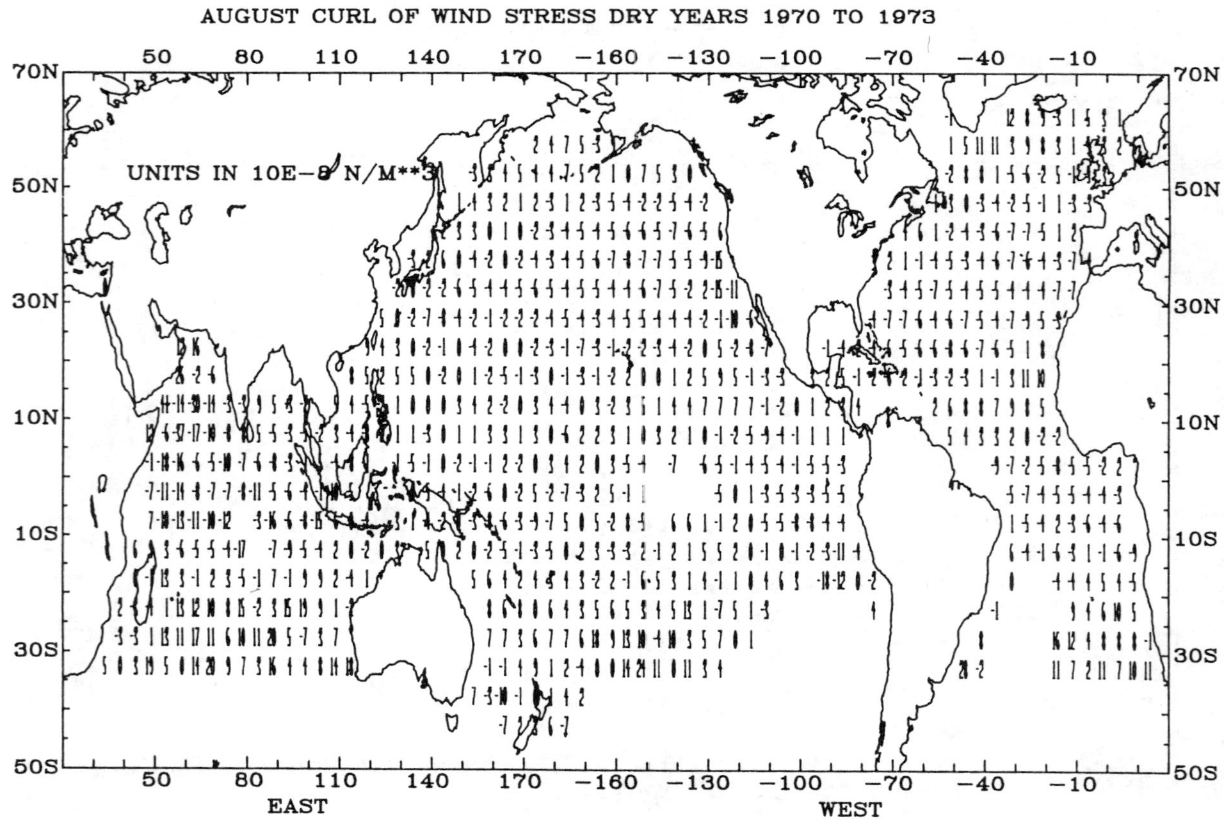

Figure 7b. Curl of August wind stress for dry years (1970-1973). Units: 10^{-8} N/m^3.

surface anomalies extend to the depth of the thermocline then from maps of this level by Levitus (1982) and Fig. 5 we estimate that the change in energy content of the Atlantic between the wet and dry periods amounts to about 7.7 x 10^{21}J; if the change occurs in two years the energy flux change is 1.2 x 10^{14}W. The northward flux of energy crossing the equator in the Atlantic sector is 5.7 x 10^{14}W (Bunker, 1976; Hsiung, 1985) so a 20% decrease could account for this change. Hsiung (1983) computed the change of energy entering the Indian Ocean through the surface between the two periods and found a value of 7 x 10^{14}W less in the second period. A decrease of 10^{14}W passing through the Atlantic can then be adequately accounted for on the basis of a reduction in the energy supply to the South Atlantic from the Indian Ocean. As far as the North Atlantic is concerned then, there is indirect evidence that three items in the chain - energy entering the Indian Ocean, energy passing northwards towards the tropical Atlantic and energy passing polewards across the North Atlantic - were reduced in the second period.

IMPLICATIONS FOR PALEOCLIMATE

One implication is that if the surface wind stress pattern changes and thereby the pattern of the curl of the wind stress, the oceanic energy transport will change and concomitantly high latitude ocean temperature. The extreme case is probably the period prior to 18,000 BP when the zero line of the wind stress curl in the North Atlantic gradually pointed more towards the east eventually becoming east-west (Ruddiman and McIntyre, 1981). With this alignment there is little poleward energy flux by the ocean although partial compensation by an increased atmospheric energy flux may be expected (Newell, 1974). But smaller amplitude changes may have a large difference on local regions such as Britain where large abrupt temperature changes have occurred as shown by Coope (this volume). Why should these changes occur? First of all during the development and decay of ice sheets the topographic forcing, which is now fixed, was highly variable. The standing wave pattern to the lee of the Rockies can, for example, be described in terms of the height of the mountains and the velocity of the mean east-west flow (Gill, 1982). As the north-south temperature gradient changes under the influence of changing Milankovitch radiation patterns so does the downstream wave pattern. During the ice age variable topography occurred, with up to 3km of ice on Hudson Bay. The wind patterns set up by this topography coupled with the winds from the Rockies to form a changing interference pattern which, we hypothesize, altered the orientation of the Atlantic Ocean energy flow vectors with time so that the abrupt temperature changes observed for northwest Europe, such as during the Younger Dryas, are quite easily explained. Another possibility is that the changing circulation in the Northern Hemisphere alters the cross-equatorial currents, as may have occurred for the African drought case. Of course, it remains to translate these ideas into a quantitative theory - not a trivial task.

Another implication concerns salinity and though based on recent data may be relevant to paleoclimate studies. A decreased poleward energy transport in the recent dry period would give decreased salinity by evaporation, because the surface waters were cold and evaporation increases with positive temperature anomaly (Warren, 1983; Hsiung, 1983) so that the colder northern oceans should have a lower salinity. There is some experimental evidence for such a trend (Brewer et al., 1983).

A further implication concerns the differences in deep water formation in the glacial and interglacial modes. Duplessy and Shackleton (1985) find evidence of a substantial reduction of North Atlantic Deep Water in the last ice age and Broecker et al. (1985) suggest that deep water may be formed in the North Pacific rather than the North Atlantic at that time. As noted earlier the second eigenvector of SST shows a change between the two periods 1949-64 and 1965-79 in which the North Atlantic and North Pacific alter their relative phase, being out-of-phase with each other in the first period and in phase with the second. If the wind stress pattern changes are the cause of the SST changes as argued above, one can ask if more exaggerated wind pattern changes may have been responsible for the glacial-interglacial changes which apparently occurred. The link again would be the evaporational control of salinity as the main control of deep water formation. Insofar as oceanic meridional temperature gradients influence the zonal wind speed, there is considerable potential for ocean-atmosphere feedback.

A final implication concerns the El Niño mode in SST patterns. It is tempting to argue that this is the present day analogue for the climatic optimum conditions of 6000 B.P. or 1200 A.D. with the opposite situation (more upwelling in the eastern equatorial Pacific) as an analogue to the Pleistocene Ice Ages or the Little Ice Age of the 17th century. This El Niño mode is also the mechanism governing atmospheric CO_2 (Newell, Hsiung, and Navato, 1978: Flohn, 1984; Newell and Hsiung, 1984) and here too there are potential ocean-atmosphere feedback processes.

ACKNOWLEDGEMENTS

This paper was written at the "Institut fur Umweltphysik der Universitat Heidelberg" while the first author was supported by the Alexander von Humboldt Stiftung; he is grateful for this support and for the gracious surroundings provided through the courtesy of Professor K.O. Münnich of Heidelberg. The data was processed with support from the U.S. Department of Energy under Contract DE-AC02-76EV12195 and the Climate Dynamics Section of the U.S. National Science Foundation under grant ATM-8211645. We thank Rennie Selkirk for updating Figures 1 and 3 and Susan Black for her help with the manuscript.

Helpful comments on the vector heat transport idea were made by G. Russell Coope and F. Alayne Street-Perrott when it was presented at this meeting. During seminars at which the relationship between wind patterns and possible salinity trends was proposed, Wolfgang Roether of Heidelberg and Edward Boyle of MIT drew our attention to observations of the recent freshening of the North Atlantic.

REFERENCES

Angell, J.K. and Korshover, J., 1975. Estimate of the global change in tropospheric temperature between 1958 and 1973, *Mon. Wea. Rev.*, 103: 1007-1012.

Boer, G. and Higuchi, K., 1980. A study of climatic variability, *Mon. Wea. Rev.*, 108: 1326-1332.

Brewer, P.G., Broecker, W.S., Jenkins, W.J., Rhines, P.B., Rooth, C.G., Swift, Takahashi, T. and Williams, R.T., 1983. A climatic freshening of the deep Atlantic north of 50°N over the past 20 years, *Science*, 222: 1237-1239.

Broecker, W.S., Peteet, D.M. and Rind, D., 1985. Does the ocean-atmosphere system have more than one stable mode of operation?, *Nature*, 315: 21-26.

Bunker, A.F., 1976. Computations of surface energy flux and annual sea-air.interaction cycles of the North Atlantic Ocean, *Mon. Wea. Rev.*, 104: 1122-1140.

Chiu, L.S. and Newell, R.E., 1983. Variations of zonal mean sea surface temperature and large-scale air-sea interaction, *Quart. J. R. Met. Soc.*, 109: 153-168.

Climate Analysis Center, 1985. *Climate Diagnostics Bulletin*, September 1985, Global Analyses and Indices. Figure Al. NOAA/National Weather Service, National Meteorological Center, Washington, D.C. 20233.

Coope, G.R., 1987. Fossil beetle assemblages as evidence for sudden and intense climatic changes in the British Isles during the last 45,000 years, this volume.

Douglas, A., Cayan, D. and Namias, J., 1982. Large-scale changes in North Pacific and North American weather patterns in recent decades, *Mon. Wea. Rev.*, 110: 1851-1862.

Duplessy, J-C. and Shackleton, N.G., 1985. Response of global deep-water circulation to Earth's climatic change 135,000-107,000 years ago, *Nature*, 316: 500-507.

Flohn, H., 1984. A possible mechanism of abrupt climatic change. In: *Climatic Change on a Yearly to a Millenial Basis*, eds. N.-A. Morner and W. Karlen, D. Reidel Publishing Company, 521-531.

Folland, C.K., Palmer, T.W. and Parker, D.E., 1986. Sahel rainfall and worldwide sea temperatures, 1901-85, *Nature*, 320: 602-607.

Gill, A.E., 1982. *Atmosphere-Ocean Dynamics*, Academic Press, 662 pp.

Gordon, A.L., 1986. Interocean exchange of thermocline water, *J. Geophys. Res.*, 91: 5037-5046.

Horel, J.D. and Wallace, J.M., 1981. Planetary scale atmospheric phenomena associated with the Southern Oscillation, *Mon. Wea. Rev.*, 109: 813-829.

Hsiung, J., 1983. Large-scale sea-air energy fluxes and global sea surface temperature fluctuations, Ph.D. Thesis, Massachusetts Institute of Technology, 240 pp.

Hsiung, J., 1985. Estimates of global oceanic meridional heat transport, *J. Phys. Oceanogr.*, 15: 1405-1413.

Hsiung, J. and Newell, R.E., 1983. The principal nonseasonal modes of variation of global sea surface temperature, *J. Phys. Oceanogr.*, 13: 1957-1967.

Kidson, J.W., 1975. Tropical eigenvector analysis and the Southern Oscillation, *Mon. Wea. Rev.*, 103: 197-216.

Kidson, J.W., 1977. African rainfall and its relation to the upper air circulation, *Quart. J. R. Met. Soc.*, 103: 441-456.
Kukla, G.J., Angell, J.K., Korshover, J., Dronia, H., Hoshiai, M., Namias, J., Rodenwald, M., Yamamoto, R. and Iwashima, T., 1977. New data on climatic trends, *Nature*, 270: 573-577.
Lamb, P.J., 1983. Sub-Saharan rainfall update for 1982: continued drought, *J. Climatol.*, 3: 419-422.
Levitus, S. 1982. *Climatological Atlas of the World Ocean*, NOAA Professional Paper 13, U.S. Government Printing Office, Washington, D.C. 20402, 173 pp.
Navato, A.R., Newell, R.E., Hsiung, J., Billing, C.B., Jr. and Weare, B.C., 1981. Tropospheric mean temperature and its relationship to the ocean and atmospheric aerosols, *Mon. Wea. Rev.*, 109: 224-254.
Newell, R.E., 1974. Changes in the poleward energy flux by the atmosphere and ocean as a possible cause for ice ages, *Quaternary Res.*, 4: 117-127.
Newell, R.E., 1979. Climate and the ocean, *American Scientist*, 67: 405-416.
Newell, R.E. and Hsiung, J., 1979. Fluctuations in zonal mean sea surface temperature, *Riv. Ital. Geofis. Sci. Affini*, 5: 121-125.
Newell, R.E. and Hsiung, J., 1984. Sea surface temperature, atmospheric CO_2 and the global energy budget: some comparisons between the past and present. In: *Climatic Changes on a Yearly to Millenial Basis*, eds. N.-A. Morner and W. Karlen, D. Reidel Publishing Company, 533-561.
Newell, R.E. and Kidson, J.W., 1984. African mean wind changes between Sahelian wet and dry periods, *J. Climatol.*, 4: 27-33.
Newell, R.E., Hsiung, J. and Navato, A.R., 1978. Long-term global sea surface temperature fluctuations and their possible influence on atmospheric CO_2 concentrations, *Pure and Applied Geophysics*, 116: 351-371.
Newell, R.E., Selkirk, R. and Ebisuzaki, W., 1982. The Southern Oscillation: sea surface temperature and wind relationships in a 100-year data set, *J. Climatol.*, 2: 357-373.
Newell, R.E. and Weare, B.C., 1976a. Ocean temperatures and large-scale atmospheric variations, *Nature*, 262: 40-41.
Newell, R.E. and Weare, B.C., 1976b. Factors governing tropospheric mean temperature, *Science*, 194: 1413-1414.
Nicholson, S.E., 1981. Rainfall and atmospheric circulation during drought periods and wetter years in West Africa, *Mon. Wea. Rev.*, 109: 2191.
Pan, Y.H. and Oort, A.H., 1983. Global climate variations connected with sea surface temperature anomalies in the eastern equatorial Pacific Ocean for the 1958-73 period, *Mon. Wea. Rev.*, 111: 1244-1258.
Parker, D.E., 1985. The influence of the Southern Oscillation and volcanic eruptions on temperature in the tropical troposphere, *J. Climatol.*, 5: 273-282.
Ruddiman, W.F. and McIntyre, A.F., 1981. The North Atlantic Ocean during the last deglaciation, paleogeography and paleoclimatology, *Paleoecology*, 35: 145-214.
Warren, B.A., 1983. Why is no deep water formed in the North Pacific?, *J. Marine Res.*, 41: 327-347.

Weare, B.C., Navato, A.R. and Newell, R.E., 1976. Empirical orthogonal analysis of Pacific Ocean surface temperatures, *J. Phys. Oceanogr.*, 6: 671-678.

SUMMER TEMPERATURE CHANGES FROM TREE RINGS IN THE MEDITERRANEAN AREA DURING THE LAST 800 YEARS

F. Serre-Bachet and J. Guiot

Laboratoire de Botanique historique et
Palynologie, UA 1152 du CNRS,
Faculté des Sciences et Techniques
St-Jérôme, 13397 Marseille Cedex 13
France

ABSTRACT. Reconstructions of the mean summer temperature (June to September) for Rome, Marseille and Grand St-Bernard from 1150 to 1972 are based on four tree-ring series of the Alps and of the Mediterranean region. They indicate a succession of alternately cold and warm episodes, the longest and most recent being the so called "Little Ice Age".

INTRODUCTION

In dendroclimatology, ring-widths or parameters derived from densitometric measurements of wood are used to reconstruct the temperature and precipitation history. For a few years tentative reconstructions of climatic parameters, based on tree rings and regarding the preinstrumental time, have been developing in Europe (Schweingruber et al., 1978; Eckstein and Aniol, 1981; Bircher, 1982; Aniol and Eckstein, 1983; Briffa et al., 1983; Bednarz, 1984; Hughes et al., 1984; Jones et al., 1984; Guiot et al., 1982 a, b, 1984 a, b, 1985, 1986). The tree-ring series are provided by several taxa, mainly conifers from relatively high altitudes: *Pinus silvestris* L., *Pinus cembra* L., *Picea abies* Karst, *Larix decidua* Mill., *Cedrus atlantica* Manetti. However, for the reconstruction of the riverflow of three river catchments in southern Britain, Jones et al. (1984) used oak (*Quercus petraea* Liebl. and *Quercus robur* L.). The reconstructions mainly concern mean "summer" temperatures, the temperature of 1, 2 or 3 months from June to September, on different periods and in different meteorological stations (Table 1).

Here we are concerned with the reconstruction of the mean temperature of the four summer months June to September, during the period 1150-1972, at Rome (Italy), Marseille (France) and Grand St-Bernard in the Pennine Alps. This period is particularly interesting because it includes the so-called "Little Ice Age" which was marked by an unusual number of cool summers in western Europe and elsewhere (Pfister, 1980; Le Roy Ladurie and Baulan, 1981; Lamb, 1982, 1984).

W. H. Berger and L. D. Labeyrie (eds.), Abrupt Climatic Change, 89–97.

TABLE 1. Length of the estimated periods and meteorological stations involved in climatic reconstructions in Europe and North Africa

---

94 yrs (1880-1973 : Schweingruber et al., 1978),
54 and 107 yrs (1903-1960; 1860-1976 : Bircher, 1982),
131 yrs (1840-1970 : Guiot et al., 1982b),
226 yrs (1755-1980 : Jones et al., 1984)
225 yrs (1741-1965 : Bednarz, 1984),
255 yrs (1721-1975 : Hughes et al., 1984),
300 yrs (1680-1979 : Aniol and Eckstein, 1983),
371 yrs (1590-1960 : Guiot, 1984a),
498 yrs (1471-1968 : Eckstein and Aniol, 1981).
775 yrs (1200-1974 : Guiot, 1984b, 1985, 1986) : the longest period

Meteorological stations in

- Scotland (Edinburgh : Hughes et al., 1984),
- northern Sweden (Kiruna : Aniol and Eckstein, 1983),
- the Tatra mountains (Zakopane : Bednarz, 1984)
- the Austrian or Swiss alpine region (Vent : Eckstein and Aniol, 1981; Zurich, Rigi, Grand St-Bernard and Chaumont : Schweingruber et al., 1978; Grand St-Bernard and St-Gothard : Bircher, 1982; Grand St-Bernard : Guiot, 1984a; Bivio : Guiot et al., 1982 a,b), - the South of France (Nice : Guiot, 1984b, 1986; Marseille : Guiot, 1985, 1986)
- Morocco (Tetouan and Marrakech : Guiot et al., 1982b).

Four tree ring series, labelled A, B, C and D (Fig. 1) are used. They are of various length and derived from three different species, as follows:

A. *Larix decidua* Mill., 1150-1970, from "Les Merveilles" near Tende (44°03'N, 7°27'E) at 2150-2180m elevation (Serre, 1978, 1979)
B. *Larix decidua* Mill., 1353-1970, from "l'Orgère", near Modane (45°13'N, 6°40'E) at 1900-2300m elevation (Tessier, 1981; Tessier et al., 1986)
C. *Pinus leucodermis* Ant., 1150-1970, from Calabria, near Cosenza (39°26'N, 16°12'E) at 1900m elevation (Serre-Bachet, 1985)
D. *Abies alba* Mill., 1660-1970, from Mont-Ventoux, near Avignon (44°10'N, 5°15'E) at 1750-1800m elevation (Serre-Bachet, 1986).

When reconstructing climate it is of interest to use different species from widely dispersed sites, because each species reacts in its own way to climatic influences; the combination of different species and different responses to climate allows a better reconstruction of the climate itself.

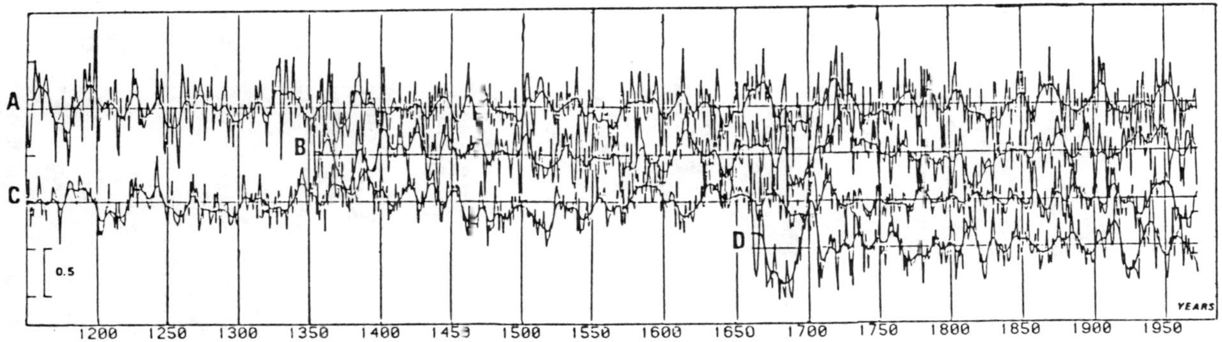

Figure 1. The indexed tree-ring master chronologies: A : Les Merveilles; B : L'Orgère; C : Calabria; D : Mont-Ventoux.

Figure 2. Reconstructions of summer temperature (°C departure from mean) in 1) Rome, 2) Marseille, 3) Grand St-Bernard, 4) decennial index of summer (June+July) rainfalls in Belgium, Germany and Northern France from Alexandre (1977).

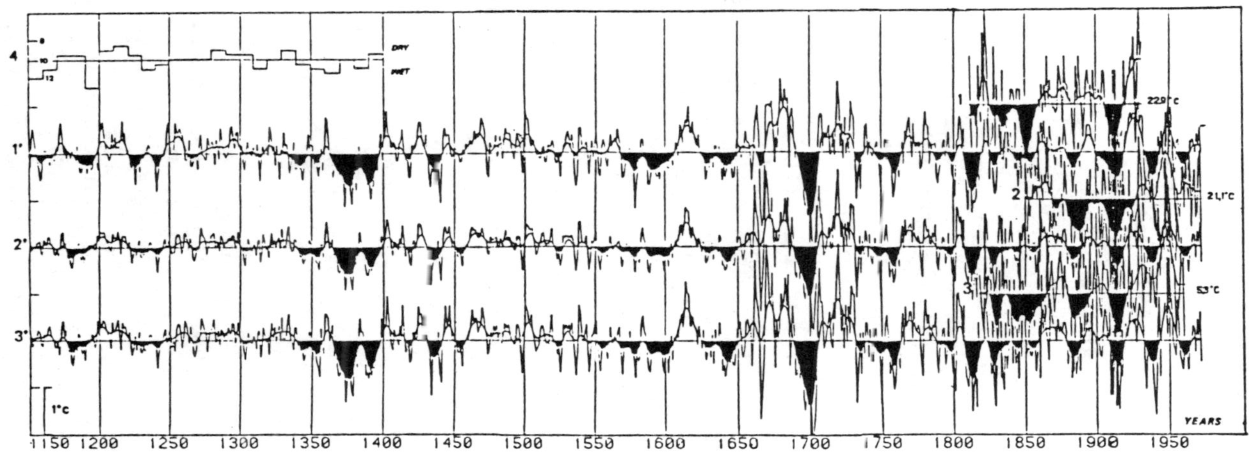

THE DATA

There are about 10 trees per sites and 3 cores per tree. In order to allow for the effect of age on growth, ring-widths were standardized (Fritts, 1976) and then the index series were averaged to obtain the ring width chronology of each site. The curves on Fig. 1 show the annual variation (high frequency variations) of these chronologies. Superimposed are the low frequency variations calculated on a digital filter with a cut-off period of 8 years.

Because of their proximity, records A and B are positively correlated (r=0.42), both correlate negatively with D, albeit weakly (-0.18 and -0.21 respectively); record C is not correlated with the three others.

The three meteorological stations of Rome (elev. 63m), Marseille (elev. 75m) and Grand St-Bernard (elev. 2470m) were chosen for calibration of the records because they provide the longest instrumental data series (Rome : 1811-1930 and 1951-1970, Marseille : 1851-1970, Grand St-Bernard : 1818-1960). As is often the case, the meteorological stations are at quite a distance from the sites studied, except for Grand St-Bernard and l'Orgère, and at a far lower elevation, except again Grand St-Bernard. Therefore we chose to reconstruct temperature, as it does not differ as much as precipitation from one site to another.

THE RECONSTRUCTIONS

For each meteorological station, correlations between the 12 monthly temperatures and records A, B, C and D showed a strong and coherent influence of summer temperature (June to September). Therefore, the mean temperature of the four summer months was used as predicted. The predictors were chosen according to the standard methods used in dendroclimatology (Fritts, 1976; Hughes et al., 1984); they are the four standardized tree ring series at time t and t+1 - so as to include direct and delayed effect - and at time t-1 -in order to correct for autocorrelation.

Twelve, nine or six predictors were then used for regressions: Regression 1, with 12 predictors, enables one to extrapolate back to 1660 A.D., Regression 2, with 9 predictors, back to 1353 A.D. and Regression 3, with 6 predictors, back to 1150 A.D.

For the three regressions and for each meteorological station, the calibration period is 1851-1930; 60 years (1811-1850/1951-1970) of actual temperature data at Rome are retained for the independent verification period, 42 years (1931-1972) at Marseille, 63 years (1818-1850/1931-1960) at Grand St-Bernard.

The regression coefficients calculated on the calibration period show that 1) for populations A (Les Merveilles) and B (l'Orgère), the June to September mean temperature has a positive and unlagged effect on growth, 2) for population C (Calabria), the effect is negative and continues the following year, and 3) for population D (Ventoux), the effect is also negative but somewhat delayed.

The correlation between estimated and actual values on the calibration and on the verification periods (Table 2) are highly significant and stable for Regression 1, but less so, or not at all, for Regressions 2 and 3. In those cases, the apparent instability is caused by high frequencies prevailing in a too short verification period (63 years or less).

There are few proxy series available to check the reliability of the reconstructed summer temperature trend before 1400 A.D.; among them the decennial index of summer rainfall for Belgium, Rhenanie and northern France by Alexandre (1977) is plotted in Figure 2. Assuming that precipitation is negatively related to temperature in summer, this index is consistent with the reconstructions based on tree rings. Therefore the long-term variations of these reconstructions appear reliable. Regression 3 may be used before 1353 A.D., at least for the low frequencies.

TABLE 2. Correlations between estimated and actual values for the three regressions on the calibration period (1851-1930), on the independent period (1811-1850 and 1931-1972), on the total period (1811-1972), significant at level 0.10 (*), 0.05 (**) or 0.01 (***).
The number of degrees of freedom is indicated in brackets.

correlations	regression 1 A-B-C	regression 2 A-B-C	regression 3 A-C
calibration			
number of P.C	12	9	6
Rome	0.66 ***	0.49 ***	0.37 *
Marseille	0.55 ***	0.36	0.28
Grd St-Bernard	0.56 ***	0.39	0.28
verification			
<u>1811-1850/1931-1972</u>			
Rome (60)	0.29 **	-0.04	-0.06
Marseille (42)	0.49 ***	-0.08	-0.35
Grd St-Bernard (630	0.35	0.19 *	-0.15
<u>1811-1972</u>			
Rome (140)	0.50 ***	0.24 ***	0.12 *
Marseille (122)	0.48 ***	0.24 ***	0.09
Grd St-Bernard (143)	0.42 ***	0.27 ***	0.08

DISCUSSION

The reconstructions, and more especially the below-average values of the low frequency curves plotted in black (Fig. 2) have been compared with those made previously for Marseille, Nice and Grand St-Bernard on one or several tree ring-series by Guiot et al. (1982a), Guiot (1984a, b, 1986) and for the Austrian Alps by Eckstein and Aniol (1981) as well as with independently recorded meteorological data (Guiot, 1984b). These comparisons show a fairly good agreement between the main under-mean episodes of the corresponding periods. The cold periods of 1570-1600, 1630-1650, 1690-1700, 1740-1760, 1810-1840 characterize the so-called Little Ice Age. The 1460-1550 period differs from the preceding and from the next one by rather weak variations. The medieval period of 1150 to 1450 is marked by the cold episodes of 1180-1190, 1220-1240, 1340-1360, 1370-1400, 1430-1440.

It is worth noting that the main cold episodes in the reconstructed mean summer temperature (Fig. 2) correspond to the periods characterized by a high mean sensitivity (a statistic which measures the average relative difference of one ring-width to the next (Fritts, 1976)) (Fig. 3). This has already been stated by Serre (1979), Tessier (1981) and Tessier et al. (1986).

An analysis of the climate changes from 1150 A.D. was made for Marseille using the Mann-Kendall rank statistics (Goossens and Berger, in this volume). Figure 4 shows for the three regressions used for this station (Table 2): 1) a cooling from 1335 A.D. to 1456 A.D. significant from 1369 A.D. (Fig. 4.3); 2) a warming from 1456 A.D. to 1551 A.D. significant from 1511 A.D. (Fig. 4.3); 3) a second cooling from 1551 A.D. to 1809 A.D. with a warm episode at the beginning of the 18th century (Fig. 4.3 and 4.2); 4) a final warming from 1860 A.D. to now with a maximum around 1950 A.D. (Fig. 4.1)

The coherency between Figures 4.1, 4.2 and 4.3 suggests again the reliability of the reconstructions. There appears no continuous trend from 1150 A.D. up to now but rather a succession of alternately cold and warm episodes lasting 120 to 300 years, the last one being the so-called Little Ice Age.

CONCLUSION

The mean summer temperature reconstructions presented here for the period 1150-1972 can be considered as reliable. However the use of a small number of tree-ring series (2) for the oldest part of the analyzed period is responsible for less reliability as compared with the more recent part of the period for which a greater number of tree-ring series (3 or 4) is available.

Using Mann-Kendall statistics for the reconstructed Marseille record, four main climatic periods can be distinguished: first a warm period before 1335, followed by a marked cooling down to 1456; then, from 1551, another cold period corresponding to the Little Ice Age followed just after 1860 by a warming that is still prevailing now.

Figure 3. Variation of the mean sensitivity calculated on 20 years overlapping periods for the four populations A, B, C, D.

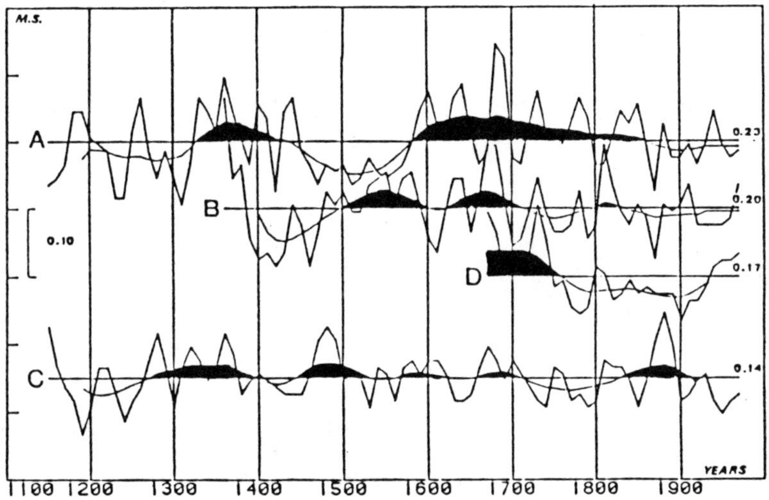

Figure 4. Mann-Kendall rank statistics (Goossens and Berger, this volume) cumulated in the chronological direction and in the inverse direction for the three reconstructions from 4 populations (1), 3 populations (2) and 2 populations (3).

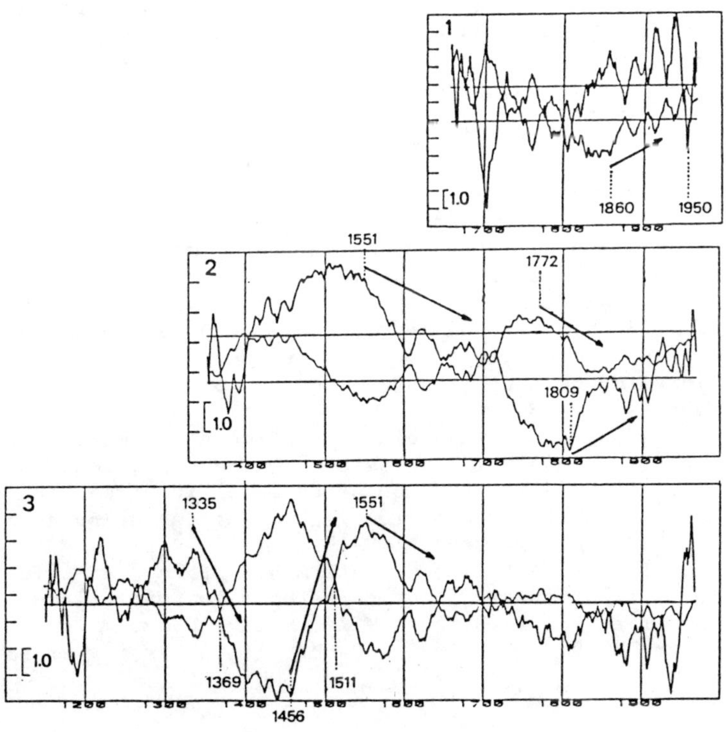

The greater variability observed in ring-width during the cold summer periods seems to indicate that cold summer temperatures are more variable than warm ones. Within the framework of the present analysis there is no evidence for notable differences between the long term climatic changes recorded in the Mediterranean area at Marseille or Rome and those found in the Alps at Grand St-Bernard. Thus it is once more verified that there are coherent temperature changes over large regions, especially in the "Little Ice Age".

REFERENCES

Alexandre, P., 1977. Les variations climatiques au Moyen Age (Belgique, Rhénanie, Nord de la France). *Annales*, 2, 183-197.

Aniol, R.W. and Eckstein, D., 1984. Dendroclimatological studies at the northern timberline. In: *Climatic Changes on a Yearly to Millenial Basis*. N.A. Morner and W. Karlen (eds.), Dordrecht : D. Reidel Publ. Co., 273-279.

Bednarz, Z., 1984. The comparison of dendroclimatological reconstructions of summer temperatures from the Alps and Tatra mountains from 1741-1965. *Dendrochronologia*, II, 63-72.

Bircher, W, 1982. Zur Gletscher und Klimageschichte des Saastales, glazialmorphologische und dendroklimatologische Untersuchungen. *Pysische Geographie*, Universitat Zurich, 9, 233 p.

Briffa, K.R., Jones, P.D., Wigley, T.M.L., Pilcher, J.R. and Baillie, M.G.L., 1983. Climate reconstruction from tree rings. Part 1, Basic methodology and preliminary results for England. *Journal of Climatology*, 3, 233

Eckstein, D. and Aniol, R.W., 1981. Dendroclimatological reconstruction of summer temperatures for an alpine region. *Mitt. Forst. Bundesversuchsanst.*, Wien, 142, 391-398.

Fritts, C., 1976. *Tree rings and Climate*. Academic Press, 567 p.

Goossens, C. and Berger, A., 1987. How to recognize an abrupt climatic change?, this volume.

Guiot, J., 1984a. Deux méthodes d'utilisation de l'épaisseur des cernes ligneux pour la reconstruction de paramètres climatiques anciens, l'exemple de leur application dans le domaine alpin. *Palaeogeography, Palaeoclimatology, Palaeoecology*, 45, 347-368.

Guiot, J., 1984b. Mise en oeuvre d'une méthode d'analyse multivariée de séries chronologiques dans un but de reconstructions dendroclimatiques concernant les Alpes du sud. *Statistique et Analyse des données*, 9, 35-59.

Guiot, J., 1985. The extrapolation of recent climatological series with spectral canonical regression. *Journal of Climatology*, 5, 325-335.

Guiot, J., 1986. ARMA techniques for modelling tree ring response to climate and for reconstructing variations of palaeoclimates. *Ecological modelling*, 33, 149-171.

Guiot, J., Berger, A.L. and Munaut, A.V., 1982a. An illustration of alternative transfer function methods in Switzerland. In: *Climate from Tree Rings*, Hughes, M.K., Kelly, P.M., Pilcher, J.R. and LaMarche, V.C. Jr, 1982 , Cambridge University Press, 160-162.

Guiot, J., Berger, A.L., Munaut, A.V. and Till, C., 1982b. Some new mathematical procedures in dendroclimatology, with examples from Switzerland and Morocco. *Tree Ring Bull.*, 42, 33-48.

Hughes, M.K., Schweingruber, F., Cartwright, D. and Kelly, P.M., 1984. July-August temperature at Edinburgh between 1721 and 1975 from tree ring density and width data. *Nature*, 308, 341-344.

Jones, P.D., Briffa, K.R. and Pilcher, J.R., 1984. Riverflow reconstruction from tree rings in southern Britain. *Journal of Climatology*, 4, 461-472.

Lamb, H.H., 1982. *Climate, History and the Modern World*. London (Methuen), 387 pp.

Lamb, H.H., 1984. Climate and history in northern Europe and elsewhere. In: *Climatic Changes on a Yearly to Millenial Basis*. N.A. Morner and W. Karlén (eds). D. Reidel Publ. Co., 225-240.

Le Roy Ladurie, E. and Baulant, M., 1981. Grape harvests from the fifteenth through the nineteenth centuries. In: *Climate and History : Studies in Interdisciplinary History*, R.I. Rotberg and T.K. Rabb (eds), Princeton: Princeton University Press, 259-269.

Pfister, C., 1980. The climate of Switzerland in the last 450 years. *Geographica Helvetica*, 35, 15-20.

Schweingruber, F.H., Fritts, H.C., Braker, O.U., Drew, L.G. and Schar, E., 1978. The x-ray technique as applied to dendroclimatology. *Tree-ring Bull.* 38, 61-91.

Serre, F., 1978. Résultats dendroclimatiques pour les Alpes méridionales françaises. Colloque International du Centre National d'Etudes Spatiales: "Evolution des atmosphères planétaires et climatologie de la terre". Nice, 16-20 Octobre 1978, ed. CNES, Toulouse, 381-385.

Serre, F., 1979. The dendroclimatological value of the European larch (*Larix decidua* Mill) in the French Maritime Alps. *Tree Ring Bull.* 38, 25-34.

Serre-Bachet, F., 1985. Une chronologie pluriséculaire du sud de l'Italie. *Dendrochronologia*, 3, 45-66.

Serre-Bachet, F., 1986. Une chronologie maîtresse du sapin (*Abies alba* Mill.) du Mont Ventoux (France). *Dendrochronologia*, 4, 87-96.

Tessier, L., 1981. Contribution dendroclimatologique à la connaissance écologique du peuplement forestier des environs des chalets de l'Orgère (Parc National de la Vanoise). *Trav. Scient. Parc Nat. Vanoise*, XI, 29-61.

Tessier, L., Couteaux, M. and Guiot, J., 1986. An attempt at an absolute dating of a sediment from the last glacial recurrence through correlations between pollenanalytical and tree-ring data. *Pollen et Spores*, 28, n°1, 61-76.

EVIDENCE OF ABRUPT CLIMATIC CHANGE DURING THE LAST 1,500 YEARS RECORDED IN ICE CORES FROM THE TROPICAL QUELCCAYA ICE CAP, PERU

Lonnie G. Thompson and Ellen Mosley-Thompson

Institute of Polar Studies
The Ohio State University
Columbus, Ohio 43210
U.S.A.

ABSTRACT. Reliable observations of climatic variation prior to 1850 are limited or absent for many regions, especially *South America*. Ice sheet and ice caps can provide a valuable supplement to historical and other proxy records. The *tropical* Quelccaya ice cap in Peru (5,670m) has provided a *1500-year* record of *climatic variability*. Abrupt climatic events are recorded in this ice by *microparticle concentrations*, *conductivities* and *oxygen isotopes* which are discussed in detail for three periods: A.D. 1928-1947, 1864-1905 and 1452-1550. These data suggest that both the onset and termination of the "*Little Ice Age*" were abrupt in tropical South America.

INTRODUCTION

Climate fluctuations over the last 1,000 years represent the most probable range of natural variability for the next century and are of great interest. Therefore historical records chronicling the extent of past climatic changes over the last 1,000 years are ideal for assessing the time scales and amplitudes of change. Unfortunately, long historical records are geographically restricted to the Orient (China and Japan) and West Europe (Gribbin and Lamb, 1978). In the Southern Hemisphere, and especially in South America, the reconstruction of global climate changes depend almost entirely upon proxy records such as ice cores, tree rings and lake sediments.

Tropical and subtropical conditions inhibit the formation of seasonal rings (Stockton et al., 1985), thus, dendroclimatological records from Ecuador, Peru, and Brazil and southward along the eastern flank of the Andes to about 23°S are very limited. Fortunately, some ice fields in the Andes contain records of tropical and subtropical climatic variability with high temporal resolution over several millennia. In an effort to obtain such a record, a field program has been conducted on the Quelccaya ice cap each summer for the period 1976 through 1984 with one central objective - to recover an ice core to bedrock. It was anticipated that a 1,000-year climatic history for tropical South America could be reconstructed. The Quelccaya ice cap program was designed as a

W. H. Berger and L. D. Labeyrie (eds.), Abrupt Climatic Change, 99–110.

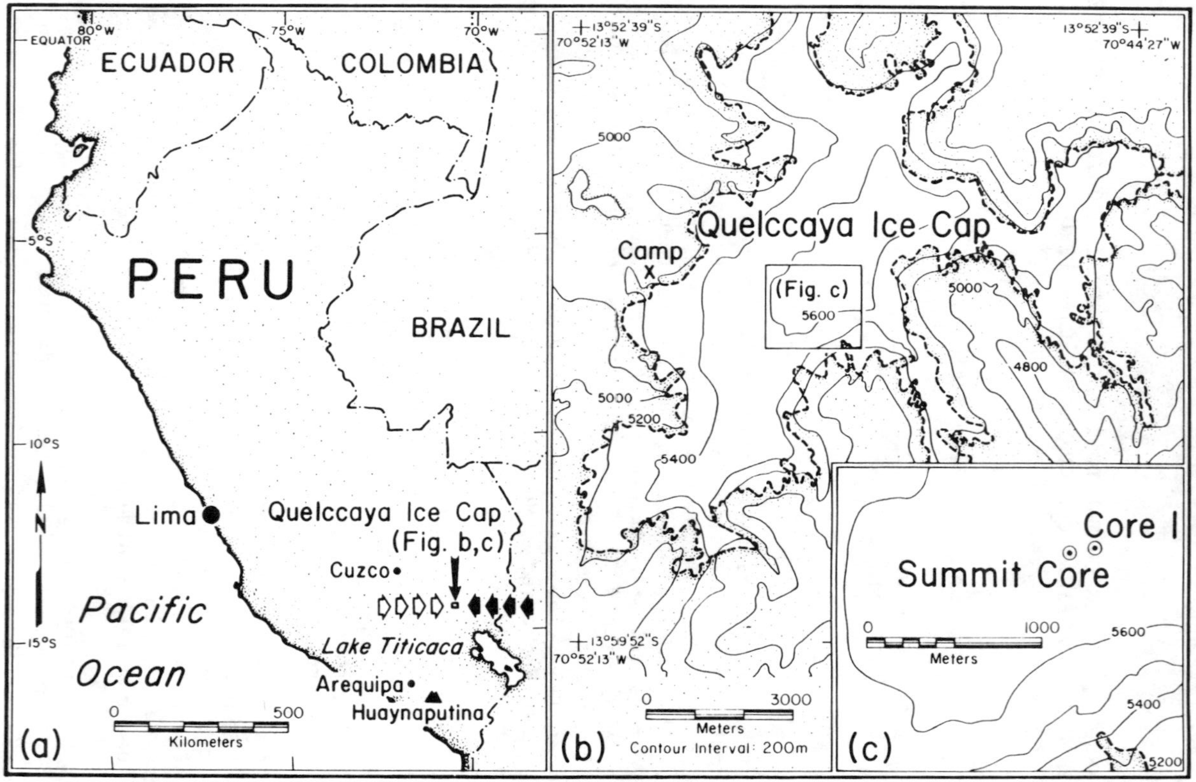

Figure 1a. Location of the Quelccaya ice cap. Black arrows indicate dominant wet season wind direction while the open arrows indicate dominant dry season wind directions (Schwerdtfeger, 1976), 1b shows map of ice cap and 1c illustrates the locations of two long cores drilled in 1983.

field and laboratory study to acquire data for the reconstruction of this climatic record. Results from snow pit and shallow ice core studies have been published (Thompson et al., 1979, 1984a, b; Thompson, 1980) These studies demonstrate seasonal cycles in microparticle concentrations, conductivities and oxygen isotope ratios.

In 1983, the central objective was accomplished by recovering one 154.8m summit core (SC) containing 1,350 years and another 163.6m core (C-1) containing 1,500 years. The cores were dated using a combination of visible annual dust layers, microparticle concentrations, conductivity and oxygen isotope ratios. Thompson et al. (1985) present a synthesis of the 1,500-year precipitation record from both cores, while Thompson et al. (1986) present the decadal variations of microparticle concentrations, conductivity and oxygen isotopes for the last 1,000 years. Here we focus on three abrupt events recorded within the last 1,000 years in the visible stratigraphy, microparticle concentrations, conductivity variations and oxygen isotopes preserved in the summit core.

The Quelccaya ice cap (13°56′S; 70°50′W) rests on the easternmost rise of the Andes (Fig. 1) on the western margin of the Amazon Basin. On this ice cap there is a distinct seasonality in snowfall which is ultimately preserved in the ice stratigraphy. The elevation (5,670m) results in a near zero net surface energy balance and an annual mean temperature of -3°C, and consequently there is an insignificant amount of melting and percolation. This has been confirmed by dry season field observations using *in situ* percolation pans. Distortion of the preserved record by ice flow is minimized as the ice cap rests on an extensive plateau with gentle topography. The plateau has been mapped using a monopulse radar system. A maximum thickness of 180 meters and an annual net balance of 1.15 meters $(H_2O)a^{-1}$ result in a long and temporally well-resolved, proxy climate record.

Precipitation on the Quelccaya ice cap falls as snow throughout the year. The major moisture source for wet season snowfall (November-April) is from air masses advected from the Amazon Basin (Thompson et al., 1984b). The 500mb cloud flow patterns across South America during the wet season indicate the major moisture source is ultimately the Atlantic Ocean (Schwerdtfeger, 1976; Virji, 1979). Meteorological data from an automatic weather station maintained on the summit of Quelccaya from 1976 to 1984 indicate a dominant wind direction out of the east during the wet season and wind direction from the west during the dry season. Thus, the main source for the dust is from the west from the dry altiplano of Southern Peru and the snowfall from the east from the vast Amazon Basin and ultimately the Atlantic Ocean.

In this paper dealing with abrupt climatic events we focus upon three different time intervals in which abrupt deviation from average microparticle concentrations, conductivities and oxygen isotopes are recorded contemporaneously in all three parameters. The periods discussed include the greatest drought recorded in southern Peru (A.D. 1933 to 1945) and the onset (A.D. 1452 to 1550) and termination (A.D. 1864-1905) of the "Little Ice Age".

MATERIALS AND METHODS

The two long ice cores (see Fig. 1b and c for location) were recovered without contaminating the pristine environment or the core samples by use of a newly-developed, sun-powered drilling system. Approximately 6,000 samples were collected from pits, shallow cores, and the two long cores. The samples were melted by passive solar-heating in a laboratory tent, placed in polyethylene bottles, sealed in wax, and shipped to The Ohio State University. In the laboratory, samples were divided so that microparticle concentrations and size distributions, conductivity, and oxygen isotope measurements could be made on the same samples.

The analysis of microparticle concentration and size distribution (Thompson, 1977) is conducted under Class 100 clean room conditions using two Model TA II Coulter Counters, which electronically separate particles into 15 size ranges between 0.4 and 16μm in diameter. The microparticle concentration (MPC) data are presented as total particles $\geq$ 0.63μm diameter per milliliter (ml) sample. Electrolytic conductivity (S), measured under Class 100 clean room conditions using an Altex RC-16C Conductivity Bridge, furnishes an excellent estimate of the soluble impurities in the meltwater. Conductivity (S) measurements, in microsiemens cm^{-1}, have an accuracy within 1% of the reading. All measurements were conducted under temperatures of 21 $\pm$ 1°C, and with replicate analyses. The oxygen isotope analyses for the SC were conducted at the University of Copenhagen (Denmark) and at the University of Washington for the C-1. The oxygen isotope results are expressed as $\delta^{18}O$, the relative difference in isotopic abundance ratio $^{18}O/^{16}O$ between the sample and the Vienna Standard Mean Ocean Water (VSMOW) expressed in per mil ($^{o}/oo$). Control samples for interlaboratory comparison indicate a maximum difference of 0.30 $^{o}/oo$.

RESULTS AND CONCLUSIONS

"Little Ice Age" in Peru

Figure 2 illustrates the decadal averages of MPC (both >.63 and >1.59μm diameter particles), conductivity, *accumulation* and $\delta^{18}O$ isotope records for the last 1,000 years in the summit core. These records suggest that in the last millenium the most significant climatic event in tropical South America coincided with the "Little Ice Age" here recorded between A.D. 1490 and 1880 (Thompson et al., 1986).

In the Northern Hemisphere the "Little Ice Age" (LIA) lasted from approximately the early 1500's to the late 1800's and was characterized by colder-than-usual temperatures and expanded glaciers. The dates of the LIA determined from historical and proxy climate records vary depending on location and the observed parameter, for example 1550-1850 (Flohn and Fantechi, 1984) and 1430-1850 (Gribbin and Lamb, 1978). This period, corresponding to the Northern Hemisphere LIA, is a prominent feature in the microparticle concentrations, conductivities and $\delta^{18}O$ features of the Quelccaya ice core record. Microparticle and conductivity averages are as much as 30% higher during the LIA than during the 14th, 15th and 20th centuries (Fig. 2). In addition, the average $\delta^{18}O$

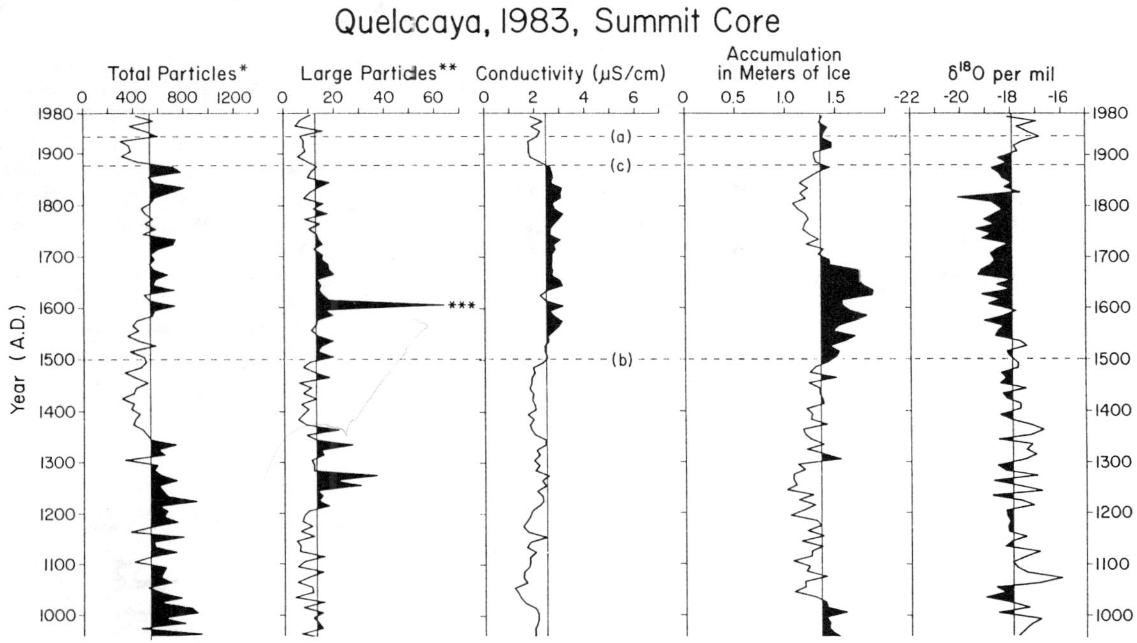

* Total Particles ≥ 0.63 to ≤ 16.0μm in Diameter per ml of Sample (10^3)
** Large Particles > 1.59μm in Diameter per ml of Sample (10^3)
*** Huaynaputina (Peru) eruption of Feb. 19, - March 6, 1600 A.D.

Figure 2. Decadal variations of total microparticles (>0.63μm and >1.59μm diameter), liquid conductivities, net accumulation and $\delta^{18}O$ over the last 1,000 years (modified from Thompson et al., 1986). The solid line represents the 1,000 year averages. The "Little Ice Age" appears as the largest climatic event in the last 1,000 years in tropical South America. The periods interpreted as representing abrupt climatic changes which are discussed in this paper are indicated as a, b and c. These are enlarged in Figures 3, 4 and 5, respectively.

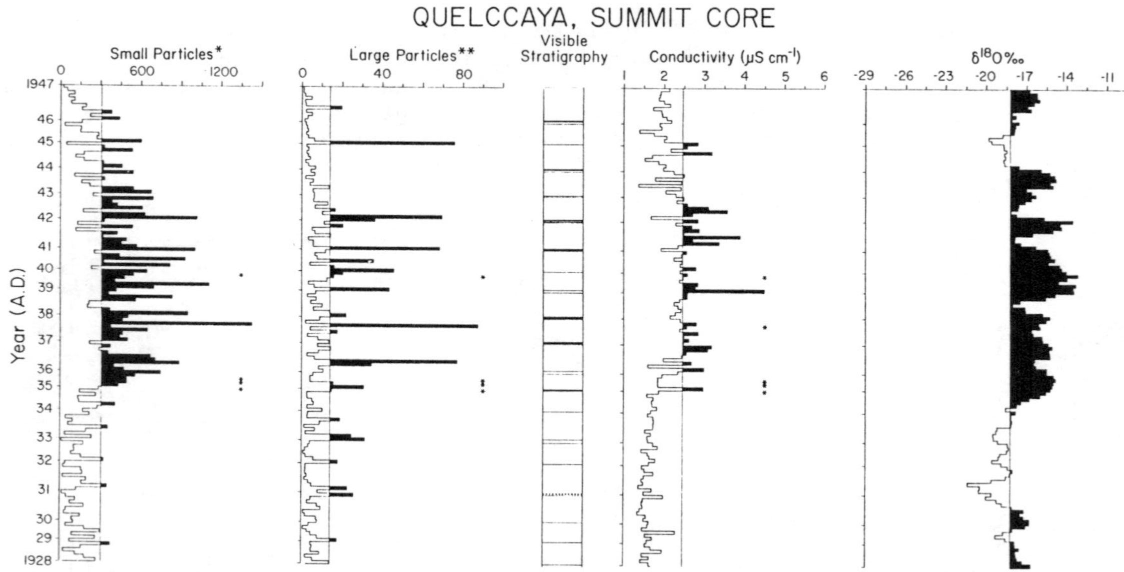

* Small Particles ≥ 0.63 to ≤ 0.80μm in Diameter per ml of Sample (10^3)

** Large Particles ≥ 1.59 to ≤ 16.0μm in Diameter per ml of Sample (10^3)

Figure 3. Ice core parameters measured in the summit core for the period A.D. 1928 to 1947. Shaded areas here and in Figures 4 and 5 represent positive deviations from the averages calculated over the entire core record. The drought in southern Peru, A.D. 1933 to 1945, is recorded by increases in both soluble (conductivity) and insoluble (microparticle concentrations) particulates and less negative $\delta^{18}O$ ratios. The ice core visible stratigraphy is illustrated on the right of Figures 3, 4 and 5 where: a single solid line represents a normal dry season dust layer; a single dashed line and double dashed line represent very light to light dust layers, respectively; and a triple dashed line and quadruple dashed line represent distinct and very distinct dust layers, respectively. The drought of the 1930's and 1940's is characterized by as many as five closely spaced dry season layers. The dots identify missing samples.

values average 0.9 $^{o}/oo$ more negative than for the last 100 years or the 1,000 years prior to 1500 A.D. The Quelccaya ice core records of precipitation and conductivity suggest that the onset of the LIA occurred rather abruptly (Fig. 2b) although the δO^{18} change is more gradual. However, the termination of the LIA is abrupt in all the ice core parameters measured (Fig. 2c).

As this paper deals with abrupt climatic events, emphasis is placed upon three time intervals in which contemporaneous, abrupt deviations from the average are recorded in microparticle concentrations, conductivities, precipitation and oxygen isotopes. Special attention is given the first period, 1928 to 1947 (Fig. 2a, Fig. 3) as it may provide a modern analog facilitating better interpretation of older parts of the ice core records.

Twentieth Century Drought

The period A.D. 1933 to 1945 encompassed the greatest drought recorded in southern Peru. The drought is reflected by a large drop in Lake Titicaca water levels over this period (Newell, 1949) and in the precipitation record from El Alto in Bolivia. Lake Titicaca, the highest (3,812m) large lake in the world, covers 8,446km. From 1933 to 1945 water levels dropped almost 5m (Newell, 1949; see pages 10-17) causing extreme difficulties for the Peruvian Corporation which operated a fleet of steamboats. The lake was closed to traffic for several years and installations were abandoned because of the shallowing.

Evidence of this drought is excellently preserved in the Quelccaya ice core records. It is characterized by reduced accumulation, by an increase in the concentration of both large and small particles, and by higher conductivities for the years from 1934 to 1945 (see Fig. 3). In association with this drought the $\delta^{18}O$ ratios are distinctly less negative than under the wetter conditions prevailing several decades prior to and since that time. The substantial increase in particle concentrations and conductivities probably reflect the increased dryness of the altiplano in association with the drought. The drought is also evident in the visible stratigraphic record as an increase in the frequency of multiple dust bands (produced during the dry season) within a single year (Fig. 3). It should be possible to isolate other extensive drought periods in this region from the longer ice core using distinct characteristics of the microparticles, conductivities, $\delta^{18}O$ and net annual accumulation.

Onset of the "Little Ice Age"

The onset of the LIA in Peru, inferred from the increase in microparticles and the sharp increase in conductivities (Fig. 2b; Fig. 4) occurred around 1490 A.D. The high and typically more extreme values which characterize the $\delta^{18}O$ signal throughout the LIA, did not begin until the 1520's, 30 years after the initial increase in both soluble and large-diameter insoluble particulates. From these ice core records it appears that a profound and lasting change in climate took place within a few decades. The atmosphere became far more dust laden, both with insoluble particles but particularly with soluble particulates,

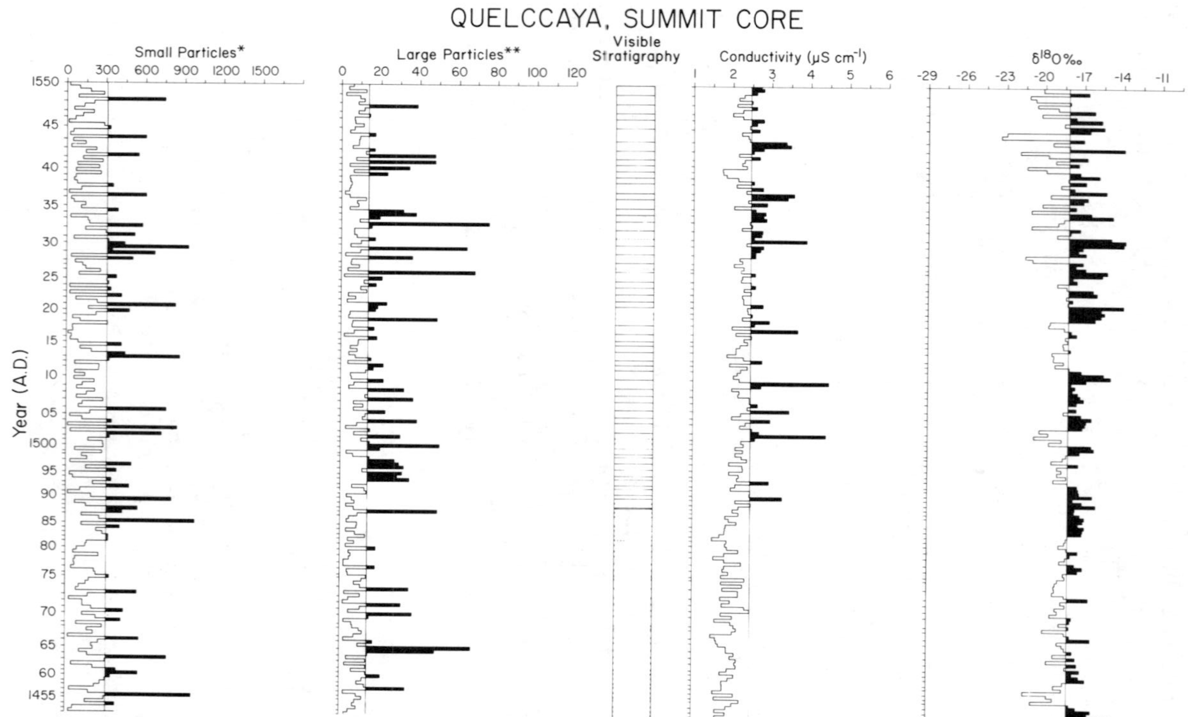

*Small Particles ≥ 0.63 to ≤ 0.80μm in Diameter per ml of Sample (10^3)
**Large Particles ≥ 1.59 to ≤ 16.0μm in Diameter per ml of Sample (10^3)

Figure 4. Ice core parameters measured in the summit core for the period A.D. 1452 to 1550. The increase in soluble (conductivity) and insoluble (microparticle concentrations) particulates characterize the "Little Ice Age" (LIA) inferred as beginning about 1490 A.D. The visual stratigraphy recorded in the field reveals a change in the characteristics of the annual dust layers at this time. High seasonal oscillations and more negative $\delta^{18}O$ values which characterize the LIA begin around 1520 A.D.

precipitation increased abruptly around A.D. 1490. Thirty years later the lower oxygen isotopes suggest colder temperatures, while the greater interannual variability is due at least in part to greater seasonal variability during the Little Ice Age. The initial increase in the concentration of microparticles and the sharp increase in conductivity, characterizing the entire LIA, were very abrupt at about 1490 A.D. The dry season dust layers are very distinct in the visual stratigraphy before 1490 A.D. The general increase in soluble and insoluble particles at the onset of the "Little Ice Age" may have reduced the overall clarity of the ice core making the dry season dust layers appear less distinct.

Termination of the "Little Ice Age"

One of the most abrupt events recorded in the Quelccaya ice cores is the termination of the LIA (Fig. 2c; Fig. 5). Centered nearly on A.D. 1880 there is a sharp reduction in both the insoluble and soluble particulate concentrations from the higher levels characterizing the LIA. These reduced levels are more characteristic of the last 100 years. Additionally, the amplitudes of the seasonal $\delta^{18}O$ variations which characterized the LIA period rapidly decrease just before A.D. 1880, while mean oxygen isotope values become less negative. These data indicate that for the last 100 years: (a) reduced mean wind velocities resulted in a sharp drop in insoluble and soluble particles, (b) an increase in annual snowfall and, (c) an increase in annual mean temperatures (as inferred from less negative oxygen isotope values).

SUMMARY

Many geological records of climate variability covering the last 1,000 years lack good temporal resolution. This reinforces the conventional expectation of gradual climatic change. The relatively large annual accumulation on ice sheets and ice caps and especially on Quelccaya (1.15 meters water equivalent a^{-1}) allows reconstruction of both high frequency (annual) and longer-term climatic variations. Some evidence of abrupt changes within the climate system have been reported from polar ice cores, e.g., Camp Century (Thompson, 1977) and Dye 3 in Greenland (Dansgaard et al., 1982; Herron and Langway, 1985). These results and those from the tropical Quelccaya ice cap strongly suggest that atmospheric conditions can change substantially over only a few years. Ice cores offer the most direct evidence of the past variations in atmospheric aerosols. The dust records provide a unique opportunity to acquire the information essential for the reconstruction of conditions of drought, volcanic activity and variations in past circulation patterns. The insoluble and soluble particulates from the Quelccaya ice cores have been used to identify a historical volcanic eruption (Thompson et al., 1986) and in this paper increases in particle concentrations have also been shown to be one of the ice core indicators of drought conditions. The oxygen isotope data suggest a significant cooling in the tropics during the "Little Ice Age" and the large seasonal range during the LIA may indicate greater sublimation of surface snow during the dry season and/or greater seasonal variability in moisture source during this period.

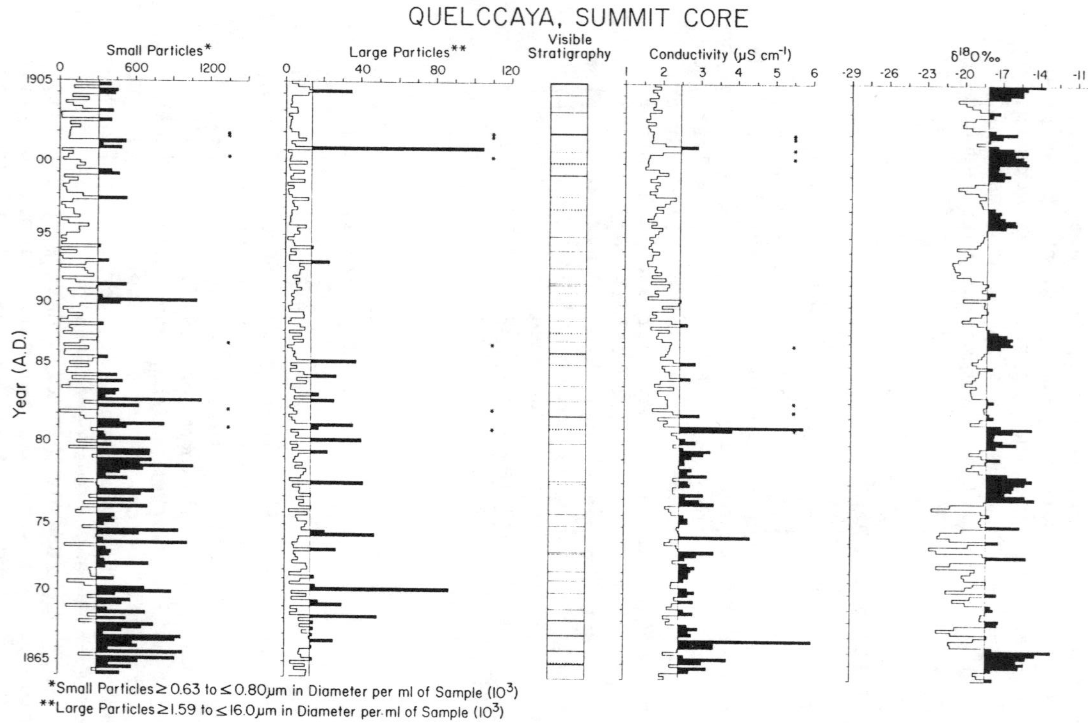

Figure 5. Ice core parameters measured in summit core for the period A.D. 1864 to 1905. The soluble (conductivity) and insoluble (microparticle concentrations) particulates drop abruptly from the elevated LIA levels around A.D. 1880. From A.D. 1875 to 1880 there is an abrupt reduction in the annual amplitude of the $\delta^{18}O$ signal from the large seasonal ranges characterizing the LIA to lower amplitudes and less negative average values characteristic of the last 100 years.

ACKNOWLEDGEMENTS

Supported by NSF grants ATM75-15513A02, ATM78-1609A01, ATM81-05079A02, and ATM82-13601A02. The NSF Division of Polar Programs supported 1974 field investigations (GV1411) and development of the solar-powered drill. We thank the Polar Ice Coring Office for the design and development of the drill. We thank Prof. W. Dansgaard for the oxygen isotope analysis of Summit Core and for many valuable discussions. We thank B. Morales Arnao as well as numerous scientists, engineers, and technicians from Electroperu for scientific and logistical assistance and the Inter-American Geodetic Survey office in Lima for logistical support. Many people have participated in the ten field programs, and their efforts are gratefully acknowledged. Contribution 561 of the Institute of Polar Studies, Ohio State University.

REFERENCES

Dansgaard, W., Clausen, H.B., Gundestrup, N., Hammer, C.U. Johnsen, S.F., Kristindotter, P.M. and Reed, N., 1982. A New Greenland Deep Ice Core, *Science*, 218 (4579): 1273-1277.

Flohn, H. and Fantechi, R., 1984. *The Climate of Europe: Past, Present and Future*, Reidel Pub. Co., Dordrecht, 24-64.

Gribbin, J. and Lamb, H.H., 1978. Climatic change in historical times. In: *Climatic Change* edited by J. Gribbin, Cambridge University Press, 68-82.

Herron, M.M. and Langway, C.C., Jr., 1985. Chloride, nitrate and sulfate in the Dye 3 and Camp Century, Greenland ice cores. In: *Greenland Ice Core: Geophysics, Geochemistry, and the Environment*, edited by C.C. Langway, Jr., H. Oeschger and W. Dansgaard, American Geophysical Union Monograph, 33, 77-84.

Newell, N.D., 1949. *Geology of the Lake Titicaca Region, Peru and Bolivia*, The Geological Society of America Memoir 36, p. 11, Waverly Press, Inc.

Schwerdtfeger, W., 1976. The Climate of Peru, Bolivia and Ecuador, In: *Climates of Central and South America* edited by W. Schwerdtfeger, World Survey of Climatology, Elsevier Scientific Publishing Company, 12, 192-218.

Stockton, C.W., Boggess, W.R. and Meko, D.M., 1985. Climate and tree rings. In: *Paleoclimate Analysis and Modeling*, edited by A. D. Hecht, 71-161, John Wiley and Sons.

Thompson, L.G., 1977. Variations in microparticle concentration, size distribution and elemental composition found in Camp Century, Greenland, and Byrd Station, Antarctica, ice cores, Proceedings of the Grenoble Symposium, 1975, *IAHS AISH Publ. No. 178*, 351-364.

Thompson, L.G., 1980. Glaciological investigations of the tropical Quelccaya ice cap, *J. Glac.*, 25 (91): 69-84.

Thompson, L.G., Hastenrath S., and Arnao, B.M., 1979. Climatic ice core records from the tropical Quelccaya Ice Cap, *Science*, 203 (4386): 1240-1243.

Thompson, L.G., Mosley-Thompson, E. and Arnao, B.M., 1984a. El Niño-Southern Oscillation Events Recorded in the stratigraphy of the tropical Quelccaya ice cap, Peru, *Science*, 226 (4670): 50-53.

Thompson, L.G., Mosley-Thompson, E., Grootes, P.M., Pourchet, M. and Hastenrath, S., 1984b. Tropical glaciers: potential for ice core paleoclimatic reconstructions, *J. Geophys. Res.*, 89 (3): 4638-4646.

Thompson, L.G., Mosley-Thompson, E., Bolzan, J.F. and Koci, B.R., 1985. A 1500-year record of tropical precipitation records in ice cores from the Quelccaya ice cap, Peru, *Science*, 229 (4717): 971-973.

Thompson, L.G., Mosley-Thompson, E., Dansgaard, W. and Grootes, P.M., 1986. The "Little Ice Age" as recorded in the stratigraphy of the tropical Quelccaya ice cap, *Science*, 234 (4774), 361-364,

Virji, H., 1979. Summer circulation over South America from satellite data, PhD thesis, The University of Wisconsin, 146.

Part III.

Abrupt Climatic Change during the Holocene

THE LATE-QUATERNARY CLIMATE OF THE WESTERN AMAZON BASIN

Paul A. Colinvaux[1] and Kam-biu Liu[2]

[1]Departments of Zoology and Anthropology
and Byrd Polar Research Center
The Ohio State University
Columbus, Ohio 43210, USA

[2]Department of Geography and Anthropology
Louisiana State University
Baton Rouge, Louisiana 70803, USA

ABSTRACT. Claims that Amazon climates were arid at glacial maxima are without radiocarbon dated evidence to justify them. Although aridity has been demonstrated for areas peripheral to South America, like the Galapagos Islands or northern Venezuela, these sites are separated from the Amazon by such climatic divides as the Andes mountains and the ITCZ. Radiocarbon dated evidence from a single site in the western Amazon basin near the Andean foothills suggests descent of montane forest of more than 700m at 33,000-26,000 yr BP, requiring temperature depression of at least 4.5°C in the Amazon lowlands. Cooling rather than aridity might have been a determining factor of glacial climate in the Amazon. Within the Holocene an episode of greatly increased precipitation has been demonstrated for the western Amazon from 1300 to 800 yr BP. Scattered data suggest that this millennial time-scale event might have been one of a long succession in late-Quaternary times.

INTRODUCTION

Recent opinions on the late-Quaternary history of Amazonia hold that climates were more arid in glacial maxima than in interglacials like the present (Haffer, 1969, 1974; Prance, 1982). The Holocene is thought to have been characterized by only modest climatic changes, possibly including a series of drier intervals (Absy, 1979). These opinions are based on extraordinarily few data. Proponents of an arid Amazon at glacial maxima have not produced a single radiocarbon dated sample in support of their hypothesis and accounts of Holocene events in the basin are as yet based on histories of *varzea* lake formations subject to annual flooding which might be expected to be stochastically intermittant.

The concept of an arid Amazon has its strongest support from biogeographers seeking to explain disjunct distributions of modern Amazonian taxa. Bird, butterfly and other species distributions are

W. H. Berger and L. D. Labeyrie (eds.), Abrupt Climatic Change, 113–122.

found to be clumped round regions of high relief, generally at the margins of the Amazon basin. These regions of high relief are, not unreasonably, among the wettest in the modern Amazon, letting Haffer (1969, 1974) put forward the hypothesis that these wetter areas should have remained wet also throughout the Pleistocene. The modern lowlands, like the wetter elevations themselves, support tropical rain forest. If, however, times of northern glaciation were periods of aridity in the Amazon basin, then it should be possible that the elevated areas of wettest modern climates should have remained moist. These areas might then have been able to support tropical rain forest. The biogeographical hypothesis to explain disjunct species distributions in the Amazon, therefore, postulates their occupying elevated refugia in arid glacial times, essentially islands of rain forest in a sea of savanna.

Very little direct evidence from the Amazon basin has been put forward to test this hypothesis, though there have been numerous discussions to marshall data that might be consistent with the hypothesis. Among these data have been descriptions of dune landforms and stone pavements in various parts of Amazonia, none of them with independent dating (Prance, 1982). A single pollen diagram from Rondonia in the southwestern Amazon, a part of the basin with strongly seasonal climates at present, shows a savanna interval at some undated time in the past (Absy and van der Hammen, 1976). The savanna interval is as likely to be Holocene as of glacial age, since the section is in an ecotonal region within a hundred kilometers of modern savannas.

Only one Amazon data set gives independent evidence for aridity with glacial radiocarbon ages, that of the history of mineral deposition on the continental shelf off the Amazon mouth by Damuth and Fairbridge (1970). These authors demonstrate that included in glacial age sediments discharged from the Amazon mouth were minerals suggestive of erosion of arid land surfaces that are not included in Holocene sediments, yet the hypothesis they put forward to explain this is not aridity in the lowlands but aridity in the Guiana Highlands to the north. These are drained by southward flowing rivers that feed into the Amazon system. The Guiana Highlands are today an ecotonal region of savannas, suggesting that only modest climatic changes were necessary to start arid erosion cycles there. The neccessary changes in climate for this ecotonal region might have followed displacements of the Intertropical Convergence Zone (ITCZ), or have resulted from a diminution of monsoonal rains of the kind modeled when Milankovitch forcing is applied to the Community Circulation Model (CCM) (Kutzbach and Guetter, 1984, 1986). Thus the Damuth and Fairbridge (1970) results are parsimoniously explained by changes in the ecotonal climates of the north flank of the Amazon basin and do not require aridity in the Amazon lowlands.

The biogeographic hypothesis of past aridity in the Amazon is made plausible by a general belief in tropical aridity as a property of the ice-age earth. This is brought into a South American context by citing evidence from various authors who demonstrate aridity in various parts of South America peripheral to the Amazon basin. The Galapagos Islands have been shown to have been more arid than now throughout the time of the classical Wisconsin maximum (Colinvaux, 1972). Late-glacial aridity in northern Venezuela has been demonstrated from studies at Lake Valencia (Bradbury et al., 1981) and on the coast of British Guiana (Wijmstra and van der Hammen, 1966). But the aridity demonstrated in these and kindred studies is for climatic systems quite outside the Amazon basin. The Galapagos history, for instance, undoubtedly has been influenced by patterns of El Niño and the Southern Oscillation, a climatic system separated from the Amazon basin by the Andes chain, and the Lake Valencia record is from the other side of the path of ITCZ from Amazonia as well as being separated from the basin by major relief. Demonstration of aridity in local parts of South America outside Amazonia at various times in the last glacial cycle is in no sense evidence that the whole continent was subject to aridity.

Recent models applying Milankovitch orbital parameters to the Community Circulation Model (CCM) for the ice-age earth predict a diminution of monsoon rains of about 10% (Kutzbach and Guetter, 1984). Monsoonal climates are, by definition, seasonal. A general 10% reduction of monsoon rains will produce noticeable aridity only in habitats with strongly seasonal climates, and then only in ecotonal areas. Possibly this mechanism did cause significant aridity in the northern and southwestern portions of Amazonia where savannas are included in the modern vegetation mosaic. But in the wetter parts of modern lowland Amazonia precipitation is over 3000mm annually and may approach 5000 - 7000mm locally. Neither are the climates in the western Amazon basin strongly seasonal, suggesting that Milankovitch forcing would not result in arid climates there.

Thus arguments for an arid Amazon in glacial maxima are tenuous and without hard, radiocarbon dated evidence.

TEMPERATURE DEPRESSION IN THE AMAZON LOWLANDS

At Mera at an elevation of 1100m at the base of the Andes mountains in western Amazonia (Fig. 1), pollen and megafossil evidence suggests a lowering of vegetation belts consistent with a temperature depression of 4.5°C between 33,000 BP and 26,000 BP (Liu and Colinvaux, 1985). These are the only radiocarbon dated data for glacial times yet retrieved from the Amazon basin. If correctly interpreted, the data suggest, not only temperature depression in lowland equatorial regions, but also that postulated rain forest refugial areas of biogeographic theory did not in fact support rain forest continuously throughout the last glacial maximum. This result would make plausible the concept of a wet rain forest climate in lower-lying regions of Amazonia in place of the aridity of current concepts.

The Mera site is at the upper limits of lowland tropical rain forest in Amazonia (Grubb et al., 1963). Liu and Colinvaux (1986) used surface pollen samples to show that the modern rain forest near the Mera elevation of 1100m is indeed comparable to that at elevations down to 250m. Pollen from the fossil sections differed from these modern rain forest spectra in ways that suggested affinities with forest types not found at high elevations in the Andes. As with other pollen sections

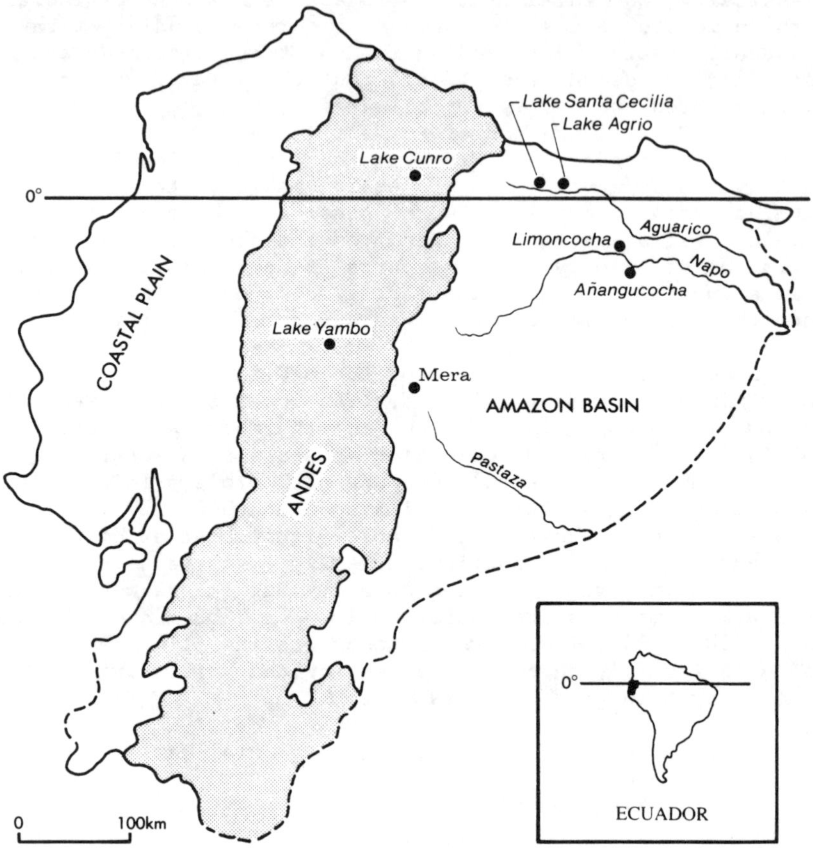

Figure 1. Map of Ecuador showing paleoecological study sites mentioned in the text.

from the Amazonian lowlands, the pollen spectra included numerous rare pollen taxa which could not be given botanical names. In addition, the pollen spectra are not directly comparable to any modern plant associations. That the vegetation differed markedly from modern Amazonian rain forest, however, is evident. This shows that the rain forest refugium postulated for the River Napo watershed by refugial theorists did not persist throughout the glacial cycle, unless displaced largely to lowlands well to the east.

The Mera pollen spectra also differ from a series of surface samples from the Inter-Andean Plateau of Ecuador in being without large proportions of grass and other pollen types of open or lakeside habitats (Liu and Colinvaux, 1985). The presence of these taxa in modern Andean samples certainly reflects the aridity of many modern Andean sites resulting from rain shadows of the cordilleras to the east and west.

That they were reduced in the Mera pollen assemblage is strong evidence that the vegetation was moist; a reasonable conclusion considering that Mera is in the Amazonian foothills of the Andes in one of the wettest climates of the region today. Thus the fossil Mera pollen spectra suggest a complex, moist, montane forest with many elements now prominent in the Andes at elevations of 2000m or higher.

Wood samples included in the polliniferous deposits included softwoods thought to be of Andean species of *Podocarpus* (Liu and Colinvaux, 1986). It has since been pointed out that some species of *Podocarpus* exist in lowland forests, though none have been recorded from Ecuador (T. Stone and A.H. Gentry; personal communications). But these lowland podocarps appear to be constituents of lowland sites with pronounced dry seasons as in Rondonia of the southern edges of the Amazon basin. The pollen assemblage from the Mera deposits, in which *Podocarpus* pollen appears in association with appropriate taxa of moist Andean forests, seems to confirm the original interpretation that the logs are of Andean *Podocarpus* species now prominent at elevations above 2000m.

Since the lowest record for *Podocarpus* in Ecuador is at 1800m, Liu and Colinvaux (1985) concluded that forest types were lowered by at least 700m at 33,000 and 26,000 yr BP, implying a minimal temperature depression of 4.5°C during times of glacial maxima. This temperature depression is minimal, both because the data give no reason for thinking that the Mera site was at the lowest elevation of the ancient *Podocarpus* stands and because the dating is consistent with the fossil forests having been of interstadial rather than full glacial age (op. cit.). No other calculations of temperatures at glacial maxima appear to have been made in lowland Amazonia, or indeed anywhere in the wet equatorial lowlands, though there is ample evidence for temperature depressions at higher elevations along the equator.

The closest relevant calculation is that for the Colombian Andes at about 5° N latitude and above 2000m elevation where pollen data suggest a temperature depression of about 6°C in full glacial times (van der Hammen and Gonzalez, 1960). Temperature depressions of this order in the high Andes are also suggested by depression of firn lines (Hastenrath, 1971) and by the budget for an extended Quelccaya ice cap (Thompson et al., 1984).

These results conflict with the widespread belief that temperature depression at low latitudes and low elevations was modest. CLIMAP (1976) shows many areas of tropical oceans with surface temperatures within 2°C of modern temperatures at 18,000 yr BP, letting many conclude that temperature depression was also minimal at low elevation over equatorial lands. Close inspection of the CLIMAP results, however, does not support the contention that all equatorial ocean masses maintained near modern temperatures at glacial maxima (Rind and Peteet, 1986). The true pattern is one of changed oceanic circulation in which water masses were displaced so that some equatorial water masses were colder than others. Furthermore, the actual data coverage of equatorial oceans is still sparse so that the map of ocean surface temperatures is far from complete. Thus our calculation of a minimal temperature depression of

4.5°C over the western Amazon during glacial maxima is fully in accord with what little circumstantial data exist and is in conflict with none. Very likely the true temperature depression in the lowland Amazonian forests of Ecuador at the last glacial maximum was in the order of 6°C as in the Colombian Andes.

A MILLENNIAL TIME-SCALE PRECIPITATION EVENT

Lake sediment records from close to the equator in Ecuador have been taken to show regional flooding in the western Amazon 1300-800 yr BP (Colinvaux et al., 1985). Current research on sedimentary histories of endorheic lakes in the adjacent Andes mountains now show that the flooding event was contemporaneous with episodes of high lake level. Together, these records suggest enhanced precipitation on the equatorial Cordillera Oriental during the 500 yr period ending 800 years ago.

The prime evidence for flooding in the lowlands comes from four sedimentary records of riverine lakes in the Oriente Province of Ecuador which suggest that the parent rivers reentered old abandoned channels between these dates. Lakes Agrio and Santa Cecilia lie along the River Aguarico and Limoncocha and Añangucocha (n.b. the Quechua suffix cocha means "lake") along the River Napo (Fig. 1). All four lakes occupy deep sections of channels abandoned by the neighboring river. Lakes Santa Cecilia and Agrio lie close to the Andean foothills, although well down in the lowland rain forest at 330m elevation. Añangucocha and Limoncocha are about 100km further away from the foothills to the east at about 230m elevation. The sediment cores show that all four lakes have deposited organic rich gyttja continuously for the last 700-800 radiocarbon years, although at different rates. Underlying the gyttja in all four lakes is grey mineral sediment comparable to sediment in the modern river beds, showing that all four channels were occupied by the parent rivers until the sediment boundary with gyttja, which is dated to near 800 BP in all four lakes. In Añangucocha the old riverine sediment is itself underlain by peaty gyttja, showing that the River Napo had reentered a still older channel at about 1300 BP. A pollen record from the three part sequence at Añangucocha gives strong evidence of local vegetation before the reentry event being displaced by simpler vegetation suggestive of flooded localities during the reentry period. Vegetation closely similar to the preexisting forest was then restored during the last 800 yr (Frost, 1983).

Only in Añangucocha is reentry of a previously existing channel demonstrated unequivocally, though a radiocarbon age of 1230 ±120 BP towards the bottom of the riverine sediment in Limoncocha shows that the span of deposition of river sediment was similar to the span of reentry beginning at 1310 ±140 BP at Añangucocha (Table I).

Synchroneity of termination and duration of the reentry event in four lakes, in two river valleys, and on an east-west traverse of >100 km, suggests a regional climatic event as the primary mechanism directing the flows of the rivers. Colinvaux et al. (1985) concluded that this event was probably enhanced precipiation in watersheds of the western Amazon system and on the flanks of the Cordillera Oriental.

TABLE I. Dated episodes of Amazon flooding and high Andean precipitation of the 1300-800 yr BP event.

	ONSET	END
Western Amazon lakes:		
Limoncocha:	>1230 +/- 120	755 +/- 135
Añangucocha:	1340 +/- 140	830 +/- 160
L. St. Cecilia:	u/d	>785 +/- 100
L. Agrio:	u/d	>675 +/- 190
Andean endorheic lakes:		
L. Cunro:	1580[2]	710[2]
L. Yambo:	1600[1]	800[1]

[1] = interpolated age, [2] = age based on close-interval radiocarbon dating, (u/d) = undated.

This hypothesis has now been tested with pollen and sedimentary histories from two lakes of the Inter-Andean Plateau in Ecuador, Lakes Cunro and Yambo (Fig. 1). In both these endorheic lakes we find evidence for high water levels at ages that correlate closely with the duration of the reentry event in the riverine lakes (Table I).

Lake Cunro occupies a maar at 2800m elevation on the west-facing flank of the eastern Andean chain (Cordillera Oriental), just across the divide from the watersheds of the Napo and Aguarico rivers (Steinitz-Kannan et al., 1983). The (unpublished) reconstruction of the history of pollen from aquatic plants in sediments spanning 6000 yr suggests that water level was at its highest, resulting in a lake maintained by overflowing the crater rim between 1580 to 710 BP. This result appears to require heavy precipiation on the Cordillera Oriental during this interval.

The second unpublished record, from Lake Yambo, is from a site further out onto the floor of the Inter-Andean Plateau in the of the most arid parts of the plateau, the local vegetation of which is desert. Bare ground between desert shrubs is deeply dissected by rare storm rains. In this landscape, Lake Yambo occupies a graben, the lake apparently being fed from underground aquifers, supplemented by rare rains (Steinitz-Kannan et al., 1983). A pollen diagram spanning all the 2700 yr of Lake Yambo's history shows only one significant event represented by an *Alternanthera* pollen zone spanning from extrapolated ages of 1600 to 800 BP. The pollen taxa of this zone suggest a local marsh bordering the lake, which would be provided if lake level rose to flood a flat bench that borders the modern lake at one end. The pollen data, however, give no hint of significant changes in the vegetation of the broader region, which remained that of desert similar to that of the present day. The *Alternanthera* pollen zone, therefore, is parsimoniously explained if a raising of Andean water tables, following increased precipitation on higher slopes or on the eastern flank of the cordillera, raised the level of the lake.

Dating of high lake levels in the Andes shows the onset of increased precipitation as predating somewhat the first evidence for river reentry of old channels on the Amazonian watershed (Table I). This could merely reflect inaccuracies in the dating resulting from extrapolations between radiocarbon determinations. An alternative explanation is that levels of endorheic lakes responded more quickly to increased precipiation than could changes in the river system. The precipitation event probably would not have been detected by our Amazonian lake cores until after a threshold had been passed which let the rivers spread so far as to reenter their ancient channels.

DISCUSSION

These few data from the western Amazon region suggest local climate with an interesting and complex history. This should not be surprising, because the region is at the conjunction of major climatic systems; separated from climates of the Eastern Pacific Ocean by the Andean chain, from the climates of northern South America by the ITCZ and a mountainous massif, and from southern South America by still more mountains. Small readjustments of the relationships of the outside climatic systems should have large local consequences in the Amazon. For the western Amazon these readjustments included temperature depression in glacial times and at least one extended and significant change in precipitation during the Holocene.

Apart from showing how interesting a full climatic history of the region might be, this study demonstrates the importance of direct paleoecological data for reconstructing past Amazonian climates. The first radiocarbon dated evidence for glacial age vegetation and climate of Amazonia seems utterly at variance with current interpretations of biogeographic theory. Although it must be at least as rash to extrapolate to the whole Amazon basin from a single sample at its western edge as to infer ice age climates from modern bird distributions, these, the only dated paleoecological data available, allow the hypothesis of a cooler Amazon basin with wet tropical rain forest restricted to those bottom lands thought to hold only savanna or seasonal woodland by much contemporary opinion.

Just as we have as yet insufficient data to describe glacial period climates of Amazonia with any confidence, so also we cannot reconstruct the Holocene in detail. But our data do show that significant climatic changes on millennial time-scales do occur. The precipitation event on the equatorial Cordillera Oriental and adjacent Amazonian watersheds between 1300 and 800 yr BP is the most clearly documented and dated event now available from the entire Amazon basin. Other events requiring seasonal aridity have been proposed (Absy, 1979), and wildfires at irregular intervals over the last 6000 yr have been demonstrated from southern Venezuela (Sanford et al., 1985). Changing pathways of river meanders on century time-scales have been documented from Peru (Salo et al., 1986), and these conceivably could themselves reflect changes in patterns of precipitation. Besides the geological evidence for massive flooding from 11,000 to 5000 yr BP in Peru (Campbell and Frailey, 1984), we still lack sufficient, well-dated, long records of the Holocene in Amazonia to test the hypothesis that other, secular, regional, or long term, climatic changes took place in the earlier Holocene. On the basis of the few records that we have, however, it seems likely that they did.

Our final thought is that there is an urgent need to reexamine the hypothesis that temperature depression at glacial maxima was modest over equatorial land masses at low elevation. The single datum available from the Amazon basin suggests otherwise.

ACKNOWLEDGMENTS

M. Steinitz-Kannan, I. Frost, M. Miller, and K. Olson contributed greatly to this research both in field and laboratory. We give special thanks to H. and T. Steinitz of Quito, who provided our base in Ecuador. The Ecuadorian work was supported by grants from the General Ecology Program of NSF to the Ohio State University.

REFERENCES

Absy, M.L., 1979. *A palynological study of Holocene sediments in the Amazon Basin*. Ph.D. dissertation, University of Amsterdam, Amsterdam, Netherlands.

Absy, M.L. and van der Hammen, T., 1976. Some paleoecological data from Rondonia, southern part of the Amazon basin, *Acta Amazonica*, 6: 293-299.

Bradbury, J.P., Leyden, B., Salgado-Labouriau, M.L., Lewis, W.M., Schubert, C., Binford, M.W., Frey, D.G., Whitehead, D.R. and Weibezahn, F.H., 1981. Late Quaternary environmental history of Lake Valencia, Venezuela, *Science*, 214: 1299-1305.

Campbell, K.E. and Frailey, D., 1984. Holocene flooding and species diversity in southwestern Amazonia, *Quaternary Research*, 21: 369-375.

CLIMAP Project Members, 1976. The surface of the ice-age Earth, *Science*, 191: 1131-1137.

Colinvaux, P.A., 1972. Climate and the Galapagos Islands, *Nature*, 240: 17-20.

Colinvaux, P.A., Miller, M.C., Liu, K-b., Steinitz-Kannan, M. and Frost, I., 1985. Discovery of permanent Amazon Lakes and hydraulic disturbance in the upper Amazon Basin, *Nature*, 313: 42-45.

Damuth, J.E. and Fairbridge, R.W., 1970. Equatorial Atlantic deep-sea arkosic sands and ice-age aridity in tropical South America, *Bulletin of the Geological Society of America*, 81: 189-206.

Frost, I., 1984. *A paleclimnological and palynological investigation in the Equadorian rain forest: evidence of regional flooding and paleohydrological disturbance in the Amazon rain forest.* MS Thesis. The Ohio State University, Columbus, Ohio, U.S.A.

Grubb, P.J., Lloyd, J.R., Pennington, T.D. and Whitmore, T.C., 1963. A comparison of montane and lowland forests in Ecuador 1: the forest structure, physiognomy, and floristics, *Journal of Ecology*, 51: 576-601.

Haffer, J., 1969. Speciation in Amazon forest birds, *Science*, 165: 131-137.

Haffer, J., 1974. *Avian speciation in tropical South America*, Nuttal Ornithological Club, Cambridge, Mass., U.S.A.

Hastenrath, S., 1971. On the Pleistocene snowline depression in the arid regions of the South American Andes, *Journal of Glaciology*, 10: 225-267.

Kutzbach, J.E. and Guetter, P.J., 1981. The sensitivity of Monsoon Climates to Orbital Parameter Changes for 9000 years B.P.: Experiments with the NCAR General Circulation Model. In: Berger, A.L. Ed, *Milankovitch and Climate*, part 2, D. Reidel, 801-820.

Kutzbach, J.E. and Guetter, P.J., 1986. The influence of Changing Orbital Parameters and Surface Boundary Conditions on Climate Simulations for the past 18,000 Years, *Journal of Atmospheric Sciences*, 43: 1726-1759.

Liu, K-b. and Colinvaux, P.A. 1985. Forest Changes in the Amazon Basin during the last Glacial Maximum, *Nature*, 318: 556-557.

Prance, G.T., 1982 (ed.). *Biological Diversification in the Tropics*, Columbia University Press., N.Y., U.S.A.

Rind, D. and Peteet, D., 1986. Terrestrial conditions at the last glacial maximum and CLIMAP sea-surface temperature estimates: Are they consistent? *Quaternary Research*, 24: 1-22.

Salo, J., Kalliola, R., Häkkinen, L., Käkinen, Y., Niemelä,P., Puhakka, M. and Coley, P.D., 1986. River dynamics and the diversity of Amazon lowland forest. *Nature*, 322: 254-258.

Sanford, R.L., Saldarriaga, J., Clark, K.E, Uhl, C. and Herrera, R., 1985. Amazon rain-forest fires, *Science* 227: 53-55.

Steinitz-Kannan, M., Colinvaux, P.A. and Kannan, R., 1983. Limnological studies in Ecuador: 1. A survey of chemical and physical properties of Equadorian lakes, *Archive fur Hydrobiologie/ Suppl.*, 65: 61-105.

Thompson, L.G., Mosley-Thompson, E. and Arnao, B.M., 1984. El Niño-Southern Oscillation events recorded in the stratigraphy of the tropical Quelccaya ice cap, Peru, *Science*, 226: 50-53.

Wijmstra, T.A. and van der Hammen, T., 1966. Palynological data on the history of tropical savannas in northern South America, *Leidse Geologische Mededelingen*, 38: 71-83.

van der Hammen, T. and Gonzalez, E., 1960. Upper Pleistocene and Holocene climate and vegetation of the "Sabana de Bogota" (Colombia, South America), *Leidse Geologische Mededelingen*, 25: 262-315.

LATE HOLOCENE MORPHOLOGICAL CHANGES IN A JAMAICAN LAND SNAIL: EVIDENCE FOR CHANGES IN RAINFALL

Glenn A. Goodfriend

Isotope Department
The Weizmann Institute of Science
76100 Rehovot, Israel*

and

Department of Zoology
University of Florida
Gainesville, FL 32611

ABSTRACT. The occurrence of giant late Holocene subfossil shells of the Jamaican land snail *Pleurodonte lucerna* is interpreted as representing a period significantly wetter than today. Sometime in the last c. 2000 yr, a major decrease in precipitation appears to have occurred in Jamaica. Extinctions of some land snail species in drier areas of the island may also be related to this climatic change.

INTRODUCTION

Land snails often show morphological variation that correlates with climatic variation (Goodfriend, 1986). By extrapolation of these relationships to fossil land snails, paleoclimatic information can be obtained (e.g., Magaritz and Heller, 1980).

The endemic Jamaican land snail *Pleurodonte lucerna* (Müller) shows a dramatic pattern of geographic variation in size and other morphological characteristics which relates to rainfall (Goodfriend, 1983). Subfossil specimens of *P. lucerna* found in soils often show major morphological differences from populations now living at the same sites. In particular, giant specimens, far outside the size range of the living snails, are found at most sites that have been sampled. Morphological comparisons were made between these subfossil specimens and live-collected specimens of the same species from the same sites. Dating of selected subfossil specimens was carried out by amino acid epimerization analysis (alloisoleucine/isoleucine ratio, or A/I) (Harmon et al., 1983).

---

* present address

W. H. Berger and L. D. Labeyrie (eds.), Abrupt Climatic Change, 123–126.

METHODS AND MATERIALS

Geographic variation in living *P. lucerna* was documented by morphological analysis of 20 specimens from each of 15 populations along a N-S transect in north-central Jamaica. The relationship of shell form to rainfall was documented from four transect populations and four north-central Jamaican populations from outside the transect, all collected within several kilometers of a rainfall station. Subfossil material was collected from some of these same sites, from the humus-root mat that overlies the mineral soil and from under loose limestone rocks. Amino acid epimerization analysis was carried out by Dr. R.M. Mitterer (University of Texas at Dallas), according to methods given in Kriausakul and Mitterer (1980).

RESULTS AND DISCUSSION

In north-central Jamaica, living *P. lucerna* are very small at the coast and large at the top of the coastal slope (Fig. 1). Southward across the interior plateau, they become smaller (in the rainshadow) and then larger again further south. The mean shell diameter of populations correlates strongly with mean annual rainfall ($r = 0.89$; Fig. 2).

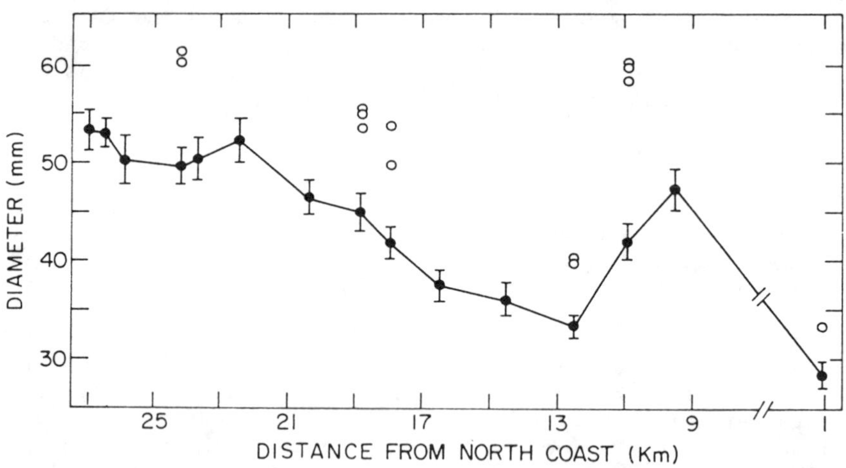

Figure 1. Shell diameter of *Pleurodonte lucerna* along a north-south transect in north-central Jamaica. Solid circles = mean shell diameter of living populations (± 2 S.D.); open circles = diameter of individual late Holocene subfossil shells (modified from Goodfriend, 1983)

Giant subfossil *P. lucerna* specimens were found to have diameters 3-9 standard deviations above the mean of populations living at the same sites (Fig. 1). Amino acid epimerization analysis yielded A/I ratios of 0.02-0.03 (n=8) for the giant shells from most sites. At Coco Ree, Jamaica, land snails radiocarbon dated at 13,400±180 yr BP have a mean A/I=0.16±0.02 (Goodfriend and Mitterer, submitted for publication).

Using the epimerization rate calculated from the Coco Ree sample, the giant subfossil *Pleurodonte lucerna* shells are estimated to average c. 2000 yr old. A giant subfossil from another site (km 12.5, Fig. 1) has A/I=0.06 and is estimated to be c. 5000 yr old.

Although the *mean* shell size of the late Holocene subfossil populations is not known (only the giants were dated), it must have been significantly larger than that of modern populations. Brown (1911) also reported giant subfossils of this species from west-central Jamaica. These data indicate a major diminution of shell size throughout the region sometime in the last c. 2000 yr. Since size variation in living snails is very closely related to mean annual rainfall, this suggests that Jamaica has experienced a major drop in precipitation during this time.

This recent desiccation phase may also have been responsible for the apparently recent extinctions of some species of land snails in north-central Jamaica. For example, *Zaphysema tumida* (Pfeiffer) occurs abundantly throughout the plateau region of north-central Jamaica (St. Ann Parish) as bleached, subfossil shells (collected from 46 sites) but occurs alive today only in the wetter areas to the west (the Cockpit Country of southern Trelawny and northern St. Elizabeth, and in northern Clarendon). A similar pattern is seen in *Alcadia major*, with bleached shells common in eastern St. Ann, but living populations occurring only to the west.

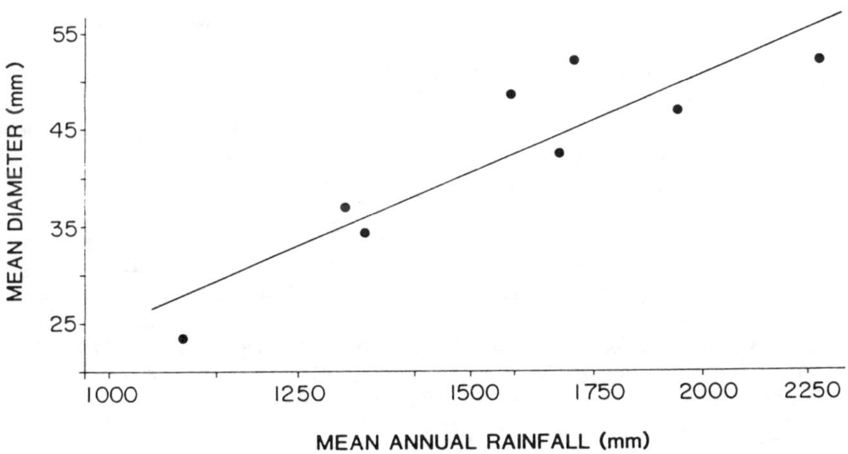

Figure 2. Mean shell diameter of living *Pleurodonte lucerna* populations in relation to mean annual rainfall. Note that rainfall is plotted on a log scale.

Further study is needed to quantify late Holocene morphological changes in *P. lucerna* and to obtain more precise chronological information on these variations as well as on local species extinctions. Unfortunately, dating is hindered by the occurrence of variable radiocarbon age anomalies in shell carbonate of Jamaican land snails from limestone areas (Goodfriend and Stipp, 1983), making radiocarbon dating of late Holocene shell carbonate very imprecise (Goodfriend, in press).

ACKNOWLEDGEMENTS

The author is indebted to Dr. R.M. Mitterer for amino acid analyses.

REFERENCES

Brown, A.P., 1911. Variation in some Jamaican species of *Pleurodonte*, *Proc. Acad. Nat. Sci. Phila.*, 63: 117-164.

Goodfriend, G.A., 1983. Clinal variation and natural selection in the land snail *Pleurodonte lucerna* in western St. Ann Parish, Jamaica, Ph.D. dissertation, Univ. of Florida, Gainesville.

Goodfriend, G.A., 1986. Causes of variation in land snail shell form and size: a review, *Syst. Zool.* 35: 204-223.

Goodfriend, G.A., in press. Radiocarbon age anomalies in shell carbonate of landsnails from semi-arid areas, *Radiocarbon*, 29 (2).

Goodfriend, G.A. and Mitterer, R.M., (submitted for publication). Late Quaternary land snail faunal history and chronostratigraphy of cave sediments at Coco Ree, Jamaica.

Goodfriend, G.A. and Stipp, J.J., 1983. Limestone and the problem of radiocarbon dating of land-snail shell carbonate, *Geology*, 11: 575-577.

Harmon, R.S., Mitterer, R.M., Kriausakul, N., Land, L.S., Schwarcz, H.P., Garrett, P., Larson, G.J., Vacher, H.L. and Rowe, M., 1983. U-series and amino-acid racemization geochronology of Bermuda: implications for eustatic sea-level fluctuation over the past 250,000 years, *Palaeogeogr., Palaeoclimat., Palaeoecol.*, 44: 41-70.

Kriausakul, N. and Mitterer, R.M., 1980. Some factors affecting the epimerization of isoleucine in peptides and proteins. In: Hare, P.E., Hoering, T.C. and King, K., Jr. (eds.), *Biogeochemistry of Amino Acids*, John Wiley and Sons, N.Y., pp. 283-296.

Magaritz, M. and Heller, J., 1980. A desert migration indicator - oxygen isotopic composition of land snail shells, *Palaeogeogr., Palaeoclimat., Palaeoecol.*, 32: 153-162.

INDICATIONS FOR ABRUPT HOLOCENE CLIMATIC CHANGE: LATE HOLOCENE OXYGEN ISOTOPE STRATIGRAPHY OF THE GREAT SALT LAKE, UTAH

J. A. McKenzie and G. P. Eberli

Department of Geology
University of Florida
Gainesville, Florida 32611
U.S.A.

ABSTRACT. Oxygen-isotope stratigraphy in sediment cores together with carbonate mineralogy and percent carbonate stratigraphies have been used to reconstruct salinity changes, that is lake-level changes of the Great Salt Lake, Utah. For the late Holocene (last 5,500 years), the data suggest the following:

1) the lake level fluctuated frequently with abrupt changes above and below the modern mean elevation showing high and low extremes greater than those recorded in the historic lake-level curve,

2) there were two periods (minima) of approximately 750 years duration each, when salinities were probably greater and more constant than the modern mean, interpreted as periods of lower average lake levels with fresh-water inflow predominantly from summer precipitation,

3) there were three periods (maxima) of approximately 1500 years duration each, when probable salinities were often less than now but experienced large excursions around the modern mean. These are interpreted as periods of higher lake levels with fresh-water inflow predominantly from winter precipitation,

4) the maxima and minima show an apparent cyclicity and are considered to be controlled by climatically induced changes in weather patterns. The transitions between periods appear to be abrupt occurring on the order of about 100 years.

INTRODUCTION

In his monumental monograph on Pleistocene Lake Bonneville, G. K. Gilbert (1890) recognized the hydrologic significance of ancient, elevated shorelines encircling the modern Great Salt Lake, Utah and postulated that water levels had varied dramatically being up to about 300m higher than the modern average. With the historic lake-level record available at that time, Gilbert also foresaw the flood potential of the lake in recognizing that only slight changes in water depth would result

W. H. Berger and L. D. Labeyrie (eds.), Abrupt Climatic Change, 127–136.

in a tremendous change in the area covered by the lake. Recently, Gilbert's observations have assumed renewed significance with the sudden rise in the lake level beginning in 1982/1983, which has inundated much of the low-lying developed areas along the fluctuating shoreline and remains a major threat to higher standing properties as the present trend continues (Kay and Diaz, 1985). In May, 1986, the lake level surpassed its 1873 historic high of 1283.8m above sea level (Weisburd, 1986). This most recent rise in lake elevation raises questions about the climatic control on lake-level fluctuations: Is it a short-term phenomenon or does it represent a longer term trend? Do lake-level fluctuations reflect climatic change?

Recent studies of the Great Salt Lake and its precursors have concentrated on the ancient sedimentary record in order to evaluate the relationship between lake-level fluctuations and climatic change. Geomorphologic studies of littoral zone features around the lake together with radiocarbon dating have been used to reconstruct the past history and magnitude of longer term lake-level variations (Currey et al., 1984; Currey and Oviatt, 1985). They prove to be inadequate to decipher shorter term or abrupt changes, which must have occurred, however, throughout the Holocene as exhibited by fluctuations in the historic lake-level curve (Austin, 1980). More detailed continuous Holocene records of salinity changes and, hence, lake-level fluctuations are contained in high-resolution statigraphies of cored sediments (Spencer et al., 1984; McKenzie and Eberli, 1985).

The Great Salt Lake is situated topographically at the bottom of a closed basin. The lake level is controlled by a balance between an inflow of fresh water from rivers, groundwater and direct precipitation versus an evaporative outflow. Abrupt climatic changes can dramatically affect this balance by altering the amount and timing of the fresh-water inflow, which then results in salinity changes. Variations in salinity are stratigraphically reflected in the authigenic carbonate mineralogy, percent carbonate and oxygen-isotope composition of cored sediments. The sediments, therefore, directly record salinity changes which can be correlated with lake-level fluctuations. For example, the Pleistocene/Holocene evaporative drawdown of Lake Bonneville to Great Salt Lake levels was recorded in the sediments as a progressive linear change from calcite to Mg-calcite to aragonite with a steadily increasing percentage of carbonate and oxygen-18 values from -8.0 to -1.7^{o}/oo (Spencer et al., 1984). This paper presents the results of high-resolution analytical studies of cored sediments representing the late Holocene, approximately the last 5500 years, and interprets the changes as a function of the overall hydrologic budget which is probably ultimately controlled by climate.

METHODOLOGY

In 1979 sediment cores were secured from the south arm of the Great Salt Lake for a multi-disciplinary study of lake-level history. Stratigraphic correlations of the various types of data (sedimentology, stable isotope, organic geochemistry, pollen, ostracode assemblages, etc.) were used to interpret latest Pleistocene/Holocene lake elevations (Spencer et al., 1984). For this study of late Holocene lake levels, we chose cores from Site I, which lies west of Antelope Island at about 41°N, 112°20'W. Sedimentary correlations showed that cores from this site contained an expanded sequence above the highest tephra layer, which is correlated with the Mazama ash dated at 6,900 years B.P. The sediments comprise laminated, aragonitic alterations of sorted pellets, ripple laminae, unsorted parallel flat laminae and disrupted laminae (Spencer et al., 1984). It was anticipated that greater stratigraphic resolution could be obtained from such a laminated, expanded section.

A piston core (GSL-79-05P2) was sampled every 5cm over 0.5cm intervals in order to achieve an unbiased delineation of isotopic variations with depth in the core. After the analysis of these initial samples was successfully completed, the upper 0.45m of the piston core was resampled at approximately 2cm intervals to obtain a more detailed isotope record for the latest Holocene sediments. The upper 0.44m of a gravity core (GSL-5G2) with less coring disturbance than the piston core was likewise densely sampled. With this dense sampling of two cores, we obtained an isotope record with enough continuity and detail that it could be correlated with the historic record of lake-level fluctuations.

Each sample was treated first with a 6% sodium hypochlorite solution to remove any organic matter and soluble salts. Afterwards, the washed and dried samples were analyzed by X-ray diffraction techniques. The total weight percentage of carbonate in the treated samples was also measured using a combustion-titration apparatus, which in combination with the X-ray data enabled the calculation of the percentages of aragonite, calcite, dolomite, and quartz in the samples. Stable isotope analyses of CO_2 gas released from the samples during reaction with phosphoric acid at 25°C under vacuum were made with a triple-collector mass spectrometer, VG Micromass 903. The oxygen-isotope ratio of the bulk carbonate in the samples is reported as the per mil ($^{o}/oo$) deviation from the international PDB isotopic standard. The pertinent data are shown in Figures 1 and 2.

RESULTS AND DISCUSSIONS

Source of Oxygen-18 Variations

The first test of the applicability of oxygen-isotope stratigraphy in the study of lake-level fluctuations was to determine the sedimentary source of the oxygen-isotope variations. Our data for the late Holocene, as shown in Fig. 1 for the gravity core, consistently showed that the weight percentage of the detrital carbonates, calcite and dolomite, remained constant while the weight percentage of aragonite and total carbonate varied in a parallel manner, exclusive of periods of very low salinities when calcite is the stable authigenic precipitate. Further, the oxygen-18 composition tended to vary correspondingly, i.e. higher percent aragonite (or percent carbonate) occurs with more enriched oxygen-18 values and vice versa. If we assume that the oxygen-18 composition of the authigenic aragonite reflects the isotope ratio of the lake water at the time of precipitation, we can then correlate periods of increasing oxygen-18 and percent aragonite (or percent carbonate) with periods of decreased fresh-water inflow, and periods of decreasing oxygen-18 and percent aragonite (or percent carbonate) with periods of increasing fresh-water inflow.

Determination of Stratigraphic Ages

The second test of our proposed relationship between oxygen-isotope stratigraphy and lake-level fluctuations was to compare the isotopic record with the historic record. To date, the age control in the cores has been based on very limited data. The presence of the Mazama tephra layer at 4.38m in the 5.01m piston core places a maximum age of 6900 years B.P. near the base of the core. The minimum age (1978 A.D.) was set at the top of the gravity core. A third, very approximate age of 5500 years B.P. is set at 1.825m, where the last appearance of authigenic dolomite occurs in the sediments. A mid-Holocene playa stage (7000 to 5500 years B.P.) has been recognized for the region containing the Great Salt Lake Basin (Currey and James, 1982). We correlate this playa stage with the frequent appearance of authigenic dolomite in the sediments. Above 1.825m the authigenic carbonate is dominantly aragonite with the exception of a brief interval around 0.95m where calcite becomes the dominant authigenic carbonate. We associate this latter mineralogic change with the mid-neolacustral episode which occurred between 3,500 and 2,200 years B.P. and is recognized by an incursion of lacustrine facies in a sequence of salt marsh sediments (Mehringer, 1977; Currey and James, 1982).

Placing approximate ages of 5,500 and 2,200 years B.P., respectively, at these two points of mineralogic change (1.825 and 0.95m), an average sedimentation rate of 0.26mm/yr was used to extrapolate our stratigraphic ages between 5,500 and 1,000 years B.P. (Fig. 2). Since compaction of the younger sediments will not be as extensive as in the older sediments, we distribute the last 1,000 years between 0.65m and the top of the core taking into account our correlation of the isotope stratigraphy with the historic record.

In Fig. 1, the historic lake-level record is correlated with the oxygen-isotope record contained in the upper 45cm of cored sediment. Although this correlation remains only very approximate due to the lack of absolute dating of the cores, a strong similarity between the oscillations in the isotopic and historic curves apparently exists. In particular, the shape of the isotopic curve for the uppermost 22cm of the gravity core shows a remarkable resemblance to that from the historic record. The most positive oxygen-18 value at 2.5cm could record the historic low stand in 1963 A.D. It is undoubtedly significant that the intervening isotope data between 20 and 5cm indicate two low and two high stands between the 1873 and 1963 A.D. extremes. These isotopic fluctuations could be a manifestation of the 1905 and 1945 A.D. lower stands and 1925 and 1952 A.D. higher stands, as seen in the historic lake-level record. Although not as complete or distinctive, the isotope data for the piston core show comparable trends. The average oxygen-18 value for all the points correlated with the historic record is $-6.06^{o}/oo$, which would correspond to an average modern lake elevation of 1280.8m. Together, the data from lake cores strongly suggest that the

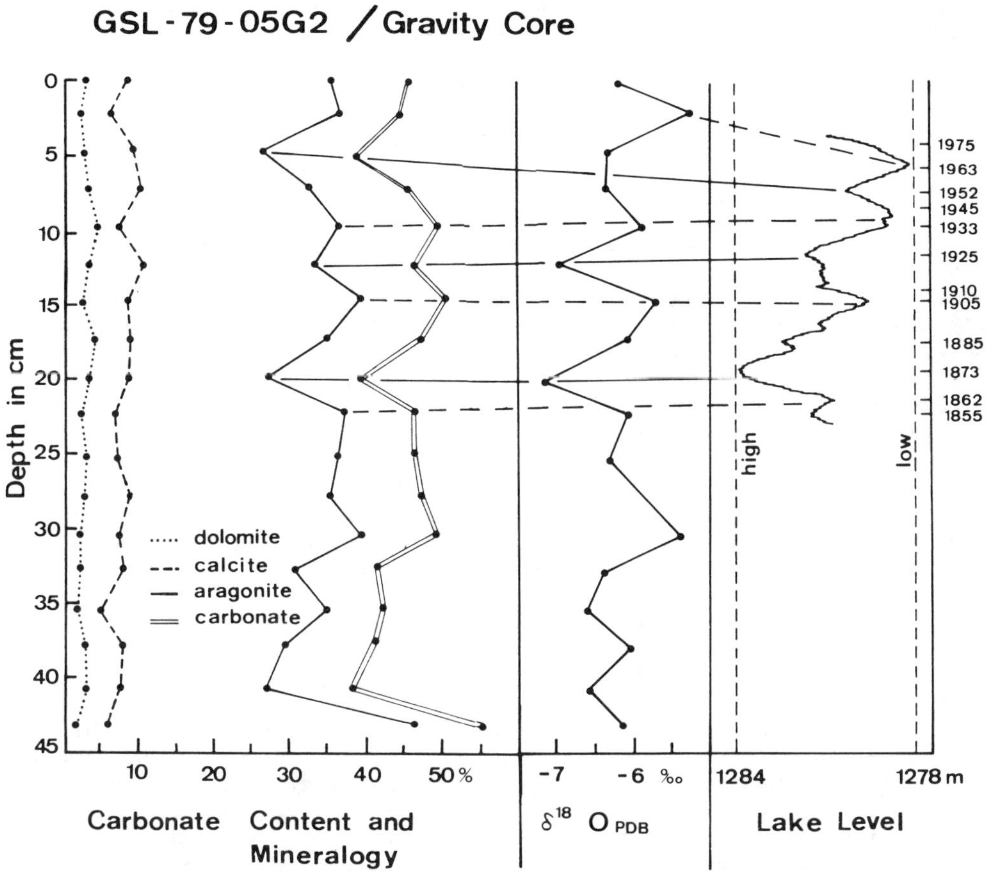

Figure 1. Carbonate mineralogy, carbonate content and oxygen-isotope stratigraphies of gravity core (GSL-79-05G2) correlated with historic lake-level curve (after Austin, 1980). See text for discussion.

oxygen-isotope signal recorded in the authigenic carbonate does reflect changes in salinity and, hence, lake elevation.

The Late Holocene Isotope Record

Holocene paleoenvironments of the northeastern Great Basin which includes the modern Great Salt Lake have been extensively analyzed and correlated utilizing litho- and morphostratigraphies, pollen and macrofossil stratigraphies and radiocarbon dating (summarized in Currey and James, 1982). These correlations have led to a consistent interpretation of the paleoclimatic changes in the region during the late Holocene. After a long dry period (7000-5500 years B.P.) when the Great Salt Lake experienced near-desiccation, the climate of the region apparently became wetter with the onset of a neopluvial episode, which has been subdivided into three glacial-lacustrine-pluvial maxima and two minima (Currey and James, 1982). The early, mid- and late neopluvial stages or maxima are correlated with periods of higher lake levels and cooling coinciding with increased westerly flow and winter precipitation beginning at about 5000, 3500 and 600 years B.P., respectively. The minima are associated with warmer periods of strengthened Mexican monsoon circulation, increased summer precipitation and wide-spread mesic conditions, beginning at about 4500 and 1500 years B.P. (Currey, 1976).

The late Holocene oxygen-isotope stratigraphy in sediment cores from the Great Salt Lake (Fig. 2) is basically in agreement with the climatic record compiled from the numerous studies, as summarized by Currey and James (1982). Combining both the isotopic and carbonate data for the lacustrine sediment record and using the approximated sedimentation rate, the elevation of the Great Salt Lake relative to the modern average (1280.8m) for the past 5500 years can be divided into three periods of higher stands (maxima) with two intervening periods of lower stands (minima). The estimated stratigraphic ages placed on these intervals are consistent with the general climatic record for the region.

Fundamentally, the oxygen-isotope composition of the authigenic carbonates, a reflection of the isotopic composition of the water from which precipitation occurred, could be controlled primarily by two isotope-fractionation processes, both of which would mutually reinforce the tendency towards maxima or minima, as seen in the isotope curve (Fig. 2). The first process produces isotopic fractionation during the evaporation of natural waters; in general, evaporated water becomes enriched in the heavier isotope (Epstein and Mayeda, 1953; Craig, 1961). If the minima represent lower stands of the lake due to relatively decreased rates of fresh-water inflow, the oxygen-isotope value of the authigenic carbonates could indicate an oxygen-18 enrichment in the lake waters, a reflection of the effect of evaporation on a decreased water volume. On the other hand, at lower lake levels, less fresh-water inflow is required to maintain the lake at a given level (Williams, 1985). Perhaps, an isotopic steady-state is more readily achieved at lower lake levels, i.e. a hydrologic balance is reached under conditions of decreased fresh-water inflow and decreased evaporative outflow, which results from the decreased surface area. This scenario could account for the relatively steady, but enriched, oxygen-isotope values of the proposed minima recorded in the late Holocene isotope curve (Fig. 2).

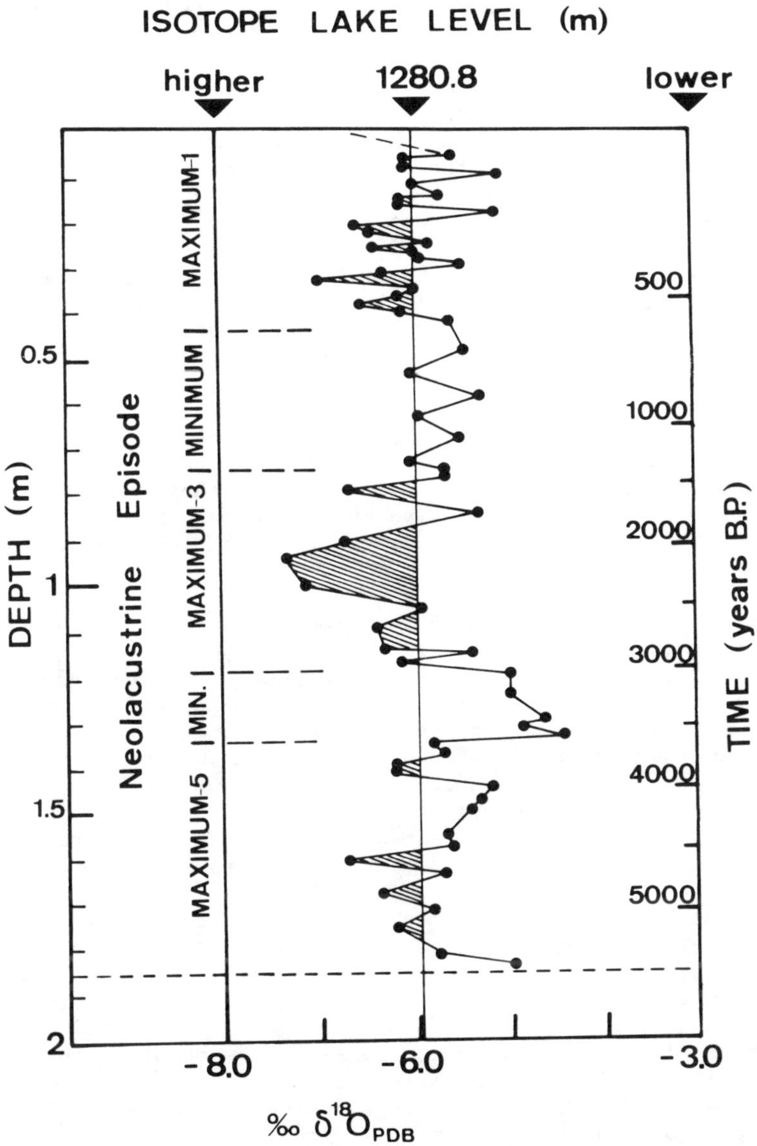

Figure 2. Oxygen-isotope stratigraphy of upper 1.8m of piston core (GSL-79-05P2) showing fluctuations of the data about an average historic value of -6.06°/oo (straight line), which represents the historic mean lake elevation of 1280.8 m. Periods when the isotope values are more positive than the historic value are associated with minima and on the average lower lake levels, while periods when isotope values tend to be generally more negative are associated with maxima and on the average higher lake levels. See text for discussion.

With relatively increased rates of fresh-water inflow, on the contrary, the lake level would rise (maxima) and the oxygen-isotope value of the authigenic carbonates could show an oxygen-18 depletion indicative of decreased salinity. To maintain the lake at higher levels, greater amounts of fresh-water inflow are required due to the relatively increased surface area which leads to increased evaporative outflow (Williams, 1985). If the rate of increased inflow is not maintained, evaporative outflow would tend to lower the lake level as well as enrich the oxygen-18 concentration of the waters. The rapid and large isotope fluctuations in the high resolution record for the most recent maximum-1 could be a consequence of periodically increased inflow producing relatively rapid rises in lake level (oxygen-18 content decreases) which is followed by greater amounts of evaporation (oxygen-18 content increases) as a consequence of the relatively increased surface area. If this interpretation is correct, the isotope record for the mid-neolacustral maximum-3 suggests that increased rates of fresh-water inflow allowed the lake level to rise to the late Holocene high of 1287m. The increased rates of inflow must have been maintained sufficiently long in order to have produced the shoreline deposits at this topographic level. Evaporative enrichment of the isotopic composition of the water only became effective at the end of this maximum period.

The second isotope-fractionation process involves the dependence of the isotope composition of precipitation on temperature (or seasonal) variations and atmospheric circulation patterns. During warmer months with average higher temperatures, precipitation is enriched in the heavier isotopes relative to water precipitated during cooler months with average lower temperature (Yurtsever, 1975). Also, the mean annual isotopic composition of precipitation is dependent upon latitude, e.g. precipitation falling in regions to the southwest of the Great Salt Lake is considerably more enriched in the heavier isotopes (Yurtsever, 1975). If the minima represent periods dominated by both summer precipitation and strengthened Mexican monsoon circulation, as suggested by other lines of evidence (Currey and James, 1982), the fresh water entering the lake could have been more enriched in oxygen-18 than during maxima, which could have been dominated by winter precipitation and increased westerly flow as represented by the modern conditions. It is interesting to note that the modern mean annual oxygen-18 value for precipitation in the region lies between -8 and -10 ^{o}/oo, relative to SMOW (Yurtsever, 1975). The most negative oxygen-18 value measured for the late Holocene is -7.39^{o}/oo, an averaged value for sediments from the maximum-3 period. Values equal to or slightly less than -8°/oo were recorded in the Lake Bonneville sediments (Spencer et al., 1984; McKenzie, unpublished data).

CONCLUSIONS

Oxygen-isotope stratigraphy in conjunction with carbonate mineralogy and percent carbonate stratigraphies indicates that the late Holocene was a period with frequent, significant fluctuations in the water level of the Great Salt Lake. Climate appears to have played a dominant role in controlling lake level by regulating the rate of fresh-water inflow relative to evaporative outflow. The record further suggests possible major changes in weather patterns in the region, i.e. periods (minima) dominated by summer precipitation due to strengthened Mexican monsoon circulation and decreased fresh-water inflow alternating with periods (maxima) dominated by winter precipitation with modern westerly flow patterns and increased fresh-water inflow. The transitions between the maxima and minima are quite abrupt occurring on the order of 100 years. For the past 5500 years representing the neo-lacustrine episode following the mid-Holocene playa stage, an apparent cyclicity of about 2250 years can be observed in the record with maxima having about 1500-years duration and minima spanning about 750 years. The most recent maximum-1 began about 700 years ago and weather patterns typifying maxima continue to dominate in the Great Salt Lake region, as illustrated by recent increases in winter precipitation and sharp rises in the lake level.

REFERENCES

Austin, L. H., 1980. Lake level predictions of the Great Salt Lake. In: *The Great Salt Lake - a Scientific, Historical, and Economic Review*, Salt Lake City, edited by J. W. Gwynn, Utah Geological and Mineral Survey Bulletin, 116: 273-277.

Craig, H., 1961. Isotopic variations in meteoric water, *Science*, 133: 1702-1703.

Currey, D. R., 1976. Late Quaternary geomorphic history of Pint-size Shelter. In: *Pint-size Shelter*, Salt Lake City, edited by L-M. W. Lindsay and C. K. Lund, Utah Division of State History, Antiquities Section Selected Papers, 3(10): 59-65.

Currey, D. R. and James, S. R., 1982. Paleoenvironments of the northeastern Great Basin and northwestern basin rim region: a review of geological and biological evidence. In: *Man and Environment in the Great Basin*, edited by D. B. Madsen and J.F. O'Connell, Washington, D.C. Society for American Archaeology, SAA Paper No. 2, p. 27-52.

Currey, D. R., Atwood, G., and Mabey, D. R., 1984. Major levels of Great Salt Lake and Lake Bonneville, *Utah Geological and Mineral Survey Map 73*.

Currey, D. R. and Oviatt, 1985. Duration, average rates, and probable causes of Lake Bonneville expansions, standstills, and contractions during the last deep-lake cycle, 32,000 to 10,000 years ago. In: *Problems of and Prospects for Predicting Great Salt Lake Levels*, Salt Lake City, edited by P. A. Kay and H. F. Diaz, Center for Public Affairs and Administration, University of Utah, p. 9-24.

Epstein, S. and Mayeda, T., 1953. Variations in ^{18}O contents of waters from natural sources, *Geochim. Cosmochim. Acta*, 27: 213-224.

Gilbert, G. K., 1890. Lake Bonneville, *U.S. Geological Survey Monograph 1*, 438 p.

Kay, P. A. and Diaz, H. F., 1985. *Problems of and Prospects for Predicting Great Salt Lake Levels*, Salt Lake City, Center for Public Affairs and Administration, University of Utah, 309 p.

McKenzie, J.A. and Eberli, G. P., 1985. Late Holocene lake-level fluctuations of the Great Salt Lake (Utah) as deduced from oxygen isotope and carbonate contents of cored sediments. In: *Problems of and Prospects for Predicting Great Salt Lake Levels*, edited by P. A. Kay and H. F. Diaz, Salt Lake City, Center for Public Affairs and Administration, University of Utah, p. 25-39.

Mehringer, P. A., 1977. Great Basin Quaternary environments and chronology. In: *Models and Great Basin Prehistory: A Symposium*, edited by E. B. Fowler, Reno Desert Research Institute Publications in the Social Sciences, 12: 113-167.

Spencer, R. J., Baedecker, M. J., Eugster, H. P., Forester, R. M., Goldhaber, M. B., Jones, B. F., Kelts, K., McKenzie, J., Madsen, D. B., Rettig, S. L., Rubin, M. and Bowser, C. J., 1984. Great Salt Lake, and precursors, Utah: the last 30,000 years, *Contributions to Mineralogy and Petrology*, 86: 321-334.

Weisburd, S., 1986. Lake at record high, *Science News*, 129: 325.

Williams, G., 1985. Some relationships of precipitation and Great Salt Lake elevations, past, present and near future. In: *Problems of and Prospects for Predicting Great Salt Lake Levels*, edited by P. A. Kay and H. F. Diaz, Salt Lake City, Center for Public Affairs and Administration, University of Utah, p. 189-204.

Yurtsever, Y., 1975. Worldwide survey of stable isotopes in precipitation, Rept. Sect. Isotope Hydrology, *IAEA*, November 1975, 40 p.

POLLEN TIME SERIES AND HOLOCENE CLIMATE VARIABILITY OF THE MIDWEST UNITED STATES

Jonathan T. Overpeck

Lamont-Doherty Geological Observatory
Palisades, New York 10964
USA

ABSTRACT. Time series of pollen data are important sources of information on climate variability within the century and millennium frequency bands. Twenty pollen records from the midwestern United States were investigated using two methods of time series analysis in order to characterize Holocene climatic variability in this region. Event correlation among the twenty pollen records suggested that change in vegetation, and thus in climate, was episodic during the Holocene and was most substantial during the period between 11,500 and 10,000 yr B.P. Spectral analysis, cross-spectral analysis, and band-pass filtering of six of these pollen records revealed that a portion of the Holocene climate variability in the Midwest may have been quasi-periodic with a central frequency of approximately 1 cycle/1100 years. These preliminary results suggest that vegetation change in the century and millennium frequency bands must be considered in the context of systematic century- and millennium-scale climatic change.

INTRODUCTION

Sediments from mid-latitude lakes provide continuous records of late Quaternary climate change and can be dated using radiocarbon or annual laminations. In particular, fossil pollen from lake sediments have been used to document 9000-year trends in climate (Bartlein et al., 1984) as well as the magnitude and timing of short (100-1000 year), and in some cases abrupt, climate events such as the Little Ice Age (Bernabo, 1981). In order to gain a keener understanding of climate change on the time scales of centuries and millennia, two quantitative methods were used to study time series of Holocene pollen data from the midwestern United States. The rates and timing of pollen change, and hence vegetation change, were examined first using a new event correlation method to determine whether climate change was episodic or gradual during the late Pleistocene and Holocene. Spectral analysis was then employed to study the distribution of Holocene pollen variance as a function of frequency (Overpeck, 1986). The midwestern United States region was selected for study because it contains a high density of published radiocarbon-dated pollen records that are located across steep

W. H. Berger and L. D. Labeyrie (eds.), Abrupt Climatic Change, 137–143.

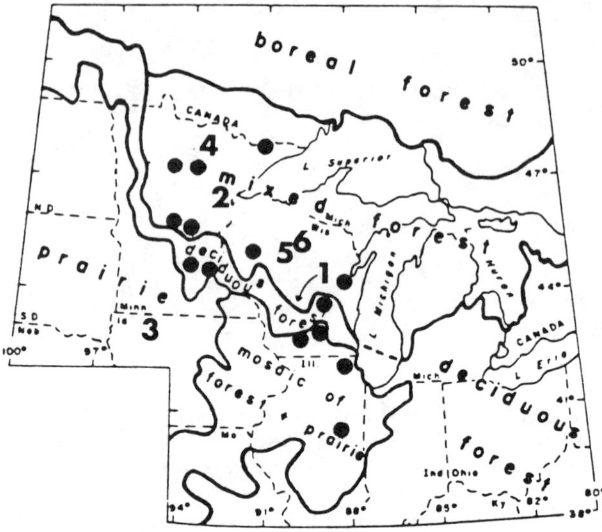

Figure 1. Map showing the pollen sites (numbers and black dots) and major vegetation regions of the Midwest. Spectral analysis was used to study the pollen time series from the numbered sites: 1. Devils Lake (Maher, 1982), 2. Jacobson Lake (Wright and Watts, 1969), 3. Lake West Okoboji (Van Zant, 1979), 4. Myrtle Lake (Janssen, 1968), 5. Spruce Lake (Overpeck, 1986), and 6. Wood Lake (Heide, 1984; Overpeck, 1986). Map and references for all black dots after Webb et al., 1983.

regional vegetation and climate gradients. The forests and grasslands of the Midwest are also characterized by relatively high rates of natural disturbance (by fire and windthrow), and thus have the potential to respond rapidly to climate change.

DATA

Two new Holocene pollen records from the mixed conifer-hardwood forests of Wisconsin were generated (Overpeck, 1986) and 18 additional radiocarbon-dated records were selected from the literature (Fig. 1). The total length of these records ranged from 9000 to 14,000 years. The results discussed here were produced using radiocarbon dates that were corrected to remove the effect of the Holocene-long trend in secular ^{14}C variation (Klein et al., 1982). Age models were developed for each pollen record using one of several interpolation and approximation schemes (Overpeck and Fleri, 1982), and were used to create time series with equally spaced observations. Duplicate results, generated using uncorrected dates, did not change the primary conclusions given here.

RESULTS

Event Correlation

Quantitative "event correlation" of the 20 pollen records was used to identify times of abrupt (i.e. rapid with respect to average rates of change) vegetation change in the Midwest. Squared chord distances between stratigraphically adjacent samples were calculated and normalized at each site. High numerical dissimilarity (greater than one standard deviation above the mean for a given pollen record) between adjacent samples identified intervals that were characterized by rapid change in the surrounding vegetation. The results, summarized in Figures 2a and 2b, suggested that rapid vegetation change in the Midwest was not randomly distributed over the past 11,800 years but instead was most prevalent at certain dates.

Event correlation among the seven longest pollen records suggested that vegetation change was most substantial during the period of deglaciation before 10,000 years B.P. (Fig. 2a). During this time, the recently deglaciated Midwest was colonized by plant populations that moved from unglaciated regions, but the nature of regional climate change in the Midwest at the end of the Pleistocene is as yet unclear. Climate change at this time may have been abrupt and unidirectional or episodic and oscillatory. The probability of vegetation change at sites in different vegetation regions across a large area (e.g. the Midwest) is greater during periods when regional climate change is faster and/or larger than average. The results shown here suggest that vegetation, and therefore climate, changes most dramatically during the period 10,900 to 10,600 years B.P.

A second set of event correlation results, involving all 20 of the pollen records, was generated for the past 8300 years. The method used was the same as above, but the period before 8300 years B.P. was excluded from the analysis in order to highlight the rapid vegetational changes within the Holocene. This second set of results (Fig. 2b) is preliminary and supports the earlier contention that rapid vegetation change was not randomly distributed during the Holocene, but was concentrated in certain time periods. In addition, these results suggest that intervals with high numbers of vegetational events occurred at approximately 1000-year intervals. Midwestern Holocene climate variability may have contained an episodic or quasi-periodic component.

Spectral Analysis

The distribution of pollen and climate variance, as a function of frequency, was examined using spectral analysis (Jenkins and Watts, 1968; Overpeck, 1986). Only six pollen records (Fig. 1) were analyzed because other available pollen records did not contain enough samples for analysis. In order to use time series that were as stationary as possible, only the period from 8300 to 400 years B.P. was examined. The variance spectra for several pollen variables at all six sites were red, indicating that pollen variance, and hence climatic variance, was preferentially distributed at increasingly lower frequencies, and that the pollen variability was not random white noise. To a large degree, this

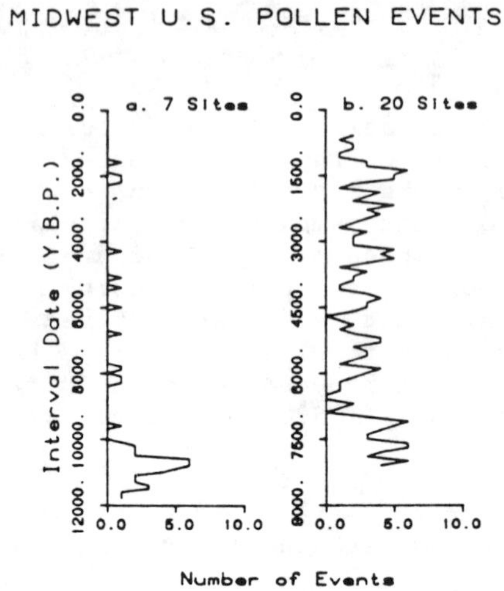

Figure 2. Time series of the number of pollen (vegetation) events (intervals of rapid change). Figure 2a was generated using seven 11,800-year pollen records, and Figure 2b was generated using twenty 8300-year records. Note that the time axes are scaled differently.

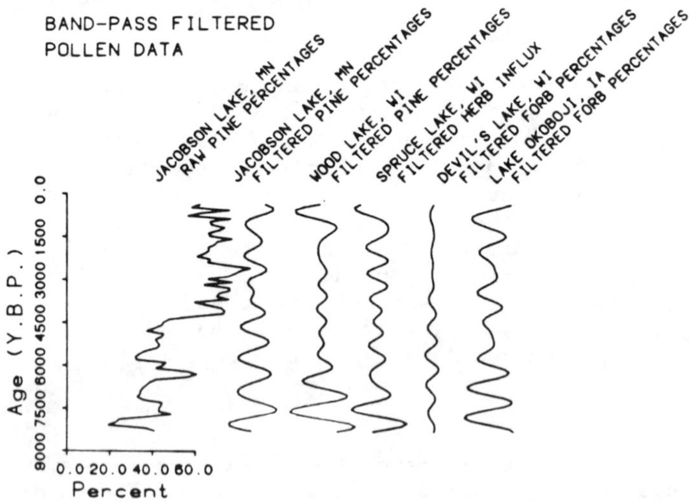

Figure 3. Time-domain perspective of the 1100-year broad-band quasi-periodic component of pollen variability in the Midwest. A band-pass filter was used to produce all but the left-most curve, which is a plot of the unfiltered pine pollen percentages for Jacobson Lake.

redness was due to the general Holocene-long trends in vegetation and climate change that are related to astronomical forcing in the Milankovitch frequency band (Imbrie, 1985; Overpeck, 1986).

Superimposed on the red continuum, however, were concentrations of variance in a frequency band bounded by 1 cycle/1200 years and 1 cycle/1000 years. These broad-band spectral peaks were present in the univariate spectra for pine pollen variability at Wood, Spruce, and Jacobson Lakes in the mixed conifer-hardwood vegetation region and in the spectra for forb (all herbs except Gramineae and Cyperaceae) pollen at Devils Lake and Lake West Okoboji in the deciduous forest and prairie regions. Cross-spectral analysis between these time series suggested that this broad-band component of pollen variability was coherent between the different sites. This component of pollen variability therefore appears to have been related to a quasi-periodic component in the regional climate of the Midwest (Overpeck, 1986).

Band-pass filtering of the 1100-year quasi-periodic component revealed that the component was most attenuated during the mid-Holocene and that it was developed best in the early- and late-Holocene (Fig. 3). In addition, the filtered signal suggested that the quasi-periodic behavior of the pollen time series was related to alternating wet and dry periods in the regional climate of the Midwest. Increased abundances of forb pollen at Devils Lake and Lake West Okoboji and increased abundances of pine pollen at Spruce and Wood Lakes marked periods of increased aridity. The abundance of pine pollen at Jacobson Lake was approximately 180 degrees out of phase with the other pine pollen series because decreased amounts of pine marked greater aridity in the regions (e.g. around Jacobson Lake) where pine trees competed with more dry-adapted plants. Change at Devils Lake and Lake West Okoboji may have led change at other sites. This phase relation is reasonable because the higher rates of disturbance in the deciduous forest and prairie allows these vegetation types to respond more rapidly to climate change than the less fire-adapted mixed conifer-hardwood forests.

CONCLUSIONS

Taken together, the preliminary results of event correlation and spectral analysis of Holocene pollen time series reveal a possible new scenario for vegetation and climate change in the Midwest United States (Overpeck, 1986). Periods of rapid vegetation change appear to relate systematically to intervals of time when the band-pass filtered pollen data suggest that climate was changing from more arid to moister conditions. The Little Ice Age manifests itself in the Midwest possibly as the last period of increased moisture, and it appears that other climate events occurred within the Holocene. Stream-flow data from Wisconsin (Knox, 1985) and lake level data from Minnesota (Dean et al., 1984) support the idea that multiple climate events, with durations less than 1000 years, occurred in the Midwest during the Holocene.

Fossil pollen from lake sediments has an important role to play in expanding our knowledge of mid-latitude Holocene climate variability. In order to use fossil pollen to study the spectrum of climate variability, the precision and accuracy of age models need to be improved. Sound statistical arguments require absolute dating control that exceeds the present limits of radiocarbon dates of lake sediments. Given present age models, it is difficult to resolve the frequency band over which an hypothesized quasi-periodic component operates. Present radiocarbon-based age models are also too inexact to constrain phase relations accurately between different sites. The use of annually-laminated lake sediments appears to be a promising way to avoid the measurement error and secular variation problems that are associated with radiocarbon dates.

REFERENCES

Bartlein, P.J., Webb, T., III, and Fleri, E., 1984. Holocene climate changes in the northern Midwest: pollen derived estimates, *Quaternary Research*, 22: 361-374.

Bernabo, J.C., 1981. Quantitative estimates of temperature over the last 2700 years in Michigan based on pollen data, *Quaternary Research*, 15: 143-159.

Dean, W.E., Bradbury, J.P., Anderson, R.Y., and Barnosky, C.W., 1984. The variability of Holocene climate change: evidence from varved lake sediments, *Science*, 226: 1191-1194.

Heide, K., 1984. Holocene pollen stratigraphy from a lake and small hollow in north-central Wisconsin, USA, *Palynology*, 8: 1-19.

Imbrie, J., 1985. A theoretical framework for the Pleistocene ice ages, *Journal of the Geological Society of London*, 142: 417-432.

Janssen, C.R., 1968. Myrtle Lake: A late and postglacial pollen diagram from northern Minnesota, *Canadian Journal of Botany*, 46: 1397-1410.

Jenkins, G.M. and Watts, D.G., 1968. Spectral Analysis and its Applications, Holden-Day, San Francisco.

Knox, J.C., 1985. Responses of floods to Holocene climatic change in the upper Mississippi Valley, *Quaternary Research*, 23: 287-300.

Klein, J., Lerman, J.C., Damon, R.E., and Ralph, E.K., 1982. Calibration of radiocarbon dates, *Radiocarbon*, 24: 103-150.

Maher, L.J., 1982. The palynology of Devils Lake, Sauk County, Wisconsin. In: *Quaternary History of the Driftless Area* (Knox, J.C., Clayton, L., and Mickelson, D.M., eds.), Wisconsin Geological and Natural History Survey Field Trip Guide Book Number 5, 119-135.

Overpeck, J.T., 1986. Time Series Analysis of Holocene Pollen Data: Paleoclimatological and Paleoecological Applications. Ph.D. Thesis, Brown University.

Overpeck, J.T. and Fleri, E.C., 1982. The development of age models for Holocene sediment cores: northeast North American examples, *American Quaternary Association Abstracts*, 7: 152.

Van Zant, K., 1979. Late glacial and postglacial pollen and plant macrofossils from Lake West Okoboji, northwestern Iowa, *Quaternary Research*, 12: 358-380.

Webb, T., III, Cushing, E.J., and Wright, H.E., Jr., 1983. Holocene changes in the vegetation of the Midwest. In: *The Late-Quaternary of the United States - Volume 2. The Holocene* (Wright, H.E., Jr., ed.), University of Minnesota Press, Minneapolis, 142-165.

Wright, H.E., Jr. and Watts, W.A., 1969. Glacial and vegetational history of northeastern Minnesota, *Minnesota Geological Survey Special Publication* 11.

Part IV.

Glacial-Holocene Transition: Land Records

FOSSIL BEETLE ASSEMBLAGES AS EVIDENCE FOR SUDDEN AND INTENSE CLIMATIC CHANGES IN THE BRITISH ISLES DURING THE LAST 45,000 YEARS

G. R. Coope

Department of Geological Sciences
University of Birmingham
PO Box 363
Birmingham B15 2TT
England

ABSTRACT. Insect species can be shown from fossil evidence to have reacted promptly and intensely to episodes of abrupt climatic change during the Quaternary. Their response has not been by evolution but by tracking suitable conditions across the continents as the climatic zones shifted. Within the last glacial-interglacial cycle alone some species have altered their geographical ranges by thousands of kilometers. Since coleoptera (beetles) are common as fossils and they can be identified to the species level, large assemblages can be obtained from single localities representing short time intervals. By using a computer-based program the overlap of species ranges can be used to reconstruct past thermal environmental changes in terms of mean July temperatures and mean January temperatures which can then be plotted against radiocarbon time. It can be shown that summer temperatures in central Britain, at about 13,000 BP, rose suddenly by 7°C and that winter temperatures may have risen by 20°C at the same time. A similar event occurred at about 10,000 BP. This reconstruction of climatic changes differs considerably from the traditional view of European climatic changes towards the close of the last glacial period.

Insect remains, chiefly of coleoptera (beetles) are abundant in sediments that accumulated in shallow freshwater or in waterlogged deposits on land such as peats. Many of these fossils are sufficiently well preserved to permit identification to the species level. It can be shown that during Quaternary time the great majority of insects within the temperate zones have maintained morphological constancy. There is, at the moment, almost no information of Quaternary fossil insects from tropical latitudes. Although physiological stability cannot be demonstrated directly, there is good reason to believe from past community structures that this has been equally constant and that the fossils can therefore be used as indicators of past environments. The coleoptera inhabit a wide spectrum of habitats, with the exception of marine conditions, though each species is often narrowly restricted to some well defined set of living conditions.

W. H. Berger and L. D. Labeyrie (eds.), Abrupt Climatic Change, 147–150.

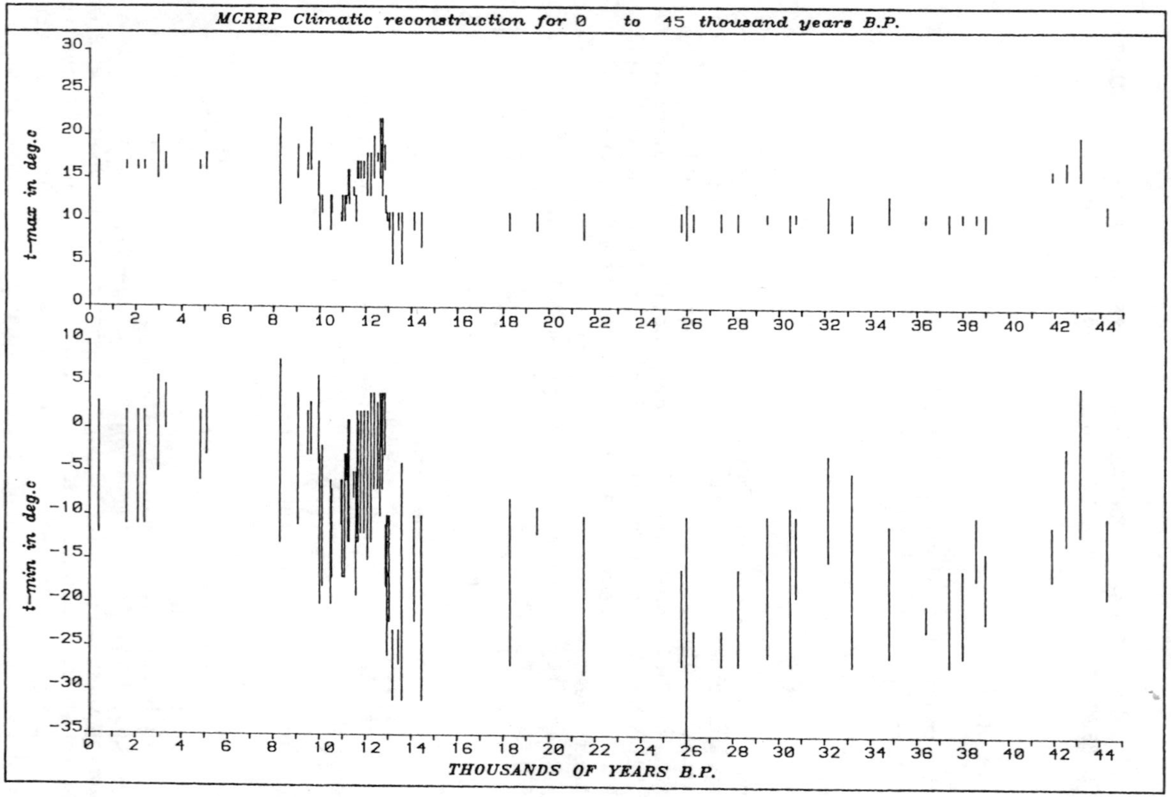

Figure 1. Reconstructions of the mean July temperatures (t-max) and mean January temperatures (t-min) for the last 45,000 years in central Britain based on fossil coleoptera (beetle) assemblages.

One of the most important factors governing the acceptability of habitat is the suitability of the thermal regime, particularly the temperature. However, although the details of the environmental parameters in which the animal actually lives are those of the microclimate, it is apparent from the broad scale geographical distributions of species that appropriate microclimatic conditions are to a large extent determined by macroclimatic factors. The way in which the distributions of coleoptera reflect overall climates is highlighted by the great changes in the geographical ranges of many species in response to the Quaternary climatic oscillations. These have involved shifts of thousands of kilometers for some species during the latest glacial/interglacial cycle alone. It is apparent that the coleoptera respond to abrupt climatic changes more promptly and in a more extreme fashion, than any other component of the terrestrial biota.

In the past, interpretations of paleotemperatures from insect data have been based upon a consideration of range limits of species, their abundance and their associates. Because of the large number of species in the fossil assemblages, temperature estimates have been arrived at by the subjective summation of all these factors. Figure 1 shows an attempt at objective reconstruction of paleotemperatures for central Britain during the past 45,000 years using the computer based Mutual Climatic Range Method (Atkinson et al., 1986 and this volume). This analysis has involved the use of almost four hundred species from about forty different sites. In this reconstruction, t-max = the mean temperature of the warmest month and t-min = the mean temperature of the coldest month. The bars shown indicate that the mean monthly temperatures lay somewhere within its limits; they do not indicate that the mean temperatures ranged between these limits. The ages of the various sites are given in radiocarbon years BP.

Figure 1 shows considerable differences from the traditional pictures of climatic changes in northwestern Europe during this period. Several sudden and intense episodes of climatic changes are evident. Unfortunately the short but very warm climatic event at 43,000 BP is very close to the acceptable limit of radiocarbon dating techniques and apart from inferring suddenness on stratigraphical grounds (Morgan, 1973; Girling, 1974) and intensity on entomological grounds (Coope and Angus, 1975). it is difficult to quantify the rates of change involved. If anything, it is likely that this warm episode was rather older than the radiocarbon dates indicate. There are no indications of the separate Mid-Weichselian interstadials of the Netherlands (Hengelo and Denekamp Interstadials of van der Hammen et al., 1967). The dramatic climatic warming at about 13,000 BP can be quantified with much more accuracy. Figure 1 shows a rise of about 7°C in the mean July temperature and about 20°C in mean January temperatures in an immeasurably short period of time that may have been as short as a century (Coope and Brophy, 1972). Subsequent deterioration in temperature took place in two stages; firstly at just before 12,000 BP and secondly at about 11,000 BP, ushering in the Younger Dryas cold period during which temperatures returned to glacial severity. The climatic improvement at the beginning of the present interglacial (the Flandrian) was as abrupt and as intense as that at 13,000 BP with temperatures as warm as those of the present day established by 9,500 BP (Osborne, 1975; Bishop and Coope, 1977).

A clear relationship emerges between "warm" periods and climatic oceanicity and, conversely, between "cold" periods and climatic continentality. The transitions from one climatic mode to the other were so sudden at times that the whole terrestrial biota was disrupted giving strange non-analogue communities of animals and plants that arose, not because of non-analogue conditions but as a direct result of the abruptness and intensity of climatic changes.

REFERENCES

Atkinson, T.C., Briffa, K.R., Coope, G.R., Joachim, M.J. and Perry, D.W., 1985. Climatic calibration of coleopteran data. In: B.E. Berglund (ed.) *Handbook of Holocene Palaeoecology and Palaeohydrology*, J. Wiley & Son, London, 851-858.

Bishop, W.W. and Coope, G.R., 1977. Stratigraphical and faunal evidence for Lateglacial and Early Flandrian environments in Southwest Scotland. In: J.M. Gray and J.J. Lowe (eds.) *Studies in the Scottish Lateglacial Environment*, Pergamon, Oxford, 61-88.

Coope, G.R. and Angus, R.B., 1975. An ecological study of a temperate interlude in the middle of the last glaciation based on fossil Coleoptera from Isleworth, Middlesex, *Journal of Animal Ecology*, 44: 365-391.

Coope, G.R. and Brophy, J.A., 1972. Late Glacial environmental changes indicated by a coleopteran succession from North Wales, *Boreas*, 1: 97-142.

Girling, M.A., 1974. Evidence from Lincolnshire of the age and intensity of the mid-Devensian temperate episode, *Nature*, 250: 270.

Morgan, A., 1973. Late Pleistocene environmental changes indicated by fossil insect faunas from the English Midlands, *Boreas*, 2: 173-212.

Osborne, P.J., 1974. An insect assemblage of Early Flandrian age and its bearing on contemporary climate and ecology, *Quaternary Research*, 4: 471-486.

van der Hammen, T., Maaleveld, G.C., Vogel, I.C. and Zagwijn, W.H., 1967. Stratigraphy, climate succession and radiocarbon dating of the Last Glacial in the Netherlands, *Geologie en Mijnbouw N.S.*, 46: 79-95.

ON THE DURATION OF THE INTERGLACIAL TO GLACIAL TRANSITION AT THE END OF THE EEMIAN INTERGLACIAL (DEEP SEA STAGE 5 E): BOTANICAL AND SEDIMENTOLOGICAL EVIDENCE

Burkhard Frenzel and Wolfgang Bludau

Botanisches Institut
Universität Stuttgart-Hohenheim
Garbenstr. 30
D-7000 Stuttgart 70
F.R.G.

ABSTRACT. According to Woillard (1979), the last interglacial might have ended within only 150 years. It can be shown by pollen analysis of annually laminated lake sediments that this transition lasted for more than 3000 years. Thus this transition is much less abrupt than envisioned by Woillard.

INTRODUCTION

In Europe, at sites north of the Alps, the last interglacial period is defined by the pollen sequence typical of the Eemian proper. It was only recently that the pollen record was directly correlated with the $\delta^{18}O$-curve of deep-sea sediments (Turon, 1984) proving that the Eemian is the equivalent of Isotopic Stage 5e. So the typical Eemian pollen sequence or its East European equivalent, i.e. the Mikulino interglacial, may reliably serve as marker horizon for Stage 5e in terrestrial sediment profiles, provided that the pollen flora met with is of undoubtedly Eemian or Mikulino age.

As to this interglacial period, Woillard (1979) had published the stimulating hypothesis that the transition from the end of the *Carpinus-Abies*-phase to the beginning of the *Pinus-Betula-Picea*-phase, rich in various heliophilous herbs of the ensuing last glaciation, might only have lasted for some 150±75 years. This conclusion was derived from calculations of the length of various phases in the interglacial development of vegetation at the Grande Pile peat bog, Southwestern Vosges Mts., Eastern France. The calculations were based on the assumption that the sedimentation rate within the interglacial lake of Grande Pile had remained constant over the interglacial period, the duration of which was derived from a paper by Müller (1974b) on a fossil lake of the Eemian interglacial in the Lüneburger Heide, Northern Germany. Here, it could be shown by an analysis of annually layered lake sediments that this interglacial might have lasted approximately 11,000 years.

W. H. Berger and L. D. Labeyrie (eds.), Abrupt Climatic Change, 151–162.

Figure 1. Laminated lake sediments of the interglacial Krumbach section, a) before the hiatus at the end of the interglacial sequence (coarse laminae), b) after the hiatus (fine laminae).

Seret (1983) criticized Woillard's (1979) approach, since it is neither proved that the sedimentation rate in the interglacial Grande Pile lake has been constant all over the interglacial period, nor that the finely-layered sediments occurring there show annual features. On the other hand, Müller (1965, 1974a, 1979) has repeatedly pointed out, based on pollen studies, that even during the Holsteinian and during one preceding interglacial, climate seems to have changed several times abruptly, from approximately present-day conditions in Central Europe to those of the modern polar forest-limit in Fennoscandia. This would mean an extremely high instability of climate, at least during interglacial phases. Models constructed by Flohn (1979, 1984) give a theoretical background for the understanding of such extremely abrupt climatical changes.

In view of the extremely important scientific and practical consequences of such putative abrupt changes of climate, G. Woillard had already tried to check her hypothesis on laminated lacustrine sediments together with B. Frenzel, as a Humboldt Scholar at the Botanisches Institut der Universität Hohenheim. After her sudden death B. Frenzel and W. Bludau have continued her work. This paper is dedicated to her memory.

THE GEOLOGICAL SETTING

The Krumbach section is situated between the terminal moraine systems of the classical Würmian and Rissian glaciations in the northernmost regions of the former Rhein-Glacier, Southern Germany. It consists of lake sediments of the preceding glaciation, of the Eemian interglacial proper, and of the beginning of the ensuing glaciation. All these phases are palynologically well documented. This sequence is overlaid by minerogenic sediments, the origin of which is a matter of debate. According to Frenzel (1978a), it is covered by basal till (type basal lodgement till) and by younger solifluction layers and loesses, whereas Schreiner and Werner (1980) and Schreiner and Ebel (1981) suggest that all these sediments overlying the lacustrine ones are solely of periglacial origin. Nevertheless, the interglacial proper can palynologically be easily correlated with the Eemian type-sequence in the Netherlands and in Northern Germany, and it can be proved paleomagnetically that the transition from lake sediments to the minerogenic sediments just mentioned dates from about 100,000 b.p. (Koci, in a letter to Frenzel).

During the end of the Krumbach Eemian interglacial and during the ensuing period of a deteriorating climate leading to pleniglacial conditions, laminated sediments were formed in the then-existing lake (Fig. 1). The laminae have a thickness of between 15mm and less than 0.5mm. They consist of lighter and darker brown layers, and it was assumed that one light and one dark layer were formed during only one year. The correctness of the assumption is shown. On the other hand, not all parts of the inter- and anaglacial sequence of the Krumbach section seem to be layered. We used two approaches for the determination of the duration of the interglacial to glacial transition: a) as far as possible, the annual layers were counted; b) if it turned out not to be possible to

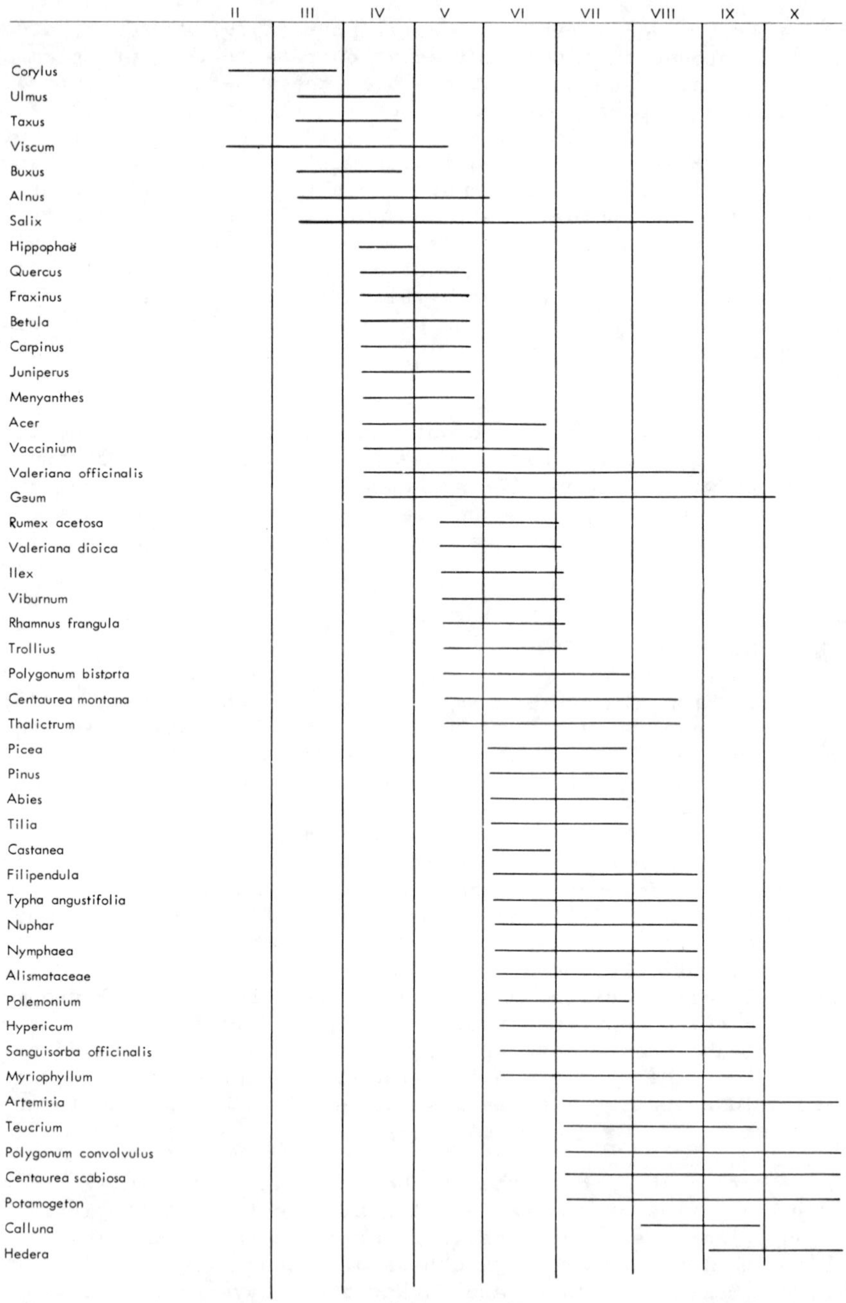

Figure 2. Phenological calendar of the interglacial Krumbach flora. Times of blossoming are indicated (months of the year).

rely on annual layers, the maximum velocity of spontaneous changes in the prevailing forest communities of that time was used, provided that climate had remained more or less constant. Moreover, this calculation was based on a comparison with well-dated spontaneous changes in nearby Holocene sites (Bludau, 1986; Frenzel, 1978b, 1983).

METHODS

At Krumbach, the interglacial and glacial sediments can only be obtained by coring. A piston corer, type "Mobile Drill," was used, with an inner diameter of 6cm and a length of 90cm. Before coring, the former lake basin had been investigated seismically and geoelectrically by the Niedersächsische Landesamt für Bodenforschung, Hannover. We are greatly indebted to them for this help.

In fresh cores, it is generally very difficult to recognize the laminae. So the cores were dried for several weeks; drying time depended on the best visibility of the laminae achieved (in general 4 to 5 weeks under normal room conditions). Thereafter each lamina was divided into its light and dark brown section under the binocular lense by thoroughly scraping off either the light or the dark brown layer. The scraped-off material was collected and prepared for pollen analysis, using the method described by Frenzel (1964) and its improvements by Schedler (1981) and by Urban (1978). Since the pollen content was always a very high one, at least 400 to 500 pollen grains were counted in each scraped-off layer. Because it had to be proved that the laminae investigated (always consisting of the lighter and the darker brown layer) dated from only one year, the pollen flora was divided into plants blossoming in early spring, in late spring, in summer, in autumn, (and perhaps winter; Fig. 2). Findings confirmed that both of these layers were formed during one year (Fig. 3). It can be seen that the light layers were formed during summer and early autumn, the dark ones during early spring. From this firm starting point the duration of the interglacial to glacial transition might be extrapolated; yet we have analyzed several hundred layers in order to be sure that a) we are right in ascribing these layers to single years only, and b) that hiatuses could be reliably well recognized (see below).

We also thoroughly investigated the interglacial vegetation history, primarily at the very end of the interglacial proper, registering the occurrence of indicator-plants which can prove whether ecological conditions remained constant or not. If the conditions remained constant, the mean duration of the establishment of comparable Holocene forest-types, before the impact of man had begun, was used for the evaluation of the minimal time necessary for the change from one prevailing forest-type to another. Here minimal ages are given only in order not to exaggerate the time necessary for this change. Of course, this is no direct determination of time, yet for an estimate of minimum time, this method may suffice.

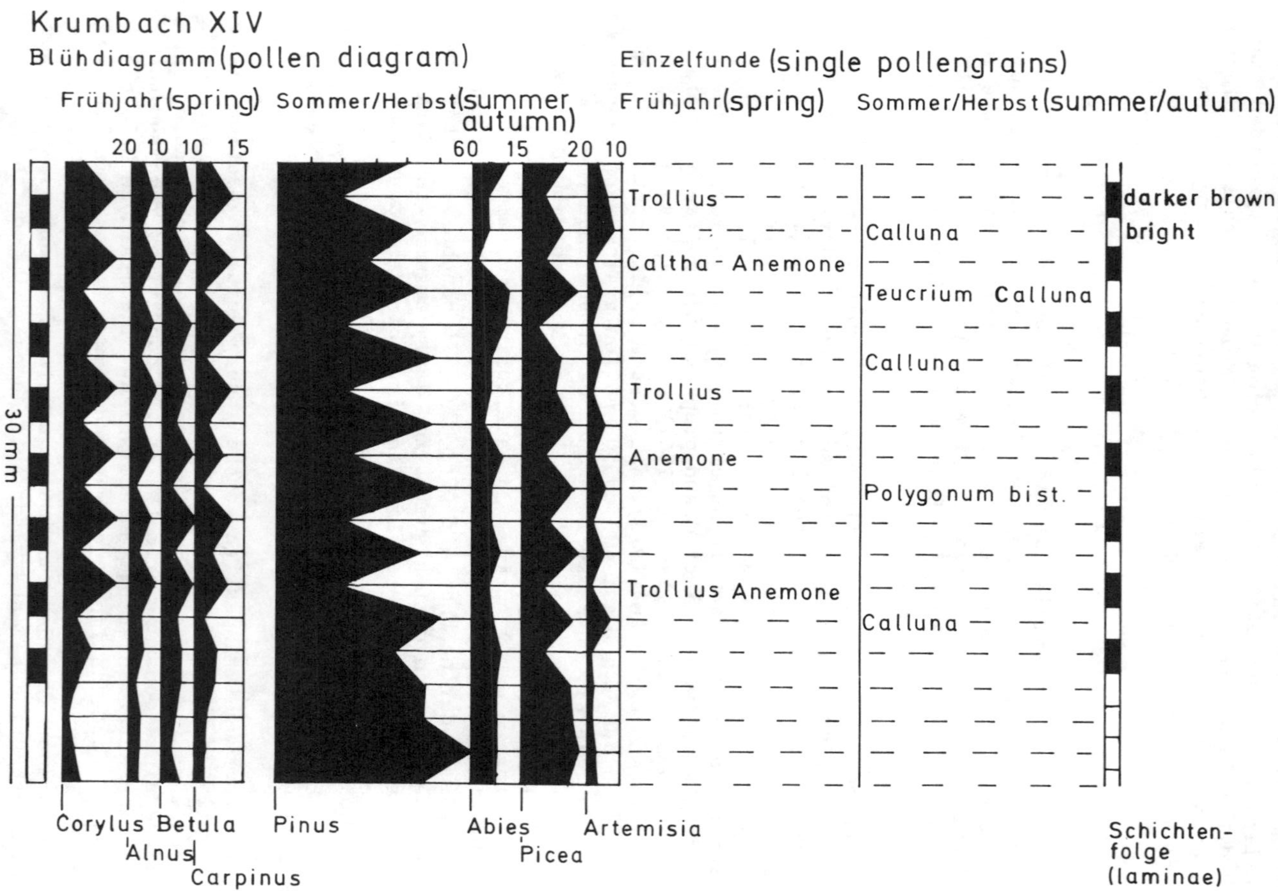

Figure 3. The time of formation of light or darker brown laminae, respectively, in the interglacial Krumbach section at times of blossoming of various plant species (see Fig. 2).

RESULTS

Up to the end of the *Carpinus-Corylus* (hornbeam-hazel) warm phase, demanding plant taxa like *Taxus* (yew), *Acer* (maple), *Hedera* (ivy), and *Ilex* (holly) prove optimal climatic conditions at that time. When *Abies* (fir) began to spread into the hornbeam forests, the same thermophilous and hygrophilous plants persisted, together with *Buxus* (box). Thus very favorable climatic conditions must have continued uninterruptedly. If this is accepted and if these changes in vegetation are compared with related ones during the Holocene in Southern Germany, it must be concluded that the immigration of *Abies* at the expense of *Carpinus* should have lasted for at least 500 years. As far as we know from several investigations on the Holocene vegetation history under comparable climatic and pedological conditions, this is a minimal age only. Furthermore, it may be suggested that the ensuing *Abies-Carpinus*-phase should have lasted for at least one tree-generation, i.e. for about 250 to 300 years.

Later on these forests were replaced by *Picea* (spruce), yet even then *Carpinus* and *Abies* acted as major constituents of the then-existing vegetation. At the same time, such thermophilous and hygrophilous plants like *Taxus*, *Buxus*, *Hedera*, and *Ilex* slowly disappeared. Yet the amount of *Pinus* (pine) was negligible and the same held true for heliophilous herbs. On the other hand, *Alnus* (alder) and fen-plants began to spread. Therefore it must be concluded that there was not a strong deterioration of climate causing the retreat of the thermophilous plants and opening the forests; the forests remained as dense as they had been formerly and only a weak paludification seems to have occurred. Since *Abies* and *Carpinus* did not disappear, it may be concluded that spruce had immigrated onto waterlogged sites as it did during the Holocene in the nearby Ammergebirge (Bludau, 1986). From this it may be concluded that the immigration of Picea during this later part of the interglacial period should have lasted for at least 500 years, and one would suggest that the ensuing *Picea-Abies-Carpinus*-phase should have had a duration of at least one tree-generation, i.e. about 250 to 300 years.

One gets the impression then that the development of the interglacial forest vegetation at Krumbach, beginning with the spreading of *Abies* within the formerly dominating hornbeam forests, up to the domination of spruce forests, with some admixture of Abies and of *Carpinus*, lasted for at least 1500 to 1600 years (Fig. 4).

The further development of the interglacial forests at Krumbach is interrupted within the *Picea-Abies-Carpinus*-phase by a hiatus, the duration of which is unknown (Fig. 5). It camouflages the further development of this vegetation to the phase of prevailing pine-spruce forests, with only a small admixture of herbs. This points to the problem of hiatuses in the evaluation of time, being complicated here by the fact that this hiatus cannot be observed sedimentologically.

The ensuing *Pinus-Picea*-phase was characterized by an extremely small percentage of herbs in the then-existing forest. Yet the presence of these herbs points to an increasing drought and coldness of climate. During this phase beautifully laminated lake sediments were formed,

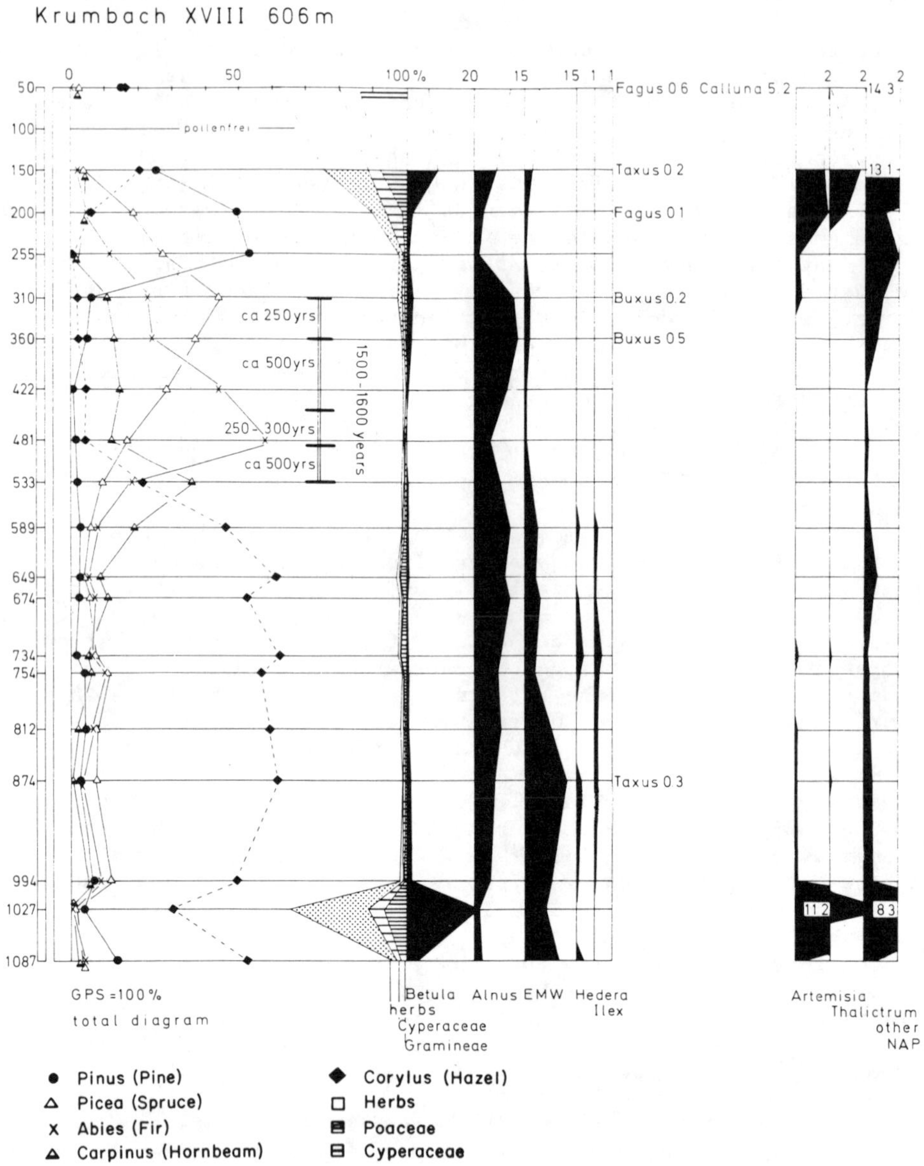

Figure 4. Duration of the *Abies-Carpinus*-phase with thermophilous plants and of the *Picea-Abies-Carpinus*-phase, respectively, in the interglacial Krumbach section (see text).

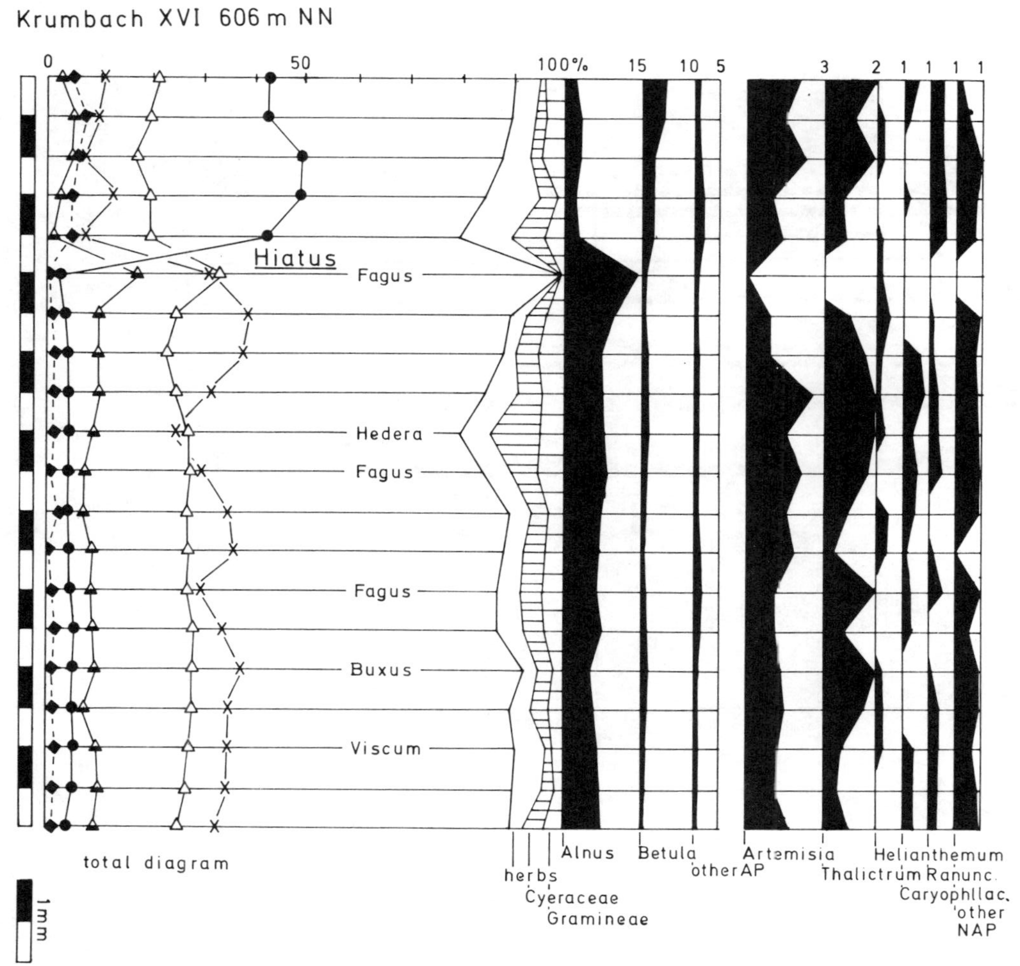

Figure 5. Position of a hiatus in the interglacial Krumbach lake sediments at the end of the *Picea-Abies-Carpinus*-phase. Key as in Fig. 4.

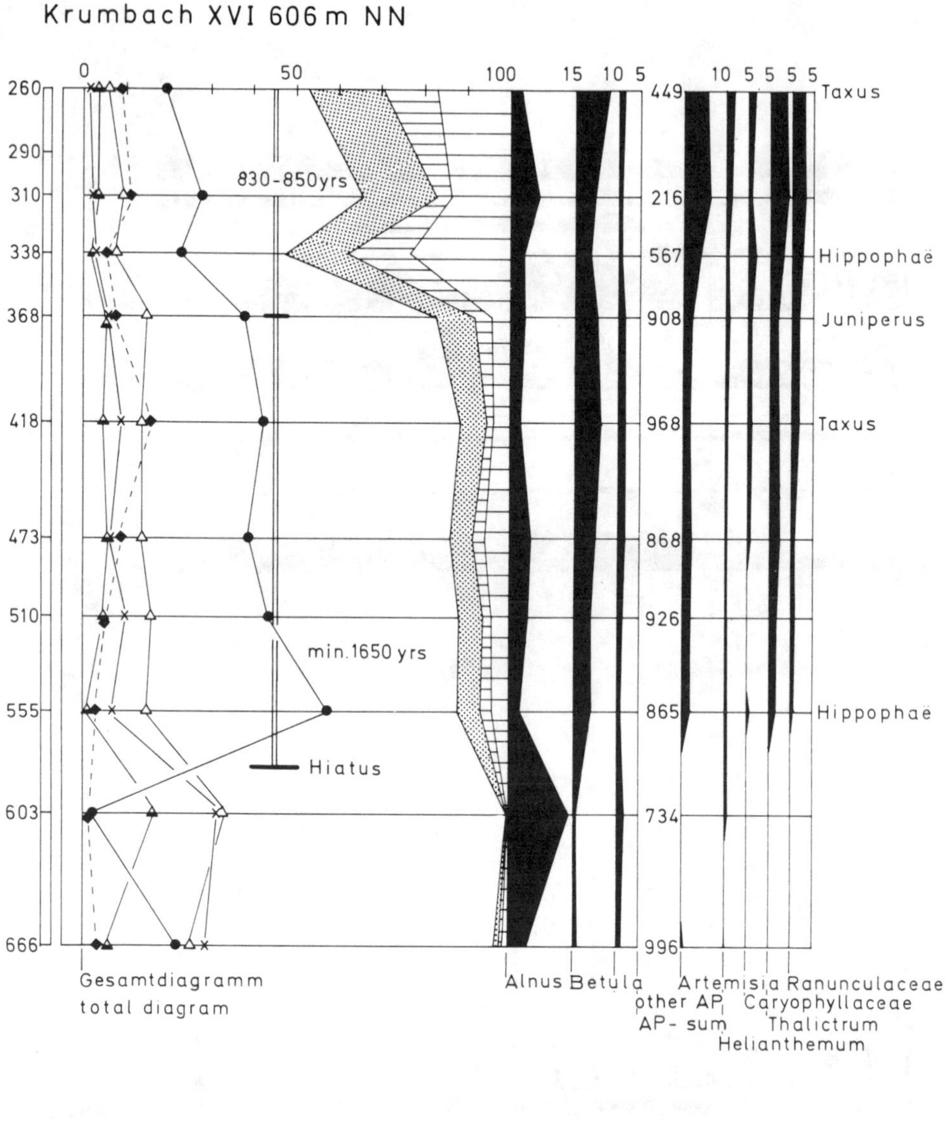

Figure 6. Duration of the very end of the Krumbach interglacial (*Pinus-Picea*-phase and *NAP-Pinus*-phase) according to the number of annual laminae of the lake sediments. Key as in Fig. 4.

which contained rich pollen floras. So, by counting these annual layers, it can be deduced that this phase, excluding the hiatus mentioned, had lasted for at least 1650 years. Only after this long span of time, during the next 820 to 850 years (annually laminated sediments!) glacial conditions began (Fig. 6). This was characterized by a strong increase in nonarboreal pollen up to more than 25 to 50% of the total pollen sum.

DISCUSSION

From the preceding, it follows that in the Krumbach section, at least 1500 to 1600 years elapsed between the end of the *Carpinus-Corylus*-phase and the end of the *Picea-Abies-Carpinus*-phase. It must be stressed that the exact length of time cannot be given due to the hiatus mentioned, which cuts off the last steps in the evolution of the vegetation of this phase. Müller (1974b) calculated about 2000 years for the last interglacial in Northern Germany and for the same series of events, whereas Woillard (1979) determined about 150±75 years.

According to our investigations of the annually layered sediments of the Krumbach section, the ensuing *Pinus-Picea*-phase, which still belonged to the last interglacial proper, lasted for at least 1650 years, excluding the unknown length of the hiatus. Müller (1974b) had calculated approximately 2000 years for the same phase in interglacial vegetation history. Woillard (1979), on the other hand, suggested that about 20 years might have sufficed.

Our results and the calculations done by Müller point in the same direction; a minimal value of about 3000 years can be given for the development of the interglacial vegetation, extending from the end of the hornbeam-hazel-phase to the very beginning of the last glaciation. Müller (1974b) suggested 4000 years. This may be a better value, if the hiatus in our section is taken into consideration. Since the hypothesis of Woillard (1979) is based on the fallacious assumption of a constant sedimentation rate during the last interglacial in the Grand Pile lake alone, without this hypothesis having been checked by counting annually layered sediments, it must be concluded that Müller's calculations and our calculations and countings approach reality much better than the assumption of an abrupt end of the last interglacial.

ACKNOWLEDGEMENTS

This work was supported by a grant of the Bundesministerium für Forschung und Technologie, Project-Group "Terrestrische Paläoklimatologie". We are much indebted to Mrs. Gaby Eisele and Mrs. Susanne Liner for their skillful laboratory work.

REFERENCES

Bludau, W., 1986. *Zur Paläoökologie des Ammergebirges im Spät- und Postglazial*, Schäuble Verlag Rheinfelden, 365 pp.

Flohn, H., 1979. Climatic change, ice sheets and sea level, *Sea Level, Ice, and Climatic Change* (Proceedings of the Canberra Symposium, December 1979), IAHS Publ., 131, 431-440.
Flohn, H., 1984. A possible mechanism of abrupt climatic changes. In: *Climatic Changes on A Yearly To Millenial Basis*, N.-A. Mörner and W. K. Karlén (eds.), 521-531.
Frenzel, B., 1964. Zur Pollenanalyse von Lössen, *Eiszeitalter u. Gegenwart*, 15: 5-39.
Frenzel B., 1978a. Das Problem der Riss/Würmwarmzeit im deutschen Alpenvorland, *Führer zur Exkursionstagung des IGCP-Projects* 73/1/24, "Quarternary Glaciations in the Northern Hemisphere", B. Frenzel (ed.), 103-114.
Frenzel, B., 1978b. Landschaftsgeschichte und Landschaftsökologie des Kreises Freudenstadt, *Der Kreis Freudenstadt*, Theiss-Verlag Stuttgart und Aalen, 52-76.
Frenzel, B., 1983. Die Vegetationsgeschichte Süddeutschlands im Eiszeitalter. In *Urgeschichte in Baden-Württemberg*, H.J. Müller-Beck (ed.), Konrad Theiss-Verlag, Stuttgart, 91-166.
Müller, H., 1965. Eine pollenanalytische Neubearbeitung des Interglazial-Profils von Bilshausen (Unter-Eichsfeld), *Geol. Jb.*, 83: 327-352.
Müller, H., 1974a. Pollenanalytische Untersuchungen und Jahresschichtenzählungen an der holstein-zeitlichen Kieselgur von Munster-Brehloh, *Geol. Jb.*, A 21: 107-140.
Müller, H., 1974b. Pollenanalytische Untersuchungen und Jahresschichtenzählungen an der eem-zeitlichen Kieselgur von Bispingen/Luhe, *Geol. Jb.*, A 21: 149-169.
Müller H., 1979. Climatic changes during the last three Interglacials. In: *Man's Impact on Climate. Developments in Atmospheric Science*, Bach, Pankrath & Kellog (eds.), 10, Amsterdam, Oxford, New York, 29-39.
Schedler, J., 1981. Vegetationsgeschichtliche Untersuchungen an altpleistozänen Ablagerungen in Südwestdeutschland, *Diss. Bot.*, 58, J. Cramer, Vaduz, 157 pp.
Schreiner, A. and Ebel, R., 1981. Quartärgeologische Untersuchungen in der Umgebung von Interglazialvorkommen im östlichen Rheingletschergebiet (Baden-Würtemberg), *Geol. Jb.*, A 59: 3-64.
Schreiner, A. and Werner, J., 1980. Zur Quartärgeologie in der Umgebung des Eem-Interglazials von Krumbach/Saulgau (Baden-Württemberg), *Geol. Jb.*, A 56: 5-43.
Seret, G., 1983. Rather long Duration of the Transient Climatic Events in the "Grande Pile" (Vosges - France). *In Paleoclimatic Research and Models*, A. Ghazi (ed.), D. Reidel Publ. Company, Dordrecht, Boston, Lancaster, 139-143.
Turon, J.-L., 1984. Direct land/sea correlations in the last interglacial complex, *Nature*, 309: 673-676.
Urban, B., 1978. Vegetationsgeschichtliche Untersuchungen zur Gliederung des Altquartärs der niederrheinischen Bucht, Sonderveröff. Geol. Inst. Univ. Köln, 34: 165 pp.
Woillard, G., 1979. Abrupt end of the last interglacial s.s. in North-East France, *Nature*, 281: 558-562.

THE ALLERÖD/YOUNGER DRYAS BOUNDARY

Jan Mangerud

University of Bergen
Department of Geology, Sec. B
Allegt. 41
N-5007 Bergen
Norway

ABSTRACT. From former cirque glaciers in western Norway, it is calculated that the summer (1.May to 30.September) temperature dropped 5-6°C during less than two centuries, probably within decades, at the Alleröd/Younger Dryas transition, some 11,000 years ago. From the same data the Younger Dryas summer temperature is estimated to have been 8-10°C lower than at present, and from fossil ice wedges the mean annual temperatures 13°C lower than at present in the same area. The amplification of the climatic change in western Europe at the Alleröd/Younger Dryas transition appears due to a step function in the change in the path of the North Atlantic Current.

INTRODUCTION

During the Alleröd Chronozone, 11,800 to 11,000 years ago, western Europe approached the present day environmental and climatic situation, after having suffered the last glacial maximum some 20,000 to 18,000 years ago. However, the climatic deterioration 11,000 years ago led to nearly fully glacial conditions on this continent for some few hundreds of years during the Younger Dryas. This change is completely out of phase with the Milankovitch (orbital) forcing as this is understood today, and therefore its cause is of major interest.

Since the Alleröd/Younger Dryas boundary is well inside the period of radiocarbon dating, it is also of great interest as a model for the mode of response, time lags, and feedback mechanisms of different biota, glacial systems, periglacial features, etc., to this type climatic change, which in Europe was very similar to the onset of a glaciation, even though the forcing was different.

In this paper I will describe some of the geological records of this boundary. Description of other features, and some more general considerations, are recently given by the NATO/NSF working group (Mangerud et al., 1986b), and other abstracts in that volume. Finally, I provide some speculations on the cause of the Younger Dryas.

W. H. Berger and L. D. Labeyrie (eds.), Abrupt Climatic Change, 163–171.

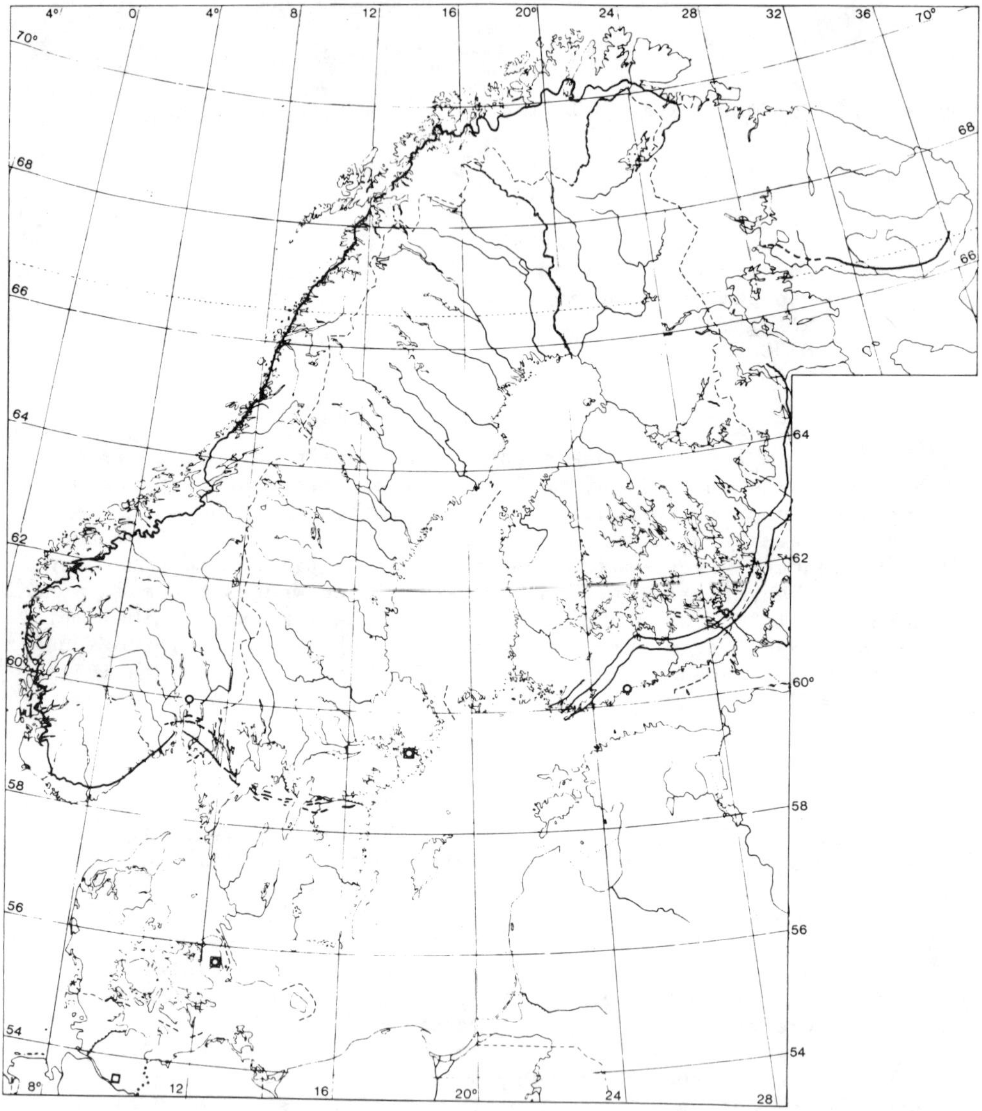

Figure 1. Map of Younger Dryas end moraines around the Scandinavian Ice Sheet. Compiled from various sources.

EVIDENCE

The Scandinavian Ice-sheet

At the time of the Alleröd/Younger Dryas transition, the Scandinavian ice-sheet was still a major element in the climate system. The record from the Younger Dryas is distinct, consisting of ice-marginal deposits that are mapped nearly continuously around Scandinavia (Fig. 1) (e.g. Andersen, 1981), showing that the climate turned to a more glacial regime in both the continental climate area of USSR/Finland and the oceanic climate area of Western Norway. This suggests that lower summer temperatures, and not increased winter precipitation, was the climatic factor that determined the major pattern of glacial response. Superimposed on this major trend are smaller variations, e.g. an amplification of the re-advance in Western Norway compared to the easterly areas, due to higher winter precipitation along the western flank of the ice-sheet, and topographical and glaciological factors (Mangerud, 1980). The re-advance also caused a relative rise in sea level in Western Norway through the combination of increased gravitational attraction and a halt in glacio-isostatic uplift (Anundsen, 1985). Concerning the northern limit for the Younger Dryas glacial response, it may be mentioned that in Svalbard (78°N) a re-advance for one valley glacier is also stratigraphically demonstrated (Mangerud et al., 1985; Elgersma and Helliksen, 1986).

Permafrost

Evidence of permafrost during the Alleröd is not known from Europe. From the Younger Dryas, ice wedge casts are found in the entire area from Finland to NW Germany, and frost cracks are found in the Netherlands (Maarleveld, 1976; Larsen et al., 1984). The main requirements for formation of ice-wedges are permafrost, and some extremely low temperatures during the winter. Mean annual temperatures are below -6°C in areas where they form today (Pewe, 1966). For Germany, this would mean an annual temperature 15°C lower than today. The permafrost in the mentioned area was probably due to cold NE-winds, especially during the winters. These winds may partly have been caused by the high pressure over the Scandinavian ice-sheet. However, as a topographical element, the ice-sheet did not change much from the Alleröd to the Younger Dryas, and the NE-winds can therefore not be merely a feedback result of increased size of the ice-sheet.

We (Larsen et al., 1984) have earlier pointed out that Younger Dryas ice-wedge casts are missing along the western coast of Norway, except in the far north, and have attributed this to the snow cover during winter. Later I have found one cast in a former shore-line on the island of Vigra, in a position where the snow must have blown off the ground during formation. With the assumptions given above, this suggests a Younger Dryas annual temperature 13°C lower than today in Western Norway.

Cirque glaciers

Cirque glaciers provide a simpler system for paleoclimatic reconstructions than larger glaciers (Sutherland, 1984). Distinct end moraines of Younger Dryas age that allow precise reconstruction of each individual glacier, are common both in the British Isles (Sissons, 1980a, b) and along the coast of Norway outside the Scandinavian icesheet (Andersen, 1975; Larsen et al., 1984). We (Mangerud et al., 1979; Larsen et al., 1984) cored a lake at Krakenes in western Norway which was fed by a stream from a cirque glacier. The sediment sequence proved that no glacier existed in the cirque during the Alleröd, and that the cirque glacier was formed at the Alleröd/Younger Dryas transition.

Rates of change. A cirque glacier starts to form when the climate passes certain thresholds. The geological signal is therefore generally of an on-off type, and rates of climatic deterioration are impossible to determine. One way of solving this would be to investigate when glaciers were formed at different elevations, because glaciers would form first at higher elevations, and progressively later in lower terrain. We have not yet had time to pursue this idea, however we tried a different approach; in the lake at Krakenes we (Larsen et al. 1984) found indication of a climatic deterioration in the pollen record from a level corresponding to 50-180 years before the start of glacio-lacustrine sedimentation. In another lake we (Kristiansen, Mangerud and Lömo, manuscript in preparation) found that the pollen influx (production) of *Betula* decreased faster as a response to the climatic deterioration than the composition of the vegetation. The difference in response time is estimated as maximum 100 years. However, both in that pollen diagram, as in most others from western Norway, the Alleröd/Younger Dryas transition is seen as one steep step. As a first approximation, I conclude that in this area the full amplitude of change, from the warmest Alleröd to the coldest Younger Dryas, occurred as one step within less than two centuries; possibly within some few decades.

Amplitude of change. Cirque glaciers can climatically be characterized by one single parameter; the equilibrium line altitude (ELA)- the altitude of the boundary between the area of net yearly accumulation, and the area of net ablation. A map of the ELA for present-day glaciers in Norway is given by Östrem and Liestöl (1964), from which we calculated the present day ELA at Krakenes to 900m a.s.l. (Fig. 2). The Younger Dryas ELA could also be precisely determined from 14 reconstructed glaciers (Larsen et al., 1984) to 180m a.s.l., that is 700m lower than today. So far we have not been able to directly determine the ELA for the Alleröd, because we have not yet identified a cirque glacier that survived the Alleröd. As an approximation we have used the lowering of the tree-limit, since the ELA and tree-limit today are roughly parallel. This method suggests an Alleröd ELA of 600m a.s.l., a result that appears slightly too low from other considerations. These results give a lowering of the ELA of at least 400m at the Alleröd/ Younger Dryas boundary (Fig. 2) in western Norway.

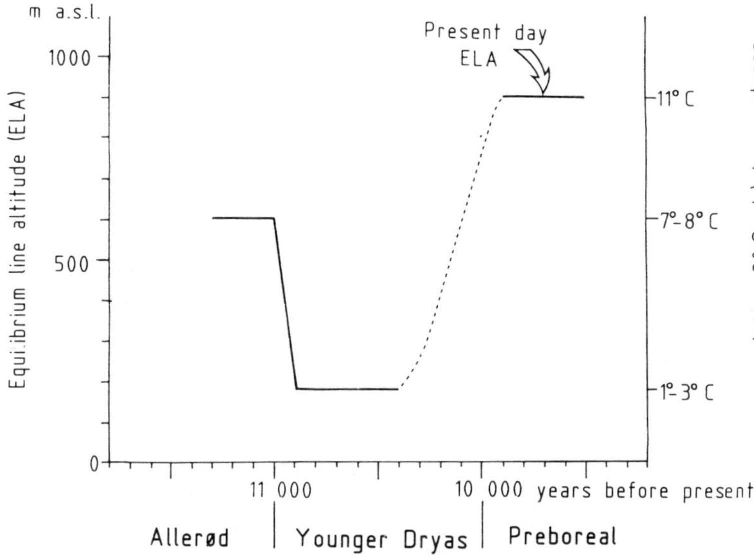

Figure 2. Variation of the equilibrium line altitude (ELA) for cirque glaciers in western Norway for the period 11,500 to 9000 years ago. For simplicity the present day ELA is plotted at 9500 Years B.P., as that certainly is close to the correct value. At the right axis the ELA variations are translated to summer temperatures, under certain assumptions discussed in the text.

The next step will be to translate the ELA changes to more direct climatic parameters. In Fig. 3 we have plotted the winter accumulation versus summer temperatures at the equilibrium line for present day glaciers in Norway and Svalbard. We have also plotted present day summer temperatures and winter precipitation for the cirques around Krakenes that housed the Younger Dryas glaciers. The idea is that the climatic conditions for the Younger Dryas glaciers were similar to the conditions for glaciers in this area today. Before comparing it should be mentioned that present-day precipitation for the Younger Dryas cirque glacier sites probably is underestimated, because the local precipitation is lower at the recording stations than beneath the high-walls in the cirques. However, even with that caution it appears evident from the plot that the Younger Dryas winter precipitation could not have been much lower than today. One may also argue that with the much colder climate a significant higher precipitation was unlikely, leading to the conclusion that winter precipitation was of the same order as today. With that assumption Fig. 3 suggests that the Younger Dryas summer temperature was 8-10°C lower than today. Extending that assumption also to the Alleröd (Fig. 2), the conclusion is that the summer (1.May to 30.September) temperature dropped 5-6°C during less than two centuries around 11,000 years ago.

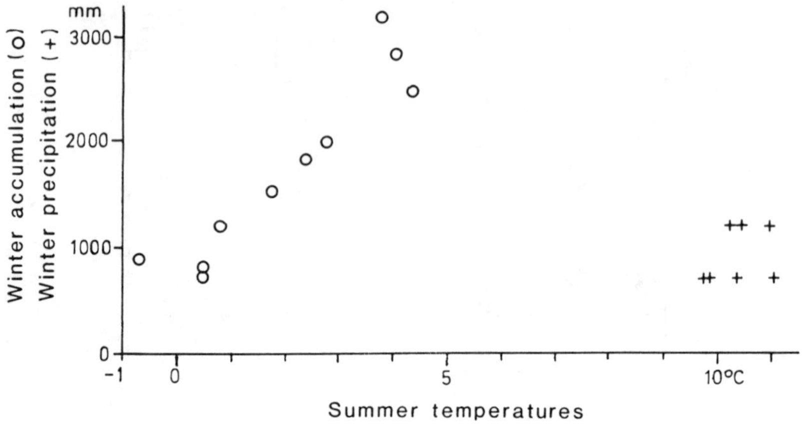

Figure 3. Circles: Measured winter accumulation versus summer temperatures at the equilibrium line (ELA) at present day glaciers in Norway and Svalbard. Winter accumulation is close to winter precipitation, since hardly any melting takes place during winter. However, in cirques winter accumulation is normally larger than the precipitation, because snow blows in from adjacent areas.
Crosses: Present day summer temperatures and winter precipitation in the cirques that housed the Younger Dryas glaciers. If the main pattern of the climate was the same during the Younger Dryas as today, the crosses should match the circles, and the relative difference in summer temperature and/or winter precipitation can be read. Constructed from Figs. 13 and 14 in Larsen et al.(1984).

THE CAUSE FOR THE YOUNGER DRYAS

In the following I will list some facts and thoughts on the Alleröd-Younger Dryas shift, without reaching a final conclusion.

1. The North Atlantic Ocean Current and the atmospheric circulation in the area are dynamically linked in a way that a change in either will cause a change in the other.

2. Following the work of Ruddiman and McIntyre (1973, 1981), the simplified path of the North Atlantic Current can be considered, on the time scale we are discussing here, as having a hinge east of North America (Lowe and Gray, 1980). The current may turn north of the British Isles into the Norwegian Sea, as it does today, and where it has a sector of freedom. It may also turn south, passing Portugal like the present day Canary Current, where it also has a large open sector for variations. The important point is that surface currents hitting the coast of Europe are forced either to the north or to the south. The mentioned hinge thus has sectors of continuous variations, but with one major step. Due to its heat transport, the North Atlantic Current is a major element in the climate of Europe, and a change in the form of a step

function of the current will probably cause a step in climatic change. Superimposed on the change of the path of the current, might be more gradual changes in its surface temperature, and volume of water.

3. During the Alleröd a branch of the North Atlantic Current entered the Norwegian Sea (Ruddiman and McIntyre, 1973, 1981). Jansen and Bjorklund (1985) and Jansen (this volume) demonstrated that this was only a narrow current in the extreme eastern part of the Norwegian Sea, while Mangerud (1977) showed that it was a warm surface current with the characteristics of the present day Atlantic Water, but which ended far south of present northern limit of Atlantic Water. However, Vorren et al. (1978, 1984) found evidence of warming as far north as Troms (70°N), and Jansen (this volume) concluded that no deep water was formed in the area during the Alleröd, implying that the current continued north along the Norwegian coast, as a progressively diluted and cooled surface and intermediate water mass.

I make the point that during the Alleröd, the Atlantic Current was in a very sensitive position; a minor change in its direction would, as a step function, move the flow of warm water from the coast of Norway towards the coast of Portugal, causing a sudden and large climatic deterioration in western Europe.

4. Because of the major climatic signal in Europe and the North Atlantic, and apparently small changes if any elsewhere in the world, the cause for the cooling has also been sought in this area. Point 1 supports this idea. Possibly a major outflow of icebergs between Iceland and Greenland, as postulated but not proven (Mercer, 1969; Ruddiman and McIntyre, 1981), was a strong enough driving force to the change. I have, however, two objections (Points 5 and 6 below) against searching for the ultimate cause only in the North Atlantic area.

5. The large Alleröd/Younger Dryas amplitude in Western Europe is, according to Points 2 and 3 above, due to an amplification caused by a geographical pattern. Thus the large signal is in itself not an argument for the location of the cause. In this "chicken and egg problem" it is just as probable that a minor change in atmospheric circulation initially caused a small shift of the North Atlantic current, and that the major climatic change was a feedback result of that change. Adjustment in atmospheric circulation to changing ice-sheet configuration (Ruddiman and McIntyre, 1981), or change in atmospheric CO_2 content (Berger and Killingley, 1982) have been proposed as possible initial causes.

6. Simply by common sense, I have often maintained that the Younger Dryas cannot be an exclusively European event. Even if the cause should be some "random" physical process within this area, the amplification led to such major changes in both atmospheric and oceanic circulation within a significant enough portion of the earth's surface that it had to cause some kind of change (cooler, warmer, drier, moister climate) in other areas. For a long time many scientists have also claimed to have found climatic changes which correspond in time to the Younger Dryas (refs. in e.g. Broecker et al., 1985). An extremely interesting task now would be to separate which changes are direct responses to the ultimate forcing of the North Atlantic changes, and which are "only" responses to the amplification in the North Atlantic.

7. Obviously the relative timing of events is important for solving many of the involved problems. We are therefore investigating volcanic ash beds from this period, that can be traced both in continental and marine sequences. Fortunately such ash beds occur both within and just after the Younger Dryas (Mangerud et al., 1984, 1986a).

REFERENCES

Andersen, B.G., 1975. Glacial geology of Northern Nordland, North Norway, *Norges Geologiske Undersökelse*, 320: 1-74.

Andersen, B.G., 1981. Northern Europe, southern Europe, Barent Sea and Arctic islands. In Denton, G.H. and Hudges, T.J. (eds), *The Last Great Ice Sheets*, New York, Wiley, 3-45.

Anundsen, K., 1985. Changes in shore-level and ice-front position in Late Weichsel and Holocene, southern Norway, *Norsk Geogr. Tidsskr.*, 39: 205-225.

Berger, W.H. and Killingley, J.S., 1982. The Worthington effect and the origin of the Younger Dryas, *Jour. Marine Res.*, 40, Suppl: 27-38.

Broecker, W.S., Peteet, D.P. and Rind, D., 1985. Does the ocean-atmosphere system have more than one stable mode of operation? *Nature*, 315: 21-25.

Elgersma, A. and Helliksen, D., 1986. Kvartaergeologiske undersökelser i Van Mijenfjordomradet, Spitsbergen, Svalbard. Univ. Bergen. Unpublished thesis.

Jansen, E., 1987. Rapid changes in the inflow of Atlantic water into the Norwegian Sea at the end of the last glaciation, this volume.

Jansen, E. and Björklund, K., 1985. Surface ocean circulation in the Norwegian Sea 15,000 B.P. to present, *Boreas*, 14: 243-257.

Larsen, E., Eide, F., Longva, O. and Mangerud, J., 1984. Alleröd-Younger Dryas climatic inferences from cirque glaciers and vegetational development in the Nordfjord area, western Norway. *Arctic and Alpine Research*, 16: 137-160.

Lowe, J.J. and Gray, J.M., 1980. The stratigraphic subdivision of the lateglacial in NW Europe: A discussion. In Lowe, J.J., Gray, J.M. and Robinsen, J.E.(eds.), *Lateglacial of North-west Europe*, Oxford, Pergamon, 157-175.

Maarleveld, G.C., 1976. Periglacial phenomena and the mean annual temperature during the last glacial times in the Netherlands, *Biuletyn Periglacjalny*, 26: 57-78.

Mangerud, J., 1977. Late Weichselian marine sediments containing shells, foraminifera, and pollen, at Agotnes, western Norway, *Norsk Geologisk Tidsskrift*, 57: 121-148.

Mangerud, J., 1980. Ice-front variations of different parts of the Scandinavian ice-sheet, 13,000-10,000 years BP. In *Lateglacial of North-west Europe*, Lowe, J.J., Gray, J.M. and Robinsen, J.E. (eds.). Oxford, Pergamon, 23-30.

Mangerud, J., Furnes, H., and Johansen, J., 1986a. A 9000-year-old ash bed on the Faroe islands, *Quaternary Res.*, 26: 262-265.

Mangerud, J., Larsen, E., Longva, O., and Sönstegaard, E., 1979. Glacial history of western Norway 15,000-10,000 B.P., *Boreas*, 8: 179-187.

Mangerud, J., Lie, S.E., Furnes, H., Kristiansen, I.L. and Lömo, L., 1984. A Younger Dryas ash bed in western Norway, and its possible correlations with tephra in cores from the Norwegian Sea and the North Atlantic, *Quaternary Research*, 21: 85-104.

Mangerud, J., Elgersma, A., Helliksen, D., Landvik, J. and Salvigsen, O., 1985. The Late Weichselian glaciation in Isfjorden a*Climate in the Arctic*, 16-17. Report from "Voksenasendagene 1984". Oslo.

Mangerud et al. (NATO/NSF working group), 1986b. Land and ice records of abrupt climatic changes between the last glacial maximum and the Holocene. In *The book on abstracts and reports from the conference on: Abrupt Climatic Change*. Univ. California. Scripps Inst. Ocean. SIO Reference 86-8.

Mercer, J.H., 1969. The Alleröd oscillation: A European climatic anomaly? *Arctic and Alpine Research*, 1: 227-234.

Östrem, G., and Liestöl, O., 1964. Glaciologiske undersökelser i Norge 1963. *Norsk Geogr. Tidskr.*, 18: 281-340.

Pewe, T.L., 1966. Paleoclimatic significance of fossil ice wedges. *Biul. Peryglac.*, 15: 65-73.

Ruddiman, W.F. and McIntyre, A., 1973. Time-transgressive deglacial retreat of Polar waters from the North Atlantic. *Quaternary Research*, 3: 117-130.

Ruddiman, W.F. and McIntyre, A., 1981. The North Atlantic Ocean during the last deglaciation, *Palaeogeogr. Paleoclimat. Paleoecol.*, 35: 145-214.

Sissons, J.B., 1980a. Paleoclimatic inference from Loch Lomond advance glaciers. In Lowe, J.J., Gray, J.M. and Robinson, J.E. (eds.), *Lateglacial of north-west Europe*. Oxford. Pergamon, 31-43.

Sissons, J.B., 1980b. The Loch Lomond advance in the Lake District, northern England. *Transactions of the Royal Society of Edinburgh, Earth Sciences*, 71: 13-27.

Sutherland, D.G., 1984. Modern glacier characteristics as a basis for inferring former climates with particular reference to the Loch Lomond Stadial. *Quaternary Science Reviews*, 3: 291-309.

Vorren, T.O., Hald, M., and Thomsen, E, 1984. Quaternary sediments and environments on the continental shelf off northern Norway. *Marine Geol.*, 57: 229-257.

Vorren, T.O., Strass, I., and Lind-Hansen, O.W., 1978. Late Quaternary sediments and stratigraphy on the continental shelf off Troms and west Finnmark, northern Norway. *Quaternary Res.*, 10: 340-365.

MOVEMENT OF THE DESERT BOUNDARY IN THE LEVANT FROM LATEST PLEISTOCENE TO EARLY HOLOCENE

Mordeckai Magaritz and Glenn A. Goodfriend

Isotope Research Department
The Weizmann Institute of Science
76100 Rehovot, Israel

ABSTRACT. Evidence of moisture conditions in the semi-arid region of the Levant during the deglaciation period, 15,000 to 8,000 yr BP, is reviewed. This evidence is derived from a wide variety of sources: sediment characteristics, dune movement, lake levels, soil calcic horizons and their isotopic composition, vertebrate faunas, the isotopic composition and size of land snail shells, pollen, and human population density. From 15,000-11,000 yr BP, relatively wet conditions prevailed, apparently as a result of a more southerly location of the subtropical high pressure belt. A short (<1000 yr) period of extremely dry conditions apparently followed. From c. 10,000-8,000 yr BP, conditions were wet again. Very little information is available on shorter-term (decadal) climatic fluctuations.

INTRODUCTION

The period from latest Pleistocene (since c. 15,000 yr BP) to early Holocene represents a time of global climatic changes referred to as the "deglaciation period." The environmental changes recorded in marine sediments, ice cores and glaciated land surfaces in high latitudes reflect the melting of the ice sheets associated with increased temperature (Ruddiman and McIntyre, 1981; Duplessy et al., 1981; Berger et al., 1985a, b). In land areas in the low to mid latitudes, the climatic changes during this period are seen primarily as changes in the amount of precipitation. Such changes are most dramatic in the desert boundary zone. Because of the steep climatic gradient at the desert boundary, even small geographic shifts in rainfall patterns can have a tremendous impact on local precipitation.

Variation in precipitation occurs on various time scales. Yearly fluctuation of rainfall in Israel, for example, can be ±50% of the mean. Pre-record precipitation is usually estimated from data derived from sediments or materials contained therein (e.g., pollen, faunal remains, carbonate nodules). Therefore usually only longer-term averages (on the order of hundreds of years) of rainfall can be estimated. Moreover, the response time of the various paleoclimatic proxy variables may differ.

W. H. Berger and L. D. Labeyrie (eds.), Abrupt Climatic Change, 173–183.

The sedimentary record in the arid and semi-arid zone is generally temporally discontinuous. In order to obtain a complete time sequence, information has to be obtained not only from various sections, but from various types of materials and by various methods (Table 1). Periods of transition from one set of climatic conditions to another may be missing in a given record. Despite these problems of temporal discontinuity, spatial continuity may be excellent, as in the case of loess accumulation or soil formation. Good geographic resolution is therefore sometimes possible. This is especially critical in a zone of strong rainfall gradients.

Most of the dates available for the deglaciation period are ^{14}C ages or are calibrated against ^{14}C (e.g., prehistoric archeological dates). It has been shown for the middle to late Holocene (Klein et al., 1982), that ^{14}C ages do not exactly correspond to calendarial ages due to fluctuation in the atmospheric concentration of ^{14}C (Suess, 1978). Corrections are available for this period (Klein et al., 1982; Stuiver, 1982) but not for earliest Holocene or Pleistocene ^{14}C dates. Large changes in atmospheric CO_2 have been documented for the deglaciation period (Berner et al., 1980; Delmas et al., 1980). If these changes are due to changes in oceanic circulation such as in upwelling of relatively ^{14}C-depleted deep ocean water, then large deviations between ^{14}C years and calendarial years may be expected (Keir, 1983).

Additional problems exist in ^{14}C dating of some terrestrial materials, for example, incorporation of ^{14}C-depleted carbonate into lake sediments (Olsson, 1974), land snail shells (Goodfriend and Stipp, 1983), and soil carbonate nodules (Williams and Polach, 1971; Magaritz et al., 1981) and incorporation of recent carbonate into soil carbonate nodules.

In this paper we summarize the evidence for climatic changes at the boundary of the desert belt in the Levant. Most of the evidence comes from the Northern Negev and Judean Desert of Israel, where a number of studies on the climate during the deglaciation period have recently been carried out. Comparisons are made with coeval conditions in North Africa in order to obtain a broader picture which permits interpretation of the meteorological basis for climatic changes.

The Northern Negev (31°15 N, 34°45E) represents the wet extreme of the Sinai-Negev deserts (Fig. 1) and receives an average of 300mm rainfall per year at its northern edge as defined here. Rainfall occurs almost entirely during the winter half of the year.

THE EARLY DEGLACIATION PERIOD (15,000-11,000 YEARS BP)

Following the dry period of the last glacial maximum (LGM) at c. 18,000 yr BP (Gat and Magaritz, 1980), a wet period is recorded at various localities. The widespread development of a soil profile during the period of c. 15,000-11,000 yr BP is indicated by ^{14}C-dating of the top calcic horizon in the northern Negev (Magaritz, 1986) and in the coastal plain of Israel (Magaritz et al., 1981), and by an archeologically-dated calcic horizon in northern Sinai (Goldberg, 1977). Formation of calcic horizons is generally interpreted as occurring under

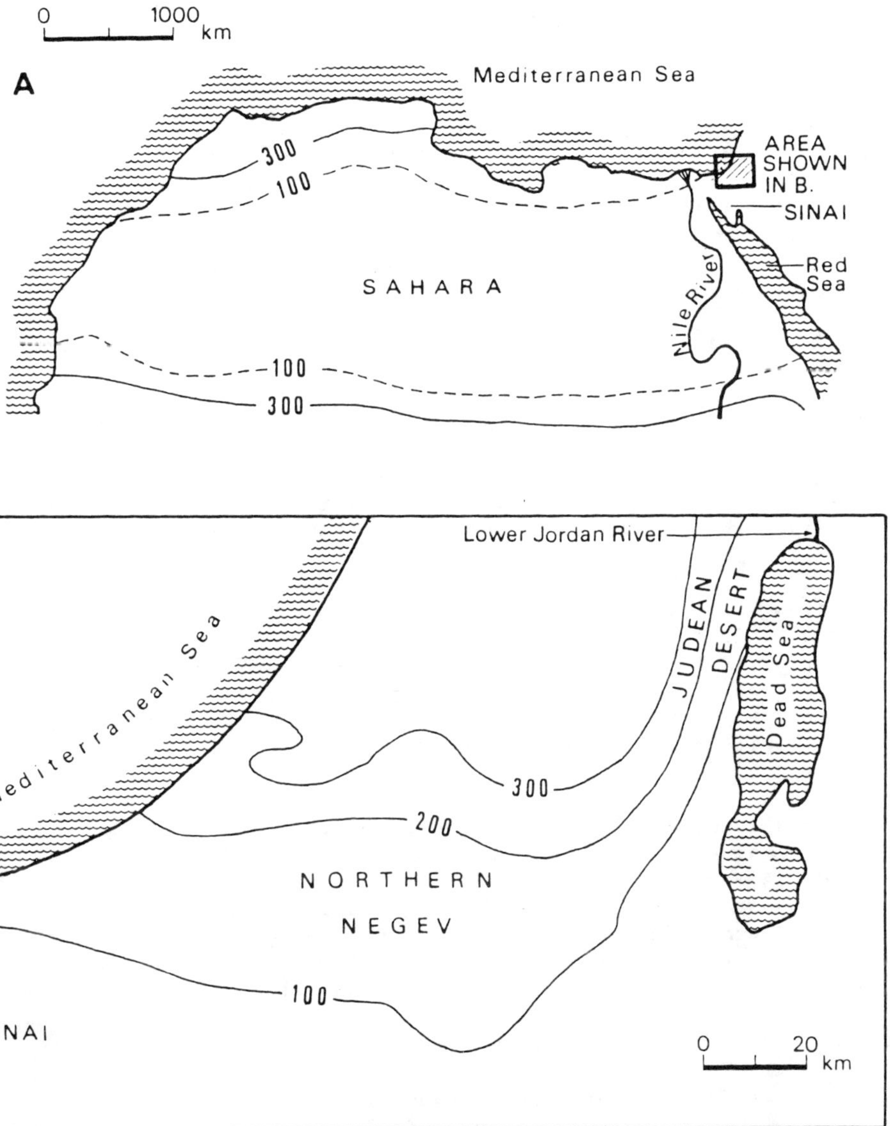

Figure 1. A. North Africa and adjoining area of the Levant, showing present-day 300mm and 100mm mean annual rainfall isohyets. B. Southeastern Mediterranean area (southern Levant) showing areas mentioned in text and mean annual rainfall isohyets.

relatively higher (but still semi-arid) rainfall conditions. Under these circumstances, rainfall is sufficient to leach carbonate from the upper layer of sediments, but not sufficient to wash it out of the sediments altogether, so that it reprecipitates at a lower depth in the sediments (Dan and Yaalon, 1982). While ^{14}C ages of soil carbonates may have significant anomalies (see above), supporting evidence regarding the age of the top soil in the coastal plain of Israel was obtained from ^{14}C analyses of organic matter in the soil (Scharpenseel, 1971) and from finds of tools of Epipaleolithic age (c. 12,000 yr BP) (Bar-Yosef, 1975). Gleying of sediments dated to 14,000 yr BP in northeastern Sinai indicates a considerable rise of the water table, which probably dates to this early deglaciation period (Goldberg, 1984).

In northern Sinai a fresh water lake occupied a large region at Gebel Maghara (Goldberg, 1977). An increase in number of human occupation sites in the present-day desert of the Negev and Sinai was documented by Goldberg and Bar-Yosef (1982) and suggests relatively moist conditions. The similarities of the tool industries in the Negev and North Africa led Goldberg and Bar-Yosef (1982) to suggest amelioration of the climate during the Epipaleolithic; the arid zone separating these areas may have been sufficiently moist at this time to permit increased contact between the cultures of these areas.

Carbon isotope analysis of carbonate nodules of the calcic horizon formed at this time indicates that a strong north-south moisture gradient existed in the northern Negev. Values of ^{13}C range from -8 to -9°/oo at the present northern boundary of the desert, increasing to 0°/oo further south (Magaritz and Goodfriend, unpublished data). Quantitative paleoclimatic interpretation of nodule ^{13}C values is not yet possible.

Lake Lisan, the Pleistocene terminal lake of the Jordan River system (covering what is now the Dead Sea and the Lower Jordan Valley) reached its maximum level at this period (Begin et al., 1985). This implies pluvial conditions in the region (Copeland and Vita-Finzi, 1978). Pollen analyses indicate a high percentage of arboreal pollen north of the Dead Sea (Bar-Yosef et al., 1974). Abundant arboreal pollen is also seen in the Central Negev at this time (Horowitz, 1979). However, it is not known to what degree this pattern may be due to low temperatures rather than to high rainfall.

In North Africa north of the Sahara, this early deglaciation period was generally one of progressively drier conditions, whereas wet conditions prevailed south of the Sahara (Rognon, 1976). However, evidence for humid phases, perhaps of short duration, also exists for the north. For example, in the early part of this period (17,000-13,000 yr BP), intense wadi activity occurred in southern Egypt (Butzer and Hansen, 1968, as cited by Nicholson and Flohn, 1980), which suggests wet conditions (Nicholson and Flohn, 1980). And at c. 12,000 yr BP, the Nile reached a high level (Adamson et al., 1980) and a calcic horizon formed in Tunisia (Scharpenseel et al., 1980). Calcic horizons in the Nile area also date to within ±c. 1000 yr of this time (Adamson et al., 1980). Given the uncertainties of ^{14}C dates on calcic horizons (see above), the North African and Middle Eastern calcic horizons all fall within a rather narrow time range. Thus they may all represent a single wet period that occurred throughout the region.

So while the semi-arid regions of the Levant experienced rather wet conditions at this time, North Africa north of the Sahara experienced generally dry conditions with wet episodes. These dry conditions apparently resulted from periodic northward movement of the desert high pressure cell with the general northward shift of the climatic zones during deglaciation (Rognon, 1976). Continuing wet conditions in the Middle East indicate that these northward incursions did not reach into this area at this time.

ARID INTERVAL (C. 11,000-10,000 YR BP)

Although the exact dates and the length of the arid interval are not known, available data suggest that it began 11,000 to 10,500 yr BP and lasted less than 1000 yr. Evidence for this period is meager, since it represents a time of erosion and non-deposition of sediments. No ^{14}C dates from this period have been obtained from sediments in northern Negev (Goodfriend, unpublished data).

Dune migration characterized this arid phase. The time of accumulation of dunes is usually hard to determine precisely, since they only rarely contain datable materials. Consequently, the minimum age of dunes is often obtained from the sediments they cover. However, in the northern Negev, dune migration blocked the drainage of Nahal Sekher and consequently created at this locality a water body in which evaporite minerals were formed (Enzel, 1984). The ^{14}C age of the carbonate sediments (11,680-10,300 yr BP) thus indicates the time of dune movement (Magaritz, 1986). In this case, formation of a lake represented an arid period. The lake sediments overlie the paleosol discussed earlier.

Land snail shells from archeological sites provide evidence of migration of the desert boundary northward (Magaritz and Heller, 1980). Arid conditions at c. 11,000-10,500 yr BP (archaeologically dated - Late Natufian) in a now-moist area of Israel were indicated by the enrichment of the shell carbonate in ^{18}O and by the smaller size of the shells. Analysis of pollen from a Late Natufian site in the semi-arid region of Syria also indicates relatively dry conditions from 10,500-10,200 yr BP, with a slight increase in moisture between 10,200-10,000 yr BP (Leroi-Gourhan, 1981).

It was apparently at this time that Lake Lisan reached its lowest level. Neev and Emery (1967) have interpreted the existence of a drainage system of wadis and fan deltas deep below the present Dead Sea level as indicating a drop of lake level to -740m from the high stand at -200m. By c. 9800 yr BP (^{14}C date on organic material in lake sediments), sea level had risen to over -400m (approximately its present level).

In contrast, the extensive settlement of the Central Negev by the Harifian culture during this time (well-dated at 10,600-10,200 yr BP) (Rosen, in press; Goring-Morris, 1985) is difficult to reconcile with extreme dry conditions. A gap in the record of human settlement in this area occurs just *after* this period (to c. 9500 yr BP). It is conceivable that while conditions further north were quite dry, a weaker rainfall gradient existed during this period, such that the now-arid Central Negev (rainfall <100mm) was not then especially dry. Possibly springs, originating in the preceding wet period, may have supported human populations and their prey at this time.

North Africa also experienced dry conditions at this time, apparently as a result of a northerly position of the desert high pressure cell (Rognon, 1976).

This dry period reflects a change in atmospheric circulation over the Mediterranean Sea, which is the main source of moisture for the Near East (Gat and Magaritz, 1980). At present, air masses bringing rain to the region originate as cold air masses over Europe and move to the southeast, picking up moisture from the warm Mediterranean Sea (Rindsberger et al., 1983). The dry period should reflect conditions under which air masses did not pass over the Mediterranean. Such a situation may have occurred if a low pressure trough stabilized over the central Mediterranean and did not move eastward over the Middle East. The Middle East would then have received only dry air masses from the northeast.

THE EARLY HOLOCENE

A transition to wet conditions occurred in the Levant with the beginning of the Holocene. Around 9,000-8,000 yr BP human density in the Levant was particularly high and reflected an economy based largely on farming (Moore, 1979). By 9500 yr BP, the Central Negev was again inhabited (Goring-Morris, personal communication). Vertebrate faunal evidence from north of the Dead Sea (Tchernov, 1980) and from southern Sinai (Tchernov and Bar-Yosef, 1982) indicates the presence of freshwater bodies in these now-arid areas at c. 10,000 and 8,500 yr BP respectively. At archeological sites north of the Dead Sea, scars of endolithic lichens on rocks buried c. 9500 yr BP indicate the occurrence of relatively moist conditions, probably dating to earliest Holocene (Danin, 1985). ^{13}C analysis of organic matter in early Holocene land snails from the Northern Negev indicates that the northern limit of xeric C_4 semi-shrubs (mostly Chenopodeaceae and enriched in ^{13}C relative to C_3 plants) was more than 30km south of its present position at the 300mm isohyet (Goodfriend, unpublished data). This implies an approximate doubling of rainfall relative to the present. Wet conditions are also indicated by the very small size of the shells of the land snail *Trochoidea seetzeni* (20-25% smaller diameter than modern ones) from this period (Goodfriend, 1986); modern *T. seetzeni* show a geographic gradient in size which is negatively correlated with mean annual rainfall (the opposite of the pattern seen in *Levantina*, above). In Syria, pollen analysis indicates a significant increase in moisture, reflected by abundant arboreal pollen, from 9,800-8,800 yr BP (Leroi-Gourhan, 1981).

In North Africa north of the Sahara, moist conditions also prevailed in the east at this time (Wendorf et al., 1977; Butzer and

Hansen, 1968; Adamson et al., 1980). A sapropel layer in eastern Mediterranean sediments dates to this period and indicates a large influx of fresh water, perhaps from the Nile (Vergnaud-Grazzini, 1985). However, to the west continued existence of dry conditions is documented at a large number of sites (Rognon, 1976). Nicholson and Flohn (1980) have hypothesized that the Saharan depression intensified at this time, displacing the high pressure cell centered over northeast Africa to a more easterly position, whereas the high pressure cell over northwest Africa remained in place. This would have brought wetter conditions to the Middle East and Northeast Africa.

DISCUSSION AND CONCLUSIONS

Although sufficient data are available to indicate long-term trends (on the order of 1000yr) in the region (Table 1), information on shorter-term fluctuations is largely lacking. Land snail shell size is probably responsive to rainfall variation on a relatively short time scale, perhaps on the order of 10yr or less, but an objective estimate of response time is not available. For the early Holocene, the uniformity of the size of *Trochoidea seetzeni* (see above) is suggestive of a stable rainfall regime.

Perhaps the greatest problem in studying paleoclimate in the desert margin region is that many of the paleoclimatic indicators traditionally used reflect climate over too broad an area. For example, the Dead Sea, a desert lake receiving only 50-100mm direct rainfall per year, receives runoff from as far north as southern Lebanon via the Jordan River. The flow regime of large wadis, and consequently the nature of the sediments they deposit at a given point, may be affected by rainfall conditions anywhere upstream. In the desert margin region, wadis may pass through a wide range of rainfall regimes (e.g. Nahal Besor, 100-300mm mean annual rainfall). Pollen may blow in from hundreds of kilometers away (e.g. Van Campo, 1975); in the area of the Judean Desert, a gradient from 100 to 600mm rainfall occurs over less than 20km. These geographically-broad paleoclimatic indicators may, at a given site, be biased toward wetter or drier areas or may indicate a regional average, rather than local conditions. For these reasons, paleoclimatic indicators which reflect local conditions are to be preferred in studies of areas of strong climatic gradients. These include such materials as calcic horizons, remains of animals of limited mobility (e.g. land snails or small mammals), tree rings, and plant remains in rodent middens.

A final problem which must be emphasized is the paucity of quantitative paleoclimatic information for the Levant. It is not clear how wet the wet period of 15,000-11,000 yr BP was as compared to the early Holocene wet period in the semi-arid region of the Levant. The formation of a calcic horizon during the former period but not during the latter would suggest that the former was significantly wetter. However, the abundant supply of loess then available (due to dry conditions in the loess source region, the Sahara) may have been a factor in calcic horizon development. Other evidence permitting a comparison of these periods is presently unavailable.

TABLE 1. Moisture conditions in the present-day semi-arid zone of the Levant, from 15,000-8,000 yr BP. See text for discussion and references.

Period	Climate	Evidence
15,000-11,000 yr BP	wet	soil calcic horizons in Negev high Lake Lisan level lake in n. Sinai high arboreal pollen, Jordan Valley and Central Negev
c. 11,000-10,000 yr BP	dry	dune migration, N. Negev high ^{18}O and small size of land snail *Levantina*, n. Israel abundant chenopode and low arboreal pollen, Syria low Lake Lisan level
c. 10,000-8,000 yr BP	wet	increased human habitation vertebrate faunas, Lower Jordan Valley and s. Sinai endolithic lichen scars on rocks, Lower Jordan Valley pure C_3 plant communitiesin N. Negev small size of land snail *T. seetzeni* high arboreal pollen, Syria

REFERENCES

Adamson, D.A., Gasse, F., Street, F.A. and Williams, M.A.J., 1980. Late Quaternary history of the Nile, *Nature*, 287: 50-55.

Bar-Yosef, O., 1975. The Epi-paleolithic in Palestine and Sinai. In: F. Wendorf and A.E. Marks (eds.), *Problems in Prehistory: Northwest Africa and the Levant*, SMU Press, Dallas, Texas, p. 363-378.

Bar-Yosef, O., Goldberg, P. and Leveson, T., 1974. Late Quaternary stratigraphy and prehistory in Wadi Fazael, Jordan Valley: a preliminary report, *Paleorient*, 2: 415-428.

Begin, Z.B., Broecker, W., Buchbinder, B., Druckman, Y., Kaufman, A., Magaritz, M. and Neev, D., 1985. Dead Sea and Lake Lisan levels in the last 30,000 years: a preliminary report, *Israel Geological Survey Report*, 29/85: 1-18.

Berger, W.H., Killingley, J.S. and Vincent, E., 1985a. Timing of deglaciation from an oxygen isotope curve for Atlantic deep-sea sediments, *Nature*, 314: 156-158.

Berger, W.H., Killingley, J.S., Metzler, C.V. and Vincent, E., 1985b. Two-step deglaciation: ^{14}C-dated high-resolution $\delta^{18}O$ records from the tropical Atlantic Ocean, *Quaternary Research*, 23: 258-271.

Berner, W. Oeschger, H. and Stauffer, B., 1980. Information on the CO_2 cycles from ice core studies, *Radiocarbon*, 22: 227-235.

Butzer, K.W. and Hansen, C.L., 1968. *Desert and River in Nubia: Geomorphology and Prehistoric Environments at the Aswan Resevoir*, U. of Wisconsin Press, Madison.

Copeland, L. and Vita-Finzi, C., 1978. Archaeological dating of geological deposits in Jordan, *Levant*, 10: 10-25.

Dan, J. and Yaalon, D.H., 1982. Automorphic saline soils in Israel. In: D.H. Yaalon (ed.), *Aridic Soils and Geomorphic Processes, Catena*, Suppl., 1: 103-115.

Danin, A., 1985. Palaeoclimates in Israel: Evidence from weathering patterns of stones in and near archaeological sites, *Bull. Amer. Schools Oriental Res.*, 259: 33-43.

Delmas, R.J., Ascencio, J.-M. and Legrand, M., 1980. Polar ice evidence that atmospheric CO_2 20,000 yr BP was 50% of present, *Nature*, 284: 155-157.

Duplessy, J.C., Delibrias, G., Turon, J.L., Pujol, C. and Duprat, J., 1981. Deglacial warming of the northeastern Atlantic Ocean: correlation with the paleoclimatic evolution of the European continent, *Palaeogeogr. Palaeoclimat. Palaeoecol.*, 35: 121-144.

Enzel, Y., 1984. The geomorphology of the lower Secer Wadi, M.Sc. Thesis, Institute of Earth Sci., Hebrew Univ., Jerusalem (in Hebrew).

Gat, J.R. and Magaritz, M., 1980. Climatic variations in the Eastern Mediterranean Sea area, *Naturwiss.*, 67: 80-87.

Goldberg, P., 1977. Late Quaternary stratigraphy of Gebel Maghara, *Qedem*, 7: 11-31.

Goldberg, P., 1984. Late Quaternary history of Qadesh Barnea, Northeastern Sinai, *Z. Geomorph.*, n.f., 28: 193-217.

Goldberg, P. and Bar-Yosef, O., 1982. Environmental and archaeological evidence for climatic changes in the southern Levant, *British Archaeol. Rep. Intl. Ser.*, 133: 399-414.

Goodfriend, G.A., 1986. Holocene shifts in rainfall in the Negev Desert, as inferred from isotopic, morphological, and faunal analysis of fossil land snails. In: W.H. Berger and L.D. Labeyrie (eds.) *Book of Abstracts from the Conference on Abrupt Climatic Change*, Scripps Inst. Oceanogr. Ref. Ser., 86-8, p. 122-124.

Goodfriend, G.A. and Stipp, J.J., 1983. Limestone and the problem of radiocarbon dating of land-snail shell carbonate, *Geology*, 11: 575-577.

Goring-Morris, A.N., 1985. Terminal Pleistocene hunter-gatherers in the Negev and Sinai, Ph.D. Dissertation, Inst. of Archeology, Hebrew Univ.

Horowitz, A., 1979. *The Quaternary of Israel*, Academic Press, N.Y.

Keir, R.S., 1983. Reduction of thermohaline circulation during deglaciation: the effect on atmospheric radiocarbon and CO_2, *Earth Plan. Sci. Lett.*, 64: 445-456.

Klein, J., Lerman, J.C., Damon, P.E. and Ralph, E.K., 1982. Calibration of radiocarbon dates: tables based on the consensus data of the Workshop on Calibrating the Radiocarbon Time Scale, *Radiocarbon*, 24: 103-150.

Leroi-Gourhan, A., 1981. Diagrammes polliniques de sites archeologiques au Moyen-Orient. In: W. Frey and H.-P. Uerpmann (eds.), *Beiträge zur Umweltgeschichte des Vorderen Orients. Beihefte zum Tübinger Atlas des Vorderen Orients*, reihe A (Naturwiss.), 8: 121-133.
Magaritz, M., 1986. Environmental changes recorded in the Upper Pleistocene along the desert boundary, southern Israel, *Palaeogeogr. Palaeoclimat. Palaeoecol.*, 53: 213-229.
Magaritz, M. and Heller, J., 1980. A desert migration indicator - oxygen isotopic composition of land snail shells, *Palaeogeogr. Palaeoclimat. Palaeoecol.*, 32: 153-162.
Magaritz, M., Kaufman, A. and Yaalon, D., 1981. Calcium carbon nodules in soils: $^{18}O/^{16}O$ and $^{13}C/^{12}C$ ratios and ^{14}C contents, *Geoderma*, 25: 157-172.
Moore, A.M.T., 1979. A pre-Neolithic farmers' village on the Euphrates, *Scient. Amer.*, 241 (2): 62-70.
Neev, D. and Emery, K.O., 1967. The Dead Sea, depositional processes and environments of evaporites, *Bull. Geol. Surv. Israel*, 41: 147pp.
Nicholson, S.E. and Flohn, H., 1980. African environmental and climatic changes and the general atmospheric circulation in Late Pleistocene and Holocene, *Climatic Change*, 2: 313-348.
Olsson, I.U., 1974. Some problems in connection with the evaluation of ^{14}C dates, *Geol. Fören. Stockholm Förhand.*, 96: 311-320.
Rindsberger, M., Magaritz, M., Carmi, I. and Gilad, D., 1983. The relationship between air mass trajectories and the water isotope composition of rain in the Mediterranean Sea area, *Geophys. Res. Lett.*, 10: 43-46.
Rognon, P., 1976. Essai d'interpretation des variations climatiques au Sahara depuis 40.000 ans, *Rev. Geogr. Phy. Geol. Dyn.*, 18: 251-282.
Rosen, S.A., in press. Demographic trends in the Negev Highlands: preliminary results from the emergency survey, *Bull. Amer. Sch. Oriental Res.*, 265.
Ruddiman, W.F. and McIntyre, A., 1981. The mode and mechanism of the last deglaciation: oceanic evidence, *Quaternary Res.*, 16: 125-134.
Scharpenseel, H.W., 1971. Radiocarbon dating of soils - problems, troubles, hopes. In: D.H. Yaalon (ed.), *Paleopedology: Origin, Nature and Dating of Paleosols*, Intl. Soc. Soil Sci. and Israel Universities Press, Jerusalem, p. 77-88.
Scharpenseel, H.W., Zakosek, H., Neue, U. and Schiffmann, H., 1980. Search for pedogenic phases during the Younger Pleistocene and Holocene (Soltanien and Rharbien) of Tunisia, *Radiocarbon*, 22: 879-884.
Stuiver, M., 1982. A high precision calibration of the AD radiocarbon time scale, *Radiocarbon*, 24: 1-26.
Suess, H.E., 1978. La Jolla measurements of radiocarbon in tree-ring dated wood, *Radiocarbon*, 20: 1-18.
Tchernov, E., 1980. The faunal remains from the Gilgal site. In: T. Noy, J. Schuldenrein and E. Tchernov, Gilgal, a Pre-pottery Neolithic A site in the lower Jordan Valley, *Israel Exploration Journal*, 30: 63-82.
Tchernov, E. and Bar-Yosef, O., 1982. Animal exploitation in the Pre-pottery Neolithic B period at Wadi Tbeik, southern Sinai, *Paleorient*, 8: 17-37.

Van Campo, M., 1975. Pollen analyses in the Sahara. In: F. Wendorf and A.E. Marks (eds.), *Problems in Prehistory: North Africa and the Levant*, SMU Press, Dallas, p. 45-64.

Vergnaud-Grazzini, D., 1985. Mediterranean Late Cenozoic stable isotope record: Stratigraphic and paleoclimatic implications. In: D.J. Stanley and F.-C. Wezel (eds.), *Geological Evolution of the Mediterranean Basin*, Springer-Verlag, N.Y., p. 413-451.

Wendorf, F. et al., 1977. Late Pleistocene and Recent climatic changes in the Egyptian Sahara, *Geogr. J.*, 143: 211-234.

Williams, G.E. and Polach, H.A., 1971. Radiocarbon dating of arid zone calcareous paleosols, *Bull. Geol. Soc. Amer.*, 82: 3069-3086.

YOUNGER DRYAS IN NORTH AMERICA - MODELING, DATA ANALYSIS, AND RE-EVALUATION

Dorothy M. Peteet

NASA/Goddard Space Flight Center
Institute for Space Studies
2880 Broadway
NY, NY 10025

ABSTRACT. Questions concerning the sensitivity and mechanism of Quaternary climate change are explored using the Younger Dryas abrupt climatic event. Global climate model (GCM) experiments and re-examination of the fossil record provide a means of assessing the relative magnitude and distribution of this type of climatic fluctuation, and offer opportunities for ascertaining its origin.

INTRODUCTION

The Younger Dryas climatic deterioration of northwestern Europe approximately 11-10,000 yr B.P. provides a well-documented example of the extreme and rapid variability within the climate system. Because it is embedded in a long-term warming trend, persistent questions remain concerning the origin of the Younger Dryas cooling, as increased Northern Hemisphere insolation followed the last glacial maximum (Berger, 1978). Several characteristics of this climatic reversal make it particularly useful for studies of climate sensitivity.

1) The wide variety of climatic indicators (glacial, geomorphological, varve, faunal, floral, isotopic) which recorded its impact are valuable for inter-comparisons of the timing and magnitude of the change.
2) The 11-10,000 yr B.P. interval, within the range of radiocarbon dating, is quite accessible in organic deposits throughout the world.
3) Studies of North Atlantic sediments have linked the Younger Dryas with movement of the North Atlantic polar front (Ruddiman and McIntyre, 1981), and several causal hypotheses for this movement have been proposed (cf. Mercer, 1969; Ruddiman and McIntyre, 1981). The possibility of its linkage to CO_2 changes (Oeschger et al., 1984) has relevance to earlier (possibly similar) intervals of rapid CO_2 change, and suggests that the ocean-atmosphere system may have more than one stable mode of operation involving North Atlantic deep water change (Broecker et al., 1985).

To further explore the Younger Dryas interval, questions concerning the extra-European occurrence, magnitude, timing, and origin of the

W. H. Berger and L. D. Labeyrie (eds.), Abrupt Climatic Change, 185–193.

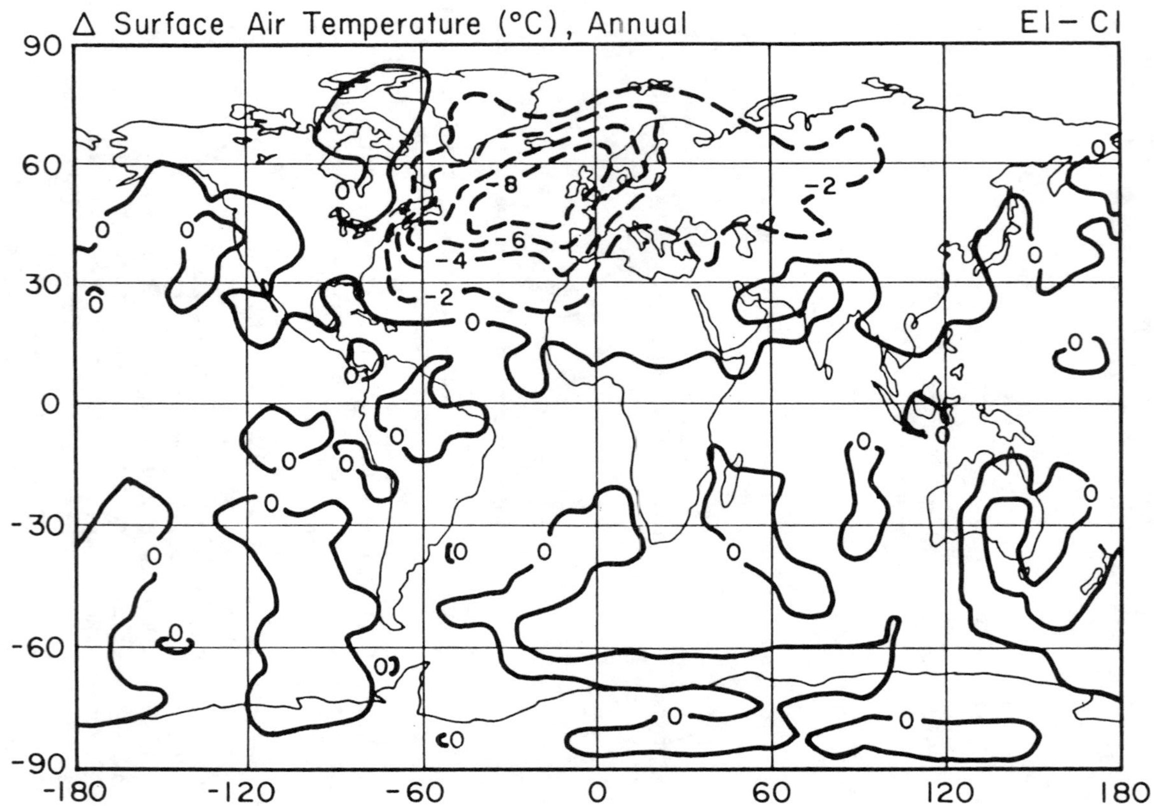

Figure 1. Changes in annual surface air temperature produced by the GMC when ice-age North Atlantic sea surface temperatures (E1) replaced modern values. (C1 - control; E1 - experiment 1)

event must be pursued. One approach to these questions involves climate modeling; the other concerns further analysis of available palynologic data.

CLIMATE MODELING

An examination of the impact of simulated late-glacial changes in ocean temperature, orbital parameters, and ice extent on global climate was performed by Lamont and Goddard Institute scientists using the Goddard Institute (GISS) GCM. The results of these experiments are compared to pollen records of rapid vegetational change 13,000 to 10,000 yr B.P. on both sides of the North Atlantic (Rind et al., 1986).

The first of these experiments was designed to ascertain the effects of a colder North Atlantic surface ocean temperature on the present global climate. Thus boundary conditions similar to those today were used along with sea surface temperature (SST) depression equal to ice age values in the North Atlantic. This experiment thus isolates the effect of colder North Atlantic temperatures during modern conditions. The resulting changes in surface air temperature are given in Fig. 1. The simulated temperature depression is similar to paleoclimatic data on both sides of the Atlantic, in that cooling effects were greater in coastal areas of Europe and diminished towards the continent (Watts, 1980). In North America oceanic effects were felt near the coast, and primarily in the Canadian maritime provinces.

In a second set of experiments, orbital values and ice conditions for 11,000 yr B.P. were included in the simulated Alleröd-Younger Dryas reconstructions, and the effect of the same North Atlantic SST depression was observed. Essentially, the annual surface air temperatures are similar to the first set of experiments. However, when the seasonal climates resulting from these changes in SST, land ice, and orbital parameters are compared to the present climate (Fig. 2), summer temperature increases of 4-10°C are recorded at continental sites (i.e. midwestern United States and western USSR), indicating that seasonal extremes were much greater during the late-glacial.

This increase in seasonality may account for anomalous (no-analog) assemblages of boreal and temperate species which characterize many late-glacial sites both in Europe and the United States. This enhanced seasonality also suggests that a re-examination of pollen records may be important in the reconstruction of vegetation and climatic interpretations.

PALYNOLOGIC DATA ANALYSIS

The extent of the Younger Dryas cooling outside of Europe is not well defined. In North America, the majority of late-glacial pollen studies do not reveal palynologic changes which can be interpreted as indicative of a climatic reversal correlative with the European sequence (cf. Davis, 1983; Watts, 1983). Sedimentary and pollen research from the maritime provinces of Newfoundland, Nova Scotia, and New Brunswick,

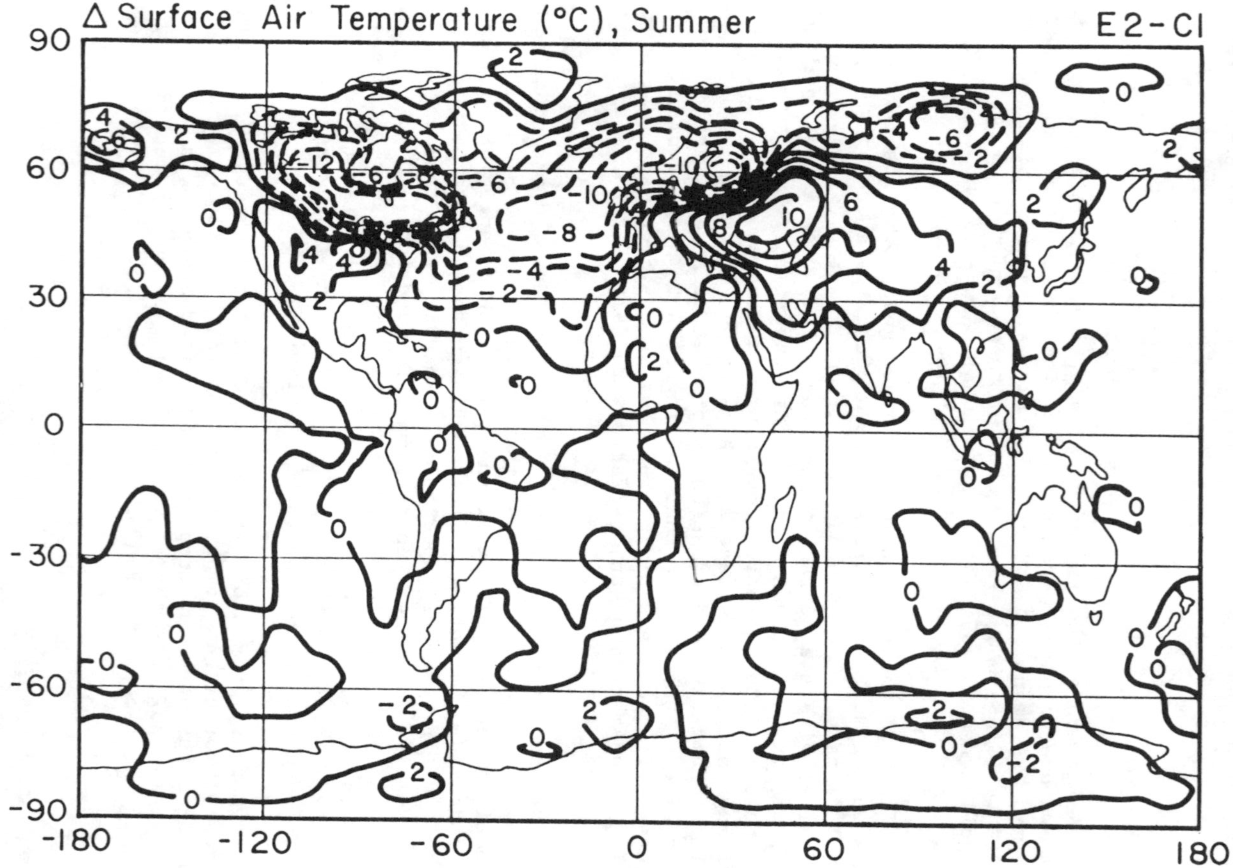

Figure 2. Changes in summer surface air temperature produced by the GCM when 11,000 yr B.P. orbital values and ice extent along with ice-age North Atlantic sea surface temperatures (E2) were compared with modern conditions (C1). (C1 - control; E2 - experiment 2)

however, suggests that the Younger Dryas cooling may have been expressed at high coastal latitudes of the N. American continent (Anderson and Macpherson, 1984; Mott et al., 1984; Mott, 1985). Questions thus arise concerning the southernmost expression of this cooling along the western North Atlantic coast.

A re-examination of northeastern U.S. late-glacial pollen sites is needed in order to determine whether this region may have actually experienced the Younger Dryas cooling during an interval of enhanced seasonality. In southern New England, late-glacial pollen stratigraphy reveals percentage increases of oak and ash (see Fig. 3) which are followed by sudden decreases of these temperate deciduous trees concurrent with increases in fir, spruce, birch, or alder. Subsequently, the temperate hardwoods again increase. This reversal in the pollen stratigraphy merits re-examination because a climatic interpretation of the reversal is controversial and has implications for the regional distribution of the Younger Dryas cooling, as well as for mechanisms involved in the climatic change. Although early interpretations of the late-glacial interval suggested that a climatic reversal took place (Leopold, 1956; Deevey, 1958; Ogden, 1959), more recent consideration of pollen accumulation rates by Davis (1967) resulted in the interpretation of the hardwood decline as the result of a dilution of allochthonous pollen by increasing numbers of autochthonous pollen. Thus the presently accepted interpretation of northeastern U.S. late-glacial pollen stratigraphy includes no climatic reversal (Bernabo and Webb, 1977; Webb, 1981; Davis, 1983; Watts, 1983, Davis and Jacobson, 1985) despite a reversal in the palynological record.

Review of available New England palynological sites (Fig. 4) suggests that the late-glacial interval reflects enhanced seasonality and should be critically re-examined in more detail for reasons concerning the oscillation in the pollen record. Pollen influx data from both Rogers Lake, Conn. in the south to Sandogardy Pond, N.H. in the north (Davis, 1983) show sharpest increases in spruce and fir from 11-10,000 yr B.P. Tannersville Bog, Pa. (Watts, 1979), Francis Lake, N.J. (Cotter, 1983), and Alpine Bog, N.J. (Nickmann, personal communication 1984) all indicate that this hardwood pollen decline was concurrent with conifer and birch/alder increases at approximately 11-10,000 yr B.P. If conifers had already colonized the landscape prior to 11,000 yr B.P., the question remains as to the reason for their sudden increase, if it was not climatic.

Re-evaluation of the North American North Atlantic coastal late-glacial climate, as inferred from the pollen record, should include the following considerations:

1) Because no-analog plant and animal species assemblages are characteristic of many late-glacial sites, extremes of seasonal temperature should be considered in formulating vegetational reconstructions from pollen assemblages. GCM model results offer one simulation of the magnitude of these changes.

2) The possible effects of atmospheric CO_2 fluctuations on species changes should be investigated.

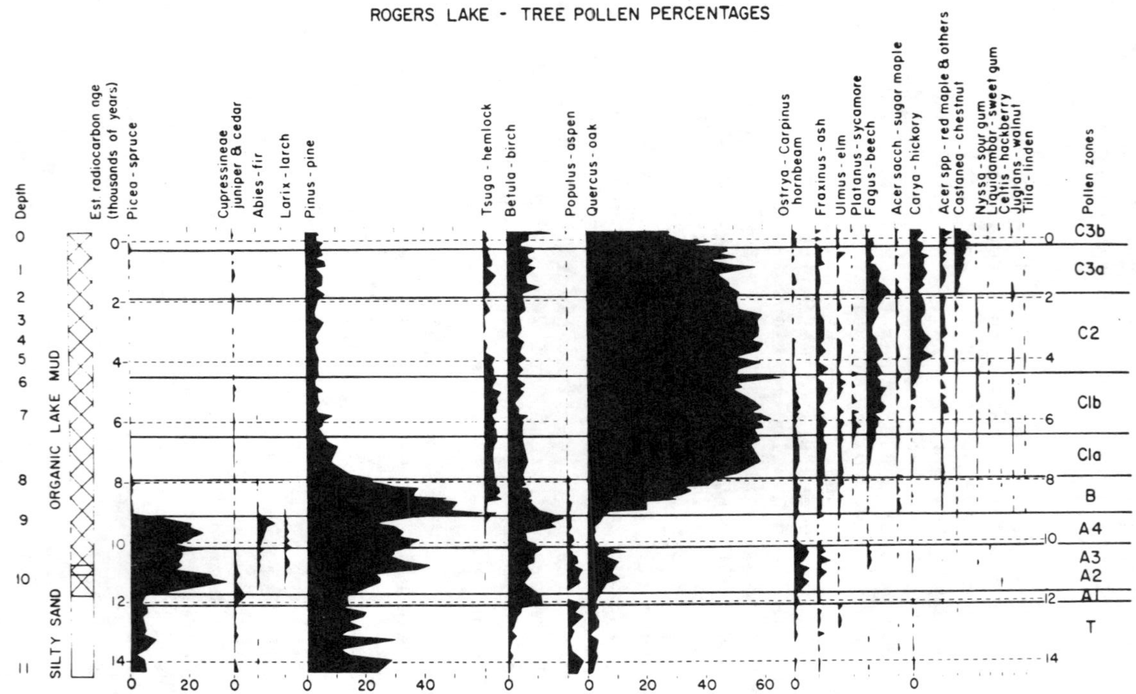

Figure 3. Tree pollen percentage diagram from Rogers Lake, Conn., calculated as percent total pollen from terrestrial plants (Davis, 1969). Recalculated ^{14}C chronology (Davis, 1983) indicates that the A-4 interval in which oak percentages drop and spruce and fir rise took place approximately 11,000 - 10,000 yr B.P.

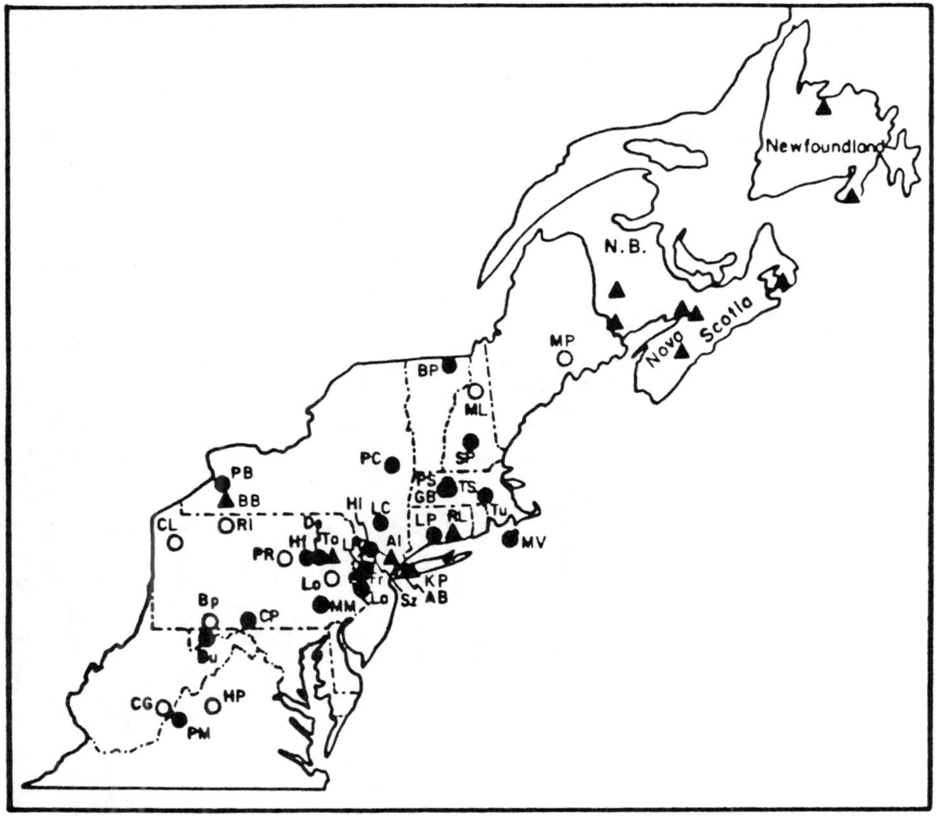

Figure 4. Location of northeastern North American pollen diagrams which include the 14,000 to 10,000 B.P. interval. Pollen assemblage hardwood-conifer oscillations ^{14}C-dated to approximately 11,000 - 10,000 yr B.P. are marked by a full triangle. Where an oscillation is apparent but insufficiently dated, sites are marked by a full circle. Sites where an oscillation is unclear are designated by an open circle.

3) Late-glacial species oscillations (see Fig. 3) are of the magnitude to be considered of climatic significance, and similar shifts in European pollen diagrams have been considered as such (c.f. Watts, 1980).

4) Detailed sampling intervals and adequate radiocarbon control for documentation of timing at present is lacking, and should be improved.

5) Pollen accumulation rates vary widely both within and between depositional basins (Pennington, 1973), and need to be considered regionally and in more detail for evidence of regional species migrations.

6) Macrofossil research is needed as an integral part of future late-glacial investigations in an attempt to firmly establish species arrival at a particular site. Accelerator dating of macrofossils should eliminate errors involved in the uncertainty of bulk sample dating (cf. Andrée et al., 1986), and aid in providing the detailed chronology necessary to identify a Younger Dryas equivalent in the U.S. if it is present.

In order to understand the phenomenon of rapid climatic change as evidenced by the Younger Dryas we must be able to reconstruct the geographic pattern and magnitude of its manifestation. The GCM study of the impact of North Atlantic cooling (Fig. 1) indicates the effects of ocean temperature change alone. To understand the applicability and limitations of the GCM in explaining the Younger Dryas effects, climate change in various parts of the world must be carefully quantified. Reevaluation of the North American east coast late-glacial climate must be accompanied by examination of terrestrial and marine data worldwide in order to assess the origin and potential regional importance of rapid climate variability.

REFERENCES

Anderson, T.W. and MacPherson, J.B., 1984. A late-glacial climatic oscillation in Newfoundland, Canada. VI Intl. Palyn. Conf., Calgary, 2.

Andrée, M., Oeschger, H., Siegenthaler, U., Riesen, T., Moell, M., Ammann, B. and Tobolski, K., 1986. ^{14}C dating of plant macrofossils in lake sediment, *Radiocarbon*, 28 (2A): 411-416.

Berger, A., 1978. Long-term variations of daily insolation and Quaternary climatic changes, *Jour. Atm. Sci.*, 35: 2362.

Bernabo, S.C. and Webb III, T., 1977. Changing patterns in the Holocene pollen record of northeastern North America, *Quat. Res.*, 8: 64-96.

Broecker, W.S., Peteet, D.M. and Rind, D., 1985. Does the ocean-atmosphere system have more than one mode of operation? *Nature*, 315: 21-26.

Cotter, J., 1983. The minimum age of the Woodfordian glaciation of northeastern Pennsylvania and northeastern New Jersey. Ph.D. Thesis, Lehigh Univ. 1983.

Davis, M.B., 1967. Late-glacial climate in northern United States: A comparison of New England and the Great Lakes region. In: *Quaternary Paleoecology*, eds. E.J. Cushing and H.E. Wright, 11-44.

Davis, M.B., 1983. Holocene vegetational history of the eastern United States. In: *Late-Quaternary Environments of the U.S.*, ed. S.E. Porter, Univ. Minn., 166-181.

Davis, R.B. and Jacobson, G.L., 1985. Late-glacial and early Holocene landscapes in northern New England and adjacent areas in Canada, *Quat. Res.*, 23: 341-368.

Deevey, E.S., 1958. Radiocarbon-dated pollen sequences in eastern North America, *Geobot. Inst. Rubel. in Zurich Veroff* 34: 30-37.

Leopold, E.B., 1956. Two late-glacial deposits in southern Connecticut, *Natl. Acad. Sci. Proceed.*, 42: 863-867.

Mercer, J., 1969. The Allerod oscillation: a European climatic anomaly? *Arctic and Alpine Res.*, 1: 227-234.

Mott, R.G., 1985. Late-glacial climatic change in the maritime provinces. In: *Climatic Change in Canada 5*, ed. C.R. Harington, 281-300.

Mott, R.G., Grant, D.R., Stea, R. and Ochietti, S., 1984. The Alleröd-Y. Dryas oscillation in North America - fact or fiction? *VI Intl. Palyn. Conf., Calgary,* 112.

Oeschger, H., Beer, J., Siegenthaler, U., Stauffer, B., Dansgaard, W. and Langway, C.C., 1984. Late-glacial climate history from ice cores. In: *Ewing Symposium Vol.* 4, eds. J.E. Hansen and T.T. Takahashi.

Ogden, J.G. III, 1959. A late-glacial pollen sequence from Martha's Vineyard, Massachusetts, *Amer. J. Sci.*, 257: 366-381.

Pennington, W., 1973. Absolute pollen frequencies in the sediments of lakes of different morphometry. In: *Quaternary Plant Ecology*, eds. H.J.B. Birks and R.G. West, 79-104.

Rind, D., Peteet, D., Broecker, W., McIntyre, A. and Ruddiman, W., 1986. The impact of cold North Atlantic sea surface temperatures on climate: implications for the Younger Dryas cooling (11-10k), *Climate Dynamics* 1: 3-33.

Ruddiman, W. and McIntyre., A., 1981. The North Atlantic Ocean during the last glaciation, *Palaeogeog., Palaeoclim., Palaeoecol.*, 35: 145-214.

Watts, W.A., 1979. Late Quaternary vegetation of central Appalachia and the New Jersey coastal plain, *Ecol. Mon.* 49: 427-469.

Watts, W.A., 1980. Regional variation in the response of vegetation of Lateglacial climatic events in Europe. In: *Studies in the Late Glacial of Northwest Europe*, eds. J.J. Lowe, J.M. Gray, and J.E. Robinson, 74-80.

Watts, W.A., 1983. Vegetational history of the Eastern United States 25,000 to 10,000 years ago. In: *Late-Quaternary Environments of the U.S.*, V. 2, ed. S.C. Porter, 294-310.

Webb, T. III, 1981. The past 11,000 years of vegetational change in eastern North America, *Bioscience*, 31: 501-508.

THE YOUNGER DRYAS IN SOUTHWESTERN EUROPE: AN ABRUPT CLIMATIC CHANGE AS EVIDENCED FROM POLLEN RECORDS

A. Pons, J.-L. de Beaulieu, J. Guiot and M. Reille

Laboratoire de Botanique historique et Palynologie,
UA CNRS n° 1152
Faculté des Sciences et Technique St-Jérôme
13397 Marseille CEDEX 13 France

ABSTRACT. In spite of some ambiguities, botanical data about the Younger Dryas in Eastern Spain and South-Eastern France highlight the abrupt and marked character of this climatic episode. A quantitative reconstruction method from pollen time-series makes it possible to document the climatic parameters of this episode (cooling of 4-5°C, decrease in precipitation of 400-500mm).

THE BOTANICAL DATA

It is generally assumed (Watts, 1980) that the transition from the last glacial to the Postglacial is marked by a brief but extreme climatic episode: the Younger Dryas. However, it seems that the botanical data available characterizing this climatic fluctuation (from the South of Spain to the North of the French Massif Central and from Oceanic regions to the maritime Alps) are not consistent with the hypothesis of a sudden and well marked climatic episode.

These apparent discrepancies are threefold:

(1) in some sites, the decrease of the sum of arboreal pollen is too slight for a supposedly marked climatic deterioration;
(2) in many cases the new rise in arboreal pollen (A.P.) frequencies takes place slowly and progressively after this episode ;
(3) lastly, the geographic variation of pollen curves during this episode leads one to question the general value of this climatic signal.

In the following these discrepancies are discussed:

First, one notes a relatively small decrease in the A.P values (Pelléautier, St Julien de Ratz, Balcère, Fig. 1); but a closer examination of the diagrams shows that the amplitude of this fall depends on the role played by *Pinus* just before that time (Fig. 2).

In regions where there used to be large forests of *Pinus* at the end of the Alleröd (Pelléautier), the climatic change is not well perceived

W. H. Berger and L. D. Labeyrie (eds.), Abrupt Climatic Change, 195–208.

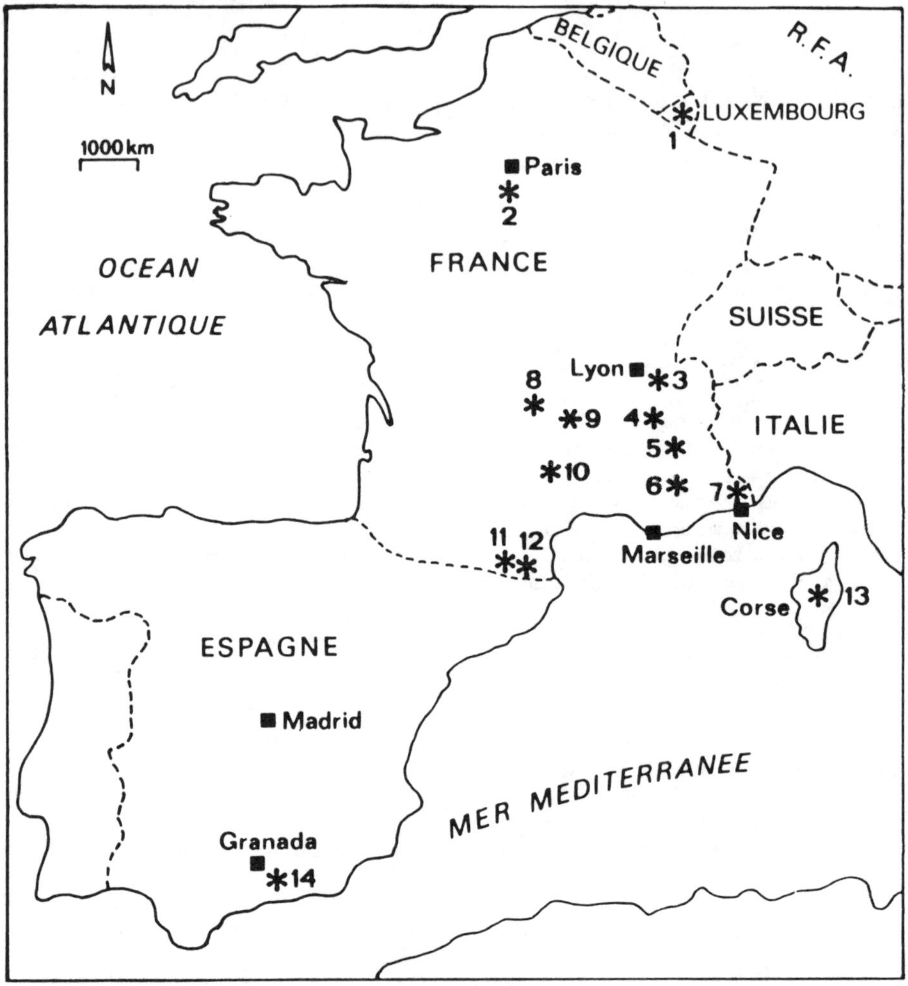

1. Echternach (Couteaux, 1970)
2. L'Archet (Jalut, 1976)
3. Hière-sur-Amby (Clerc, 1985)
4. Saint-Hilaire du Rosier (Clerc, 1985)
   Saint-Julien de Ratz (Clerc, 1985)
5. Oisans (Couteaux, 1984)
6. Pelléautier (Beaulieu and Reille, 1984)
7. Lac Long Inférieur (Beaulieu, 1977)
8. La Taphanel (Beaulieu, Pons and Reille, 1982)
9. Lac du Bouchet (Beaulieu, Pons and Reille, 1984)
10. Aubrac (Beaulieu, Reille and Pons, 1985)
11. Freychinède (Jalut et al., 1982
12. Balcère (Van Campo and Jalut, 1969)
13. Lac de Creno (Reille, 1975)
14. Padul (Pons and Reille, 1986)

Figure 1. Location of the sites.

in the diagram (fall in A.P. of only 10%), owing to the regional persistence of these forests; in regions where *Pinus* formed small local populations (Le Bouchet), these may have have been reduced leading to great changes in the diagrams (fall in A.P. of almost 50%). Lastly, in the case of a very sporadic presence of *Pinus* during the Alleröd (Brameloup : Aubrac), the pollen record of this taxon, therefore also the A.P. value, were not greatly affected (fall in A.P. of about 25%). But the expansion of *Pinus* populations just before the Younger Dryas is merely accidental (for example, among the 12 sites studied by us in the south-eastern part of the French Massif Central, only the site of Le Bouchet lake is rich in *Pinus*). The importance of the fall of *Pinus* frequencies and, as a result, of A.P. frequencies during the Younger Dryas is unequivocal and therefore climatic, whatever its amplitude.

The second apparent difficulty results from the fact that in many cases, at the end of the Younger Dryas, from the level with the first occurrences or recurrences of mesothermophilous trees that mark the inversion of the climatic trend, the A.P. slowly reaches 50%, a value which can be considered as characterizing a rather mild climate. But this slow rise comes from the fact that the mesohygrophilous herbaceous formations existing in the immediate surroundings of the site were the first to profit from the climatic improvement, and during a more or less long time they masked the development of arboreal formations: in sites with a more marked relief, herbaceous formations were rare, and, as a matter of fact, the end of the Younger Dryas is abrupt. Of course, for a valuable comparison, it is necessary to select sites with similar *Pinus* frequencies (Fig. 3).

The spatial variations of the record of the Younger Dryas constitute a third apparent difficulty. They do not contradict but rather confirm the abrupt character of the episode.

- A first variation is noted southwards: a more and more important decrease in the frequencies of arboreal pollen not belonging to *Pinus*, and their very rapid reappearance. One must admit here, as is particularly evident at Padul (Fig. 4), that the climatic deterioration has affected the pollen production of trees (a phenomenon that can be directly related to climate) as much as the size and distribution of populations (which take a longer time to change).

In Corsica, that is to say in an insular environment, the Younger Dryas is interrupted by a less arid episode (the hypothesis of a redeposition being excluded), without losing its abrupt character (Fig. 5).

Conversely, in a periglacial environment, three critical episodes have been distinguished by Coûteaux (1984): peaks of *Artemisia* related to glacial advances or peaks of *Selaginella* related to phases of stronger fluvioglacial erosion.

From the chronological point of view, our dates for these botanical data, eleven in number, place the episode between 10,700-10,300 B.P. These data, which are in good agreement with the European literature, are not consistent with the hypothesis of perturbations in the ^{14}C content during this period (Welten, 1982, p. 96). However, there arises

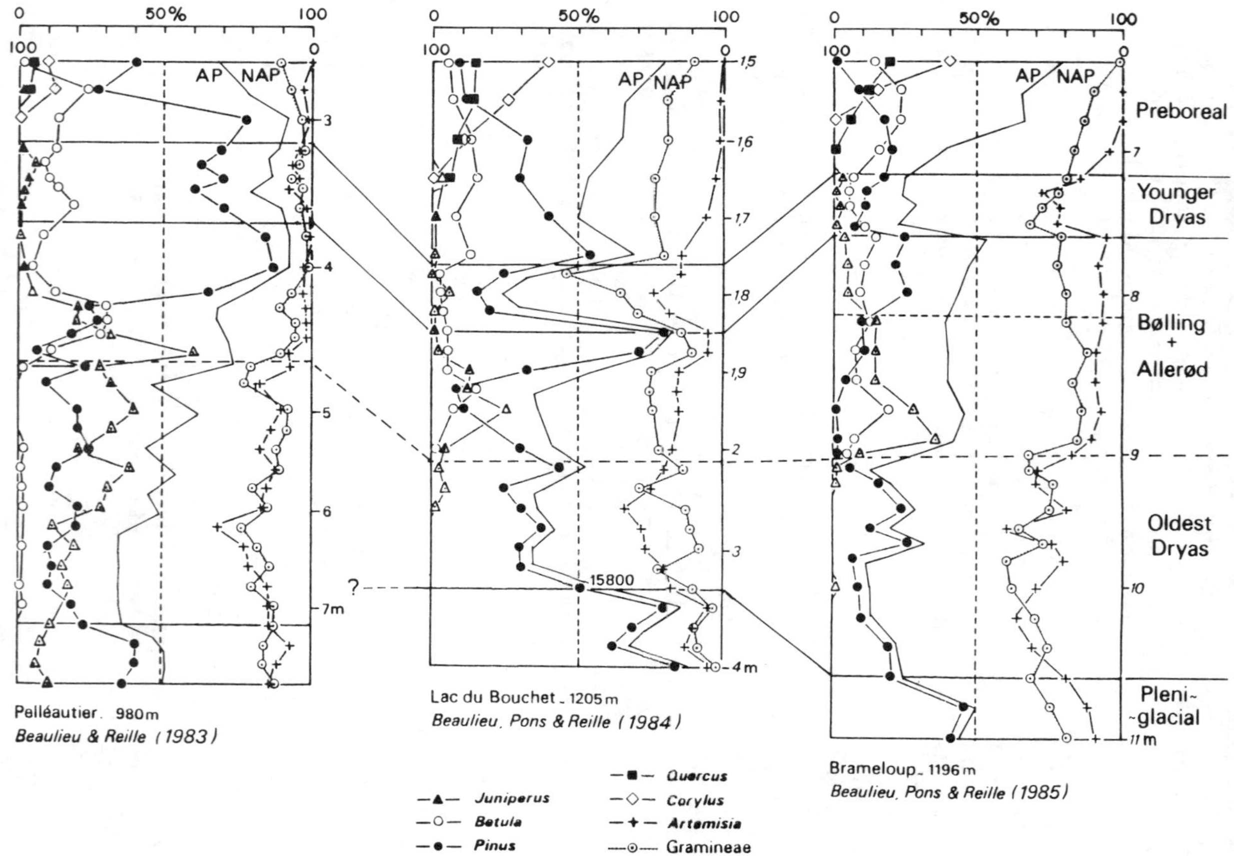

Figure 2. Variations in AP percentages during the Younger Dryas and role previously played by *Pinus*.

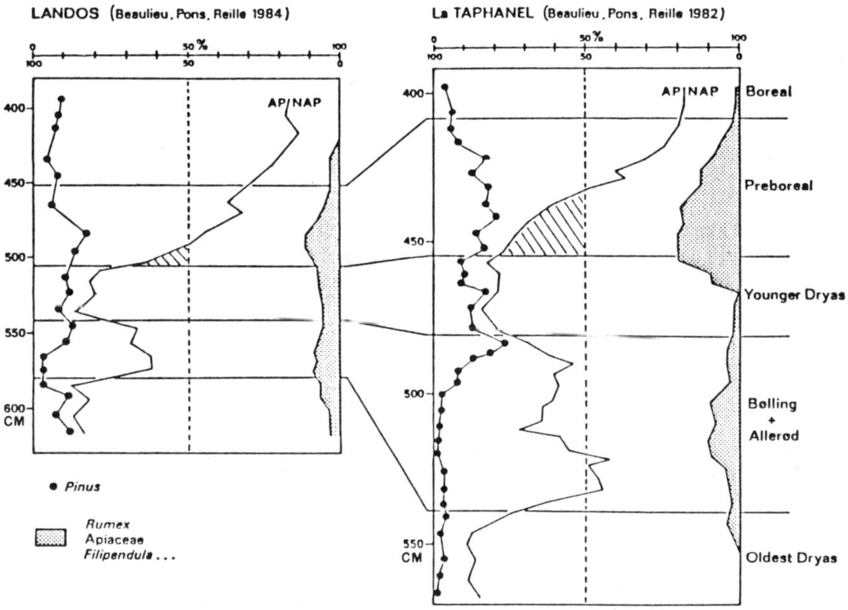

Figure 3. Major pollen curves at Landos (Velay) (Beaulieu, Pons and Reille, 1984) and at La Taphanel (Beaulieu, Pons and Reille, 1982).

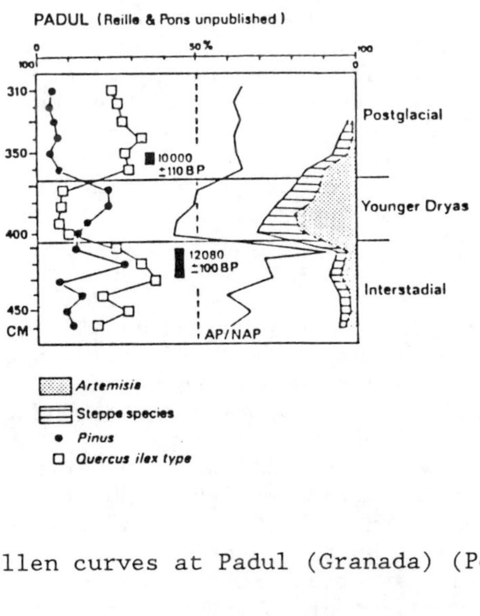

Figure 4. Major pollen curves at Padul (Granada) (Pons and Reille, 1986).

the acute problem of the characteristics of the new climate that interrupted the Lateglacial interstadium; proxiglacial data essentially suggest an increase in humidity which alone could explain the glacial advances (Coûteaux, 1984), whereas other data only point to a marked aridity, absolutely obvious in the sites.

In the detail, however, there is good agreement that the thermic factor had a major role; the eutrophic aquatic taxa that had vanished with the warming of ca 13,000 B.P. reappear and a rarefaction of the biogenic calcareous deposition is generally observed, even at low elevation. Finally pollen records lead one to admit that the Younger Dryas was quite an abrupt climatic change in southwestern Europe.

THE CLIMATIC RECONSTRUCTION METHOD

Since Imbrie and Kipp (1971), multivariate techniques have been used for evaluating deep-sea planktonic foraminifera in terms of temperature, salinity and other oceanographic variables. Webb and Bryson (1972) did similar analyses for pollen data.

The vegetation of Europe is highly complex, owing to the nature of the continent's climate and the topography. The European mountains, by contrast with those of North America, disturb the latitudinal zonation of the vegetation. In addition, the correlation of vegetation to climate is sensitive to very ancient human activities (Pons, 1984). Consequently, the calculation of transfer functions in Europe gives rise to problems that have no counterpart in North America.

The transfer function method based on modern data rests on two assumptions: 1) the relationship between climate and vegetation is constant over the time period of analysis, which implies that the ecological characteristics of the taxa used have not changed; 2) the modern observations contain all the information necessary for interpreting the fossil spectra. But this procedure is affected by three difficulties: 1) an exhaustive set of good analogues to all the fossil spectra is not always available; 2) it is difficult to evaluate the impact of local site peculiarities - such as exposure, micro-climate, human action - on the ancient vegetations and their pollen registration; 3) the vegetation, at time t, is function of climate at time t, but also of vegetation and climate at time (t-1, t-2,...).

Solutions are proposed by Howe and Webb (1983) and by Sabatier and Van Campo (1984) to solve the first and, partly, the second problem. Nevertheless, the methods used do not solve the last one: they do not take into account the character of time-process of the pollen data.

A three-stage method, based on the time-series analysis, has been proposed by Guiot and Pons (1986) slightly modified from an earlier analysis (Guiot, 1985). The first stage closely resembles the standard methods for calculating transfer functions; it is a spatial analysis or analog calibration and, after an independent verification, it involves an extrapolation to the past of an information contained in modern spatial observations. This reconstruction is called "analogue climate".

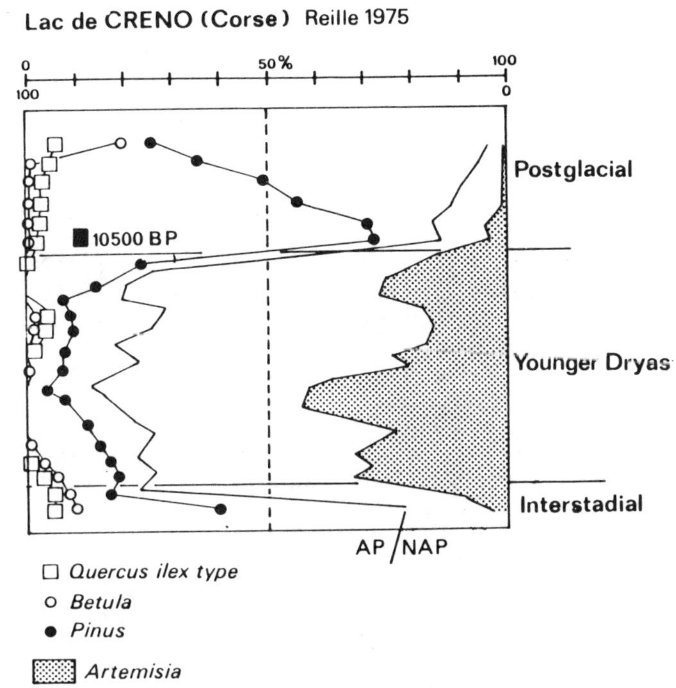

Figure 5. Major pollen curves at Lac de Creno (Reille, 1975).

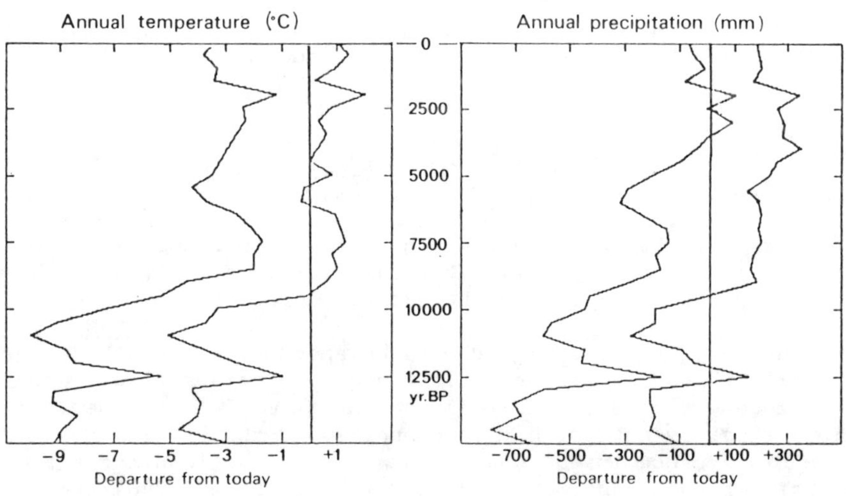

Figure 6. The 0.95-confidence interval of the reconstructions based on 9 sites (the zero reference is the corresponding present mean).

The second stage consists in finding information that is shared in the temporal variations within several pollen sequences, referred to as "palaeobioclimate". The independent reconstructions are affected by different noise sources: a Kalman filter removes these independent noises, producing a final reconstruction.

The Data

The modern data set is based on: 1) the 342 pollen spectra used by Sabatier and Van Campo (1984) reduced to 92 independent spectra; 2) the 35 spectra from southern France already used in Guiot (1985) and taken from the works of Beaulieu, Coûteaux, Reille and Laval-Triat; 3) 12 unpublished spectra sampled by Beaulieu, Coûteaux, Pons and Reille in the southern Alps; 4) 8 spectra in the eastern Pyrenées sampled by Jalut; 5) 35 spectra from USSR taken in Federova (1955) and Malgina (1973)and compiled by Jalut. All these spectra are taken from subsurface sediments. Twenty five taxa are used, a 26th variable being the varia class which represents the frequency of all the other observed taxa and a 27th variable S, the Shannon index, representing the diversity of the taxa (Guiot, 1987).

The corresponding climatological data, monthly mean air temperature and monthly total precipitation have been estimated by interpolation after taking into account an altitudinal gradient (Guiot, 1987).

The fossil data are given by the nine sites shown in Figure 1. The same 27 variables are used. Nevertheless, in the palaeobioclimatic study, it will be possible to use the *Juniperus* variable - not available in the modern data. One spectrum, at every 500 years interval, has been selected by the authors to facilitate the intersite comparison.

The Analogue Climate

This reconstruction is done using the best analogues method. The similarity between a fossil spectrum and a modern spectrum is measured by the following distance:

$$d_{ik}^2 = \sum \left(p_j \log\left(\frac{f_{ij}}{f_{kj}}\right) \right)^2 \qquad (1)$$

where f_{ij} is the relative frequency (to which 0.001 is added to remove log (0)-cases) of taxon j in the fossil record i; P_j is the loading attached to the taxon j; log is the natural logarithm.

The choice of the loadings P_j may be based either on the modern data set or the fossil data set. In the modern data set, it must reflect the goodness of fit of taxon j to the climatic data used. This selection is fairly natural but it cannot avoid the problem related to the human disturbance in the modern data. Here the loadings are chosen on the basis of the fossil data as in Guiot and Pons (1986); the loading P_j has been given by the correlation, for taxon j, between similarly dated levels of two groups of cores (Taphanel/ Pelléautier/Hières-sur-Amby and Bouchet/Lac Long/St-Julien-de-Ratz).

The distance between the fossil pollen records and each modern record is computed and a weighted mean of the climatic parameters of the selected modern analogues is used as an estimate:

$$\hat{C}_{iq} = \left(\sum_{k=1}^{a} C_{kq} / d_{ki}^{2} \right) \Big/ \left(\sum_{k=1}^{a} d_{ki}^{-2} \right) \qquad (2)$$

where $\hat{C}_{iq}$ is the estimate of the qth climatic parameter of the ith record C_{iq} and a is the number of modern analogues selected for the fossil spectrum i. The search for modern analogues of modern spectra and the corresponding estimation of the climatic parameter enables one to test the validity of the method.

The importance of a modern analogue in the estimate is a function of the inverse of the squared distance. Because the mean of the logarithm of the ratio of two frequencies tends to zero and the standard deviation to the square root of 2, the threshold for evaluating the quality of an analogue is given by:

$$d_s^2 = 2 \sum_{j=1}^{m} p_j^2 \qquad (3)$$

A dependent verification is provided by the calculation of the correlation between estimates and actual values of the climatic parameters for the modern data set. An independent verification is provided by the calculation of the same correlation but for modern data independent of the data used for analogues. The independent verification alone can check the feasibility of the climatic reconstruction from pollen. So, the first 91 records of the modern data set are used for potential analogues and the last 91 ones for the independent verification. The mean verification correlation is 0.54 for the 24 climatic parameters, which is highly significant. If the 182 modern records are used, the corresponding mean calibration correlation is 0.82.

The best analogues method has been preferred to the multiple regression used before (Guiot, 1985), because its non-linearity corresponds better to the vegetation-climate relation. Nevertheless, in the case of the summer temperature (April to September), the regression is used again. Indeed, the reference set is clearly dominated by a continental gradient from west to east, which is not at all represented by an equivalent gradient of the summer temperature. But if the region analyzed is limited, i.e. between 5°W and 10°E (to include most of the northern and high elevation locations), a preponderant altitudinal gradient is obtained, characterized by a summer temperature decrease. But in this case, the number of sites selected falls to 77, thus inducing a problematic loss of potential analogues, precisely balanced by the extrapolation possibilities of the regression.

Using the best analogues method and the multiple regression, 24 time series are produced for 9 sites, i.e. 216 series. It is not possible to present them extensively. So, the mean of these 9 reconstructions of the same parameter with a confidence interval at 95% is computed:

$$RM_t = \sum_{k=1}^{n_t} R_{tk}/n_t \qquad t = 0, 500, \ldots, 15000 \text{ yr BP}$$

$$RL_t = RM_t - t_{0.025} RS_t / \sqrt{n_t - 1} \qquad (4)$$

$$RH_t = RM_t - t_{0.025} RS_t / \sqrt{n_t - 1}$$

where R_{tk} is the reconstruction of a climatic parameter R at time t and site k; n_t is the number of sites with pollen record at time t (maximum 9); RS_t is the standard deviation corresponding to the mean RM_t and $t_{0.025}$ is the Student statistic at n_t-1 degrees of freedom and probability 0.025. The 24 time series RL_t and RH_t thus provide an envelope of the most probable climate.

The Palaeobioclimate

The palaeoecological history common to several sites close to one another, expressed by the similarity of the corresponding pollen spectra, is a picture of the regional bioclimates or eventually of the non-climatic events randomly contemporary at all the sites (these are more probably anthropogenic effects).

Canonical analysis is used to extract this common information in the case of two fossil sites (Guiot and Médus, 1985; Guiot and Pons, 1986). In order to make possible the use of the nine fossil pollen sequences, a new method, referred to as "stepwise palaeobioclimatic analysis", has been developed by Guiot (1987). It is a combination of principal component analysis and canonical analysis; the palaeobioclimate is defined by a combination of the taxa that maximizes the covariances of various pairs of fossil pollen sequences. The palaeobioclimate obtained with the two methods has a typical profile: an opposition between, on the one hand, the Oldest/Younger Dryas periods and, on the other hand, the Bölling/Alleröd/Holocene periods.

Reduction of the Noise and Final Reconstructions

Two palaeoclimatic reconstructions are now available: the analogue climate and the palaeobioclimate. The first one, absolutely quantified, is disturbed by modern noises, whereas the second, not absolutely quantified (a standardized series), is a picture of the dynamics of the climate recorded by vegetation. Because of the complementary nature of the two noise sources, the Kalman filter is very appropriate to generate a final reconstruction where the signal-to-noise ratio is increased (Guiot, 1986).

It is assumed that the analogue climate is the sum of the true climate and a white noise. It is also admitted that the palaeobioclimate has an autoregressive structure (order 1), i.e. that its state at time t depends on its state at time t-1, and that this structure reflects the structure of the true climate. These assumptions and the iterative procedure of the Kalman filter give a better estimate of the climate where the signal-to-noise ratio is increased. The deviations from the modern values of annual mean temperature and total precipitation (with the 0.95 confidence interval) are given in Figure 6.

Discussion

The analogue climate is a quantified and scaled reconstruction of the past climate. However this reconstruction is affected by noises: 1) human impact on the modern spectra disturbs the climate-pollen relationships; 2) some important taxa, such as *Juniperus*, are missing in the modern data set, so that the Bölling-Alleröd period seems too cold in Figure 6. The palaeobioclimate is a non-scaled global signal resulting from the regional climate on the record of spectra. Here the disturbances are 1) the impossibility to attribute an absolute scale to the curves; 2) the regional value of the palaeobioclimate; 3) the difficulty in separating the various climate components in the global signal.

The noise sources of the analogue climate are independent of the disturbances of the palaeobioclimate. This is why the Kalman filter improves the analogue reconstructions on the basis of the palaeobioclimate. The results are sufficiently coherent to enable us to propose a quantified scenario of the paleoclimatic evolution for the last 15 millenia. It should be pointed out that it is not possible here to discuss the time-scale because the authors selected one spectrum per 500 years (approximately). It is necessary to refer to the papers of Figure 1. The discussion proposed here is based on confidence intervals at the 0.95-level; there is a 95% chance for the temperatures and precipitations to be within those intervals. High level probability explain why these intervals are so large; one should also note that they can vary. They are function of the period but also of the number of sites available during this period. The results are validated in Guiot (1987), but in the present paper only the Late Glacial period is discussed.

It appears that these annual estimates have an error margin (at the 0.05 level) of about 1-2°C and 100-200mm (that is to say less than 20% of the total amplitude).

A sudden and marked cooling (4-5°C) as well as a fall in precipitation (400-500mm) occurred at 11,000 years BP, with a temperature and precipitation minimum that still persisted at 10,500 yr BP. At this period temperature was again similar to that of the Oldest Dryas, but precipitation was intermediate between that of the Oldest Dryas and that of the present. Nevertheless, the precipitation fall was more marked in winter than in the other seasons. It is likely that the important advance of the glaciers in the Alps was made possible as a result of a medium accumulation of snow during spring and autumn, and more importantly, weakened melting in summer.

CONCLUSIONS

The Younger Dryas appears as a relatively simple and essentially thermic fluctuation, the biological and morphological effects of which were both spectacular and diverse; thus it can be considered as one type of abrupt climatic change.

This assertion (emphasized in many other papers of this symposium!) partly rests on the fact that the climatic change that constitutes the Younger Dryas occurred as a back-current to the general warming that marked the end of the glacial times; and, as a corollary to this singularity, one notes a rapid vanishing of its effects to the South and a localization of its pulsations near the glaciers.

One can even wonder whether this singularity is not related to the fact that the very cause of this abrupt climatic change during the Younger Dryas is still unknown, as astronomical explanations are pointless in this case.

REFERENCES

Beaulieu, J.L. de, 1977. *Contribution pollenanalytique à l'histoire tardiglaciaire et holocène de la végétation des Alpes Méridionales françaises*, Thèse ès Sciences, Aix-Marseille III, 358 pp.

Beaulieu, J.L. de, Pons A. and Reille, M., 1982. Recherches pollenanalytiques sur l'histoire de la végétation de la bordure nord du Massif du Cantal (Massif Central, France), *Pollen et Spores*, 24: 251-300.

Beaulieu, J.L. de, Clerc, J., Couteaux M., Pons A. and Reille M., 1983. Pollen analysis and characters of climatic changes at the end of the Eemian and at the beginning of the Late-Würm in Western Europe. In: *Paleoclimatic Research and Models*, D. Reidel edit., 108-113.

Beaulieu J.L. de, Clerc J. and Reille, M., 1984. Late Weichselian fluctuations in the French Alps and Massif Central from pollen analysis. In: *Climatic changes on a Yearly to Millenial Basis*, N.-A. Mörner and W. Karlén (eds.),. D. Reidel Publ. Comp., 75-91.

Beaulieu, J.L. de and Reille, M., 1983. Paléoenvironnement tardiglaciaire et holocène des lacs de Pelléautier et Siguret (Hautes-Alpes). 1. Histoire de la végétation d'après les analyses polliniques, *Ecologia Mediterranea*, IX: 19-36.

Beaulieu, J.L. de, Pons, A. and Reille, M., 1984. Recherches pollenanalytiques sur l'histoire de la végétation des Monts du Velay, Massif Central, France, *Dissertationes Botanicae*, 72: 45-70.

Clerc, J., 1985. Première contribution à l'étude de la végétation tardiglaciaire et holocène du piémont dauphinois, *Documents de Cartograpie écologique*, 8: 65-83.

Couteaux, M., 1970. Etude palynologique des dépôts quaternaires de la vallée de la Sûre à Echternach et à Bedford et de la Moselle à Mertert, Institut Grand-Ducal de Luxembourg, *Archives*, 34, 297-336.

Couteaux, M., 1984. Climatic and glacial fluctuations during the Lateglacial and the Holocene in Oisans (Isère, France). In: *Climatic Changes on a Yearly to Millenial Basis*, N.-A. Mörner and W. Karlén (eds.), D. Reidel Publ. Comp., 91-94.

Fedorova, R.V., 1955. Waldphasen in der spatholozänen Vegetationgeschichte des Stavropol und Ergenigebietes, Trudy Geogr. Akad., NAUKA, 63: 57-127.

Guiot, J., 1985. A method for palaeoclimatic reconstruction in palynology based on multivariate time-series analysis, *Geographie Physique et Quaternaire*, 39: 115-125.

Guiot, J. and Médus, M., 1985. Traitement statistique de données palynologiques de l'Holocène du Lac Tanma (Sénégal), *An. Assoc. Palinol. Leng. Esp.*, 2: 273-282.

Guiot, J. and Pons, A., 1986. Une méthode de reconstruction quantitative du climat à partir de chroniques pollenanalytiques: le climat de la France depuis 15,000 ans, *C.R. Acad. Sc. Paris*, 302, sér. II: 911-916.

Guiot, J., 1987. Late Quaternary climatic change in France estimated from multivariate pollen time-series, *Quaternary Research* (in press).

Howe, S. and Webb III, T., 1983. Calibrating pollen data in climatic terms: improving the methods, *Quaternary Sciences Reviews*, 2: 17-51.

Imbrie, J. and Kipp, N.G., 1971. A new micropaleontological method for quantitative paleoclimatology: application to a late Pleistocene Caribbean core. In: *The Late Cenozoic Glacial Ages*, K.K. Turekian (ed.), New Haven, Yale University Press, 71-181.

Jalut, G., 1967. Analyse pollinique de la tourbière de l'Archet, *Pollen et Spores*, 9: 305-319.

Jalut, G., Delibrias, G., Dagnac, J., Mardones, M. and Bonhours, M., 1982. A palaeoecological approach to the last 21,000 years in the Pyrenees: the peat bog of Freychinède (alt. 1350m, Ariège, South France), *Palaeogeography, Palaeoclimatology, Palaeoecology*, 40: 321-359.

Malgina, E.A., 1973. Interpreting palynological evidence from quaternary deposits of Mongolia. In Proc. IIIrd Intern. Paly. Conf., Moscow 1971, Methodological Problems of Palynology, NAUKA, 120-127.

Pons, A., 1984. La paléoécologie face aux variations spatiales du bioclimat méditerranéen, *Bull. Soc. Bot. Fr.*, 131, Actualités Botaniques, 77-83.

Pons, A. and Reille M., 1986. Nouvelles recherches pollenanalytiques à Padul (Granada): la fin du dernier glaciaire et l'Holocène, *Quaternary climate in Western Mediterranean*, Proceedings of the Symposium on climatic fluctuations during the Quaternary in the Western Mediterranean regions. 16-21 Juno 1986. Universidad Autonomia de Madrigd, p. 405-420.

Reille, M., 1975. *Contribution pollenanalytique à l'histoire tardiglaciaire et holocène de la végétation de la montagne corse*, Thèse ès Sciences, 206 p., 44 diag., 5 pl. .t.

Sabatier, R. and Van Campo, M., 1984. L'analyse en C.P. de variables instrumentales appliquée à l'estimation des paléoclimats de la Grèce il y a 18,000 ans, *Bull. Soc. Bot. Fr.*, 131, Actualités Botaniques, 85-96.

Van Campo, M. and Jalut, G., 1969. Analyse pollinique des sédiments des Pyrénées-Orientales: lac de Balcère (1764m), *Pollen et Spores*, XI, 1: 116-126, 2 fig., 1 diagr. h.t.

Watts, W.A., 1980. Regional variation in the response of vegetation to Lateglacial climatic events in Europe. In: *Studies in the Lateglacial of North West Europe*, J.J. Lowe, J.M. Gray and J.E. Robinson (eds), Pergamon Press, Oxford, 1-21.

Webb III, T. and Bryson, R.A., 1972. Late and Postglacial change in the Northern Midwest, USA: quantitative estimates derived from fossil pollen spectra by multivariate statistical analysis, *Quaternary Research*, 2: 79-115.

ARIDIFICATION AND ABRUPT CLIMATIC EVENTS ON THE SAHARAN NORTHERN AND SOUTHERN MARGINS, 20,000 Y BP TO PRESENT

Pierre Rognon

Research group "Paleoclimates and paleoenvironments in arid regions (UA 722) Department of Earth Sciences
P. et M. Curie University 4 place Jussieu
75252 Paris Cedex 05 - France

ABSTRACT. Abundant data from paleoenvironmental studies show that the pluviometric evolution for the last 20,000 years is always more gradual and diversified in the northern than in the southern Sahara. Such a pattern is also noted from climatic measurements on the scale of the last century. On both time-scales, two types of abrupt events are much better defined in the pluviometric records of the southern fringe of the Sahara, fast acceleration of the longterm tendency and strong but brief oscillations. Oscillations are probably amplified by the high sensitivity of the monsoon rains to changes in the global climatic equilibrium or to random changes of surface water temperatures in the Atlantic Ocean. The northern boundary, on the contrary, appears to react more gradually and in a variety of ways to the fluctuations of the cyclonic trajectories.

INTRODUCTION

A new appreciation of present climate variability has resulted from recent climatic catastrophic events such as the drought in the Sahel (1968-84) or the 1982-83 El Niño in the tropical Pacific Ocean. These abrupt events show that some regions of the world are very sensitive to a minor alteration of the complex equilibrium formed by the atmosphere, the oceans, the cryosphere and the surface of continents. The Sahel, for instance, seems to be a region where the risks of climatic catastrophes are very heavy because the rapidity of these abrupt events often exceed the capacity for adjustment of natural system parameters. Long-term technological projects (dams for hydropower and irrigation or large-scale afforestation etc.) depend on the ability to forecast climate patterns. Two questions are bound to arise: Is there a predictable pattern to the occurence of series of dry years? Does the present climate, slowly or rapidly, move towards a drier one? It is, of course, very difficult to distinguish between a natural aridification and the desertification induced by man.

One can put the same questions for other desert margins of the world, and, particularly, for the northern margin of the Sahara. But

W. H. Berger and L. D. Labeyrie (eds.), Abrupt Climatic Change, 209–220.

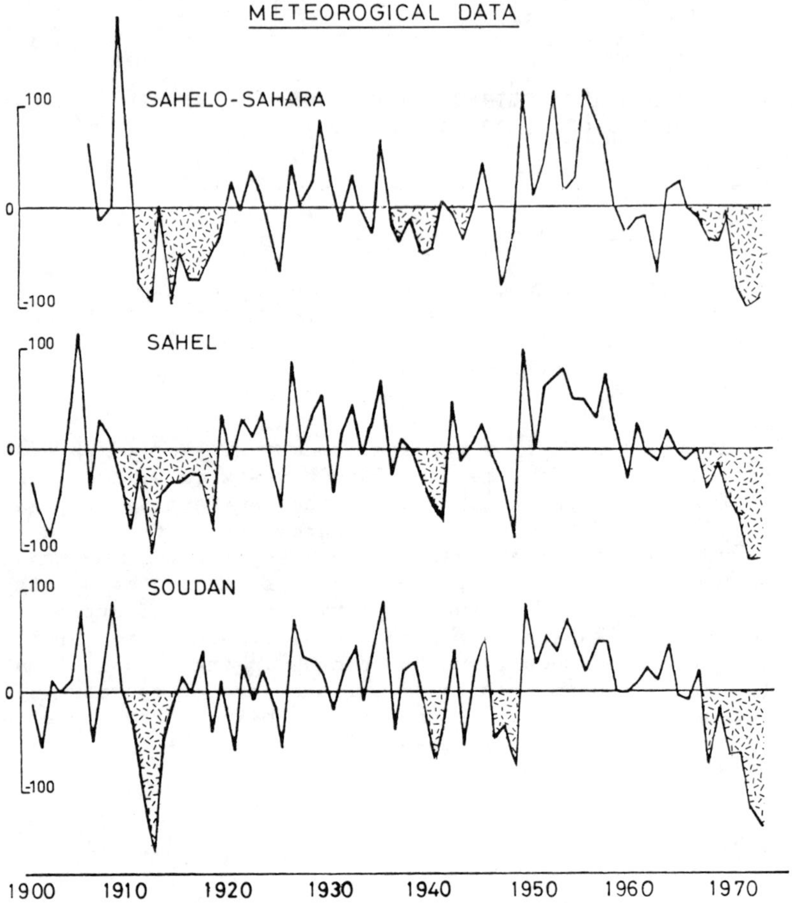

Figure 1. Trends of some Sahelian climatic indicators recorded from meteorological and hydrological data (1900-1980) (from Nicholson, 1980; Nicholson and Chervin, 1983) : 1A = rainfall.

the latter is situated in the Mediterranean zone where winter rainfall and the droughts here seem to be shorter and more localized than in the Sahel. On another time scale (20,000 BP to present), paleoclimatic reconstructions show similar contrasts between the climatic evolution of the northern and southern fringes of the Sahara. The only differences between long-term and short-term scales are in the duration and amplitudes of the phenomena. It is quite possible that every region in the world should have its own climatic patterns on different time scales. To compare these patterns is perhaps a way to understand the mechanisms which cause the abrupt events (Rognon, 1981 and 1984a).

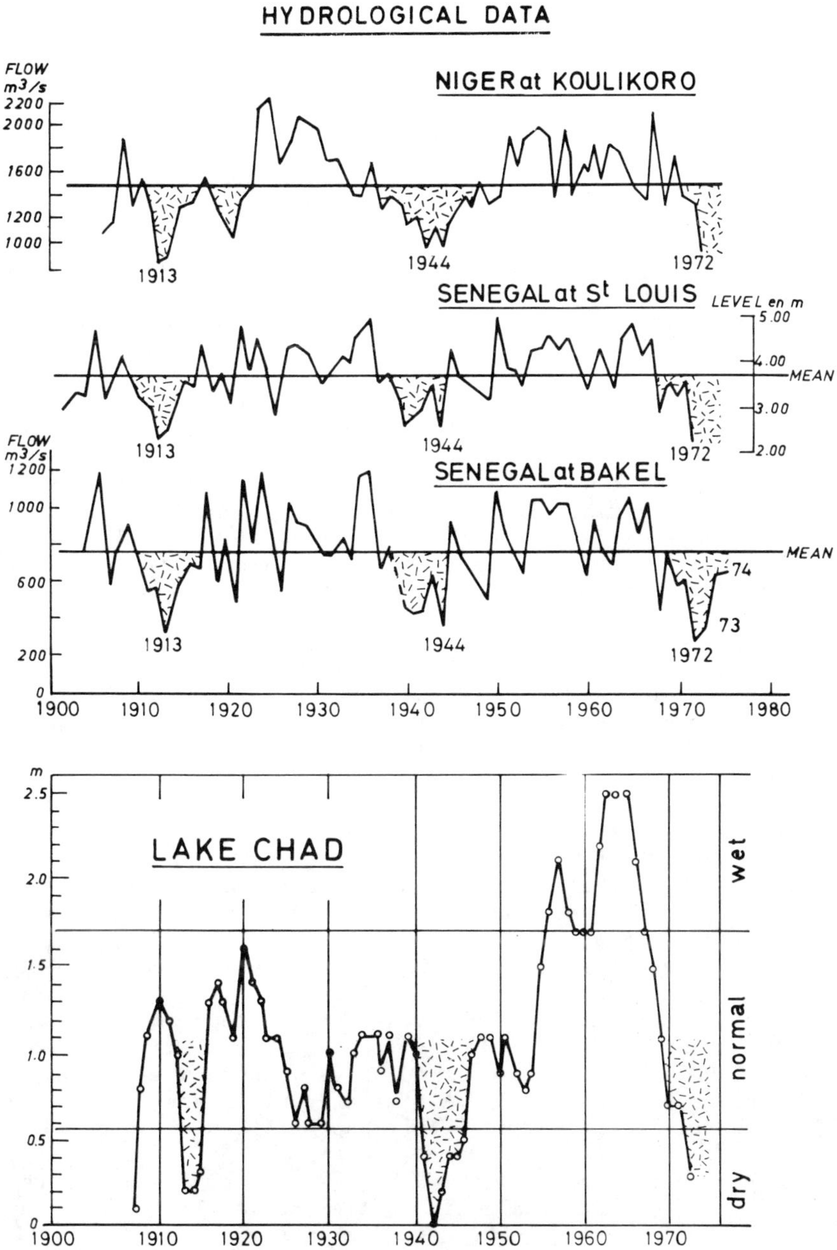

Figure 1. 1B = rivers and lake Chad.

EVIDENCE FOR TWO KINDS OF ABRUPT EVENTS DURING THE RECENT CLIMATIC EVOLUTION

The droughts appear shorter, less severe and less extensive in northern than in southern Sahara. Two sorts of abrupt events cause catastrophic effects on natural environments.

Anomalies are important but brief oscillations, followed by a return to former conditions. In the south of the Sahara, dry years cluster together during five to more than fifteen years (1910-16; 1944-48; 1968-84). They extend to all the Sahelian and Soudanian zone, where the average rainfall is about 200 to 900mm/y and the rainy season lasts two to five months (Nicholson and Chervin, 1983) (Fig. 1). This persistence of rainfall anomalies cannot be a random effect. In the Maghreb, droughts seem less catastrophic. The one from 1944 to 1948 is contemporaneous with a Sahelian drought, but less extensive. During the driest period in the Sahel, between 1968 and 1984, exceptional rainfall occurred in the eastern Maghreb in 1969, and, more localized, in 1973. The drought has been limited to the western Maghreb, but on a shorter period, 1977-84 in the West of Algeria, 1980-84 in Morocco. In North Africa, droughts result principally from abnormal seasonal rainfall distribution, for instance, too late in autumn for the period 1944-47, insufficient in spring for 1977-84 in the west of Algeria. Rainfalls are linked with cyclonic depressions of the Polar Front and the rainy period lasts from October/November to March/April, when temperatures are low. In the southern Sahara, on the contrary, the probability of monsoon summer rains extend only to 3 or 4 months. Moreover, the near oceanic moisture sources prevent a long persistence of dry atmosphere above North Africa.

Climatic steps take place between a former mean climate and another very different one, without a return to former conditions. The long Sahelian drought in 1968-84 is, perhaps, an example of this kind of abrupt events. It may be more or less linked with a feedback mechanism, the decreasing vegetation cover causes an increasing albedo and this, in turn, favors increased aridity (Charney, 1975). Its effects were the more catastrophic as it followed an abnormally wet decade (1955-65) in the Sahel and, particularly, in East-Africa. Nevertheless, this example is not unique; historical data show other rapid steps in Sahelian pluviometric fluctuations on many decadal time scale. Thus, rapid aridification after 1900 AD followed a period of very abundant and wide spread rainfall, between 1870 and 1895. A similar rapid change took place towards 1820 and aridity prevailed until about 1850 (Nicholson, 1981).

In the northern margins of the Sahara, the most abundant and extended rainfall (from Middle Algeria to Alexandria) lasted a dozen years (1884-96) and the longest drought recorded in Algiers since 1843 lasted eight years (1854-62). All available information indicates less rapid changes and shorter periods of drought in North Africa as compared with the southern fringe of the Sahara. But it is necessary to wait for more precise data from dendroclimatology studies, to confirm these different patterns of "behavior" of North and South.

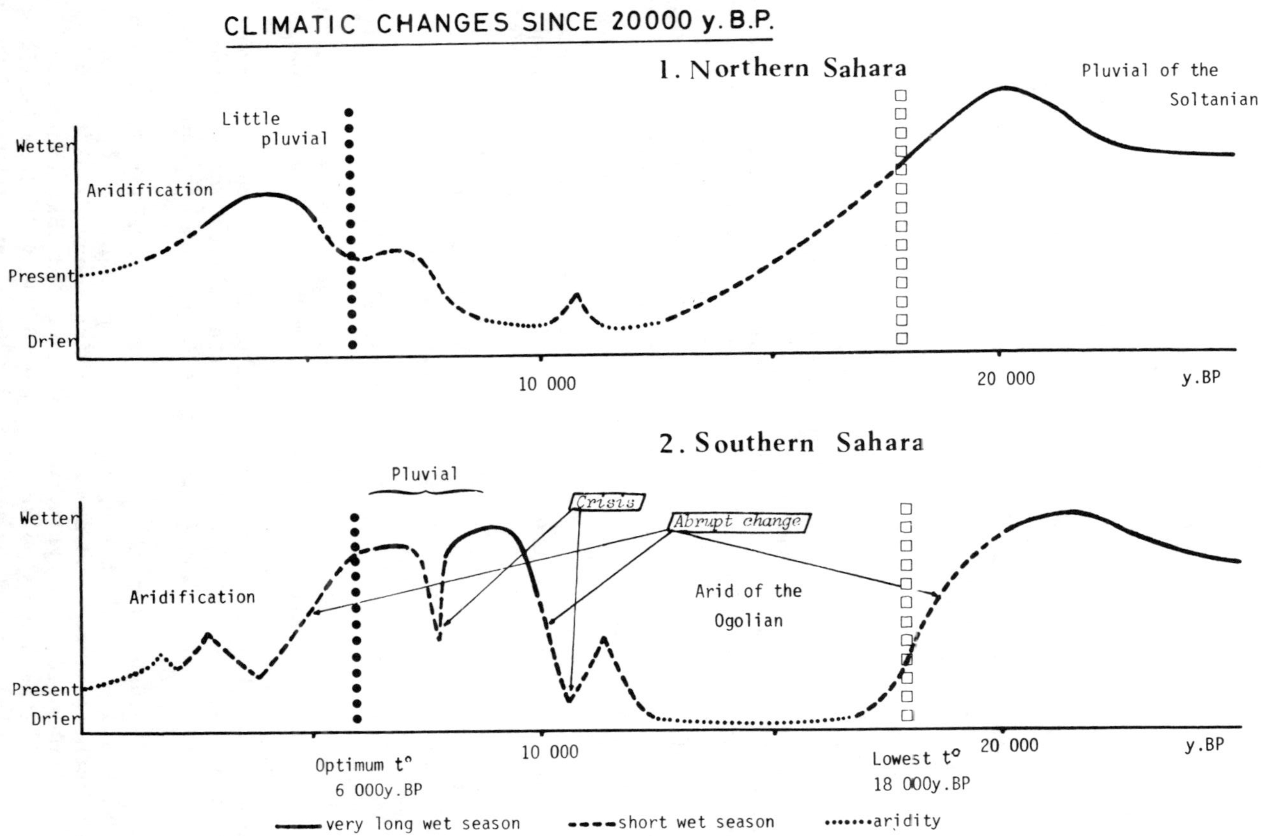

Figure 2. Summary curves of climatic changes since 20,000 y. BP both along the southern and northern margins of the Sahara.

CHANGES IN PLUVIOSITY DURING THE LAST 20,000 Y, GRADUAL IN THE NORTH AND ABRUPT IN THE SOUTHERN SAHARA.

In the Sahelian zone, the changes in pluviosity are very rapid during the last 20,000 y (Servant, 1983; Rognon, 1976; Gasse, 1980; Durand and Lang, 1986). In Fig. 2, three steps appear connecting one extreme position to another:

1. Around 22-21,000 y BP, very high lacustrine levels are recorded from Mauritania to East Africa. Two or three millenia later, these lakes were completely dry. This extremely rapid aridification corresponds to the last glacial maximum. The aridity continues until 12,500 or 12,000 y BP (Ogolian period) when temperate areas are already warming.
2. After 12-10,000 y BP, for one or two thousand years, there is a return to abundant rainfall, well distributed throughout the year, as shown by pollen-analysis, diatoms, lacustrine or alluvial deposits, paleosoils, etc. Subsequently, the Early Holocene pluvial lasts until 5000 y BP, just after the thermal optimum.
3. After 5000 y BP, the present aridification begins. This climatic change seems more progressive, probably because it is known with more details. But there is evidence for a step between 4500 and 4000 y BP, in Mauritania, North Mali, Niger, Tchad, Egypt or Ethiopia. After that, the phases of little rain or aridity appear contemporaneous on the scale of the climatic zone.

In the southern Sahara, the pluviometry curves are always characterized by abrupt changes in the climatic history which are comparable to the "terminations" in the ocean isotopic records (Fig. 2). These steps may be explained by the development of the strong high atmospheric pressure system at the Sahelian latitude. The rapid aridification, associated with reenforcement of these high pressure systems occurs immediately after the cold maximum (18,000 y BP) and after the thermal optimum (6-5,000 y BP), that is, in two completely opposite settings.

In the northern Sahara, the paleoclimatic evolution since 20,000 y BP, is known with fewer details. It appears much more gradual than in the South (Fig. 2) and also more variable within the region according to the position along the Mediterranean fringe of the Sahara. A climatic reconstruction from the Atlantic coasts to the Gulf of Gabes is to be published (Rognon, 1987). The following features can be identified:

1. *Before 20,000 y BP*, the cyclonic system was shifted far South of the Mediterranean Sea by the spreading of the North-European continental ice, both in winter and summer, bringing heavy rains throughout the year. Palynological analysis on cores revealed that the tree pollen clearly dominated the steppe pollens before 20,000 y BP in a peat bog from northern Tunisia (Ben Tiba and Reille, 1982), and around 27,000 and 21,000 y BP in marine sediments from the Gulf of Gabès (Brun, 1979). In three sites from Tunisia and one other from southern Morocco, reddish paleosoils were dated at 22-20,000 y BP (Coudé-Gaussen et al., 1983; Scharpenseel and Zakosek, 1979; Steinmann and Bartels, 1982; Rognon et al., 1984). Fine grained alluvial deposits are related to the regularity of the hydrological regime with long periods of flooding in the

valleys and the settling of silt and clay (Rognon, 1984b). In southern Tunisia, the Djerid basin was occupied by a paleolake whose shoreline is at a level of 30m above the present day bottom. Seven ^{14}C dates show this lake existed at least as early as 35,000 y BP and that its regressive stages, went on until about 22,000 y BP (Page, 1972, Richard and Vita-Finzi, 1982).

2. *Around 20,000 y BP*, there were several hydrological and biological changes in North Africa. The rain became first more seasonal and then very irregular. But this climatic evolution appears progressive over 7 to 8 thousand years. In many regions of the Maghreb, a gullying phase began only after 13,000 or at about 10,000 y BP (Conrad, 1969; Andres, 1977; Weisrock et Rognon, 1977; Rognon et al., 1983; Weisrock et al., 1985). In the South of Tunisia, loess continued to be deposited after 18,000 y BP but it was lighter in color and the isotopic composition (^{13}C and ^{18}O) of its concretions shows a distinct evolution toward dryer conditions at the end of the Upper Pleistocene. Until about 12-10,000 y BP the vegetal cover remained dense enough to trap dust and to protect the loess accumulated for tens of thousands of years (Coudé-Gaussen, 1983 and 1984).

3. *Aridity is evident only after 13-10,000 y BP*, when dunes obstructed the valleys of the Saharan Atlas. Many proofs of sanding-up or eolian deflation were mentioned for the time between the Epipaleolithic and Neolithic ages, during the Early Holocene (Estorges et al., 1969; Marmier and Trecolle, 1968, and others). But aridity has been of short durations, as at 8-9,000 y BP, traces of short, localized wetter phases mark the progressive reinstallation of wet climate in the Northern Maghreb.

4. *The reinstatement of a wet climate* is complete around 5,000-4,000 y BP, but the environments of the Middle Holocene were quite different from those existing before 20,000 BP, even though they occurred during a new period of geomorphological stability. All over North Africa, the last reddish and reddish-brown soils disappeared; the dark humic soils that succeeded were the result of a different pedogenesis and they are separated from the underlying reddish soils by a discontinuity of erosion or an eolian accumulation. They reached a sort of climax during their maximum extension in about 4,000 y BP. This extension corresponds to a return to an alluviation of silts and fine sands in many valleys of North Africa during the Middle Holocene and to the development of swamps or little lakes in closed basins along the northern Sahara margins. Since 3,000 y BP, it is difficult to differentiate natural aridification and anthropogenic desertification.

Besides being more gradual, the climatic evolution of North Africa also shows greater regional contrasts than the one along the Sahelian boundaries (Fig. 3). There is, for instance, a completely opposite pluviometric evolution in the Maghreb and in the Libyco-Egyptian desert. Each wet phase in the West (until 12,000 y BP, or after 6-5,000 y BP) corresponds to an extreme aridity in the East and the arid phase in North Africa around 12,000-6,000 y BP is opposite to the rain extension on the Libyco-Egyptian desert during the Early Holocene. These differences in "behavior" are probably explained by gradual changes in the

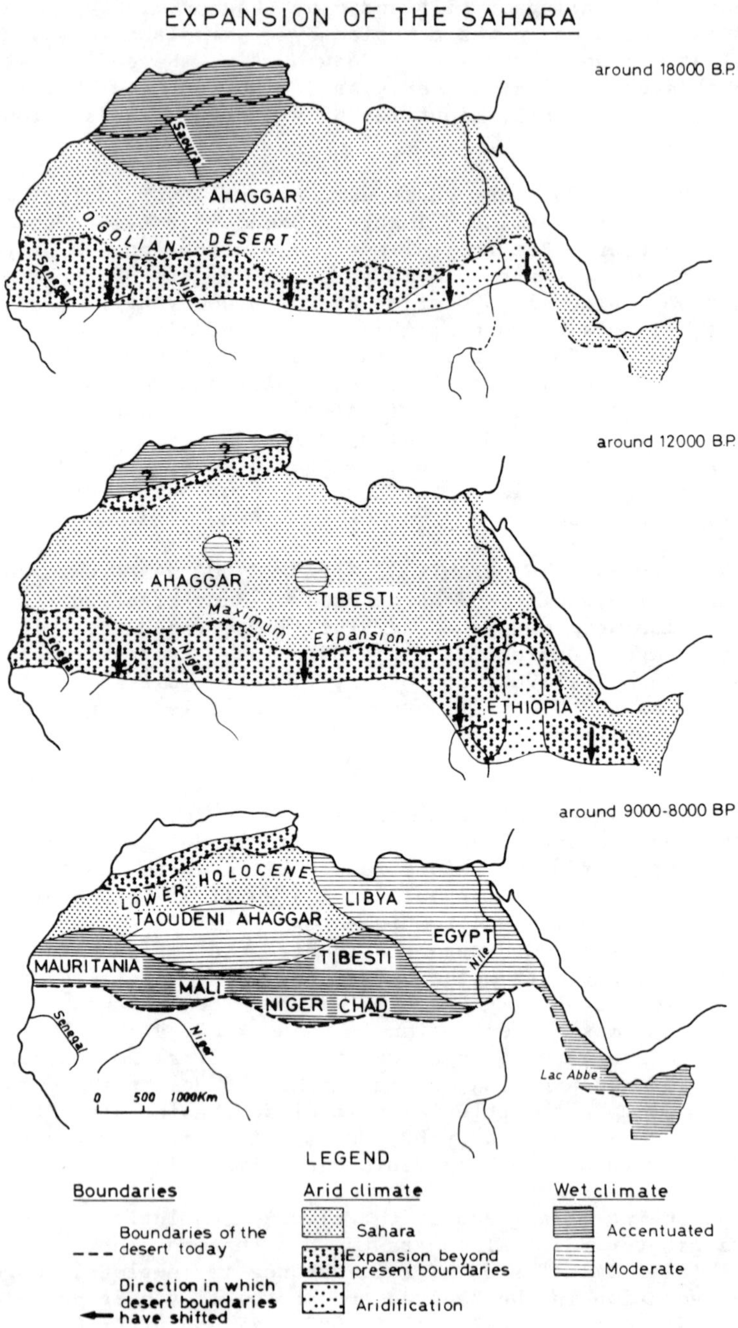

Figure 3. Schematic maps of rainfall patterns for three major stages of the paleoclimatic evolution since 20,000 y BP.

number and location of meanders in the Jet Stream, which directly control the rain system in the southern Mediterranean area.

BRIEF BUT STRONG ARIDITY CRISES IN THE SOUTHERN FRINGE OF THE SAHARA.

In the Sahelian zone, pluviometric evolution since 20,000 y BP is interrupted by very strong aridity crises lasting several centuries. These crises are analogous, on another scale, to the present catastrophic droughts, and have until now not been found in the northern Saharan paleoclimatic records. Two examples are well known (Rognon, 1982):

1. Around 13,000 and 10,000 y BP, the general tendency is for a rapid increase of pluviosity after a long arid period. However, on the more detailed pluviometric records (Nile or Niger deltas, Chad or East Africa lakes etc.), there is a short return, for several centuries, between 11,000 and 10,000 y BP, to conditions analogous to the previous aridity (Rossignol-Strick et al., 1982; Williams and Adamson, 1980; Pastouret et al., 1978; Gasse et al., 1980; and others). This crisis does not change the general tendency (Fig. 2). In fact, it is an important mark, because before this crisis, rain was still very seasonal (strict monsoon regime?) and afterward, the rain has a tendency to be evenly distributed throughout the year.

2. Around 8-7,000 y BP, during the time when pluviosity is at its maximum (early Holocene), another aridity crises is recorded, which creates a serious degradation of the environment for several centuries wherever this evolution has been precisely recorded (Senegal, Chad, Ethiopia). This crisis does not signal the end of the pluvial, but an important change in the rain distribution. After this crisis, rains are, indeed, again concentrated over a few months of the year and the dry season expands progressively (Gasse et al., 1980; Servant and Servant-Vildary, 1980; Street, 1979). Crises of this type have not been documented on the northern side of the Sahara, perhaps because of the more preliminary state of research in this area. However, when discontinuities are observed in the sections studied north of the Sahara, they are generally local or regional, but not zonal as along the southern Sahara.

The two aridity crises in the southern Sahara are generally explained by two well-defined events during the waning of the northern ice sheets after the last glacial maximum. The first one appears to have affected the northern Atlantic ocean between 11,000 and 10,000 y BP, with an increase in sea ice and icebergs as far as the Bay of Biscay (Younger Dryas); the second would be simultaneous with the rapid melting of the North American ice sheet and the resulting input of cold water over the NW Atlantic between 8,000-7,000 (Flohn, 1983). The origin of these crises would therefore be independent of the general evolution of the pluviometry, which is regulated by mechanisms such as orbital parameters, thermal gradients and pressure fields on the scale of the Northern Hemisphere. The crises are superimposed on the multimillenian tendency. When they occur during a rapid acceleration of that trend, the climatic instability reaches a maximum, as for instance, around 11-10,000 y BP (Fig. 2).

CONCLUSION

The study of the pluviometric evolution of the Saharan boundaries over the last 20,000 yrs shows a high diversity of "behavior" on a time scale which is only 1/5th of a complete paleoclimatic cycle. During this period, two types of abrupt events are much better defined in the pluviometric records of the southern fringe of the Sahara than in the North:

1. Fast acceleration of the multimillenian tendency
2. Strong but brief climatic crises.

Although they have very different origins, these events are characterized here by their large amplitudes and their quasi-zonal extension. This confirms that on a geological scale, as well as on the scale of the last century, the pluviometric evolution of the southern boundary of the Sahara is very sensitive to all the changes in the global climatic equilibrium, and, for instance, to accidental changes of surface water temperatures in the North Atlantic, which here create sharp catastrophic changes for the environment. The northern boundary of the Sahara, on the contrary, appears to react more gradually and in a more diversified way to the fluctuations of the cyclonic trajectories in the temperate zone.

REFERENCES

Andres, W., 1977. Studien zur jungquartären Reliefentwicklung des südwestlichen Anti-Atlas und seines saharischen Vorlandes (Marokko), *Mainzer Geog. St.*, 9, 147 p.

Ben Tiba, B. and Reille, M., 1982. Recherches pollenanalytiques dans les montagnes de Kroumirie (Tunisie septentrionale): premiers résultats, *Ecol. médit*, 8 (4): 75-86.

Brun, A., 1979. Recherches palynologiques sur les sédiments marins du Golfe de Gabès. *Rev. Géol. médit.*, Marseille, 4 (1): 247-264.

Charney, J.G., 1975. Dynamics of deserts and drought in the Sahel, *Quart. J. Royal Meteo. Soc.*, 101 (428): 193-202.

Conrad, G., 1969. L'évolution continentale post-hercynienne du Sahara algérien. C.N.R.S., *Centre Rech. Zones Arides*, Paris, 10, 527 p.

Coudé-Gaussen, G., 1983. Mise en place des basses terrasses holocènes dans les Matmata et leurs bordures (Sud-Tunisien), *Bull. Ass. Fr. et. Quat.*, Paris p. 173-180.

Coudé-Gaussen, G., 1984. Le cycle des poussières éoliennes désertiques actuelles et la sédimentation des loess péridésertiques quaternaires, *Bull. Elf Aquitaine*, 8 (1): 167-182.

Coudé-Gaussen, G., Olive, P. and Rognon, P., 1983. Datation de dépots loessiques et variations climatiques à la bordure nord du Sahara algéro-tunisien, *Rev. Géol. dyn et Géog. phys.*, 24: 61-73.

Durand, A. and Lang, J., 1986. Approche critique des méthodes de reconstitution paléoclimatique: le Sahel nigéro-tchadien depuis 40,000 ans, *Bull. Soc. Géol. Fr.*, Paris, 8 (2): 267-278.

Estorges, P., Aumassip, G. and Dagorne, A., 1969. El Haouita, un exemple de remblaiement fini-wurmien, *Libyca CRAPE*, Algiers, 17: 53-91.

Flohn, H., 1983. Actual palaeoclimatic problems from a climatologist's viewpoint. In: A. Ghazi (ed.). *Paleoclimatic Research and Models*, D. Reidel, Dordrecht p. 17-33.

Gasse, F., 1980. Late Quaternary changes in lake levels and diatom assemblages on the southeastern margin of the Sahara, *Palaecol. of Af.*, 12: 333-350.

Gasse, F., Rognon, P. and Street, A., 1980. Quaternary History of the Afar and Ethiopian Rift Lakes. In: *The Sahara and the Nile*, M.A.J. Williams and H. Faure (eds.) Balkema, p. 361-400.

Marmier, F. and Trecolle, G., 1968. Stratigraphie du gisement d'Hassi Mouillah, région de Ouargla (Algérie) *Bull. Soc. Préhist. Fr.*, Paris, 65 (4): 121-127.

Nicholson, S.E., 1980. Saharan climates in historic times. In: *The Sahara and the Nile*. M.A.J. Williams and H. Faure eds., Balkema, p. 173-200.

Nicholson, S.E., 1981. The historical climatology of Africa. In: *Climate and History*. Cambridge Univ. Press, p. 249-270.

Nicholson, S.E. and Chervin, R.M., 1983. Recent rainfall fluctuations in Africa. Interhemispheric teleconnections. In: *Variations in the Global Water Budget*, A. Street-Perrott (ed.) Reidel Publ., p. 221-238.

Page, W.D., 1972. The geological setting of the archaeological site at O. el Akarit and the paleoclimatic significance of gypsum soils (S. Tunisia). Thesis Doc. Univ. of Colorado, 111 p.

Pastouret, L., Chamley, H., Delibrias, G., Duplessy, J.C. and Thiede, J., 1978. Late Quaternary climatic changes in Western Tropical Africa deduced from deep-sea sedimentation of the Niger delta, *Oceanol. Acta*, 1 (2): 217-231.

Richards, G.W. and Vita-Finzi, C., 1982, Marine deposits 35,000-25,000 years old in the Chott el Djerid, southern-Tunisia, *Nature*, 295: 54-55.

Rognon, P., 1976. Essai d'interprétation des variations climatiques au Sahara depuis 40,000 ans, *Rev. Géog. Phys et Géol. Dyn.*, 18: 251-282.

Rognon, P., 1981. Les crises climatiques, *La Recherche*, 12: 1354-1364.

Rognon, P., 1982. Les crises climatiques de courte dúrée (quelques années è quelques siècles) et leur enregistrement dans la sédimentation continentale. In: *Palaeoclimatic Res. and Models*, C.E.E., A. Ghazi (ed.), Reidel Publ., p. 114-123.

Rognon, P., 1984a. Les anomalies du climat actuel et les crises climatiques, *Images de la Géographie*, Courrier du CNRS., 57: 60-64.

Rognon, P., 1984b. Signification dynamique et climatique des formations et terrasses fluviatiles en Afrique du Nord et au Proche Orient, *Bull. Assoc. Fr. Et. Quat.*, 1-2: 161-169.

Rognon, P., 1987. Late Quaternary climatic reconstitution for the Maghreb (North Africa), *Paleogeog. Paleoclim. Paleoecol.* (in press)

Rognon, P., Levy, A., Ballais, J.L., Coudé-Gaussen, G. and Riser, J., 1983. Essai d'interprétation des coupes du Quaternaire récent de l'O. el Akarit (Sud Tunisien). *Géol. Médit.*, 10 (2): 71-92.

Rognon, P., Weisrock, A., Olive, P. and Coudé-Gaussen, G., 1984. Premières datations d'un paléosol du dernier maximum glaciaire (18-20,000 BP) au Maroc, *Médit.*, Aix en Provence., 3: 65-69 .

Rossignol-Strick, M., Nesteroff, W., Olive, P. and Vergnaud-Grazzini, C., 1982. After the deluge: Mediterranean stagnation and sapropel formations, *Nature*, 295: 105-110.

Servant, M., 1983. Séquences continentales et variations climatique: Evolution du Bassin du Tchad au Cénozoique supérieur. Trav. et Doc., ORSTOM, Paris, 573 p.

Servant, M. and Servant-Vildary, S., 1980. L'environnement quaternaire du Bassin du Tchad. In: *The Sahara and the Nile*, M.A.J. Williams et H. Faure (eds.), Balkema, p. 133-162.

Scharpenseel, H. and Zakosek, H., 1979. Phasen der Bodenbildung in Tunesien, *Z. für Geomorph.*, N.F., 33: 118-126.

Steinmann, S. and Bartels, G., 1982. Quartärgeomorphologische Beobachtungen aus Nord-and Süd-Tunisien, *Catena*, 9: 95-108.

Street, A., 1979. The relative importance of climate and local hydrogeological factors in influencing lake-level fluctuations, *Palaeoecol. of Af.*, 12: 137-158.

Weisrock, A., Delibrias, G., Rognon, P. and Coudé-Gaussen, G., 1985. Variations climatiques et morphogénèse au Maroc atlantique (30-33°N) à la limite Pléistocène-Holocène, *Bull. Soc. Géol. Fr.*, 8 (1): 105-108.

Weisrock, A. and Rognon, P., 1977. Evolution morphologique des basses vallées de l'Atlas atlantique marocain, *Géologie médit.*, Marseille, IV, 4: 313-334.

Williams, M.A.J. and Adamson, D.A., 1980. Late Quaternary depositional history of the Blue and White Nile rivers in central Sudan. In: *The Sahara and the Nile*, Balkema, Rotterdam, p. 281-304.

Part V.

Glacial-Holocene Transition: Ice Core Record

ICE CORE EVIDENCE OF ABRUPT CLIMATIC CHANGES

W. Dansgaard

Geophysical Institute
Haraldsgade 6
DK-2200 Copenhagen N
Denmark

ABSTRACT. $\delta^{18}O$ profiles along five deep ice cores are compared. The two Greenland profiles reveal many abrupt climatic changes in the North Atlantic Ocean during the Wisconsin glaciation. Dust and trace element concentrations show radical environmental changes. The targets of future deep ice core drilling are discussed.

INTRODUCTION

In the last decades several ice cores to a thousand meters depth or more have been recovered from the two large ice sheets: Camp Century, Northwest Greenland 1964 (1400 meters, 1964; Hansen and Langway, 1966); Byrd Station, West Antarctica 1966 (2163 meters; Ueda and Garfield, 1969); Dome C, East Antarctica 1978 (950 meters, Lorius et al., 1979); Dye 3, Southeast Greenland 1981 (2037 meters, Gundestrup and Johnsen, 1985); and Vostok, East Antarctica 1982 (2083 meters, Barkov et al., 1977; Gordienko et al., 1983).

All of these ice cores reach back into the last glaciation (Wisconsin) in an essentially continuous sequence (Johnsen et al., 1972; Lorius et al., 1979; Dansgaard et al., 1982), the Vostok core, and perhaps that from Camp Century, even into the preceeding glaciation (Lorius et al., 1985; Dansgaard et al., 1984).

In areas and periods with an annual snow accumulation, which exceeds the surface roughness (height of sastrugies), both winter and summer snow will usually be represented in each annual layer in the snow pack and thereby in ice cores. In such cases, the individual annual layers in the core can be identified by measuring continuous profiles of one or several of the numerous chemical and physical characteristics of snow, which vary with the season, for example concentrations of heavy stable isotopes and continental dust (Hammer et al., 1978), H^+ (Hammer, 1980) and chemical constituents (Langway et al., 1977).

These techniques allow absolute dating of Holocene ice (Hammer et al., 1986), and perhaps some even older ice in relatively high accumulation areas of Antarctica (Hammer et al., 1985a). The Pleistocene ice in

W. H. Berger and L. D. Labeyrie (eds.), Abrupt Climatic Change, 223–233.

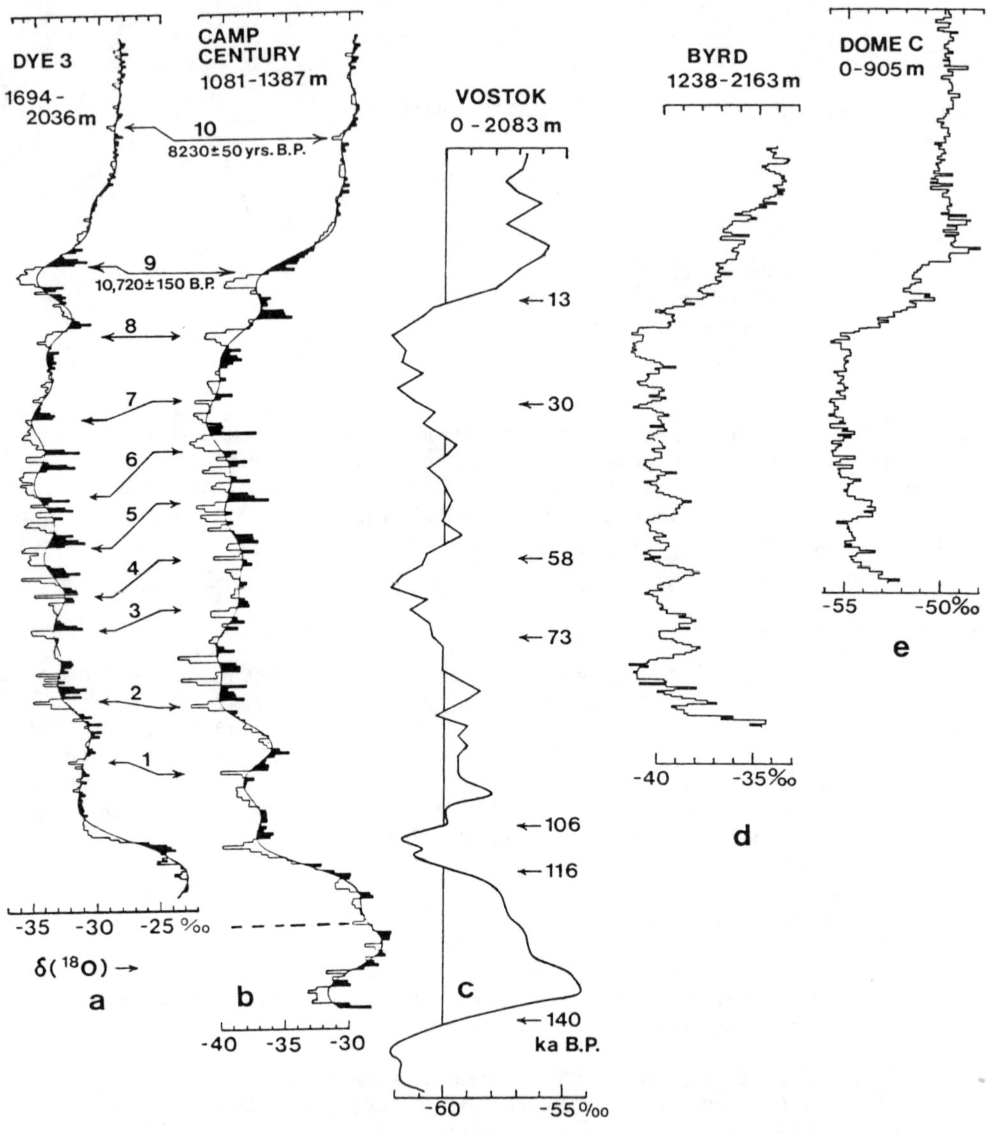

Figure 1. $\delta^{18}O$ profiles along the five deep ice cores from the polar ice sheets. They are all plotted on linear depths scales spanning the depth intervals indicated on top. δ-data from Johnsen et al. (1972), Dansgaard et al. (1982), and Lorius et al. (1979, 1985). The time scale along the Vostok record (in 1000 years before present) is not necessarily valid for the other records.

the Vostok core has been dated indirectly using glaciological models to estimate ice flow and accumulation (Lorius et al., 1985), and the Greenland deep cores by tentative correlation of climatic changes found in ice and sea floor sediment cores (Dansgaard et al., 1982). But there is a great need for checking these dating methods by radioactive or other independent methods. This will form a safer basis for estimating, if the major climatic changes occurred simultaneously in the two hemispheres, and it will allow a better estimate of the frequency and abruptness of the numerous radical climatic changes during the Wisconsin glaciation revealed by the Greenland deep ice cores in particular.

ICE CORE DATA

Fig. 1 shows $\delta^{18}O$ records (a through e) along five deep ice cores, plotted on linear depth scales and spanning the depth intervals indicated in the upper part of the figure. The individual samples represent 1, 1, 1.5-2, 4, and 3-4 meters of core increments, respectively. The δ-profiles are continuous, except for the upper two thirds of the Vostok record, which is based on one 1.5-2m sample every 25m of core. Only the Vostok and Dome C records are shown in their full length downwards from the surface. The time scale along the Vostok record (Lorius et al., 1985) is not necessarily valid for the other ones.

The arrows between the two Greenland profiles show common δ-features indicating simultaneous deposition. The Holocene part of the Dye 3 core has been dated by counting seasonal $\delta^{18}O$, dust concentration and acidity oscillations in a nearly continuous sequence downwards from the surface (Hammer et al., 1986). Notice in particular the 10,750 ± 150 yrs B.P. dating of the last abrupt δ-shift, which undoubtedly reveals the northern hemisphere transition from the Younger Dryas to Holocene (Oeschger et al., 1984). The abruptness of this transition (probably within a few decades) may be explained by the fast retreat of the sea-ice cover in the North Atlantic Ocean, which occurred by that time, associated with a drastic northward displacement of the polar front (Ruddiman and McIntyre, 1981).

Similar events may be the immediate cause of the abrupt δ-shifts from low to high values, which occurred many times during the Wisconsin glaciation, cp. Fig. 1a and b. If so, we are faced with a regional phenomenon in the North Atlantic, where the ocean-atmosphere system has oscillated between two quasi-stationary stages, dominated by extended and considerably reduced sea-ice cover, respectively. Due to the general wind circulation the consequent climatic oscillations affected primarily Europe and, obviously, Greenland, but to a lesser degree North America (as was the case during the last glacial oscillation, Bölling-Alleröd-Younger Dryas), and much less Antarctica, where no distinct evidence of these oscillations has been found so far in the Byrd and Dome C cores (the Vostok core has not yet been analyzed in sufficient detail), cp. Fig. 1.

The oscillating sea-ice hypothesis is consistent with the fact that under the mainly ice-free conditions in the North Atlantic during Holocene time no δ-oscillations seem to have occurred at Dye 3, South-

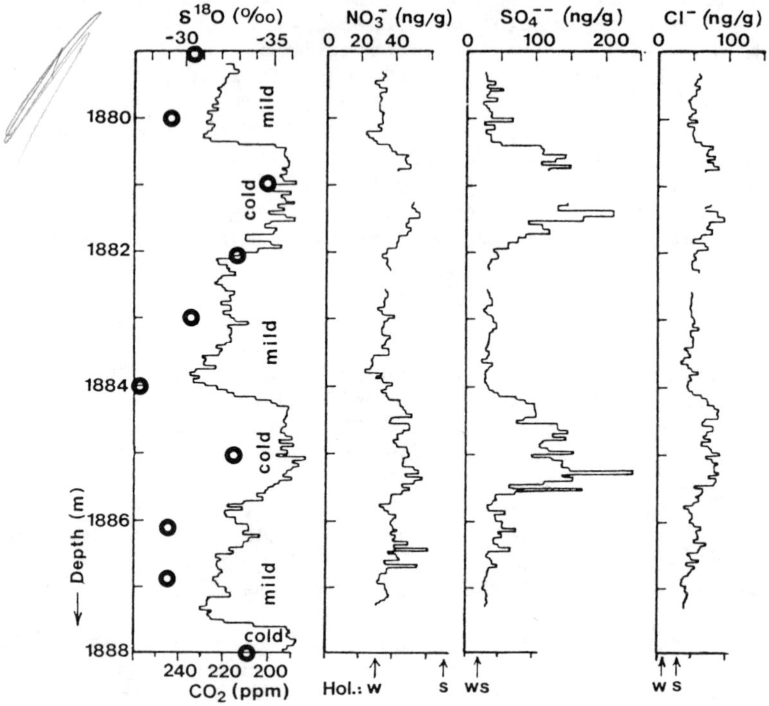

Figure 2. Abrupt environmental changes during the Wisconsin glaciation, reflected by shifts in isotopic composition (Dansgaard et al., 1984; notice the reverse scale on top), atmospheric CO_2 concentration (circles; Stauffer et al., 1984), and concentration of nitrate, sulphate and chloride (Finkel and Langway, 1985; and Finkel et al., 1986). W and S: Holocene winter and summer values.

east Greenland (the record still needs some corrections for upstream effects, though). However, weak δ-oscillations, in antiphase with worldwide glaciation changes, were found in the Camp Century record from Northwest Greenland, which is still under the influence of sea-ice in Baffin Bay. The period is approximately 2550 yrs, which may also hold true for the much stronger Wisconsin oscillations (Dansgaard et al., 1971, 1984). The same periodicity has been noticed in the isotopic composition of foraminifera in deep sea cores (Pisias et al., 1973).

The primary reason for the climatic oscillations is a problem that is still open to discussion (Dansgaard et al., 1984). But, there is no doubt that the δ-shifts represent abrupt and drastic changes in the North Atlantic environment.

Profiles of the parameters mentioned in Section 1, in addition to others, for example CO_2 (Delmas et al., 1980; Stauffer et al., 1984; Neftel et al., 1985) and ^{10}Be (Beer et al., 1985), have been measured in continuous sequences along those parts of the cores, which contain stable isotope evidence of considerable climatic changes.

Chemical composition and CO_2

Below, we shall consider a number of trace elements measured through the three δ peaks in the 1879 to 1888m depth interval, cp. Fig. 2.

The concentrations of the most important anions, NO_3^-, $SO_4^=$ and Cl^- all vary in antiphase with δ, (Finkel and Langway, 1985) yet with less abrupt shifts between cold and mild phases, which supports the conclusion that the δ shifts are not a result of discontinuities in the layer sequence (Dansgaard et al., 1984). The arrows at the bottom of Fig. 2 show Finkel et al.'s (1986) measurements of mean summer (s) and winter (w) values in recent, yet pre-industrial time (1770's and 1830's):

δ (Notice the reverse scale): The shifts from cold to mild conditions seem to have occurred more abruptly than the reverse shifts (see Dansgaard et al. (1984) for further details). In the mild phases, the annual mean δ came close to Holocene winter values (ca. -29.5 per mille).

NO_3^-: The atmospheric turnover of nitrate is still poorly understood. It is remarkable that, compared with Holocene values, the NO_3^- levels are reduced only in the mild phases of Wisconsin. One possibility is that the atmospheric NO_3^- concentration was generally lower during the glaciation, for example due to another circulation pattern, and that lower accumulation rates in the coldest phases lifted the NO_3^- concentrations in the ice to Holocene level.

$SO_4^=$: The main sources of $SO_4^=$ are volcanism and biological activity, but minor contributions originate from sea spray and continental dust. Fig. 2 shows slightly elevated $SO_4^=$ concentrations in the mild phases, perhaps due to lower accumulation rates than in Holocene, but up to ten times higher levels in the cold phases. The high peaks suggest periods of strong volcanism in the cold phases. This has previously been suggested as the cause for similar features in ice from the late and coldest part of the glaciation (Herron and Langway, 1985).

Cl^-: Sea spray is the main source of chloride. In the mild and cold Wisconsin phases, the chloride concentrations were roughly 2 and 4 times the Holocene mean value, respectively. This might be just a result of accumulation rates being 2 and 4 times lower, and this possibility must be checked by measuring annual layer thicknesses. However, the higher wind speeds, previously suggested by several authors (e.g. Lamb, 1971; Ruddiman and Glover, 1975) and confirmed by the dust measurements, must have increased the chlorine aerosol load over the ocean (cp. Petit et al., 1981; Herron and Langway, 1985), at the same time as the extended ice cover tended to reduce it over the ice sheet. Hence, the Cl^- changes observed are probably more than a simple accumulation effect.

CO_2 concentrations in entrapped air bubbles were measured in ppm by Stauffer et al. (1984), and are shown by the circles to the left in Fig. 2. The atmospheric CO_2 concentration seems to have oscillated

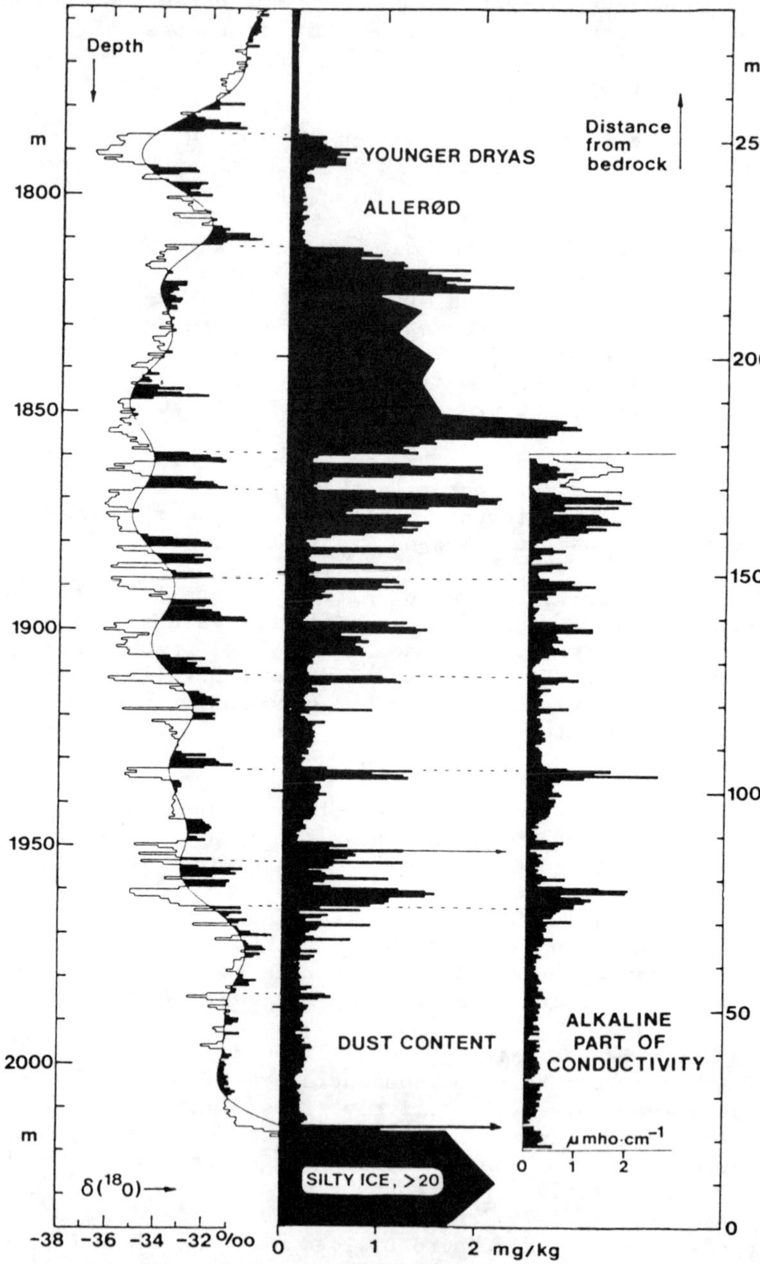

Figure 3. Schematic maps of rainfall patterns for three major stages of the paleoclimatic evolution since 20,000 y BP.

considerably and in phase with δ, but fast CO_2 concentration shifts of the order of 50 ppm are difficult to perceive. One might speculate that they could be caused by rapidly changing ocean water circulation; Duplessy and Shackleton (1985) have shown that the deep water formation in the North Atlantic Ocean is very sensitive to climatic change. If the surface water maintained relatively high temperatures during the first stages of the cold phases, as it did during the period of rapid ice growth following the Eemian interglacial (Ruddiman and McIntyre, 1979), the conditions would be ideal for intense deep water formation and, thereby, fast removal of CO_2 from the atmosphere. Conversely, the deep water formation, and thereby the removal of atmospheric CO_2 to the North Atlantic Ocean, was reduced considerably, if not stopped, due to warming and decreasing salinity of the surface water during the initial stages of the mild phases. However, if the CO_2 shifts, measured by Stauffer et al. (1984) reflect atmospheric changes, they should be of global extension, and one might expect a more jumpy course of Shackleton et al.'s (1983) record on atmospheric CO_2 based on the $\delta^{13}C$ difference between planktonic and benthic foraminifera. Anyway, the CO_2 shifts should also show up in Antarctica, and further analysis of deep Antarctic ice cores is therefore awaited with greatest interest.

Dust concentration and alkalinity

Fig. 3 (from Hammer et al., 1985b) shows the deepest 280m of the continuous Dye 3 δ-profile to the left. The insoluble particle (dust) concentration profile in the mid section is also continuous, but for the 1824-1853m depth interval. The antiphase variation between δ and dust concentration is a general feature in Greenland ice deposited during the glaciation, probably because in periods of colder climate (lower δ values), there was less precipitation, and higher flux of dust into the Arctic due to higher storminess in loess-covered areas in North America and Europe, as well as over the Atlantic Ocean.

The Holocene ice is acidic, but the Wisconsin ice in Greenland is alkaline (cp. alkalinity profile to the right in Fig. 3 expressed as HCO_3^-), mainly due to a $CaCO_3$ aerosol which probably originated from continental shelves exposed to heavy storms, particularly in the coldest periods, cp. the antiphase variation between δ and alkalinity. The ice approaches the neutral point only in the very mildest periods.

FUTURE DEEP ICE CORE DRILLING

There is still a much to learn from ice cores, and further deep drilling through both the Antarctic and the Greenland ice sheets have first priority with several international research planning groups in the fields of climatology and atmospheric physics and chemistry. Each of the ice sheets has its advantages and disadvantages: the Antarctic ice sheet is extremely cold and clean, which is favorable for measuring, i.e., the atmospheric CO_2 content in the past, and it is acid throughout, which is favorable for annual layer identification by a simple technique (Hammer et al., 1986). On the other hand, Greenland is close to the North Atlantic Ocean, which seems to be the "climatic generator" of a major part of the Northern Hemisphere. The relatively high accumulation rates reduce the risk of discontinuities in the layer sequence, and logistics is less expensive in Greenland.

An often raised question is: how old is the ice close to the bedrock of the ice sheets? Lorius et al.'s (1985) $\delta^{18}O$ profile through half of the ice sheet at Vostok clearly shows that in East Antarctica the deepest ice is at least several hundred thousand years old, even if melting occurs at the bedrock, which is the case in large areas, according to radio-echo sounding Oswald and Robin, (1973). As to Greenland, model calculations by Gregoryan et al., (1985) suggest that melting occurs around most of main ice divide, including the Dye 3 area in South Greenland. However, this is inconsistent with direct measurements in the deep hole at Dye 3, showing a bedrock temperature as cold as -13°C (Gundestrup and Hansen, 1984; Dahl-Jensen and Johnsen, 1986). Paterson and Waddington (1986) calculated that the bedrock temperature at Crête (close to the summit in Central Greenland) is, and has always been below the pressure melting point, if the geothermal heat flux is less than 48 mW/m^2. None of the modelers considered the fact that the ice deposited during Wisconsin is up to an order of magnitude softer than Holocene ice (Gundestrup and Hansen, 1984; Shoji and Langway, 1985). Including this in Paterson and Waddington's (1986) model would lower the calculated maximum bedrock temperature, which occurred late in the glaciation, when most of the ice sheet consisted of ice much softer than the present.

Hence, there are good reasons to believe that the bedrock temperature is, and has always been, below the pressure melting point in large areas in Central and North Greenland, where the age of the deepest ice may then be half a million years or more.

ACKNOWLEDGEMENTS

This work was supported by the Danish Natural Science Council, the Danish Commission for Scientific Research in Greenland, and the Commission of the European Communities (contract CLI-067-DK(G)).

REFERENCES

Barkov, N.I., Korotkewitch, E.S., Gordienko, F.G. and Kotlyakov, V.M., 1977. The isotope analysis of ice cores from Vostok Station (Antarctica), to the depth of 950m, *IAHS. Publ.*, 118: 312-321.

Beer, J., Andree, M., Oeschger, H., Stauffer, B., Balzar, R., Bonani, G., Stoller, C., Suter, M., Wolfli, W. and Finkel, R.C., 1985. ^{10}Be variations in polar ice cores, *Am. Geophys. Union, Monograph* 33, *Greenland Ice Core: Geophysics, Geochemistry, and the Environment* (Eds. C.C. Langway, H. Oeschger and W. Dansgaard), 66-70.

Dahl-Jensen, D. and Johnsen, S.J., 1986. Palaeotemperatures still exist in the Greenland ice sheet, *Nature*, 320: 250-252.

Dansgaard, W., Johnsen, S.J., Clausen, H.B. and Langway, C.C., Jr., 1971. Climatic record revealed by the Camp Century ice core, In *The Late Cenozoic Glacial Ages* (ed. K.K. Turekian), Yale Univ. Press, New Haven, Conn., 37-56.

Dansgaard, W., Clausen, H.B., Gundestrup, N., Hammer, C.U., Johnsen, S.J., Kristinsdottir, P. and Reeh, N., 1982. A new Greenland deep ice core, *Science*, 218: 1273-1277.

Dansgaard, W., Johnsen, S.J., Clausen, H.B., Dahl-Jensen, D., Gundestrup, H., Hammer, C.U. and Oeschger, H., 1984. North Atlantic climatic oscillations revealed by deep Greenland ice cores, *Am. Geophys. Union, Geophys. Monograph* 29, *Climate Processes and Climate Sensitivity* (eds. J.E. Hansen and T. Takahashi; Maurice Ewing Series), 288-298.

Delmas, R.J., Ascencio, J.M. and Legrand, M., 1980. Polar ice evidence that atmospheric CO_2 20,000 yr. B.P. was 50% of present, *Nature*, 284: 155-157.

Duplessy, J.-C. and Shackleton, N.J., 1985. Response of global deep-water circulation to Earth's climatic change 135,000-107,000 years ago, *Nature*, 316: 500-507.

Finkel, R.C. and Langway, C.C., Jr., 1985. Global and local influences on the chemical composition of snowfall at Dye 3, Greenland: the record between 10 ka B.P. and 40 ka B.P., *Earth and Planet. Sci. Lett.*, 73: 196-206.

Finkel, R.C., Langway, C.C., Jr. and Clausen, H.B., 1986. Changes in precipitation chemistry at Dye 3, Greenland, *Journ. Geophys. Res.*, 91 (09): 9849-9855.

Gordienko, F.G., Kotlyakov, V.M., Barkov, N.I., Korotkevich, Y.E. and Nikolajev, S.D., 1983. Data of glaciological studies, *Akad. of Sci. of USSR*, 46: 168-171.

Gregoryan, S.S., Buyanov, S.A., Krass, M.S. and Shumskiy, P.A., 1985. The mathematical model of ice sheets and the calculation of the evolution of the Greenland ice sheet, *Journ. of Glaciol*, 31: 281-292.

Gundestrup, N.S. and Hansen, B.L., 1984. Bore-hole survey at Dye 3, South Greenland, *Journ of Glaciol*, 30: 282-288.

Gundestrup, N.S. and Johnsen, S.J., 1985. A battery powered, instrumented deep ice core drill for liquid filled holes, *Am. Geophys. Union, Geophys. Monograph* 33, *Greenland Ice Core: Geophysics, Geochemistry and the Environment* (eds. C.C. Langway, Jr., H. Oeschger and W. Dansgaard), 19-22.

Hammer, C.U., 1980. Acidity of polar ice cores in relation to absolute dating, past volcanism, and radio-echoes, *Journ. of Glaciol.*, 25: 359-372.

Hammer, C.U., Clausen, H.B. and Langway, C.C., Jr., 1985a. The Byrd ice core: continuous acidity measurements and solid electrical conductivity measurements, *Annals of Glaciol.*, 7: 214.

Hammer, C.U., Clausen, H.B., Dansgaard, W., Gundestrup, N., Johnsen, S.J. and Reeh, N., 1978. Dating of Greenland ice cores by flow models, isotopes, volcanic debris, and continental dust, *Journ. of Glaciol*, 20: 3-26.

Hammer, C.U., Clausen, H.B., Dansgaard, W., Neftel, A., Kristinsdottir, P. and Johnson, E., 1985b. Continuous impurity analysis along the Dye 3 deep core, *Am. Geophys. Union, Geophys. Monograph* 33, *Greenland Ice Core: Geophysics, Geochemistry and the Environment* (eds. C.C. Langway, Jr., H. Oeschger and W. Dansgaard), 90-94.

Hammer, C.U., Clausen, H.B. and Tauber, H., 1986. Ice core dating of the Pleistocene/Holocene boundary applied to a calibration of the C-14 time scale, *Radiocarbon*, 28 (2A): 284-291.

Hansen, B.L. and Langway, C.C., Jr., 1966. Deep core drilling in ice and core analyses at Camp Century, Greenland, 1961-66, *U.S. Antarctic Journ.*, 1: 207-208.

Herron, M.M. and Langway, C.C., Jr., 1985. Chloride, nitrate, and sulfate in the Dye 3 and Camp Century, Greenland ice cores, *Am. Geophys. Union, Geophys. Monograph* 33, *Greenland Ice Core: Geophysics, Geochemistry, and the Environment* (eds. C.C. Langway, Jr., H. Oeschger and W. Dansgaard), 77-84.

Johnsen, S.J., Dansgaard, W., Clausen, H.B. and Langway, C.C., Jr., 1972. Oxygen isotope profiles through the Antarctic and Greenland ice sheets, *Nature*, 235: 429-434, and 236: 249.

Lamb, H.H., 1971. Climates and circulation regimes developed over the Northern Hemisphere during and since the last ice age, *Palaeogeogr., Palaeoclimatol., Palaeoecol.*, 10: 125-162.

Langway, C.C., Klouda, G.A., Herron, M.M. and Cragin, J.H., 1977. Seasonal variations of chemical constituents in annual layers of Greenland deep ice deposits, *IAHS. Publication*, 118: 302-306.

Lorius, C., Merlivat, L., Jouzel, J. and Pourchet, M., 1979. A 30,000 year isotope climatic record from Antarctic ice, *Nature*, 280: 644-648.

Lorius, C., Jouzel, J., Ritz, C., Merlivat, L., Barkov, N.I., Korotkevich, Y.S. and Kotlyakov, V.M., 1985. A 150,000 year climatic record from Antarctic ice, *Nature*, 316: 591-596.

Neftel, A., Moor, E., Oeschger, H. and Stauffer, B., 1985. The increase of atmospheric CO_2 in the last two centuries - evidence from polar ice cores, *Nature*, 315: 45-47.

Oeschger, H., Beer, J., Siegenthaler, U., Stauffer, B., Dansgaard, W. and Langway, C.C., Jr., 1984. Late-glacial climate history from ice cores, *Am. Geophys. Union*, *Geophys. Monograph* 29, *Climate Processes and Climate Sensitivity* (eds. J.E. Hansen and T. Takahashi; Maurice Ewing Series), 299-306.

Oswald, G.K.A. and Robin, G. de Q., 1973. Lakes beneath the Antarctic ice sheet, *Nature*, 245: 251-254.

Paterson, W.S.B. and Waddington, E.D., 1986. Estimated basal temperature at Crête, Greenland, throughout a glacial cycle, *Cold Regions Science and Technology*, 12: 99-102.

Petit, J.R., Briat, M. and Royer, A., 1981. Ice age aerosol content from East Antarctic ice core samples and past wind strength, *Nature*, 293: 391-394.

Pisias, N.G., Dauphin, J.P. and Sancetta, C., 1973. Spectral analysis of late Pleistocene-Holocene sediments, *Quat. Res.*, 3: 3-9.

Ruddiman, W.F. and Glover, L.K., 1975. North Atlantic subpolar circulation 9,300 yrs. B.P., *Quat. Res.*, 5: 361-389,

Ruddiman, W.F. and McIntyre, A., 1979. Warmth of the Subpolar North Atlantic Ocean During Northern Hemisphere Ice-Sheet Growth, *Science*, 204: 173-175.

Ruddiman, W.F. and McIntyre, A., 1981. The North Atlantic Ocean during the last deglaciation, *Palaeogeogr., Palaeoclimatol., Palaeoecol.*, 35: 145-214.

Shackleton, N.J., Hall, M.A., Line, J. and Shuxi, C., 1983. Carbon isotope data in core V 19-30 confirm reduced carbon dioxide concentration in the ice age atmosphere, *Nature*, 306: 319-322.

Shoji, H. and Langway, C.C., Jr., 1985. Mechanical properties of fresh ice core from Dye 3, Greenland, *Am. Geophys. Union, Geophys. Mono-*

graph 33, *Greenland ice core: Geophysics, Geochemistry and the Environment* (eds. C.C. Langway, H. Oeschger and W. Dansgaard), 39-48.

Stauffer, B., Hofer, H., Oeschger, H., Schwander, J. and Siegenthaler, U., 1984. Atmospheric CO_2 concentration during the last glaciation., *Ann. of Glaciol.*, 5: 160-164.

Ueda, H.T. and Garfield, D.E., 1969. Drilling through the Antarctic ice sheet, U.S. Army Corps of Engineers, Cold Regions Research and Engineering Laboratory, Hanover, N.H., *Technical Report* 239.

ABRUPT CLIMATIC CHANGES : THE ANTARCTIC ICE RECORD DURING THE LATE PLEISTOCENE

Jean Jouzel[1] Claude Lorius[2] Liliane Merlivat[1]
and Jean-Robert Petit[2]

[1]Laboratoire de Géochimie Isotopique/LODYC
(UA CNRS 1206) CEA/IRDI/DESICP, D.P.C, B.P. 2,
91191 Gif Sur Yvette Cedex, France

[2]Laboratoire de Glaciologie et Géophysique
de l'Environnement B.P. 96 38402
St. Martin D'Heres Cedex, France

ABSTRACT. The $\delta^{18}O$ records from Antarctic (Dome C and Byrd) ice cores allow us to investigate the possible existence of abrupt climatic changes as revealed by Greenland ice cores in the 40 to 10 ky BP time period. At the decade to century time-scale there is no firm conclusion about the presence of rapid climatic changes since seasonal oscillation is not completely averaged out.

At the century to millennium time-scale there are characteristic features of rapid warmings and coolings during the Last Glacial Maximum with a rate of temperature change of about 3°C/ka. This rate is similar to the one observed during the last deglaciation. During this period a significant reversal of the temperature trend is observed, (as already noted for Greenland ice cores) and discussed.

INTRODUCTION

Three deep ice cores reaching back to, at least, the Last Glacial Maximum, have been recovered in Antarctica at Byrd (West Antarctica), and at Vostok and Dome C (East Antarctica). Last Glacial ice can also be present in some shallow ice cores drilled in coastal regions. These cores have been extensively studied for their oxygen-18 content (denoted $\delta^{18}O$ expressed with respect to SMOW), providing a reconstruction of the Antarctic paleoclimate (Epstein et al., 1970; Johnsen et al., 1972; Barkov et al., 1977; Lorius and Merlivat, 1977; Lorius et al., 1979, 1985; Jouzel et al., 1982; Morgan, 1985) and for their chemical impurities (Delmas et al., 1980; Neftel et al.,1982; Petit et al., 1981; Legrand and Delmas, 1984). Of particular interest is the recently obtained 2083m (150 ky) Vostok $\delta^{18}O$ profile which is the first to span the Last Interglacial and recover ice from the preceeding glacial period. Byrd (Johnsen et al., 1972) and Dome C (Lorius et al., 1979) records respectively cover about the last 80 and 32 ky.

W. H. Berger and L. D. Labeyrie (eds.), Abrupt Climatic Change, 235–245.

Here, we will discuss in detail some of these data with the purpose of examining the occurrence and the rapidity of climatic changes recorded in the Antarctic ice during the latest part of the Last Glacial and during the last climatic transition.

This work is partly motivated by recent studies of the Dye 3 ice core and reexamination of the data previously obtained on Camp Century (both in Greenland). These studies revealed (Dansgaard et al., 1982, 1984): (1) that the Dye 3 record shows abrupt and vigorous $\delta^{18}O$ changes from approximately 37 to 28 ky BP; (2) that the late glacial Bölling-Alleröd and Younger-Dryas pollen sequences, well-defined in Europe from 13 to 10 ky BP, are also registered both in Camp Century and Dye 3. For this study, we will use data from Byrd and Dome C, adding to published results detailed isotopic measurements on selected ice from Dome C.

THE DECADE TO CENTURY TIME-SCALE

Details of the $\delta^{18}O$ oscillations in Greenland as given and discussed by Dansgaard et al. (1984) show the following characteristics: (1) the observed shifts are not due to discontinuities in the time scale; (2) they are more abrupt when going from "cold" towards "warm" values, in which case they can take place over a few decimeters, i.e. assuming no drastic changes in accumulation and flow conditions, in a few decades or, at most, in one century; (3) their amplitude can be as large as 5 permil, i.e. almost equivalent to that of 7 permil accompanying the Pleistocene to Holocene shift. These abrupt changes also occur (and over the same periods) for environmental parameters such as insoluble microparticles (Dansgaard et al., 1984), CO_2 concentrations (Stauffer et al., 1984) and the chemical composition of snow i.e. chloride, nitrate and sulfate concentrations (Finkel and Langway, 1985). Recent isotopic results (Dansgaard and White, Biviers symposium) revealed that the deuterium excess of snow (δD-$\delta^{18}O$) also changes abruptly. At Dome C, 22 ice core increments corresponding to comparable time and length scales (a few decades and a few decimeters) have been measured for their isotopic content. For illustration, 4 of the $\delta^{18}O$ profiles are presented on Figure 1 along with their age and the time span represented.

These data show that rapid and vigorous isotopic changes are displayed when the Dome C record is examined in such detail, whatever the depth. The isotopic shifts occur over one decade or less and their amplitude can reach 5 permil which is as in Dye 3, equivalent to that recorded between the Last Glacial Maximum and the average Holocene values (5.4 permil in $\delta^{18}O$). There is no clear relationship between the observed isotopic amplitude and age with large shifts (up to 3.6 permil) also occurring at greater depths. As no criteria were applied to select the ice core increments, it is quite probable that the phenomenon of large isotopic changes at the decadal time scale exists all along the Dome C core. This raises the question of the origin of these isotopic oscillations and their relation to climate. The rather smooth nature of the short term oscillations suggests that they may result from the obliteration of the isotopic signal initially contained in precipitation

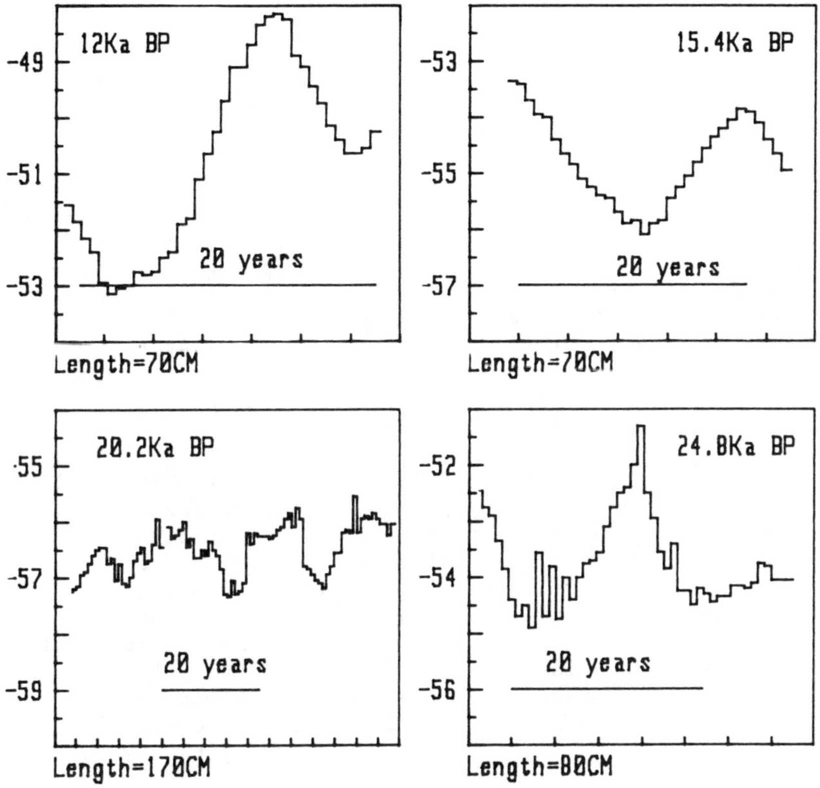

Figure 1. Detailed $\delta^{18}O$ profiles for four ice increments of the Dome C core at different depths.

with time owing to the water vapor diffusion in firn and solid ice. This smoothing effect has been discussed by many authors (Dansgaard et al., 1973; Johnsen, 1977; Johnsen and Robin, 1983; Whillans and Grootes, 1985) and was theoretically treated by Johnsen (1977) and Whillans and Grootes (1985).

According to Johnsen (1977), $\delta^{18}O$ cycles with wavelengths shorter than 20 cm of ice are obliterated during firnification, a length which would be independent of the accumulation rate and only slightly dependent on the temperature. This is contradicted by our South Pole results (Jouzel et al., 1983) showing that the seasonal oscillation is present, at least, back to the beginning of the century. Johnsen's theory, however predicts a disappearance of this oscillation after a few years. In view of these results, the Whillans and Grootes approach in which the smoothing rate is temperature dependent seems more reasonable. In both approaches, the smoothing rate decreases with increasing density and becomes much more slower in solid ice. These approaches although not in agreement, nevertheless allows us to understand better the details of the Dome C isotopic profile.

For this purpose, the characteristics of the isotopic signal in the surface snow layers must be examined. From detailed β activity, tritium and deuterium profiles, Petit et al. (1982) noted the general absence of seasonal indexes at Dome C, further confirmed by Whillans and Grootes (1985). As proposed in both articles, this is related to the low accumulation rate (3.7 cm of ice yr^{-1}). The isotopic stratigraphy is characterized by irregular accumulation events; the lack of precipitation during one season occurs relatively often and seasonal cycles are not formed. On the other hand, large oscillations (up to 10 permil in $\delta^{18}O$) exist but with a wavelength corresponding to a few years of accumulation: roughly 3 years in the example shown in Petit et al. (1982). During firnification some of the smallest oscillations are obliterated and their amplitude decreases but they do not disappear completely owing to the very low temperature (-53.5°C for present conditions). For the same reason, isotopic smoothing in solid ice is very slow, thus leading to the observed wavelength of a few decimeters and to $\delta^{18}O$ amplitudes between 1.5 and 5 permil (Figure 1). The strong variability of the amplitude with depth (more than a factor of 3) is probably due to the fact that the isotopic signal is not really seasonal in nature. It is largely irregular both for its initial wavelength and its initial amplitude. To sum up, it is very likely that the large isotopic variability exhibited at short length scales all along the Dome C core results from the smoothing of surface isotopic oscillations. Even if the recorded shifts are important in some cases (up to 5 permil), they do not correspond to short term climatic changes occurring at the scale of a few decades.

At Vostok, the same isotopic characteristics (lack of a seasonal pattern in surface and existence of decimeter wavelength oscillations at depth) are shown from the few detailed measurements existing down to the Last Glacial Maximum (Dansgaard et al., 1977). Additional data have to be obtained to check whether the relatively low $\delta^{18}O$ amplitude of these oscillations (maximum between 2 to 3 permil) is due to restricted number of ice increments studied (3 in the LGM), or to very low Vostok annual accumulation (1.5 cm of ice during the Last Glacial (Lorius et al., 1985)). At the same length scale, large (to 5 permil in the LGM) and irregular variations of seasonal origin are also present at Byrd (Johnsen et al., 1972).

At this point, one can note a difference between Greenland and Antarctic ice core isotopic profiles which is important as far as abrupt climatic changes are concerned (decadal time scale). Whereas the seasonal oscillation is completely averaged out in the Dye 3 glacial ice, this is not the case for both Antarctic ice cores owing to differences in accumulation regime and temperature. As a result, climatic changes occurring at the scale of a few decades would be more difficult to characterize in Antarctic ice than at Dye 3. Note however, that this does not definitely prove that such changes either do not exist or could not be found.

THE CENTURY TO MILLENNIUM TIME-SCALE

Isotopic smoothing explains why very abrupt isotopic shifts would be difficult to detect in Antarctic ice. However the changes taking place at the century to millennium time-scale are very well depicted in the Byrd and Dome C record. This is particularly interesting in view of a comparison with Dye 3 over the last part of the Last Glacial. Since such a comparison was previously done by Dansgaard et al. (1984), it will be only briefly discussed here.

This part of the Byrd and Dome C records along with the climatic transition and the early Holocene is presented on Figure 2. The dates are from Lorius et al. (1977) for Dome C and from Budd and Young (1983) for Byrd. The deepest part of the Dome C core is characterized by relatively low $\delta^{18}O$ followed by a general trend towards lower values between 32 and 28 ky BP (from about -53 to -55 permil excluding single extreme values). A relatively steep change (see discussion below) towards slightly higher $\delta^{18}O$ (-53.5 permil) then takes place around 27 ky BP. Over the same period, the Byrd record shows two relatively large oscillations at 28 and 35 ky BP (about 2 and 3 permil respectively).

Although dating uncertainties prevent a thorough comparison between the Dome C and Byrd records, these oscillations are different from those recorded over the same period in Dye 3 both in their amplitude (as large as 5 permil at Dye 3) and shape (Dye 3 oscillations are asymmetric and persistent unlike in Dome C and Byrd records).

More information about this last part of the Glacial will be available soon from the continuous Vostok record. However, existing data are probably sufficient to demonstrate the differences between Greenland where Camp Century and Dye 3 have common features (Dansgaard et al., 1982, 1984), and the Antarctic (Johnsen et al., 1972; Lorius et al., 1979). The oscillations contained in the Byrd and Dome C records might reflect damped versions of the largest Greenland oscillation, however, there are no indications yet of concurrent changes in environmental conditions as found in the Dye 3 record (Stauffer et al., 1984; Finkel and Langway, 1985). As a whole, this examination Antarctic data confirms the difference between the Greenland and Antarctic ice core oscillations. The specificity of Greenland oscillations is possibly linked with North Atlantic atmospheric and oceanic circulations (Dansgaard et al., 1984; Oeschger et al., 1984; Broecker et al., 1985; Rind et al., in press).

THE LAST GLACIAL TO HOLOCENE TRANSITION

Before examining the climatic transition in Antarctic ice, we first come back to Greenland cores which are the focus of much interest in recent interpretations of the Northern Hemisphere climate over this period (Dansgaard et al., 1984; Broecker et al., 1985; Rind et al., in press). The strong oscillations recorded in Camp Century and Dye 3 just before the onset of postglacial time, correspond well to the Alleröd-Younger Dryas oscillation seen in pollen records and in lake sediments throughout Europe (Siegenthaler et al., 1984; Broecker et al., 1985; Rind et

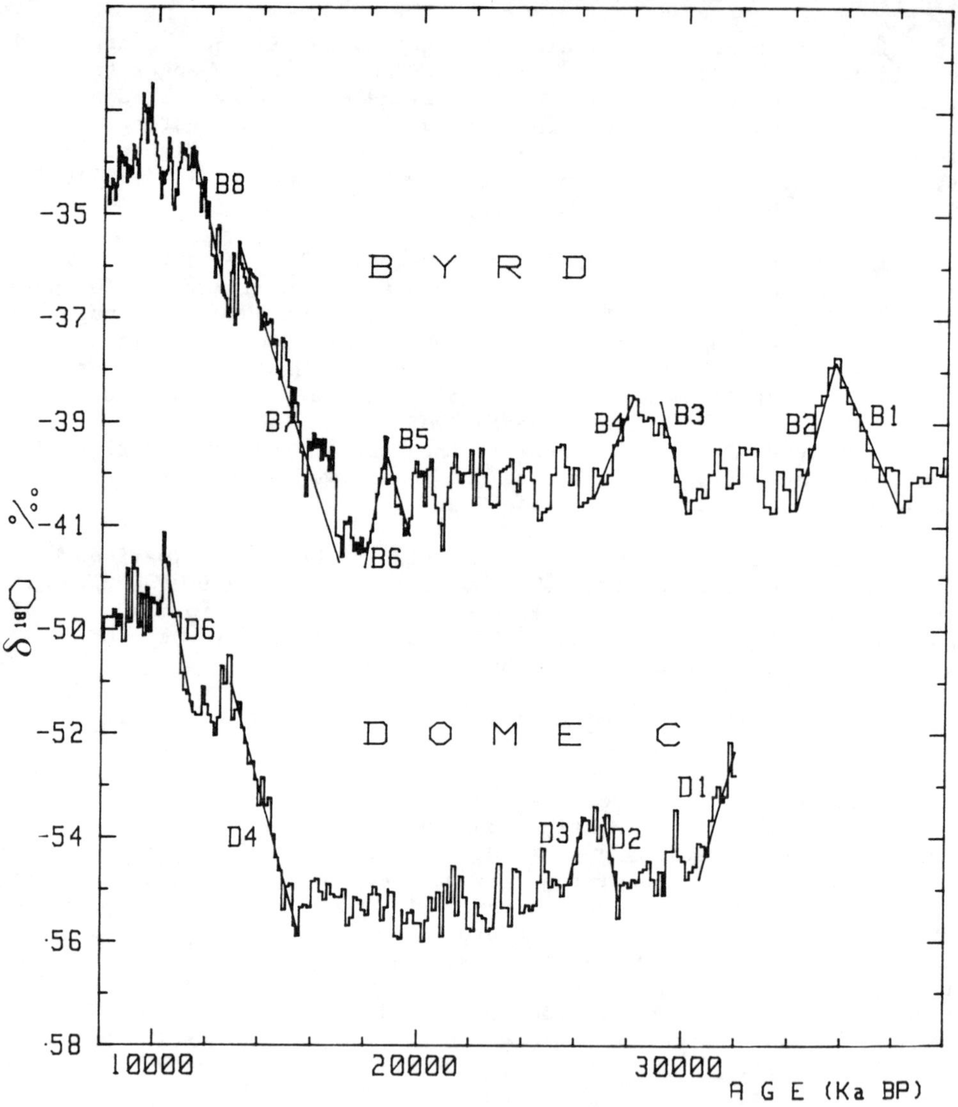

Figure 2. The Byrd and Dome C $\delta^{18}O$ records from 40 ky BP and 32 ky BP respectively to 8 ky BP as a function of age. The dates are from Budd and Young (1983) and Lorius et al. (1979) along with indication of the oscillations examined in Table I. Byrd data have been kindly provided by W. Dansgaard.

al., in press). The Younger Dryas cooling followed the warming trend of the Bölling-Alleröd, spanning approximately a millennium from 11 to 10.2 ky BP. In view of the climatic interpretation in terms of North Atlantic cooling during this period, it is important to determine if such a feature is present in Antarctic ice. Clearly, at Dome C (Figure 2) the transition is a two step process with two warming trend periods from 15.5 to 13.2 ky BP, and from 11.6 to 10.2 ky BP interrupted by a slightly colder period lasting about 1 ka (12.8 to 11.6 ky BP). The $\delta^{18}O$ amplitude of this reversal is around 1 permil, but even if much more weaker than in Greenland (4 to 5 permil at Dye 3), there is no doubt that a cold oscillation exists in the Dome C transition. The relation of this event to the European Younger Dryas seems to be contradicted by our dating which puts this "cold" period some 1.5 ka before the Younger Dryas. However, this difference is in the range of dating uncertainty. As long as absolute dating is not available, it is plausible that the Dome C reversal presently dated between 12.8 and 11.6 ky BP is a Southern Hemisphere equivalent of the Dryas. The same discussion would probably apply to the Byrd $\delta^{18}O$ profile in which a weak ($\approx$1 permil) reversal can also be noticed in the same part of the transition (Figure 2).

If real, why is this event much more weaker in Dome C and Byrd than in Dye 3? We can propose very tentatively two explanations: 1) the remoteness of the Dome C site from the ocean could lead to weakening of a climatic event possibly originating from a sea surface temperature (SST) cooling as suggested and modeled in the Northern Hemisphere for the Younger Dryas (Rind et al., in press); 2) the fact that this hypothesized SST cooling should be much more weak around Antarctica than in Northern Atlantic; in other words a damp response of the Southern Hemisphere ocean to an initial Northern Atlantic cooling.

We will not go further in this speculation since our purpose here is to present data rather than to discuss mechanisms. However, we would like to point out that the existence of this weak but convincingly recorded reversal in Antarctic ice will, when correctly dated have to be taken into account in the perspective of a global interpretation of the last deglaciation.

RAPIDITY OF CLIMATIC CHANGES RECORDED IN ANTARCTIC ICE

The Dome C and Byrd records presented on Figure 2 give a unique opportunity to analyze century and millennium climatic changes. At these time scales the isotopic signal is averaged out over a few meters of ice and the smoothing discussed above no longer operates. We have here a continuous climatic sequence, essentially undisturbed (at least for Dome C) by changes in ice flow conditions. Also, the $\delta^{18}O$ signal is an atmospheric and thus practically instantaneous response to climatic changes. Moreover, there are no known "regional" factors influencing the Antarctic climate in the sense in which North Atlantic changes influence the Greenland record, and thus the Antarctic record is of relatively large geographic significance, monitoring the conditions in the high latitudes of the entire Southern Hemisphere.

To check on rate of change we have chosen the best defined oscillations over the last part of the Glacial (40 ky BP to the LGM) i.e. those having an amplitude largely exceeding the isotopic noise, which is of the order of 0.5 permil for a 150 years sample at Dome C (Robin, 1983) and documented by numerous determinations. Three were selected at Byrd with $\delta^{18}O$ peaks around 35, 28 and 19 ka BP, and two at Dome C with peaks at 27 ka BP and around 32 ka BP (Figure 2). The length, duration and $\delta^{18}O$ amplitude of each of these oscillations are reported on Table 1 along with the $\delta^{18}O$ time gradients (with either positive or negative values).

The same calculations were done for the last transition which, owing to the cold reversal was divided in two steps. For both sites, a correction for the concurrent change in the $\delta^{18}O$ of sea surface water was applied with a maximum amplitude of 1.6 permil (Duplessy, 1978), so we assume a rate of $\delta^{18}O$ change of 0.2 permil ka for water source after 16 ky BP. For the sake of comparison with other paleoindicators, we have tentatively expressed these $\delta^{18}O$ gradients in temperature gradients assuming that even relatively minor oscillations essentially result from ground temperature variations at the site. They are obtained by multiplying the $\delta^{18}O$ time gradients by 1.3, a figure derived from theoretical considerations but which also explains well the modern East Antarctica $\delta^{18}O$ change with temperature (Jouzel and Merlivat, 1984). However, we emphasize that these figures should be used with caution for the many reasons which could obscure the δ-temperature relationship (changes in the ice thickness, change in the origin of the precipitation, changes in

Site	Feature N°	Timing Ky BP	Duration Ka	$\delta^{18}O$ permil per Ka	T°C per Ka
Byrd	B1	38.2 - 35.7	2.5	+ 1.12	+ 1.49
	B2	35.6 - 34.1	1.4	- 1.96	- 2.61
	B3	30.0 - 29.0	2.0	+ 2.19	+ 2.91
	B4	28.1 - 27.0	1.1	- 1.70	- 2.26
	B5	19.6 - 18.7	0.8	+ 1.70	+ 2.26
	B6	18.6 - 17.9	0.8	- 3.05	- 4.05
	B7	16.9 - 13.1	3.8	+ 1.80	+ 2.39
	B8	12.8 - 11.4	1.4	+ 2.53	+ 3.36
Dome C	D1	32.0 - 30.3	1.7	- 1.67	- 2.22
	D2	27.6 - 27.0	0.6	+ 2.74	+ 3.64
	D3	26.3 - 25.7	0.6	- 2.16	- 2.87
	D4	15.6 - 12.9	2.7	+ 2.04	+ 2.71
	D5	11.4 - 10.4	1.0	+ 3.06	+ 4.07

Table 1. The rapidity of climatic changes as deduced from the Byrd and Dome C records. Column 1 refers to the name reported on Figure 2 for the successive oscillations examined in each profile.

the microphysics of formation and the seasonal pattern of precipitation, change in the atmospheric circulation, etc.) as developed elsewhere (Lorius et al., 1984; Joussaume et al., 1984). We also have to keep in mind that the rates of change reported in Table 1 are not precisely determined, in particular for the oscillations for which the overall uncertainty can be as high as 50 percent. This uncertainty is lower for the transition (10 to 20 percent). Within these limits, it can be noticed that the rates of temperature change can be of the same order at the onset an oscillation as during the Glacial-Interglacial transition. The rates range from 1.5 to 4.0°C/ka.

CONCLUSION

A thorough examination of the $\delta^{18}O$ Antarctic ice record (essentially Dome C and Byrd) over the 40 to 10 ky BP period with respect to rapid climatic changes reveals the following features:

1) Very rapid changes (decadal time scale) would be difficult to characterize because the short-term original surface signal is likely to be completely smoothed out.

2) There is a weak but clear (at least at Dome C) reversal inside the climatic transition. Dating uncertainties have to be resolved in order to establish if this reversal corresponds to the Northern Hemisphere Younger Dryas event. At present, such a correspondence cannot be ruled out.

3) The maximum rate of temperature change as deduced from the isotopic profiles is of about 3°C/ka. The same range of values typically between 2 and 3°C/ka are observed along the transition and during the oscillations which took place, at the end of the Last Glacial Maximum.

To conclude, climatic oscillations exist in Antarctic ice on the millennium time scale both during the last part of the Glacial and during the transitions which are, as previously noted, muted when compared with the Greenland oscillations over the same period (Dansgaard et al., 1984; Broecker et al., 1985). When better documented (in particular from the continuous Vostok profile) correctly dated and correlated between different Antarctic sites and with oceanic and continental similar events, they certainly will contribute to a better understanding of the global climate during the late Pleistocene.

ACKNOWLEDGEMENTS

We thank Jim White for very carefully reading the manuscript.

REFERENCES

Barkov, N.I., Korotkevich, E.S., Gordienko, F.G. and Kotlyakov, V.M., 1977. The isotope analysis of ice cores from Vostok station (Antarctica) to the depth of 950 m. Isotopes and Impurities in Snow and Ice. *IAHS Publ.* 118: 127-137.

Broecker, W.S., Peteet, D.M. and Rind, D., 1985. Does the ocean-atmosphere system have more than one stable mode of operation? *Nature*, 315: 21-26.

Budd, W.F. and Young, N.W., 1983. Application of modelling techniques to measured profiles of temperatures and isotopes. *The climatic record in polar Ice Sheets*, G. de Q.. Robin ed ., Cambridge University Press, 145-150.

Dansgaard, W., Johnsen, S.J., Clausen, H.B. and Gundestrup, N., 1973. Stable isotope glaciology, *Meddel. om Grönland*, 197, N°2, 53 pp.

Dansgaard, W., Barkov, N.I. and Splettstoesser, J., 1977. Stable isotope variations in snow and ice at Vostok, Antarctica, *IAHS Publ* 118: 204-209.

Dansgaard, W., Clausen, H.B., Gundestrup, N., Hammer, C.U., Johnsen, S.F, Kristinsdottir, P.M. and Reeh, N., 1982. A new Greenland deep ice core, *Science*, 218: 1273-1277.

Dansgaard, W., Johnsen, S.J., Clausen, H.B., Dahl-Jensen, D., Gundestrup, N., Hammer, C.V. and Oeschger, H., 1984. North Atlantic climatic oscillations revealed by deep Greenland ice cores, *Geophysical Monograph*, 29: 288-298.

Dansgaard, W. and White, J.W.C. Evidence for abrupt climatic change from ice cores, presented at the NATO Workshop on Abrupt Climatic Changes (this symposium).

Delmas, R.J., Ascencio, J.M. and Legrand, M., 1980. Polar ice evidence the atmospheric CO_2 20,000 yr B.P. was 50% of present, *Nature*, 2: 155-157.

Duplessy, J.C., 1978. Isotope studies, *Climatic Change*, John Gribbin ed., 46-47.

Epstein, S., Sharp, R.P. and Gow, A.J., 1970. Antarctic ice sheet: stable isotope analysis of Byrd station cores and interhemispheric climatic implications, *Science*, 168: 1570-1572.

Finkel, R.C. and Langway Jr., C.C., 1985. Global and local influences of the chemical composition of snow fall at Dye 3, Greenland: the record between 10 ka B.P. and 40 ka B.P., *Earth Planetary Science Letters*, 73: 196-206.

Johnsen, S.J., Dansgaard, W., Clausen, H.B. and Langway Jr., C.C., 1972. Oxygen isotope profiles through the Antarctic and Greenland ice sheets, *Nature*, 235: 429-434,.

Johnsen, S.J., 1977. Stable isotope homogenization of polar firn and ice, Isotopes and Impurities in Snow and Ice, *IAHS Publ. 118*: 210-219.

Johnsen, S.J., and Robin, G. de Q., 1983. Diffusion of stable isotopes, *The Climatic Record on Polar Ice Sheets*, G. de Q. Robin ed., Cambridge University Press, 57-62.

Joussaume, S., Jouzel, J. and Sadourny, R., 1984. Water isotope cycles in the atmosphere: first simulation using a general circulation model, *Nature*, 311: 24-29.

Jouzel, J., Merlivat, L. and Lorius, C., 1982. Deuterium excess in an East Antarctic ice core suggests higher relative humidity at the oceanic surface during the Last Glacial Maximum, *Nature*, 280: 644-648.

Jouzel, J., Merlivat, L., Petit, J.R. and Lorius, C., 1983. Climatic information over the last century deduced from a detailed isotopic record in the South Pole snow, *J. Geophys. Res.*, 88 (C4): 2693-2703.

Jouzel J. and Merlivat, L., 1984. Deuterium and Oxygen 18 in precipitation: Modelling of the isotopic effect during snow formation, *J. Geophys. Res.*, 89: 11749-11758.
Legrand, M., and Delmas, R.J., 1984. The ionic balance of Antarctic snow: a detailed record, *Atmospheric Environment*, 18: 1867-1874.
Lorius, C. and Merlivat, L., 1977. Distribution of mean surface stable isotope values in East Antarctica, Isotopes and Impurities in Snow and Ice, *IAHS, Pub*. 118: 127-137.
Lorius, C., Merlivat, L., Jouzel, J. and Pourchet, M., 1979. A 30,000 yr isotope record from Antarctic ice, *Nature*, 280: 644-648.
Lorius, C., Raynaud, D., Petit, J.R., Jouzel, J. and Merlivat, L., 1984. Late glacial maximum Holocene atmospheric and ice thickness changes from ice core studies, *Annals of Glaciology*, 5: 88-94.
Lorius, C., Jouzel, J., Ritz, C., Merlivat, L., Barkov, N.I., Korotkevich, Y.S. and Kotlyakov, V.M., 1985. A 150,000 year climatic record from Antarctic ice, *Nature*, 316: 591-596.
Morgan, V.I., 1985. An oxygen isotope climate record from the Law Dome, Antarctica, *Climatic Change*, 7: 415-426.
Neftel, A., Oeschger, H., Schwander, J., Stauffer, B. and Zumbrunn, R., 1982. Ice core sample measurements give atmospheric CO_2 content during the past 40,000 yr, *Nature*, 295: 220-223.
Oeschger, H., Beer, J., Siegenthaler, U., Stauffer, B., Dansgaard, W. and Langway Jr., C.C., 1984. Late glacial climate history from ice cores, *Geophysical Monograph*, 29: 299-306.
Petit, J.R., Briat, M. and Royer, A., 1981. Ice age aerosol content from East Antarctic ice samples and past wind strength, *Nature*, 293: 391-394.
Petit, J.R., Jouzel, J., Pourchet, M. and Merlivat, L., 1982. A detailed study snow accumulation and stable isotope content in the Dome C (Antarctica) area, *J. Geophys. Res.*, 87: 4301-4308.
Rind, D., Peteet, B., Broecker, W., McIntyre, A. and Ruddiman, W., in press. Impact of cold North Atlantic sea surface temperatures on climate implications for the Younger Dryas cooling.
Robin, G. de Q., 1983. The climatic record from ice cores. *The Climatic Record in Polar Ice Sheets*, G. de Q. Robin ed., Cambridge University Press, 180-195.
Siegenthaler, U., Eicher, U., Oeschger, H. and Dansgaard, W., 1984. Lake sediments as continental ^{18}O records from the glacial-postglacial transitions, *Annals of Glaciology*, 5: 149-152.
Stauffer, B., Hofer, H., Oeschger, H., Schwander, J. and Siegenthaler, U., 1984. Atmospheric CO_2 concentrations during the last glaciation, *Annals of Glaciology*, 5: 160-164,.
Whillans, I.M. and Grootes, P.M., 1985. Isotopic diffusion in cold snow and firn, *J. Geophys. Res.*, 90: 3910-3918.

ENVIRONMENTAL CHANGES DURING LAST DEGLACIATION INFERRED FROM CHEMICAL ANALYSIS OF THE DOME C ICE CORE

Michel Legrand and Robert J. Delmas

Laboratoire de Glaciologie et
Géophysique de l'Environnement
BP 96 - 38402 St. Martin D'Heres CEDEX (France)

ABSTRACT. The global environmental changes inferred from the glaciochemical study of the Dome C ice core (Antarctica) for the last climatic transition 15-11 Ka BP are reported in detail. The methods and the calculation for interpreting the data are explained. It is confirmed that the most significant changes in the deposition of aluminum and sodium occurred during the first stage of the transition, i.e. in about 2 Ka. The ratio Cl/Na varied considerably when passing from the Last Glacial Maximum to the Holocene, most probably in relation with the stability of the atmosphere. Major changes occurred also for the origin and composition of sulfate and nitrate contributions. Since the beginning of the Holocene, the acid forms (H_2SO_4 and HNO_3) have been predominant for both anions contrary for what is observed for the glacial age.

INTRODUCTION

Polar ice cores are the most relevant records of changes in atmospheric composition. Chemical analysis of deep ice cores from Greenland and Antarctica has already provided valuable information in the atmospheric aerosol composition in the past on the scale of several millenia. This paper reports briefly on the changes observed during the last deglaciation in the composition of chemical impurities along the 905m deep ice core of Dome C (74°39'S, 124°10'E), drilled in 1977 on the central East Antarctic Plateau (Lorius et al., 1979) and spanning about the last 30,000 years.

METHODS FOR OBTAINING AND INTERPRETING GLACIOCHEMICAL DATA

To study the last large climatic transition, we selected 24 core sections between 300 and 600m depth (expressed in m. of ice equiv.). This sampling spans the time period 8,500-20,000 yr BP. Each ice sample (15cm long and covering about 5 years of precipitation) was copiously washed with double deionized water in a clean air cabinet in order to remove surface contaminants. Chemical analysis was then performed by ion chromatography (Dionex 2010) for Na, NH_4, Ca, Mg, Cl, NO_3 and SO_4

W. H. Berger and L. D. Labeyrie (eds.), Abrupt Climatic Change, 247–259.

major ions and by acid titration for acidity. The detailed analytical procedure has been published elsewhere (Legrand et al., 1984).

The determination of the amounts of all major ions listed above in each of the ice sections is of considerable interest in interpreting the raw chemical data. Additional aluminum measurements were made (by neutron activation) on aliquots of the meltwater samples (De Angelis et al., 1984; Lorius et al., 1984) in order to estimate the terrestrial input. Some difficulties arose with the sodium data, since in Antarctic ice, and particularly during the last ice-age, this element is contributed both by the oceans and the continents. It was discovered that the crustal derived sodium, which is generally considered as insoluble in meltwater, dissolved completely when passing through the ion chromatographic columns. Marine sodium (Na_m) was therefore calculated from total sodium (Na) by subtracting a crustal contribution evaluated according to the relationship:

(1) $Na_m = Na - 0.25\ Al$

where 0.25 is the value of the weight ratio Na/Al in the mean crustal composition.

Thereafter it was possible to balance satisfactorily - with less than 10% imbalance - the anion and cation concentrations:

(2) $[Na_m] + [NH_4] + [K] + [Ca] + [Mg] + [H] = [Cl] + [NO_3] + [SO_4]$

Since $[NH_4]$ has been observed to be always very low (less than 1% of the total ionic budget), it was neglected in the calculation. One part of the aerosol deposited in snow is issued from sea salt. This fraction can be easily calculated using Na_m as the marine reference element; the composition of bulk sea-water being known. The non sea-salt fraction of an ion X is called excess-X or X^* (X = K, Ca, Mg, Cl or SO_4), according to the terminology commonly used in precipitation chemistry.

For the non sea-salt fraction of the impurities, we may now write:

(3) $[K]^* + [Ca]^* + [Mg]^* + [H] = [Cl]^* + [SO_4]^* + [NO_3]$

For the Holocene time period, the excess-amounts of K, Ca and Mg are negligible and (3) may be written in a simplified form:

(3′) $[H] = [Cl]^* + [SO_4]^* + [NO_3]$

In fact, equation (3′) represents the ion balance of acid species. Then we can introduce the ratio $r = [Cl]/[Na_m]$ which has been demonstrated to vary spatially and temporally in the Antarctic (Legrand and Delmas, submitted).

The reference value of r in sea water is 1.17 (for molar concentrations or 1.8 for weight concentrations). The actual value of r (higher or lower than 1.17) provides important information on the chemical composition of snow. If $r > 1.17$, excess chloride is present. In this case Cl^* and SO_4 are both in the form of HCl and H_2SO_4. As NO_3 is also in the form of HNO_3 (Legrand and Delmas, 1986) the ice contains principally 3 mineral acids (HCl, H_2SO_4 and HNO_3) and sea salt particles as soluble impurities. If $r < 1.17$, the interpretation is less immediate. We have demonstrated that, in this case, one part of SO_4^* is in the form of $Na_2SO_4^*$ and that HCl needs not be taken into account in the discussion or snow impurities. Therefore only 2 mineral acids (H_2SO_4 and HNO_3) are present. The amount of sodium present as $Na_2SO_4^*$ is equal to 1.17 Na_m-Cl.

The two cases above with $r \neq 1.17$ actually reflect the interaction of H_2SO_4 with the sea salt particles in the atmosphere. The occurrence or the absence of this interaction allows certain deductions on the behavior of the air masses over the Antarctic continent (Legrand and Delmas, submitted).

We shall now comment briefly on our present knowledge of the chemical composition of snow impurities for the Last Glacial Maximum (LGM). As already pointed out by Petit et al. (1981), the high concentrations of sodium and aluminum for this time period can be explained by enhanced continental and marine aerosol inputs. These large amounts of primary aerosol interact with the gas-derived acid aerosols, a phenomenon which does not occur during the Holocene. Moreover the crustal input contains a significant amount of salts (such as $CaSO_4$, $MgSO_4$, $Ca(NO_3)_2$ and $Mg(NO_3)_2$) responsible of the presence of the "excess" fraction or these two metals. This fraction can be calculated using the Al values (Legrand, 1985).

The sulfate and nitrate bound to excess Ca and Mg can also be calculated using the experimental correlation linking NO_3 and Al data (Legrand, 1985). All of these complications in the interpretation of the raw data of the measured elements cease at the end of the LGM. On the other hand, the fractionation of sea salt (as reflected by a value of $r \neq 1.17$), nearly zero for the LGM, becomes observable in ice just at the end of the ice age.

Having made these preliminary statements, it is now possible to present the results of the chemical measurements and their significance with respect to our understanding of the various phenomena which modified the atmospheric environment when the Earth's climate jumped from the LGM to the Holocene.

RESULTS AND DISCUSSION

The discussion will be limited to the following elements or ions: Na, Cl, SO_4, NO_3 and Al, and will focus on the transition period (15.5-10 Ka BP). However all other chemical parameters measured in this study will be used and introduced in the calculations reported in the preceding section in order to obtain a detailed interpretation of the raw data. In particular, we have tried in this paper to demonstrate that the drawing up of the ionic balance in each ice sample opens wide perspectives for the understanding of our past environment and the behavior of the major biochemical cycles during a major climate change. We have followed the dating and climatic interpretations of the Dome C ice-core of Lorius et al. (1979). A more complete set of chemical measurements made on this ice-core can be found in Legrand (1985).

Finally we stress that the compositional variations we find in the precipitation do not necessarily represent the actual variations which occurred in the atmospheric chemistry. Snow scavenges gases and aerosol particles of various sizes differently. Moreover accumulation rates vary considerably during a climate change so that wet/dry deposition rates can be modified, particularly in Central Antarctic areas where snow accumulation rates are extremely low. The variations observed in snow must be regarded more as qualitative than truly quantitative with respect to the atmospheric aerosol.

Aluminum

Aluminum is the tracer of terrestrial dust. Concentrations of this element begin to decrease at a depth of ≈500m, (Fig. 1) i.e. around 15.5 Ka BP, simultaneously with the climatic change (Petit et al., 1981). Optical measurements of dust lead to similar conclusions (Thompson et al., 1981; Royer et al., 1983). Contrary to what occurs in Greenland, where the high values of dust typical of the LGM cease in a few tens of years (Finkel et al., 1985), the Antarctic decrease lasts several millennia. Over about 2.5 Ka (until ≈ 13 Ka BP) this contribution falls by one order of magnitude. In the profile presented here (Fig. 1) - although possibly only characteristic of this sampling since not backed up by Petit et al.(1981) - values continue to decrease after this date with a minimum reached around 9 Ka BP. In this secondary phase the decrease is of a factor of 5 over about 4 Ka. In fact, Al concentrations typical of the Holocene (mean value: 2.7 ± 2.4 ng g^{-1} from Petit et al. (1981) are already reached around 13 Ka BP, however in our profile the early Holocene is a period of extremely low Al levels. In Greenland ice, the Younger Dryas (11-10 Ka BP) is characterized by a reactivation of dust transport to the ice-cap. This short event is not recorded on Dome C profiles available so far (except if the dating used here is not correct, Jouzel et al., 1987).

Sodium

Sea salt sodium (Na_m), the marine reference element, was calculated from measured sodium by subtracting a certain amount proportional to Al (equation 1). Na_m is significantly different from Na only for the highest Al values, i.e. in the LGM and at the beginning of the transition (Fig. 2).

As already pointed out by Petit et al. (1981), the input of sea salt particles starts to decrease around 15.5 Ka BP and continues until around 11 Ka BP. In 4.5 Ka, Na_m falls by only a factor of 2, and even remains constant from 13.5 to 11.5 Ka BP (a relatively cold period, Jouzel et al., 1987). Thereafter, it reaches its Holocene level in about 1 Ka. The overall decrease of this element between the LGM and the Holocene is only 1 order of magnitude, i.e. significantly less than Al (Fig. 1).

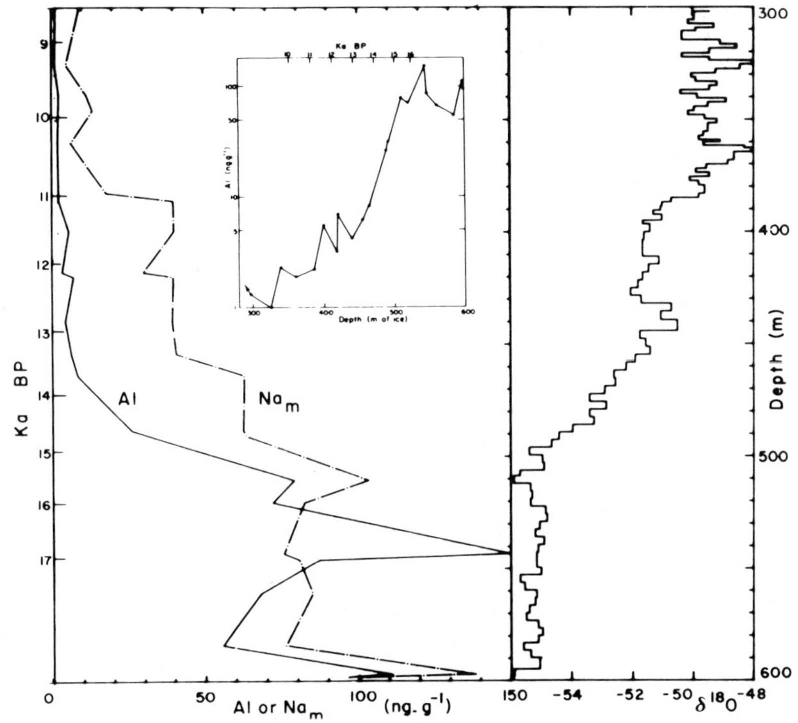

Figure 1. Aluminum (1) and marine sodium (Na_m) (2) (in ng g^{-1}) as a function of depth (in meters of ice) in the Dome C core. The inset is a representation of the aluminum decrease in a logarithmic scale, showing that the concentrations of this element seem to reach their lowest values around 9 Ka BP.

An alternative way of interpreting sodium data is to consider the amount of this element which can be bound to excess sulfate as $Na_2SO_4^*$. As explained above, this compound can be calculated only in the case where r < 1.17. Such a situation is observed throughout the transition (see below). For this time period, we have calculated a correction to obtain Na'_m, the "unfractionated" sea salt contribution:

$$(4) \quad [Na'_m] = [Na_m] - [SO_4]^* + [H] - [NO_3]$$

From measurements made in recent snow layers, it has been assumed (Legrand and Delmas, submitted) that this kind of seasalt reaches central Antarctica in winter (and also in glacial-age) meteorological conditions. The profile of Na'_m reported in Fig. 2 indicates that this contribution of sodium behaves much more similarly to Al than Na_m i.e. it decreases by about one order of magnitude between the beginning and the end of the transition. To date we can not determine whether Na_m or Na'_m is the best indicator of the sea salt condition. We can only state that each of them might be the bearer of a different glaciochemical message.

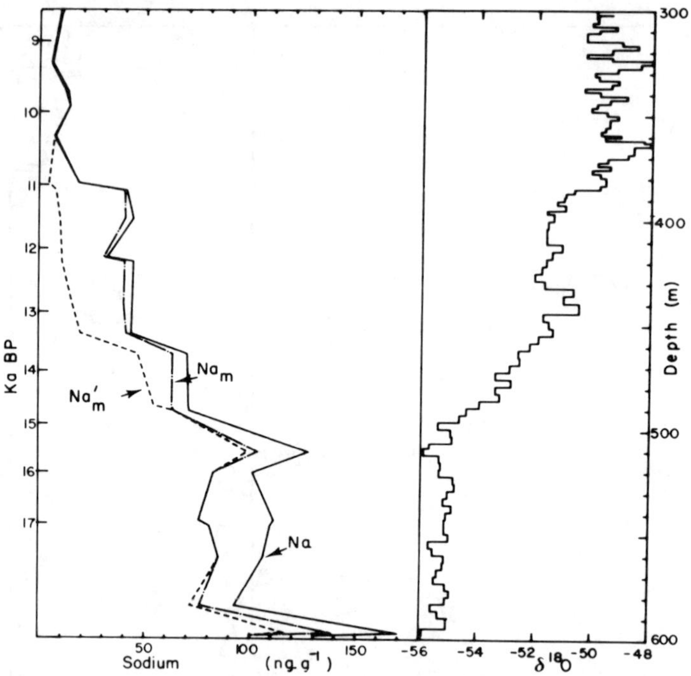

Figure 2. The total sodium (1) deposition (in ng g^{-1}) may be decomposed in marine sodium Na_m (2) or in the non-fractionated marine sodium Na_m (3), see text.

The Na_m/Al and Na'_m/Al ratios (R and R′ weight ratios)

The ratio R can be considered representative of the ratio of the marine to crustal primary aerosol contributions. The interest in calculating this ratio is to eliminate certain factors which may affect simultaneously both contributions, in particular the variations of snow accumulations rates, and, to a certain extent, the atmospheric transport (although these two aerosols are probably not transported to the Antarctic in the same air masses). Therefore the variations of R are probably mostly dependent on the intensities of the sources of Al and Na_m.

From the beginning of the climatic change around 15.5 Ka BP until the end of the transition, R increases by nearly one order of magnitude until the end of the transition (Fig. 3). Subsequently, it drops abruptly to typical Holocene values. It appears therefore that the transition values of R are about 8 times higher than the LGM values, due to the more rapid decrease of Al in comparison with Na_m.

If $R' = Na'_m$/Al is considered instead of R, a very different figure is obtained: there are no high values during the transition. Taking 1.2 as a typical LGM value for R′, the increase during the transition is

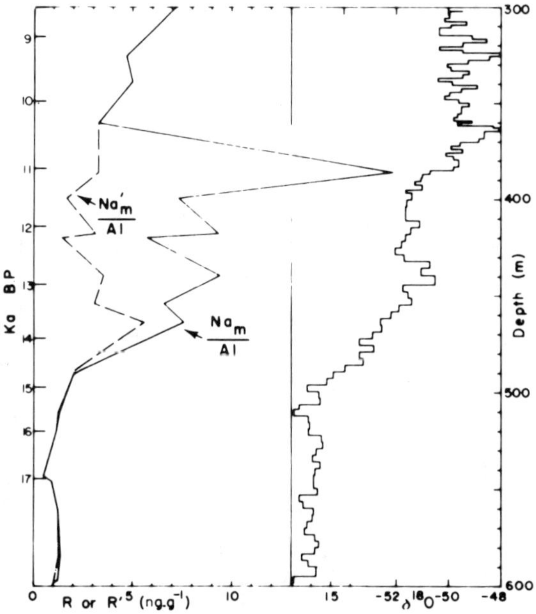

Figure 3. Na_m/Al (profile 1) and Na'_m/Al (profile 2) ratios (called R and R' in the text). The significance of the difference between R and R' during the transition is explained in the text.

only of a factor of 2 to 3. The Holocene may be characterized by still higher (x2) values.

Cl/Na_m (r ratio)

The variations of the Cl/Na_m ratio, r, for the transition period is most interesting (Fig. 4). Although r is not significantly different from the bulk sea water weight ratio (1.8) during the LGM, it begins to decrease continuously when the climate changes around 15.5 Ka until near the end of the transition. Thereafter, a very rapid variation occurs where r jumps from values lower than 1.8 to values markedly higher (up to 4), with maxima apparently corresponding to the two warm periods observed in the early Holocene. In Fig. 5 which shows r as a function of Cl, we see clearly that 3 distinct regimes may be identified: the LGM, the transition and the early Holocene.

Sulfate

Unraveling the various contributions to the total sulfate measured in the ice samples is a difficult task which necessitates taking into account all other chemical parameters entering in the ionic budget (Fig. 6). In this paper we shall describe only what is found for excess sulfate (non sea-salt sulfate, SO_4^*), the most interesting component.

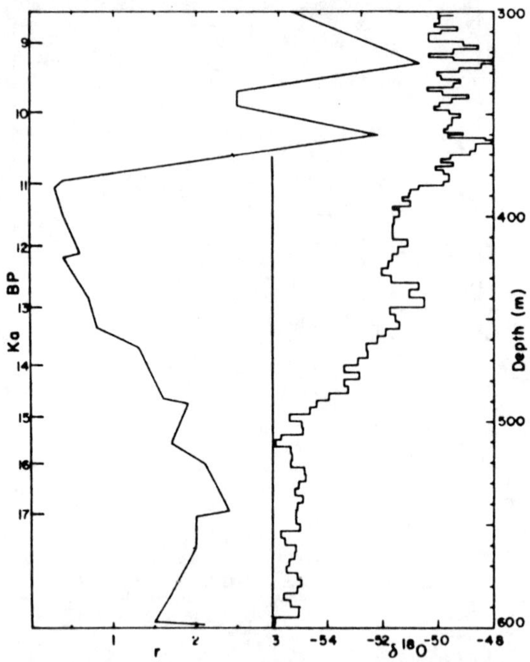

Figure 4. Variations of the Cl/Na_m ratio (r) (value of this ratio (in weight) in bulk sea water: 1.8). Note that r is close to 1.8 during the ice-age only.

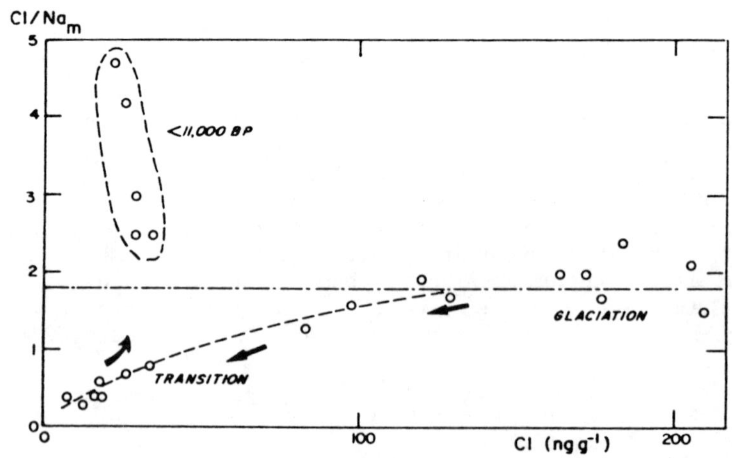

Figure 5. Values of r for different time periods as a function of Cl.

SO_4^* is high, in the range 2-4 μEq. l^{-1} during the LGM. It falls rapidly to 2 and between 1-2 μEq. l^{-1} during the transition and the Holocene, respectively. Using concentrations of all other major ionic species involved in the ion budget, the respective variations of the various forms of SO_4^* can then be calculated on the basis of equations 2, 3 and 3' according to a procedure which will be described in greater detail elsewhere (Legrand and Delmas, to be published).

1) During the LGM the major part of SO_4^* is in neutral form (SO_{4N}^*), probably linked to calcium.
2) At the beginning of the transition this neutral sulfate progressively disappears, replaced by another SO_{4N}^* (Na_2SO_4) bonded to sodium, as indicated by the decrease of the ratio Cl/Na_m to under 1.8 during the transition.
3) At the end of the transition, this last ratio rapidly becomes >1.8, which renders the calculation of $Na_2SO_4^*$ impossible (see discussion at the end of section on methods).
4) By subtracting these two forms of SO_{4N}^* we obtain the amount of excess-sulfate (SO_{4A}^*) present as an acid (H_2SO_4). Above 36m. the value of SO_{4A}^* given in Fig. 6 must be considered as an upper estimate, however, so that SO_{4A}^* in the Holocene may in fact be lower than indicated

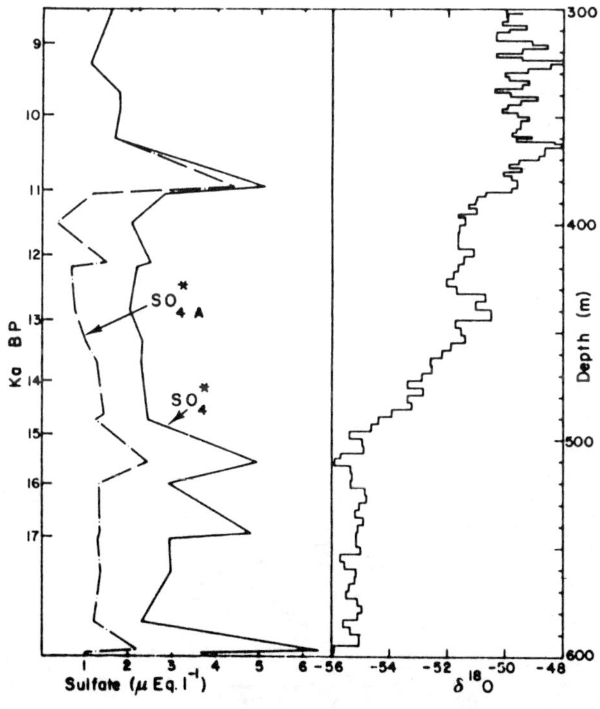

Figure 6. Total excess-sulfate (1) and acid excess-sulfate (2) as a function of depth (concentrations in μEq. l^{-1}).

here. The lower profile of Fig. 6 shows the variations of acid excess sulfate going from the ice-age to the Holocene. In comparison to the other elements or compounds examined here, the temporal variations of SO_{4A}^{*} throughout the transition are relatively weak.

Nitrate

The nitrate profile exhibits variations in the range 5-81 ng g^{-1} the highest values being encountered for the LGM, visibly in correlation with high Al values (Fig. 7). Moreover, the decrease of NO_3 starting at the beginning of the transition and parallel to the decrease of Al confirms that NO_3 has a terrestrial component at this time. From about 13 Ka BP, NO_3 no longer covaries with Al. Low values are observed until the Holocene when higher values are once again observed. The rapid augmentation of NO_3 from Transition to Holocene levels apparently occurs during the last step of the transition. Typical LGM, reversal (the cold episode around 12 ka BP in Europe and discussed by Jouzel et al., 1987, for the Dome C core) and early Holocene values would be 50, 10, and 20 ng g^{-1} respectively. When the terrestrial input of neutral NO_3 disappears (after about 13 Ka BP) it can be shown that this compound is deposited in Antarctic snow mostly as gaseous HNO_3 (Legrand and Delmas, 1986).

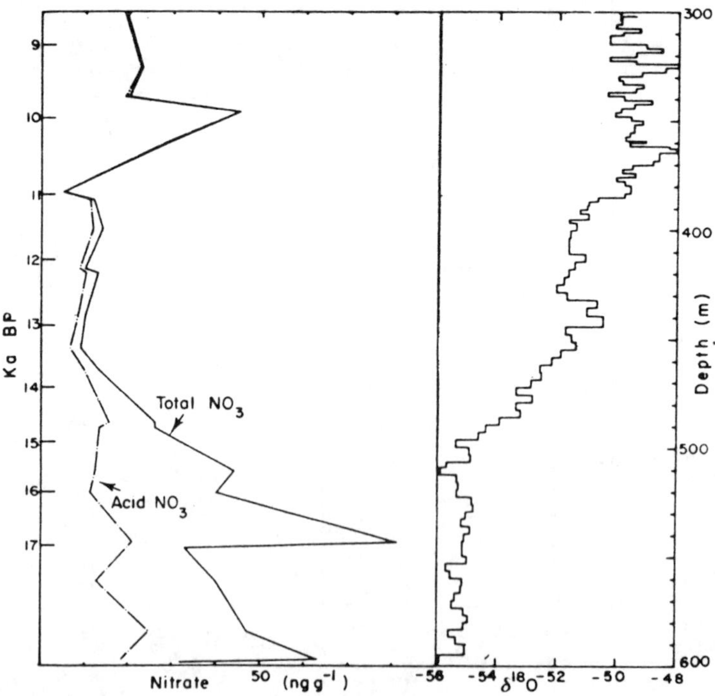

Figure 7. Total nitrate (1) and HNO_3 (2) as a function of depth (concentrations in μEq. l^{-1}). (1) is directly measured, HNO_3 is calculated from (1) by subtracting a neutral nitrate component linked with Al content (see text).

ENVIRONMENTAL SIGNIFICANCE OF THE CHEMICAL OBSERVATIONS

The more significant environmental changes appear to have occurred in the first part of the transition, between 15.5 and 13 Ka BP, when the production and transport of primary aerosols (crustal and marine material) diminished considerably. The atmosphere became progressively more stable so that long range transport to central Antarctica was more difficult than before. This is attested by the reaction of H_2SO_4 on sea salt particles which occurs only in stable atmospheric conditions (Legrand and Delmas, submitted). The cold episode called a reversal by Jouzel et al. (1987) seems to have had no effect on the increasing stability of the atmosphere, at least in the Antarctic.

The increase with time of the Na'_m/Al ratio (R') suggests that the marine aerosol input to Antarctica decreased relatively less than the crustal contribution. This could be explained by the progressive reduction of the R' ratio instead of R being a method for minimizing the influence of transport phenomena. The relatively low values of R' during the reversal stage could be due to a provisional re-extension of arid areas at this time. Thereafter the relative importance of continental primary aerosol reached a minimum during the warm period around 9 Ka BP, whereas the marine contribution remained stable. From this discussion, it seems reasonable to correct Na_m for $Na_2SO_4^*$ and therefore to consider Na'_m as more representative of the pure marine contribution than Na_m. The fractionation of sea salt, if present mechanisms were the same in the past, is essentially a summer phenomenon (Legrand and Delmas, 1984) whereas the bulk of sea salt particles arrives during winter storms. Consequently $Na_2SO_4^*$ and Na'_m could have distinct behaviors in the atmosphere before being deposited.

The extent of sea ice around the continent was probably of little influence on the amount of sea salt transported to a central Antarctic location like Dome C. This assumption is supported by the relatively weak influence of the distance from the coast on the deposition of Na found in East Antarctica (Legrand and Delmas, 1985). Finally it must be remembered that the snow accumulation rate increased during the LGM to Holocene transition. Due to the relative importance of dry deposition in central Antarctic areas, the decrease of Al and Na concentrations probably includes a term linked to this phenomenon (a factor 2 seems to be a maximum). If such a correction is made, this could finally signify that Al and Na in the Antarctic atmosphere could have decreased by less than one order of magnitude during this major climatic change. One part of SO_4^* and NO_3 was contributed by continental aerosol until the end of the cold period. Thereafter SO_4^* and NO_3 may be considered as being purely gas-derived (secondary aerosol). The evolution of SO_4^* concentrations is of considerable interest because of its possible link with marine biogenic activity. As already stated SO_4^* may be decomposed in a neutral and an acid component:

$SO_4^* = SO_{4A}^* + SO_{4N}^*$ (Legrand and Delmas, in preparation).

The exact calculation of these respective amounts is only possible when $Cl/Na_m < 1.8$, i.e. here at depths lower than 380m. This data is not sufficient to lead to firm conclusions regarding the possible change in marine biogenic activity which accompanied the climate change.

CONCLUSION

In this paper we have suggested new methods for interpreting glaciochemical profiles. Global environmental changes are probably recorded in these profiles, but a study and intercomparison of several ice-cores from different Antarctic areas and in Greenland will be needed before general conclusions can be drawn. We emphasize that the complete ion-budget must be established if chemical measurements are to be fully interpreted.

ACKNOWLEDGEMENTS

We thank J. Jouzel and L. Merlivat for having made the detailed Dome C isotopic profile available for us. We acknowledge assistance of M.C. Mieulet and J. Borde for the computation of the data. We thank M. De Angelis for the Al measurements and for helpful discussions. This work was supported by CNRS (Programme National d'Etude de la Dynamique du Climat), and the "Ministère de la Qualité de la Vie".

REFERENCES

De Angelis, M., Jouzel, J., Lorius, C., Merlivat, L., Petit, J.R. and Raynaud D., 1984. Ice age data for climate modelling from an Antarctic (Dome C) ice-core In *New Perspectives in Climate Modelling*, A.L. Berger and C. Nicolis (eds.), Elsevier Science Publ. 23-45.

Finkel, R.C. and Langway, Jr. C.C., 1985. Global and local influences on the chemical composition of snowfall at Dye 3, Greenland: the record between 10 Ka BP and 40 Ka BP, *Earth Planet. Sci. Lett.*, 73: 196-206.

Jouzel, J., Lorius, C., Merlivat, L. and Petit, J.R., 1987. Abrupt climatic changes: the Antarctic ice record during the late Pleistocene, this volume.

Legrand, M., De Angelis, M. and Delmas, R.J., 1984. Ion chromatographic determination of common ions in Antarctic snow and ice, *Analyt. Chim., Acta*, 156: 181-192.

Legrand, M., 1985. Chimie des neige et glace Antarctiques: un reflet de l'Environnement, Thèse d'Etat, Université Scientifique et Médicale de Grenoble, unpublished.

Legrand, M. and Delmas, R.J., 1984. The ionic balance of Anarctic snow: a 10 yr detailed record, *Atmospheric Environment*, 18: 1867-1876.

Legrand, M. and Delmas, R.J., 1985. Spatial and temporal variations of snow chemistry in Terre Adélie (East Antarctica), *Annals of Glaciology*, 7: 20-25.

Legrand, M. and Delmas, R.J., submitted. Formation of HCl in the Antarctic atmosphere, submitted to *J. Geophys. Res.*

Legrand, M. and Delmas, R.J., 1986. Respective contributions of tropospheric and stratospheric nitrate sources to Antarctic snow, *Tellus*, 38B: 235-249.

Lorius, C., Merlivat, L., Jouzel, J. and Pourchet, M., 1979. A 30,000 yr isotope climatic record from Antarctic ice, *Nature*, 280: 644-648.

Lorius, C., Raynaud, D., Petit, J.R., Jouzel, J. and Merlivat, L., 1984. Late-glacial maximum-Holocene atmospheric and ice-thickness changes from Antarctic ice-core studies, *Annals of Glaciology*, 5: 88-94.

Petit, J.R., Briat, M. and Royer, A., 1981. Ice age aerosol from East Antarctic ice core samples and past wind strength, *Nature*, 293: 391-394.

Royer, A., De Angelis, M. and Petit, J.R., 1983. A 30,000 yr record of physical and optical properties of microparticles from an East Antarctic ice core and implications for climatic reconstruction models, *Climatic Change*, 5: 381-412.

Thompson, L., Mosley-Thompson, E. and Petit, J.R., 1981. Glaciological interpretation of microparticle concentration from the French 905m Dome C, Antarctica core. Proc. XVII General Assembly of the IUGG, Canberra, 1979, *IAHS* Publ. 131, 227-234.

Part VI.

Glacial-Holocene Transition: Deep-Sea Record

BIOTURBATION EFFECTS ON ABRUPT CLIMATIC CHANGES RECORDED IN DEEP SEA SEDIMENTS. CORRELATION BETWEEN $\delta^{18}O$ PROFILES AND ACCELERATOR ^{14}C DATING

E. Bard[1], M. Arnold[1], J. Duprat[2], J. Moyes[2], J.C. Duplessy[1]

[1]Centre des Faibles Radioactivités, Laboratoire mixte
CNRS-CEA, P.O. Box 1
F.91190 Gif sur Yvette

[2]Département de Géologie et Océanographie
Université de Bordeaux
1 Avenue des Facultés
F. 33405 Talence

ABSTRACT. Major trends of Quaternary global climate are reflected in the continental ice volume changes which have been reconstructed by oxygen isotope analysis. $\delta^{18}O$ records from deep sea sediments show that the net glacial build-up occurs relatively slowly, but that the end of an ice age occurs quickly, in less than 10,000 yr, implying a non-linear response to simple Milankovitch theory. In order to understand the details of the last deglaciation a precise chronology of the paleoclimatic signals is needed. Carbon-14 measurements, performed by accelerator mass spectrometry, provide reliable ages on different species of foraminifera and can be used to provide a time scale for the oxygen isotope record. In addition, a bioturbation model has been used to point out the artifacts which could be falsely interpreted in terms of lead and lag in the paleoclimatic record. A comparison of the data with the restored signals enables the construction of a new deglaciation curve.

INTRODUCTION

Core CH 73-139C was recovered from the Rockall Plateau in the North-East Atlantic Ocean (54°38'N; 16°21'W). The water depth is 2209m, and is thus above the average lysocline for these areas. Thus carbonate dissolution is not important in disturbing the isotope curves.

Two oxygen isotope curves were produced: one on *Globigerina bulloides* and one using *Neogloboquadrina pachyderma* (left coiling) (Fig. 1). Both curves show, to a first approximation, the same deglaciation phenomenon marked by a clear decrease of $\delta^{18}O$ in the order of 1.5 to 2°/oo. However, detailed analysis of the two curves shows that the $\delta^{18}O$ decline for *Globigerina bulloides* (a cool water planktonic foraminifera)

W. H. Berger and L. D. Labeyrie (eds.), Abrupt Climatic Change, 263–278.

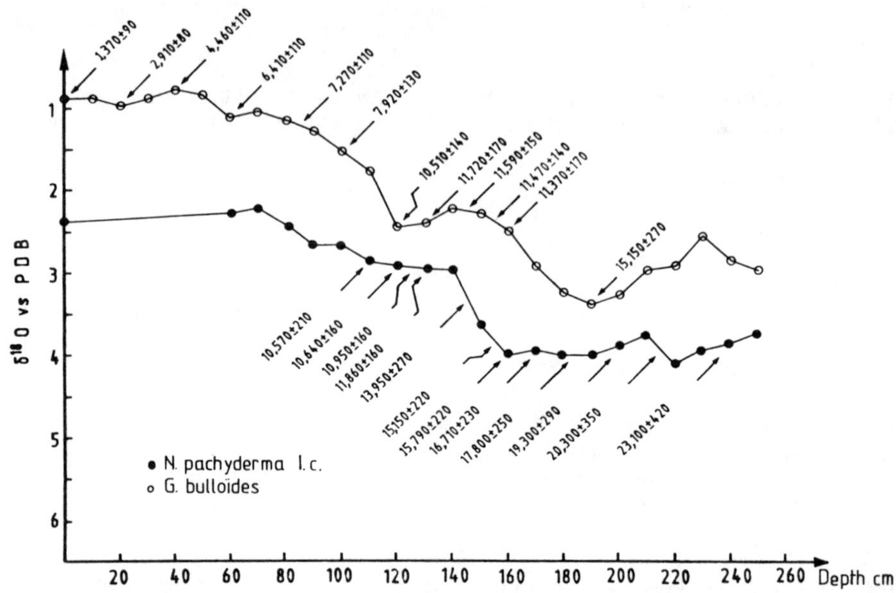

Figure 1. Oxygen isotope records and ^{14}C ages of the two foraminiferal species *G. bulloides* and *N. pachyderma* l.c. in core CH 73-139C.

precedes that for *N. pachyderma* left coiling (a cold water planktonic foraminifera) by about 30cm. While *G. bulloides* records the Bölling-Alleröd $\delta^{18}O$ minimum and the Younger Dryas maximum, *N. pachyderma* does not (for latitude ranges of the two foraminifera species, see Vincent and Berger, 1981).

In contrast to the classical method of counting beta radioactivity, the accelerator mass spectrometry enables ^{14}C ages to be obtained with only one milligram of carbon. Radiocarbon ages can thus be measured on monospecific samples of 1,000 to 2,000 foraminifera picked by hand, a technique which cuts out all risk of contamination by older detrital carbonates. This enables an age to be assigned to the "same" sample which is to be used in the oxygen isotope measurement (Broecker et al., 1984; Duplessy et al., 1986; Andree et al., 1986).

Actually the ocean surface has a $\Delta^{14}C$ of about $-50^{o}/oo$. This results in a +400 yr age shift for all oceanic ^{14}C ages relative to atmospheric conditions. Thus, in order to compare our ages with the continental chronology, all planktonic foraminiferal ^{14}C ages have been corrected for the present apparent age of surface water (i.e. 400 yr). This correction has, to a first approximation, been assumed constant for all species, despite the fact that they do not live exactly at the same water depth (Bé and Hamlin, 1967).

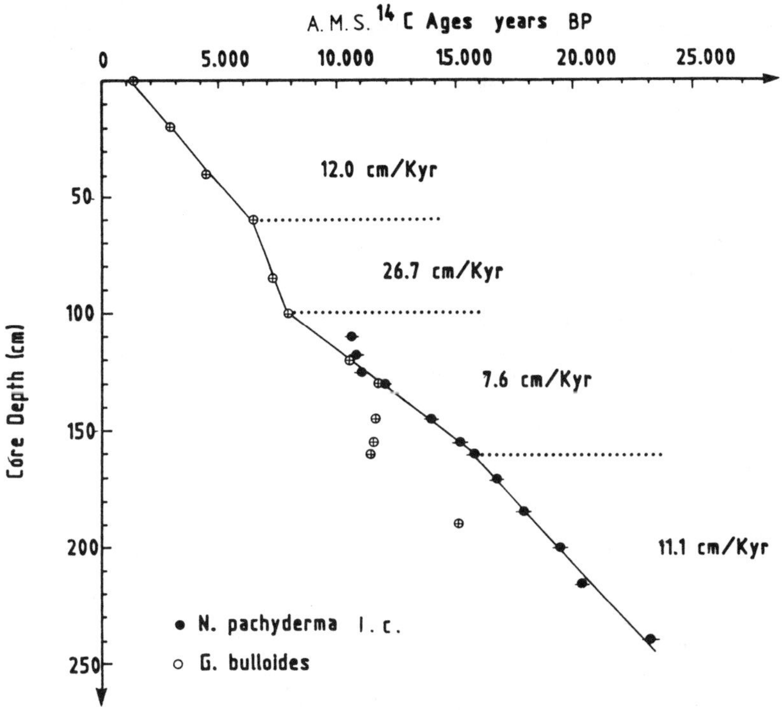

Figure 2. AMS ^{14}C records versus depth in core CH 73-139C for the two foraminiferal species *G. bulloides* and *N. pachyderma* l.c.

The sedimentation rate curve derived from the radiocarbon AMS analysis (Fig. 2) confirms the mean sedimentation rate close to 10cm/kyr found with the beta radioactivity counting method (Duplessy et al., 1981). It was quite stable during the last glacial maximum (11.1cm/kyr) and then dropped, as the deglaciation began, to a value of ca. 7.6cm/kyr. This lower rate lasted throughout Termination I (160 to 110cm in depth).

On both sides of the rate curve two "tails" of roughly constant radiocarbon ages are clearly observed. The first corresponds to an age of about 11,500 yr BP for *G. bulloides* with levels between 160 and 130cm in depth and the other to a mean age of 10,700 yr BP for *N. pachyderma* l.c. between 125 and 110cm.

Above the 100cm level only the presence of the subpolar species enables us to measure radiocarbon and this part of the curve shows that the sedimentation rate increased as the Holocene began (26.7cm/kyr; levels between 100 and 60cm) and then recovered to a value of roughly 12cm/kyr.

Assuming that the difference in depth habitats of the two foraminiferal species is relatively small and that the mixing rates within the upper layers of the ocean are rapid (that is <1kyr), the same paleoclimatological records for both species should have been observed if there were no bioturbation. Thus the signal restoration procedure should help us to recover a common climatic record for the two analyzed foraminiferal species.

Model

The approach introduced in this paper is a simple method which enables the isotopic signal to be deconvolved when the abundance of stratigraphic carriers with depth are known (Bard et al., in press).

Bioturbation is first treated as a time invariant filter whose impulse response function (IRF) is that of a first order system, and equivalent to a box model with homogeneous mixing (i.e. the well-known Berger and Heath model, 1968). In fact, the method consists of deconvolving the abundance curves of "carriers" and then restoring the isotopic signals ^{18}O and ^{14}C, taking into account the deconvolved abundances. Recurrence equations are obtained which link system output and input. Such discrete relations enable both the forward and the inverse problem to be carried out in the calculations.

The calculations reveal three limitations of the analysis:

a) The bioturbation thickness must be a multiple of the sampling step. In order to avoid this problem, the raw data can be interpolated without losing sight of the fact that the paleoclimatological significance of interpolated points is nil.

b) When the deconvolved abundance of the carrier tends rapidly towards zero, the relative error on the restored signal becomes greater and the value of deconvolved isotopes may be erroneous. In the following application the deconvolved isotopic signals will be considered as meaningless where the deconvolved abundances of the carrier tend to nearly zero.

c) The signal used is a function of sediment depth and it is assumed that proportionality exists between this variant and time. This hypothesis remains valid during periods of sedimentation rate stability but becomes false when this parameter varies.

Simulation

In order to simulate the last deglaciation we have traced schematically what could happen with two different foraminifera: a "cool" and a "cold" type (subpolar and polar species). For this purpose we have used simplified quasi-step functions for the measured changes in abundance of the different foraminiferal species. Termination I has been divided into four periods (cf. Fig. 3) correlated with the paleoclimatic evolution of the European continent (Ruddiman and McIntyre, 1981; Duplessy et al., 1981):

- the end of the glacial period characterized by high $\delta^{18}O$ values (3.5-4°/oo), a low abundance of "cool" foraminifera and a high abundance of the "cold" species.

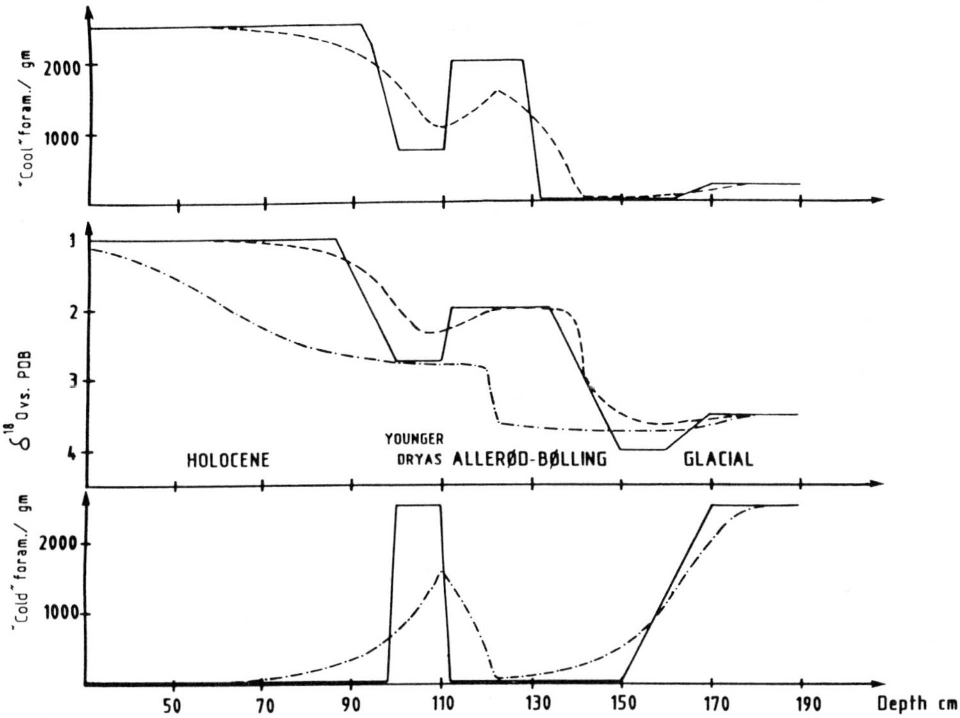

Figure 3. First attempt to simulate Termination I: $\delta^{18}O$ record (convolution sense, H = 12cm). Solid line: assumed input signal without bioturbation; short dashes: bioturbated record for a "cool" water foraminifera; dash-dot: bioturbated record for a "cold" water foraminifera.

- the Bölling-Alleröd warm event showing relatively low $\delta^{18}O$ (2°/oo), a high abundance of "cool" foraminifera and a zone nearly barren of the "cold" species.
- the Younger Dryas cold readvance characterized by intermediate $\delta^{18}O$ (2.75^o/oo), a partial disappearance of the "cool" species and a high abundance for the "cold" foraminifera.
- the Holocene showing a very low $\delta^{18}O$ (1°/oo), high abundance for the "cool" species and a very low abundance for the polar foraminifera.

For the ^{14}C age simulation the sedimentation rate has been held constant to a first approximation (10cm/kyr) and the foraminiferal abundances are the same as outlined above for the $\delta^{18}O$ simulation.

"Following" bioturbation, the $\delta^{18}O$ record of the "warm" foraminifera slightly precedes that of the "cold" type. Importantly, the latter does not record the $\delta^{18}O$ minimum corresponding to the Bölling Alleröd period (Fig. 3). When similarly convoluting an assumed ideal ^{14}C-signal (Fig. 4), the apparent sedimentation rate is not constant (fictitious drop during the entire deglaciation) and output ^{14}C ages deviate markedly from the ideal curve (output curves are traced only where the foraminifera abundances are greater than 200 shells per gram of sediment).

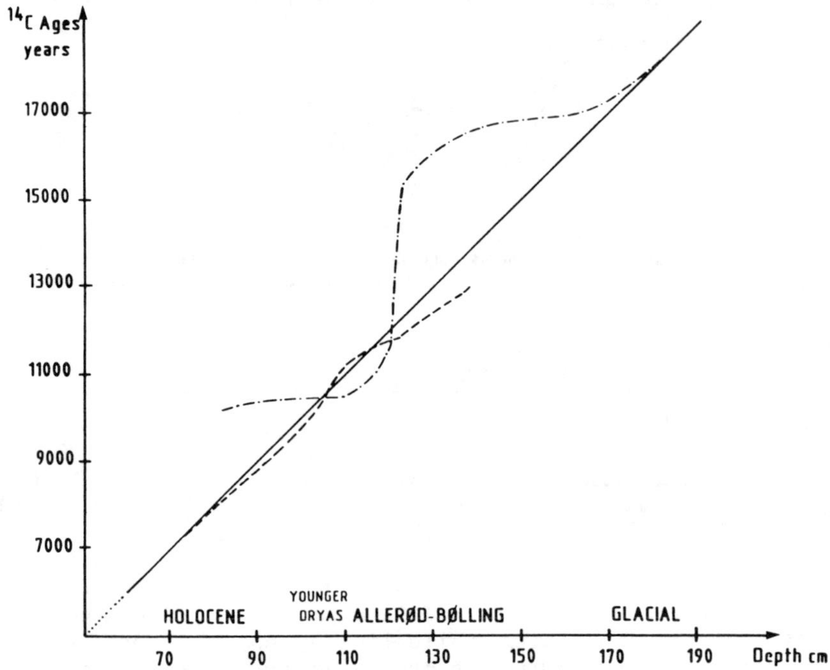

Figure 4. First attempt to simulate Termination I: ^{14}C record (convolution sense, H = 12 cm). Solid line: assumed input signal without bioturbation; short dashed: bioturbated record for a "cool" water foraminifera; dash-dot: bioturbated record for a "cold" water foraminifera.

Signal restoration

With the aim of testing the efficiency of the inverse model, deconvolved curves have been traced from available data: $\delta^{18}O$, accelerator ^{14}C, abundances of *G. bulloides* (cool water species) and *N. pachyderma* l.c. (cold water species). These paleontological signals are expressed in number of shells per unit of mass of sediment instead of the mass of shells per unit of depth of sediment. Thus the mass variation of the test of one particular foraminiferal species and the variation of the sediment density have been neglected to a first approximation. This effect may be not important as one consider other source of errors.

To minimize the effects of experimental noise amplification, the raw data have previously been smoothed using a simplified least square method (Savitzky and Golay, 1964). The absolute value of the difference between rough isotopic curves and smoothed curves is rather less than the analytical error (on the order of 0.1^{o}/oo for $\delta^{18}O$ and around 1-3% for the radiocarbon ages beyond 2000 years BP). Thus the smoothing process has a negligible effect on original climatological signals.

It should also be noted that the adjustment of smoothing parameters empirically introduces a variability in the mixing intensity (Bard et al., in press). Also, in order to test several bioturbation thickness values, the raw data have been interpolated before smoothing because the sampling step was limited to 10cm. The entire procedure is graphically summarized in Fig. 5.

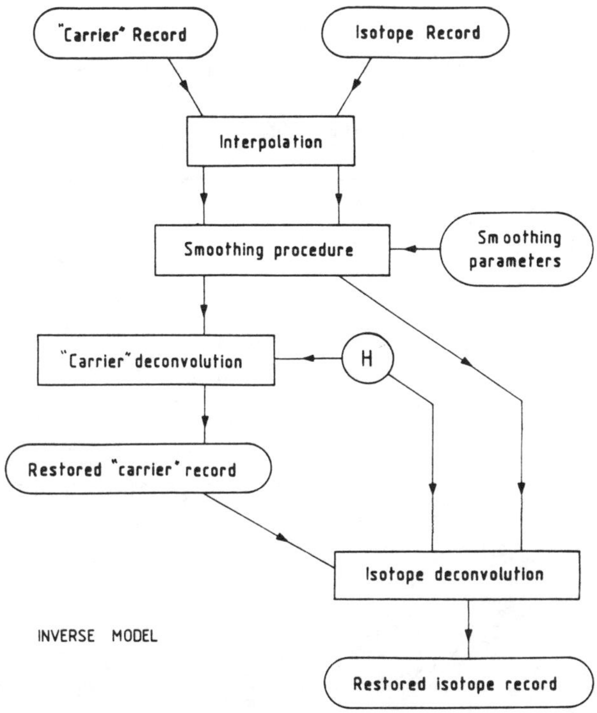

Figure 5. Schematic description of the mathematical procedure (H is the mixed layer thickness).

Description of the deconvolved curves

In order to show the sensitivity of the model, we present deconvolved $\delta^{18}O$ curves calculated with different bioturbation thicknesses (Fig. 6). The input of the model are the measured $\delta^{18}O$ and the carrier abundance data (Fig. 5). For H <10cm the stratigraphic offset is

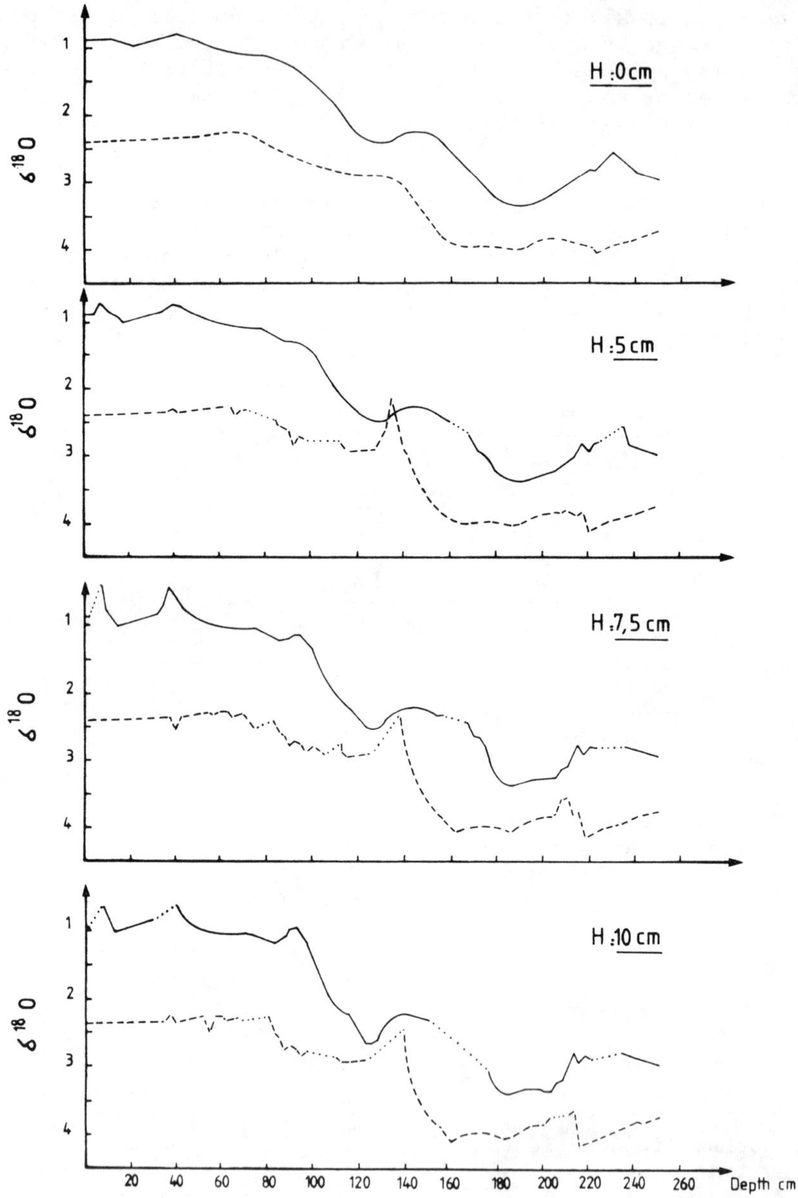

Figures 6A and 6B. Deconvolved $\delta^{18}O$ records for both foraminiferal species computed with various bioturbation thickness values. Solid line: deconvolved signal for *G. bulloides*; dashed line: deconvolved signal for *N. pachyderma* l.c.; dotted line: meaningless deconvolution intervals (no carrier).

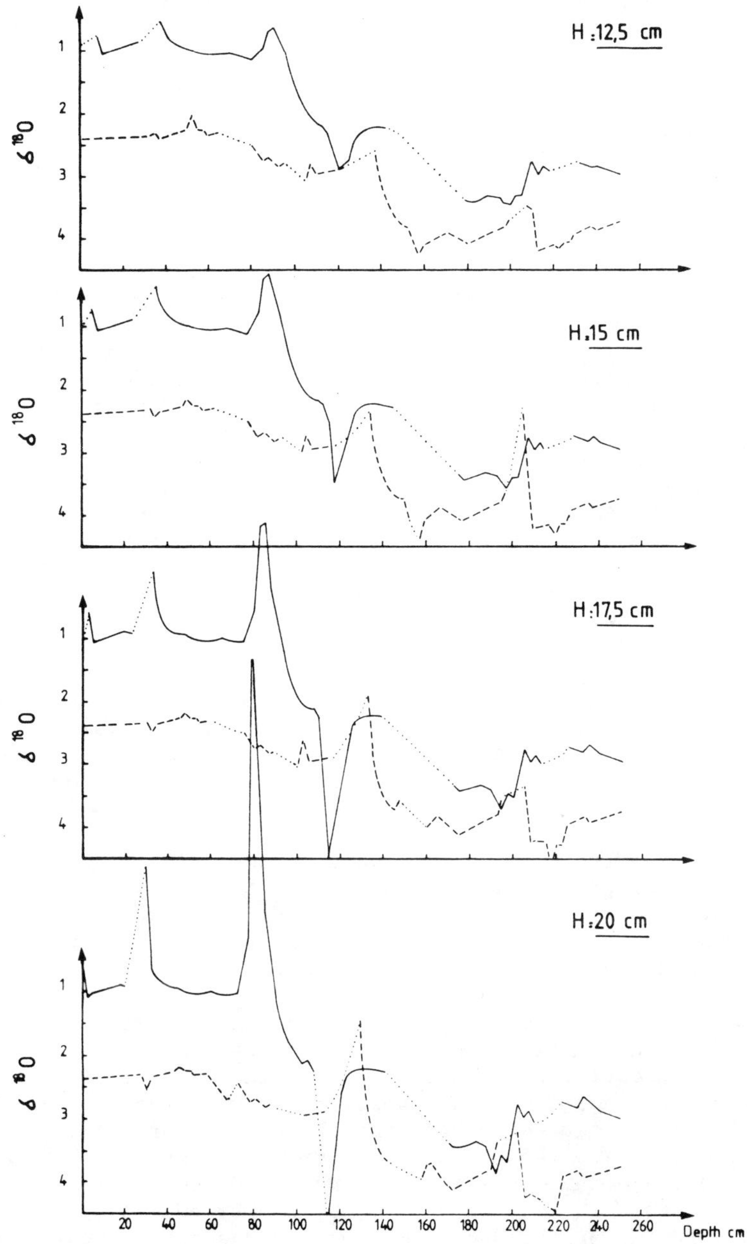
H:12,5 cm
H:15 cm
H:17,5 cm
H:20 cm
δ ¹⁸O
1
2
3
4
20
40
60
80
100
120
140
160
180
200
220
240
260
Depth cm

Figure 6B.

clearly not removed and for H >15cm great instabilities appear in the deconvolution which are not interpretable in terms of paleoclimatology.

Consequently, a thickness of H = 12.5cm appears to be the best choice for minimizing overshoots and for explaining correctly the artifacts such as lags between $\delta^{18}O$ records and ^{14}C age deviations around the sedimentation rates curve. This value has thus been used to compute the deconvolved curves for $\delta^{18}O$ and paleontological signals presented in this work (Fig. 7).

No complete deconvolved curve for ^{14}C ages has been presented in this work. This is because data are still too sparce to remove instabilities in the deconvolution procedure or perhaps because the assumption that bioturbation is the only perturbating phenomenon is not valid. Indeed during the glacial-postglacial transition some changes in the carbon cycle dynamic should have occured which could have affected the ^{14}C activities of the different carbon reservoirs (Andree et al., 1986). However, it is possible to calculate isolated deconvolved ^{14}C ages for "carrier" abundance maxima. Quite often these deconvolved ages are not different from the AMS ^{14}C ages because "carrier" abundance maxima are less susceptible to bioturbation distortion.

G. bulloides curves: A comparison of the input abundance curve with the unmixed one shows that the data signal precedes the deconvolution and that the deconvolved curve exhibits greater amplitude changes than the data curve. Additionally, there is a barren zone for *G. bulloides* between 150 and 175cm in depth. This species reappears again at about 150cm and then partly disappears until a minimum value centered at about 120cm. The presence of an abundance maximum centered near 132.5cm is remarkably clear and we shall see further why this characteristic is important in understanding the ^{14}C age deviation for *G. bulloides*.

The first conclusions are also valid when evaluating the unmixed oxygen isotope record. One exception is that the isotope deconvolution is no longer efficient when the unmixed carrier abundances drop rapidly to nearly zero. This is the case for the entire start of the deglaciation when $\delta^{18}O$ measured on G. bulloides decrease (Fig. 7, dotted line).

N. pachyderma l.c. curves: In the case of a "cold" foraminifera, the faunal abundance variations are important but almost in antiphase with those of *G. bulloides* (except for the minimum near 155-165cm which is entirely barren for the "cool" species). The intervals 165-200, 135-152.5 and 105-120cm correspond to high abundances separated by two low abundance zones at 155-165cm and 120-135cm. The latter is a virtually barren zone which coincides with a maximum of G. bulloides centered near 132.5 cm in depth. The restored $\delta^{18}O$ curve between 135 and 120cm shows an interval of meaningless deconvolution which again coincides with a carrier-barren zone (Fig. 7, dotted line). In addition, the start of the deglaciation corresponds also to an abundance minimum for *N. pachyderma* l.c. centered near 155cm.

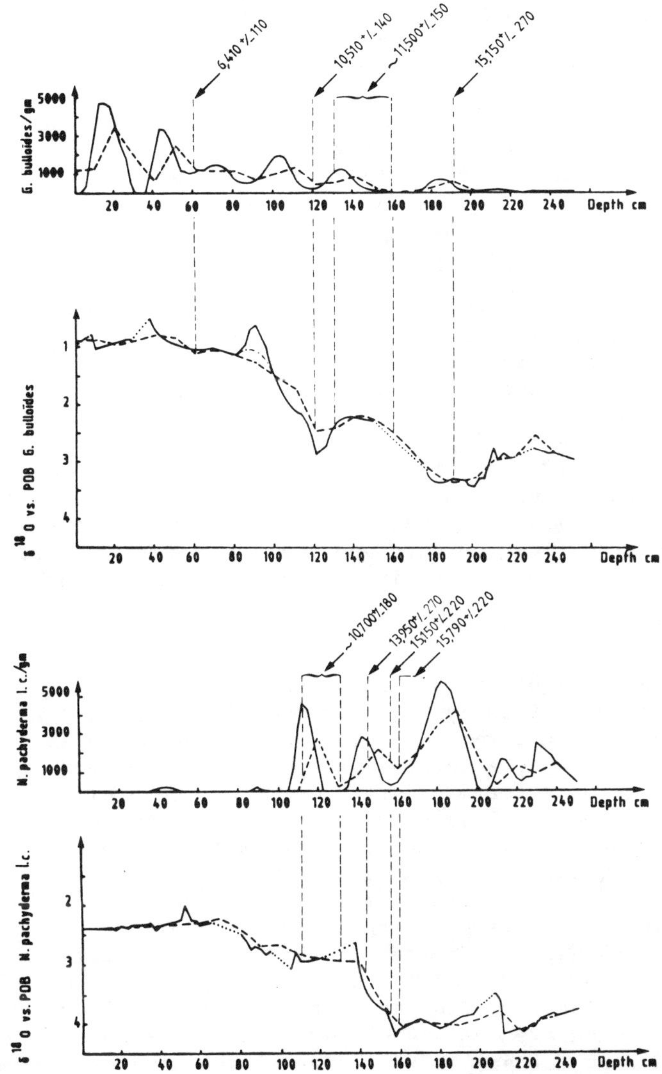

Figure 7. Data and deconvolved records for the paleontological abundances and oxygen isotopes measured on *G. bulloides* and *N. pachyderma* l.c. Dashed line: raw data records; solid line: restored signal (H = 12.5cm); dotted line: meaningless deconvolution intervals (no carrier); dash-dot line: " meltwater spike" deconvolved after a less intense mixing. The ^{14}C ages used in the text are also indicated. These ages are not deconvolved and should be considered with the raw $\delta^{18}O$ records (dashed line).

PALEOCLIMATOLOGICAL INTERPRETATIONS

In the following discussion the reader has to keep in mind that there are three inter-related components to the $\delta^{18}O$ signals: the ice volume effect, together with temperature and salinity variations. In addition, the warming estimated through the transfer function (7°C in Duplessy et al., 1981) represents a minimum value. As a matter of fact the transfer function provides temperature estimates with an accuracy of 1.5°C (at 1σ level) for non-glacial conditions. However, in the low temperature range, this transfer function is not able to estimate summer sea surface temperatures lower than 6°C.

Assuming a 0.25^{o}/oo/°C slope (Shackleton, 1974; Labeyrie et al., 1986), the warming corresponds to a 1.8^{o}/oo change in $\delta^{18}O$, not forgetting that the temperature effect on the data signals is not so obvious because our knowledge about the seasonality of the foraminifera growth is still poor. Assuming a ice effect of about 1.2^{o}/oo (Duplessy et al., 1985) the sum will be ca. 3°/oo. The deconvolved changes is about 2.3^{o}/oo for *G. bulloides* and 1.7^{o}/oo for *N. pachyderma* l.c. and therefore the difference could probably be attributed to a salinity change and/or to the problem of seasonality of foraminifera growth. Using the information derived from the data and the deconvolved records some important facts concerning the timing of deglaciation can be deduced.

Termination IA

The apparent age of 15,790±220 yr BP for the initiation of the deglaciation could be falsely based on the ^{14}C and $\delta^{18}O$ measurements at level 160cm on *N. pachyderma* l.c. This age is undoubtedly too old as this zone corresponds to a deconvolved abundance minimum centered near the depth 155cm. Most foraminiferal tests have been mixed up from the deconvolved abundance maximum situated near 185cm in depth. In addition, a *N. pachyderma* l.c. abundance maximum has been dated between 13,950±270 yr BP (level 145cm) and 15,150±220 yr BP (level 155cm). In this particular case the deconvolved age is somewhat different because the contribution of the large deconvolved maximum at about 185cm is not negligible. Consequently, the deconvolved abundance maximum centered at 145cm shows a deconvolved age of roughly 14,000 yr BP. Note that the deconvolved $\delta^{18}O$ clearly indicates that ≈40% of the first phase of deglaciation was achieved by ≈14,000 yr BP and thus we conclude that Termination IA began after 15,790±220 yr BP and somewhat before 14,000 yr BP.

The ^{14}C age of 15,150±270 yr BP determined on a small *G. bulloides* maximum precedes the beginning of Termination IA (level 190cm of the data curve). This gives another upper limit for the beginning of the climatic phenomenon.

Each foraminifera exhibits an abundance minimum for the deconvolved curve between the levels 150 and 170cm. For *G. bulloides* this is a real barren zone, already mentioned by Ruddiman and McIntyre (1981), which corresponds to the entire Termination IA. Consequently we can affirm that the $\delta^{18}O$ measured on *G. bulloides* between 165 and 185cm is typically a mixing signal due to bioturbation. As previously demonstrated deconvolution of the isotopic signal of *G. bulloides* is no longer efficient here because the restored carrier abundance is near zero.

The deglaciation pause

The constant ^{14}C ages between 130 and 160cm for *G. bulloides* (Fig. 2) also are a consequence of bioturbation. They correspond to the dating of the deconvolved abundance maximum situated near 132.5cm having been mixed up and down between 130 and 160cm in the data curve. The ^{14}C age of ca. 11,500 yr BP corresponds exactly to the Alleröd episode during which *G. bulloides* became relatively abundant and registered the low $\delta^{18}O$ values of the "warm" period.

During the Alleröd interval *N. pachyderma* l.c. disappeared totally (deconvolved curve between 120 and 135cm) explaining the absence of low $\delta^{18}O$ value in the data measured on the polar foraminifera. The absence of ^{14}C ages corresponding to the Bölling period (12,500 yr BP) suggests that both foraminiferal species were low in abundance during this early phase of the deglaciation pause.

The Younger Dryas

Following the Alleröd, the Younger Dryas "cold" period is characterized by low deconvolved abundances for *G. bulloides* and by a sharp maximum for *N. pachyderma* l.c. centered near 115cm depth.

The ^{14}C ages, roughly constant between 110 and 125cm for the *N. pachyderma* l.c. data curve might be explained by the mixing of this latter maximum which has been mixed up and down between the levels 130 and 110cm. The mean value of 10,700 yr BP corresponds exactly to the peak of the Younger Dryas period (Mangerud, 1980). This fact enables us to affirm that the data $\delta^{18}O$ between 130 and 110cm are all representative of the high Younger Dryas value.

During this cold period *G. bulloides* became scarce, as seen on the deconvolved curve between levels 110 and 125cm. However, enough foraminifera of this type remained in the water to record the Dryas high $\delta^{18}O$ values. After deconvolution the signal simply shows a much greater amplitude. The ^{14}C age of 10,510±140 yr BP measured on *G. bulloides* at the level 120cm also corresponds to the Younger Dryas. This, however, might be accidental as most of the 120cm level shells come from the bioturbation of the abundance maxima centered near 105 and 135cm on the deconvolved curve.

Termination IB

During Termination IB, *N. pachyderma* was rare but abundant enough to allow ^{18}O measurements. However, after the deconvolution procedure the abundances of the polar species are very low and the restored $\delta^{18}O$ record remains questionable. Therefore we will refer to *G. bulloides* which was sufficiently abundant to record the details of the end of the deglaciation.

The deconvolved $\delta^{18}O$ minimum situated near 90cm remains difficult to interpret using the simple model presented here. It could be either a real climatic event or a residual overshoot linked to a mixing intensity overestimation (Berger et al., 1977; Berger, 1978; Jones and Ruddiman, 1982; Schiffelbein, 1985). The smoothing parameters are of great importance to this particular "event" and this $\delta^{18}O$ minimum has been computed for the same bioturbation thickness but with a "stronger" smoothing process which corresponds to less intense mixing (Fig. 7, dash-dot line). This problem shows that more details about bioturbation parameters are needed in order to calculate the real meltwater-spike if it exists.

CONCLUSIONS

These results clearly illustrate that the complexity of the last deglaciation cannot be understood without a precise study of complementary tracers such as $\delta^{18}O$ and AMS ^{14}C dating. Following the simple deconvolution process outlined above, isotope measurements on complementary foraminiferal species enable us to precisely date the internal stages of the last deglaciation. The Termination I can be summarized as follows (with ^{14}C ages corrected for the present apparent age of the surface ocean water: 400 yr).

- The first major phase of ice melting starting after 15,150±270 yr BP and well before 14,000 yr BP. This period of intense ice-sheet disintegration has lasted roughly 1 kyr and seems to be the primary cause of the low productivity zone clearly indicated by the deconvolved abundance records (for both foraminifera species).

- A deglaciation pause occurred between 13,000 and 11,000 BP. During the warm Alleröd period *G. bulloides* reappeared and *N. pachyderma* l.c. disappeared quickly and entirely.

- The Younger Dryas cold event is dated near 10,700 yr BP on *N. pachyderma* l.c. which had reappeared. During this period *G. bulloides* dropped slightly in abundance but this species was abundant enough to record the isotopic signals of the Younger Dryas characterized by $\delta^{18}O$ values lower than those of the glacial maximum dated near 17,000 yr BP. This is also confirmed by the relatively low Younger Dryas $\delta^{18}O$ values measured on *N. pachyderma* l.c.

- The second deglaciation phase (Termination IB) started after 10,500 yr BP and ended roughly at 6,000 yr BP.

Fig. 8 schematically summarizes these important phases of the last deglaciation.

In order to obtain a clearer deglaciation record it would be useful to study cores from more southerly locations. As illustrated in this paper, even if the sedimentation rate is high, important bio-perturbations appear in cores containing "carrier" barren zones such as those at high-latitude of the North Atlantic (Ruddiman and McIntyre, 1981).

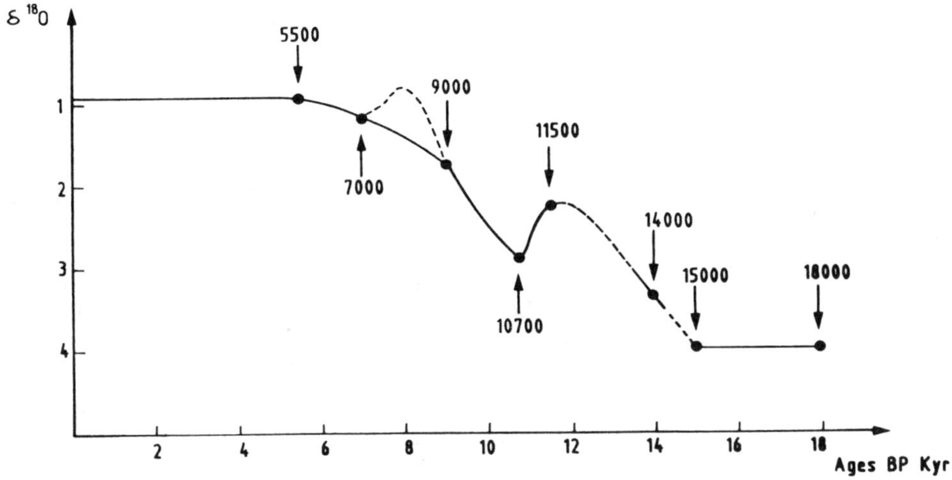

Figure 8. Summary of the new deglaciation timing.

In spite of its simplicity, the bioturbation model emphasizes some limits to signal restoration. One of the most critical is due to nearly zero restored isotope carrier abundances.

ACKNOWLEDGEMENTS

We thank J. Antignac, E. Kaltnecker, H. Leclaire, B. Le Coat and P. Maurice for help in the analyses; A. Castera and D. de Zertucha for their help in developing AMS; E. Kaltnecker for the drawing; and S. Calvert, T. Clube, J. Fayolle, L. Labeyrie, P. Monfray and D. Peteet for useful discussions and review of the manuscript.

This work was supported by CNRS, PNEDC, CEA and EEC grant CLI 085-F. CFR contribution n°788.

REFERENCES

Andree M., Oeschger, H., Siegenthaler, U., Risen, T., Moell, M., Ammann, B. and Tobolski., K., 1986. ^{14}C dating of plant macrofossils in lake sediment, *Radiocarbon*, 28 (2A): 411-416.

Andree, M., Oeschger, H., Broecker, W., Beavan, N., Klas, M., Mix, A., Bonani, G., Hofmann, H.J., Suter, M., Woelfli, W. and Peng, T.H., 1986. Limits on the ventilation rate of the deep ocean over the last 12,000 years, *Climate Dynamics*, 1:53-62

Bard, E., Arnold, M., Duplessy, J.C., Duprat, J. and Moyes, J., in press. Reconstruction of the last deglaciation: deconvolved records of $\delta^{18}O$ profiles, micropaleontological variations and accelerator mass spectrometric ^{14}C dating, *Climate Dynamics*.

Bé, A.W. and Hamlin, W.H., 1967. Ecology of recent planktonic foraminifera, Part 3 - Distribution in the north Atlantic during the summer 1962, *Micropaleontology* 13 (1): 87-106.

Berger W.H. and Heath, G.R., 1968. Vertical mixing in pelagic sediments, *J. Mar. Res.*, 2b: 134-143.

Berger, W.H., Johnson, R.F. and Killingley, J.S., 1977. Unmixing of the deep sea record and the deglacial meltwater spike, *Nature*, 269: 661-663.

Berger, W.H., 1978. Oxygen-18 stratigraphy in deep-sea sediments: additional evidence for the deglacial meltwater effect, *Deep Sea Res.*, 25: 473-480.

Broecker, W.S., Mix, A., Andree M. and Oeschger, H., 1984. Radiocarbon measurements on coexisting benthic and planktic foraminifera shell: potential for reconstructing ocean ventilation times over the past 20,000 years. *Nuclear Inst. and Meth. in Phys. Res.*, B5: 331-339.

Duplessy, J.C., Delibrias, G., Turon, J.L. Pujol C. and Duprat, J., 1981. Deglacial warming of the northeastern Atlantic Ocean: correlation with the paleoclimatic evolution of the European continent. *Palaeogeogr., Paleoclimat., Palaeoecol.*, 35: 121-144.

Duplessy, J.C., Labeyrie, L.D. and Shackleton, N.J., 1985. The oxygen isotope record of benthic foraminiferal effect of deep water temperature and ice volume changes. *EOS*, 66(18).

Duplessy, J.C., Arnold, M., Maurice, P., Bard, E., Duprat, J. and Moyes, J., 1986. Direct dating of the oxygen isotope record on the last deglaciation by ^{14}C accelerator mass spectrometry, *Nature*, 320: 350-352.

Jones, G.A. and Ruddiman, W.F., 1982. Assessing the global meltwater spike, *Quat. Res.*, 17: 148-172.

Labeyrie, L.D., Pichon, J.J., Labracherie, M., Ippolito, P., Duprat, J. and Duplessy, J.C., 1986. Melting history of Antarctica during the past 60,000 years. *Nature*, 322: 701-706.

Mangerud, J., 1980. Ice front variations of different parts of the Scandinavian ice sheet 13,000-10,000 years BP. In: Studies in the Lateglacial of North-West Europe J.L. Lowe, J.M. Gray and J.E. Robinson Ed., Pergamon Press, 23-30.

Ruddiman, W.F. and McIntyre, A., 1981. The north Atlantic ocean during the last deglaciation, *Palaeogeogr., Palaeoclimat., Palaeoecol.*, 35: 145-214.

Savitzky, A. and Golay, M., 1964. Smoothing and differenciation of data by simplified least squares procedures. *Anal. Chem.*, 36 (8): 1627-1639.

Schiffelbein, P., 1985. Extracting the benthic impulse response function: a constrained deconvolution technique, *Mar. Geol.*, 64: 313-336.

Shackleton, N.J., 1974. Attainment of isotopic equilibrium between ocean water and the benthonic foraminifera genus *Uvigerina*: Isotopic changes in the ocean during the last glacial. In: *Les méthodes quantitatives d'étude des variations du climat au cours du pleistocène*, n°219, Colloques Internationaux du CNRS.

Vincent, E. and Berger, W.H., 1981. Planktonic foraminifera and their use in paleoceanography, In: *The Oceanic Lithosphere*. The Sea VII, C. Emiliani Ed., Interscience, Wiley, New York, 1025-1119.

GLACIAL-HOLOCENE TRANSITION: CLIMATE PULSATIONS AND SPORADIC SHUTDOWN OF NADW PRODUCTION

W. H. Berger[1], S. Burke[1] and E. Vincent[2]

[1]Scripps Institution of Oceanography
University of California, San Diego
La Jolla, CA 92093

[2]Département de Géologie Dynamique
Université Pierre et Marie Curie
4 Place Jussieu
75005 Paris, France

ABSTRACT. There is evidence that climate pulsations during the Wisconsin-Holocene transition were accompanied by strong fluctuations in the intensity of North Atlantic Deep Water production. It appears likely that these fluctuations were largely a result of pulsed meltwater input. The production of NADW is intimately tied to global heat budget asymmetries (North vs. South, Atlantic vs. Pacific) and of course to the regional heat budget of the North Atlantic, which in turn acts as a climate amplifier system. Both albedo feedback and carbon dioxide feedback mechanisms would seem to depend on (and influence) NADW production. The type of arguments used to reconstruct NADW production are illustrated using isotopic, micropaleontologic and sedimentologic data from a box core in the central South Atlantic.

INTRODUCTION

The Glacial-Holocene transition (15,000-8,000 yrs B.P.) is characterized by large short-lived climatic fluctuations. The changes were especially pronounced in western Norway, where they produced substantial alternations of glacier retreat and readvance (Mangerud et al., 1979). The two main excursions are the Alleröd warm period between 12,000 and 11,000 years ago and the Younger Dryas cold spell between 11,000 and 10,000 years ago. Major abrupt changes occurred at the start of the Alleröd, at the boundary between Alleröd and Younger Dryas, and at the end of the Younger Dryas.

These climatic events are reflected in deep-sea records, especially in the North Atlantic (Caralp, 1970; Ruddiman and McIntyre, 1973, 1981a, b; Berger, 1978; Duplessy et al., 1981b; Berger et al., 1985a, b), but also in many other regions of the ocean, as well as in the ice of Greenland (Dansgaard et al., 1984) and Antarctica (Lorius et al., 1981)

W. H. Berger and L. D. Labeyrie (eds.), Abrupt Climatic Change, 279–297.

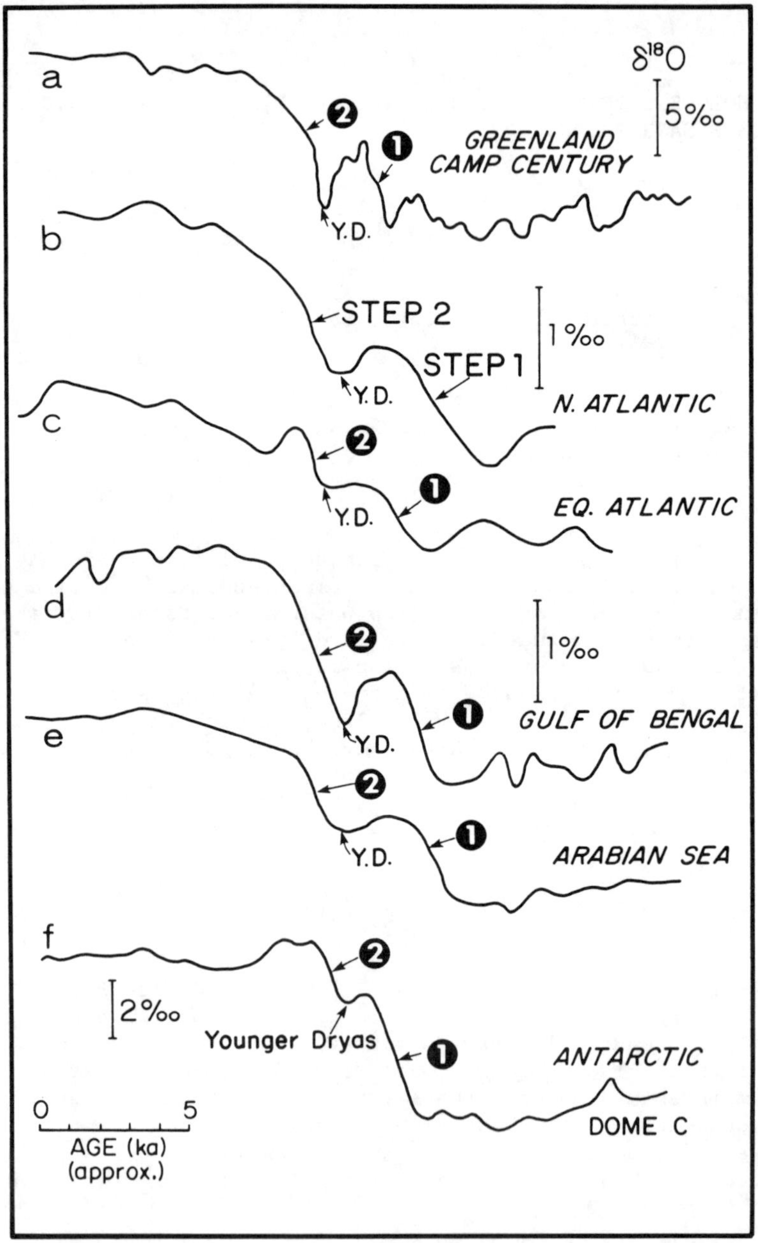

Figure 1. Evidence for major climate pulsations during the Glacial-Holocene transition as seen in ice-core data and deep-sea oxygen isotope records. Y.D., Younger Dryas. Steps 1 and 2, periods of rapid warming. Data (top to bottom): Dansgaard, 1987 (this volume); Duplessy et al. (1981a); Berger et al. (1985b); Duplessy et al. (1981b); ditto; Lorius et al.(1979). From Berger (1985), modified.

(see Fig. 1). Thus, they appear to be a worldwide phenomenon centered on the North Atlantic amplifier system..

The concept of a pulsating transition is in contrast to that of the traditional one of a sudden "termination", in the form originally proposed (Broecker and van Donk, 1970), and subsequently widely used in the interpretation of changes at the glacial-Holocene boundary. The pulsation concept calls for a unique climate regime of *transition* in which positive feedback mechanisms provide potential for runaway in opposite directions (Berger, 1982a). The stabilizing negative feedback mechanisms presumably are strongest at the end-member conditions of fully glacial or fully post-glacial configurations. In addition, stabilization must be achieved over long time spans, regardless of state, if Milankovitch forcing functions are to track climatic indices as well as they apparently do (Imbrie, this volume). Thus, the mechanisms responsible for a large non-Milankovitch climate excursion such as the Younger Dryas should be unique to the transition itself.

The one major process unique to deglaciation is the production of meltwater, and it appears reasonable, therefore, to look for meltwater effects. We here elaborate on the hypothesis that the rapid influx of meltwater to the North Atlantic, during deglaciation, resulted in sporadic shutdown of NADW-production (Olausson, 1965) and that these events were linked to abrupt climatic change (Berger and Killingley, 1982; Berger, 1985; Berger and Vincent, 1986b). The hypothesis derives from the concept that temporary decoupling of deep ocean water masses from the heat budget of the surface, through meltwater input in the North Atlantic, should increase climatic instability: the inertia of the system, is greatly decreased (Berger, 1982a).

In the following, we list evidence for and explore implications of the proposed slowdown or shutdown of NADW. As an illustration of the type of evidence available in deep-sea cores, we offer an exercise in paleoceanographic interpretation, using the record in Box Core INMD115, which was taken in the South Atlantic on the Mid-Atlantic Ridge. We realize that the low sedimentation rate of this core, combined with effects of bioturbation, puts limits of unknown severity on recovering paleoceanographic signals. Many more cores must be studied in detail before the history of NADW production can be reconstructed with some confidence. However, the data at hand appear adequate for exploring the range of possibilities and the construction of plausible scenarios.

EVIDENCE FOR VARIABLE NADW-PRODUCTION

Sources of positive feedback may be sought in albedo and greenhouse effects. The North Atlantic plays a crucial role within albedo feedback loops, providing for changes in heat budget and precipitation in sensitive latitudes, which affects land-ice and sea-ice buildup and decay, among other things (Ruddiman and McIntyre, 1984). One important aspect of such circulation changes is the production rate of North Atlantic Deep Water (NADW), which has been assumed variable on a glacial-interglacial scale for some time, on the basis of geochemical, oceanographic, and micropaleontologic evidence (Olausson, 1967; Weyl, 1968; Streeter,

1973; Schnitker, 1974, 1979; Berger, 1977; Streeter and Shackleton, 1979; Duplessy et al.,1975, 1980; Boyle and Keigwin, 1982, 1985; Corliss, 1982; Curry and Lohmann, 1982, 1983; Streeter et al., 1982; Shackleton et al., 1983; Ledbetter and Balsam, 1985; Mix and Fairbanks, 1985).

Recently, Ruddiman and McIntyre (1984) pointed to the impact, on North Atlantic climatic conditions, of variable advection of warm water northward from the low latitudes of the North and South Atlantic. Greatly expanding on this theme, Broecker et al. (1985) stressed the importance of the heat gain for the Norwegian Sea from importing warm surface waters and exporting cold deep waters. They suggested that the origin of the Alleröd-Younger Dryas fluctuation is somehow tied to changes in NADW-production during the transition. Stressing the importance of the greenhouse effect, Berger and Killingley (1982) earlier proposed that changes in the rate of deep-water production affected the partitioning of carbon dioxide between ocean and atmosphere, such that the Younger Dryas glacial readvance would have been caused by a temporary drop in atmospheric CO_2. Both mechanisms appear viable.

Evidence that the NADW-production indeed fluctuated during the transition has been presented by Jansen and Erlenkeuser (1985) based on the stable isotope composition of benthic foraminifera in cores from the Norwegian Sea. In addition, we have found evidence for major changes in deep-water properties in Atlantic and Pacific box cores (Berger and Vincent, 1986a, b) which is in agreement with their results (Fig. 2). Our data indicate that the Atlantic-Pacific difference in $\delta^{13}C$ values of benthic foraminifera is larger in the Holocene than during glacial time, as expected for increased NADW flow during the Holocene. However, between 15,000 and 10,000 years ago, this difference varies greatly and may be close to zero on occasion. We propose that this pattern represents temporary cessation or strong reduction of NADW-production. Full production, as at present, starts 10,000 years ago.

Remarkable support for these propositions derives from AMS radiocarbon data from the Pacific, published by Andree et al. (1986), and based on cores from the South China Sea, taken near 2 km depth (sill depth is near 2km also). After corrections for dissolution effects (Berger, in press), these data show a bimodal distribution of deep water ages in the North Pacific, which may be interpreted in terms of a counterpoint production of North Atlantic Deep Water and North Pacific Deep Water (Fig. 3). Furthermore, the occurrence of several exceptionally high deep water ages, centered on the second step of deglaciation, suggests the thickening of a glacial bottom water mass at that particular time. A deep saline and cold glacial bottom water mass, partially decoupled from the upper waters of the global ocean, was first predicted by Worthington (1968). Nutrients and CO_2 would have been trapped in such a water body, and nitrate would have been reduced, decreasing the productivity of the ocean in the early Holocene, with interesting consequences for climatic change.

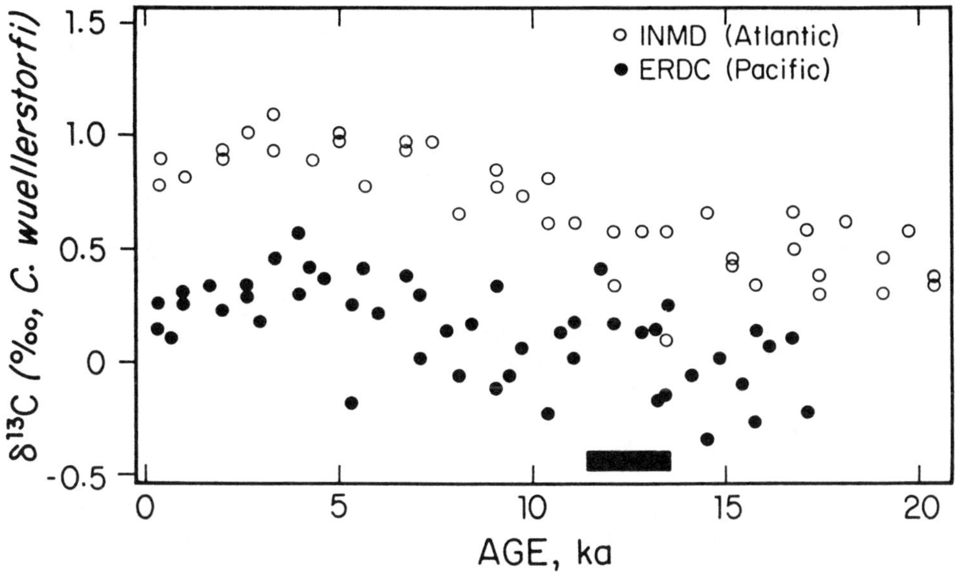

Figure 2. Carbon isotope composition of the benthic foraminifera *Cibicidoides wuellerstorfi* in four box cores, two each from the tropical South Atlantic (3400-3500m) and from the equatorial Pacific (2200-2300m). Approximate age scale based on radiocarbon dating and oxygen isotope stratigraphy. Bar: period of overlapping $\delta^{13}C$ - values. From Berger and Vincent (1986a).

NORTH ATLANTIC HEAT FLOW MODEL

The sinking of NADW in the northern North Atlantic, and the consequent advection of heat, has to be seen in the context of the overall heat budget of the Atlantic Ocean, which is highly asymmetrical (Fig. 4). The basic reason for this asymmetry is the export of water vapor, by trade winds, from the North Atlantic (whose surface waters are therefore highly saline) to the North Pacific (which has the least saline surface waters of the open ocean). As a consequence of the loss of fresh water in the North Atlantic (a) sea level is lowered and (b) the saline surface waters of the North Atlantic become very dense upon cooling, and sink to make NADW.

The lowering of sea level and the production of NADW result in water running across the equator, within the upper several hundred meters, transporting heat from south to north (about 144 x 10^{13} watts; Stommel, 1980). This heat stems both from proximal regions in the South Atlantic and from distant ones in the Pacific and Indian Ocean. About one half of the imported heat is exported as latent heat, with the moisture leaving the North Atlantic across the Isthmus of Panama.

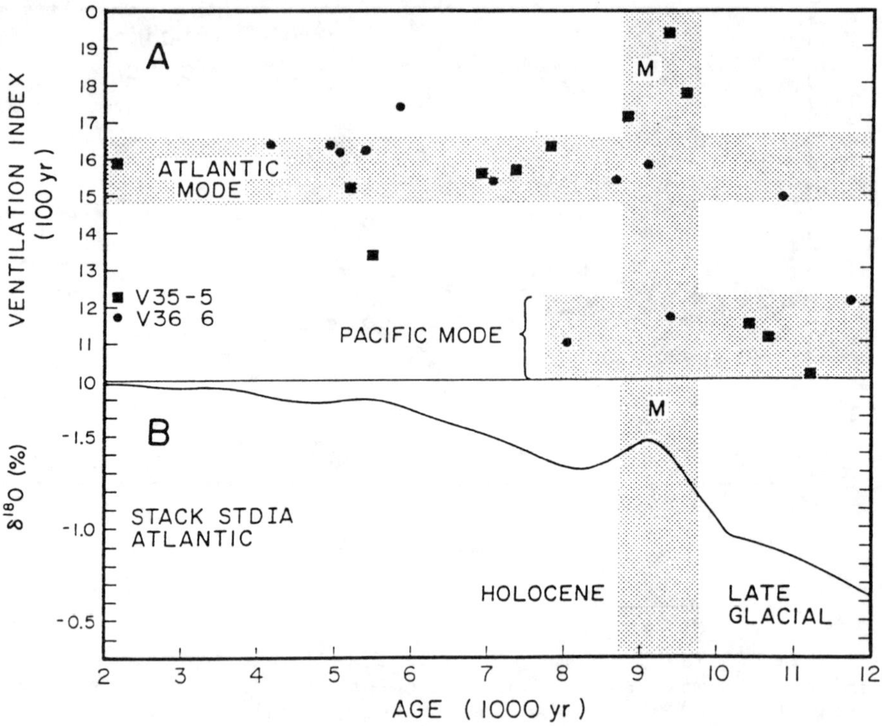

Figure 3. Bimodal distribution of deep water ages (near 2 km, western tropical North Pacific). Note the exceptional ages during maximum meltwater input, as seen in the $\delta^{18}O$ record. A, Deep water ages, based on differences in radiocarbon content of planktonic and benthic foraminifera (data in Andree et al., 1986), corrected for dissolution effects. B, $\delta^{18}O$ stratigraphy of planktonic foraminifera in the central Atlantic. Shaded zone: M-event, interpreted as rapid input of meltwater. From Berger (in press).

The (cold) NADW flows south and is given to the Circumpolar Ocean and to the deep Pacific (Reid and Lynn, 1971; Mantyla and Reid, 1983), in exchange for (warm) thermocline waters from Indian and Pacific Oceans. Gordon (1985) has estimated that as much as 47×10^{13} watts may be added from the Indian Ocean to the South Atlantic, around the Cape of Good Hope. This would represent about one third of the cross-equatorial heat flow (Fig. 4).

The net gain of heat to the North Atlantic from cross-equatorial transport, after subtracting the export of latent heat and sensible heat, is crucial to the climatic state. A fraction of the latent heat given off by the North Atlantic is available downwind to the regions of rainfall, as for example NW Europe, where precipitation warms the atmosphere. This heat is associated with the recycled portion of the fresh water, which comes back as rivers. Sensible heat moves as warm marine

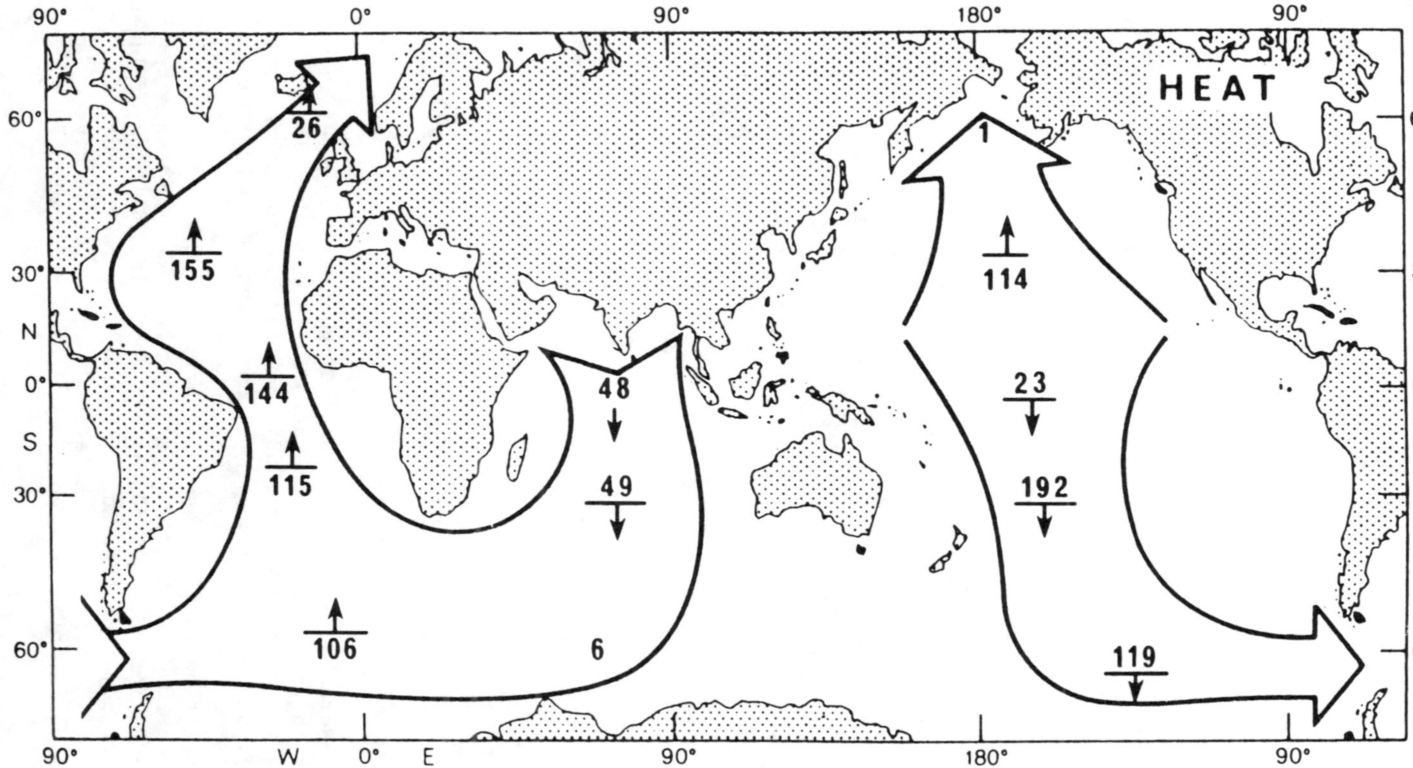

Figure 4. Meridional component of heat flow in the world ocean in 10^{13} watts. After Stommel (1980), from Wood (1981), modified.

air, of course. However, the latent heat necessary to make the North Atlantic anomalously saline must be exported to regions where fresh water will not come back to the Atlantic. It is not available to heat adjacent continents.

In sum, the overall Atlantic-Pacific deep circulation asymmetry, which is driven by NADW production (Reid, 1961), and which governs deep water chemistry (Redfield et al., 1963) and deep-sea sediment patterns (Berger, 1970), also governs an important portion of global heat flow, and hence climate (Weyl, 1968; Newell, 1974; Stommel, 1980; Wood, 1981; Gordon, 1985). When NADW is turned on, heat is sucked into the North Atlantic from the tropical regions within the Atlantic and outside of it, and some of this heat is passed on to NW Europe and other marine-air dominated regions (Fig. 5).

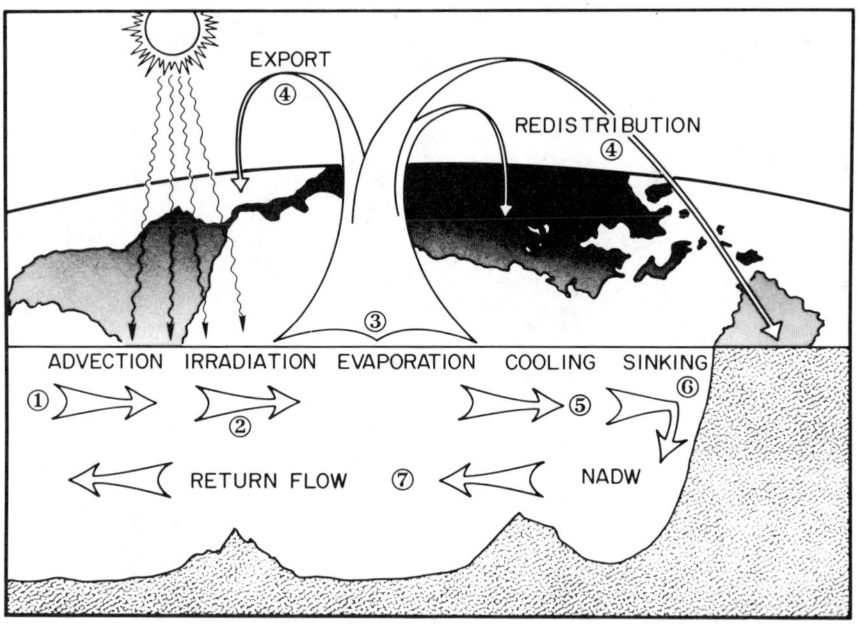

Figure 5. The North Atlantic Heat Pump. Conceptual model of import and export of heat in the North Atlantic associated with NADW formation. Based on ideas and data in Weyl (1968), Baumgartner and Reichel (1975), Stommel (1980), and Gordon (1985). (1) Advection of heat from the Indian Ocean (≈$50x10^{13}$ watts) and from subtropical South Atlantic. (2) Cross-equatorial flow (≈15sv, $144x10^{13}$ watts, see Fig. 3). (3) Evaporation of 60,000km³ (corr. to 2sv, $460x10^{13}$ W). (4) Export of 9000km³ to the Pacific (0.3sv, $70x10^{13}$W), and redistribution of latent heat to air over adjacent continents ($70x10^{13}$ W), rest to marine air ($320x10^{13}$ W). (5) Cooling in high latitudes, and (6) sinking of surface waters to make North Atlantic deep water. (7) Return of cold water to Southern Ocean.

BREAKING THE LOOP

In contemplating the machinery which produces NADW it is evident that we are dealing with a feedback loop where every link is both cause and effect. How can this loop be broken to generate an abrupt change in heat flow and hence in climate? In principle, we can (1) stop advection of heat from the south, (2) decrease heating in the tropics, or (3) decrease export of moisture to make evaporation inefficient in generating dense water. In the present context we favor a mechanism which is obviously unique to deglaciation: the addition of meltwater. It is the mechanism invoked by Olausson (1965) to shut off NADW production.

If sufficient meltwater is introduced into the North Atlantic, evaporation will be unable to produce the required density before cooling, sea level will rise relative to the global level, sinking of NADW will cease, and so will cross-equatorial heat flow. Ironically, the most favorable periods for this scenario of decreased heat import (that is, Step 1 and Step 2 of deglaciation 13,000 and 9,500 years ago), also presumably are in the greatest need for heat for melting ice.

Apparently, melting can proceed without the cross-equatorial contribution due to NADW-production. Cross-equatorial heat flow could still proceed within upper waters, however. Also much of the loss in heat import could be made up by throttling the export of latent heat to the North Pacific. In this case, no longer receiving fresh water from the North Atlantic, the North Pacific then could turn on its own source of deep water. The NADW-off/NPDW-on scenario seems to hold up for much of the deglaciation period (Berger, in press; see Fig. 3). The increased evaporation in the North Atlantic from Milankovitch warming, then, would result in more rain (and heat) for the surrounding continents, rather than increased export to the Pacific.

According to Broecker et al. (1985), a NADW shutdown would be expected to occasion a cooling (that is, the Younger Dryas), because of the decreased import of warm surface waters into the northern North Atlantic. The data of Jansen and Erlenkeuser (1985), however, indicate that a shutdown took place during major warming, just before 12,000 yrs B.P., as also suggested by our data (Fig. 2). Jansen and Erlenkeuser propose a reversal of the circulation to account for this, with the outflow of meltwater producing a "fjord ocean". We believe that this is a viable proposition, since the heat input to the Norwegian Sea could be derived from upwelling and cooling of intermediate waters and especially from the associated export of sea-ice.

If NADW-shutdown is to produce the Younger Dryas, heat import to the entire North Atlantic must be reduced, because the strengthening of trade winds (expected from the increased pole-to-equator temperature gradient) would otherwise greatly favor moisture export, and hence salinity increase and NADW-production. The effect of the still sizable ice sheet on Scandinavia, on wind patterns, must be to bring Siberian air to NW Europe (cf. Mangerud, this volume). Sea ice could then be produced (increasing the albedo) and the fjord-type Alleröd circulation of Jansen and Erlenkeuser (1985) could be choked off temporarily. A glacial mode of NADW-production could then briefly establish itself, against Milankovitch equilibrium. Due to the lack of moisture export from the Atlantic, North Pacific Deep Water could then form, as suggested by the radiocarbon data (Fig. 3).

READING THE RECORD: CORE INMD115BX

In paleoceanography one cannot hope to reconstruct the dynamics of past oceans and climate to a degree which in any sense approaches the detail of present ocean maps and models. One can aspire, however, to identify the main factors of change and to get the sign right. In the case at hand, having satisfied ourselves that NADW-production is an important ingredient in climatic change, it is desirable to establish whether a shutdown of NADW is associated with warming or with cooling, during the Glacial-Holocene transition. The problem may be reduced to one of mapping and dating the effects of changing NADW-production in deep-sea sediments.

The task is difficult because strongly contrasting climatic states (Alleröd and Younger Dryas) are separated by but 1000 years. One millenium corresponds to between 1 and 2.5cm of sediment deposition over most of the seafloor where carbonate accumulates. Bioturbation, and possibly also vertical migration of benthic foraminifera, affect depth ranges somewhat larger than this. It is evident, therefore, that the answer to the question of sign has to come from areas of high sedimentation rate. Such cores, however, have problems of their own being commonly greatly influenced by coastal ocean dynamics. In the analysis which follows, on a deep-sea core, we can only illustrate the approach, and the difficulties which arise. We do not claim that we can solve even the problem of the sign, based on this core.

Core INMD115BX was taken on the Mid-Atlantic Ridge, at 17°38'S, 16°13'W, at a depth of 3427m. Condition at retrieval was very good (i.e., the core was very little disturbed). Aragonitic pteropod shells were seen at the surface, signaling that the seafloor is bathed by NADW. Radiocarbon dates were as follows (Berger et al., 1985a): 0-2cm, 4,850 yr; 7-9cm, 7,865 yr; 12-14cm, 10,410 yr; 19-21cm, 14,165 yr. Dates are based on the Libby half-life of 5,568 yr. The sedimentation rate of 1.3cm/ky is typical of calcareous ooze in central gyre regions. Bioturbation is of low intensity here, because of low productivity.

The stable isotope stratigraphy of the core is more or less as expected as far as the long-term glacial-postglacial trends (Fig. 6): the oxygen values become lighter from glacial to postglacial time due to addition of meltwater and change in temperature (Emiliani, 1955), and the carbon values become heavier, due to buildup of the biosphere (Shackleton, 1977) and changes in deep-water circulation (Boyle and Keigwin, 1982) and also in upper water conditions. We are interested in the anomalies within the transition. For $\delta^{18}O$ these are shown shaded as difference to the global average ice effect. Here the ice effect is taken as a half-sine from 16 ka to 6 ka with an amplitude of $-1.2^{o}/oo$

(Fig. 6A) based on a review of older literature (Berger and Gardner, 1975) and on recent reconstructions (Berger et al., 1985a, 1987).

In the warm water plankton $\delta^{18}O$ record (pink *G. ruber*) we note a small warm anomaly (<1°C) at the onset of the transition followed by a substantial cold anomaly (1 to 2°C) during deglaciation proper and in the earliest Holocene. The warm anomaly may indicate diminished cross-equatorial transport, following the argument of Mix et al. (1986a, b). The cold anomalies are unexplained; they may correspondingly indicate anomalously high heat loss to the North Atlantic. It is improbable that they are due to mixing, as the abundance of pink *G. ruber* does not change much across the transition. The glacial-to-postglacial range is near 1.2°/oo, as expected for the global ice effect.

The $\delta^{18}O$ record of *C. wuellerstorfi* (Fig. 6A) follows the global curve rather nicely, suggesting that it reflects largely the meltwater effect. However, its amplitude goes beyond the ice effect of 1.2°/oo. Thus, it appears that during glaciation, and until well into the transition, the deep water was colder than now by between 1 and 2°C. A rather pronounced step occurs at the very beginning of the transition, some 16,000 or 15,000 years ago, with a warming of the deep water (that is, presumably a strengthening of the NADW signal). For three or four thousand years following this event, the temperature is quite variable, suggesting that the boundary of NADW and glacial bottom water (GBW) shifted up and down across the site a number of times. Toward the end of the transition there is a final warming to present conditions, as the NADW-GBW boundary moves below the core site for good.

The $\delta^{13}C$ record of pink *G. ruber* shows a large contrast between glacial and Holocene values, for a glacial-to-postglacial range of almost 1°/oo, clearly larger than the global range of 0.4 to 0.5°/oo. A drastic change in surface water stratification and productivity is indicated. We suggest that a substantial expansion of the subtropical surface-water lens (gyre-water) occurs in the Holocene. During the transition, the extent of this lens apparently was very restricted (as also shown on the CLIMAP 1976 reconstruction). Suprisingly, gyre-margin conditions apparently prevailed even at this central location (17.5°S, Mid-Atlantic Ridge). The planktonic and benthic foraminifera in the transition show overlapping $\delta^{13}C$ values (black areas) due to a remarkable variability of both the planktonic and the benthic signal. If this finding is verified in other cores highly unusual circulation conditions must be assumed. More nearly "normal" conditions are indicated in glacial time, *G. ruber* (pink) having distinctly heavier values than *C. wuellerstorfi*. *G. ruber* (white) values (not shown) are erratic in glacial time as well as in the transition: large vital effects may be indicated, that is, seasonal upwelling.

The $\delta^{13}C$ record of *C. wuellerstorfi* can be interpreted in terms of NADW pulses. The variability is large during the transition, and fluctuation are parallel to $\delta^{18}O$, such that warmer water (NADW) means higher $\delta^{13}C$ (more oxygenated water). Sediment mixing could conceivably simulate a variable transition in the record by injecting both glacial and Holocene materials into this zone. If mixing is responsible for the

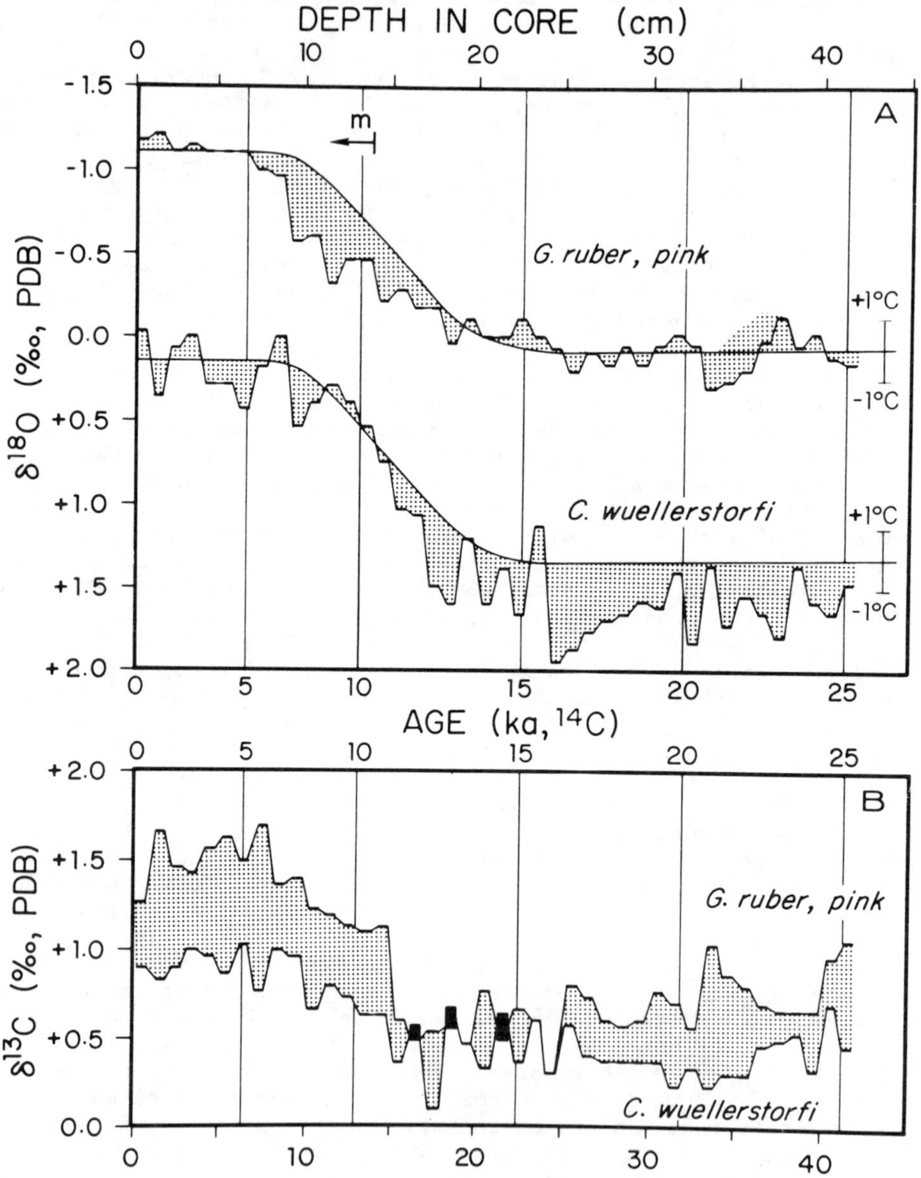

Figure 6. Stable isotopes of foraminifera in box core INMD115. A, oxygen isotopes, pink *Globigerinoides ruber* (planktonic) and *Cibicidoides wuellerstorfi* (benthic). Smooth curves: ice effect only, generalized. m, first appearance of *Globorotalia menardii*. B, carbon isotopes. Periods when *G. ruber* has higher values than *C. wuellerstorfi* are shown in black.

variability, however, it is hard to explain how glacial extremes could be retained for the $\delta^{13}C$ record but not for the $\delta^{18}O$ record.

In sum, the isotope records of both the plankton and the benthos suggest that instabilities developed in shallow and deep circulation patterns and hence in the heat flow to the North Atlantic, very early during the transition. Such heat flow instabilities might mimick the "early deglaciation" scenarios discussed in Ruddiman and Duplessy (1985). All during the transition, up to the first appearance of *G. menardii* (11,000 yr bp), instabilities persist. The *G. menardii* appearance (m in Fig. 6A) marks the reseeding of this tropical species around the Cape of Good Hope and hence may signal the onset of heat transfer from the Indian Ocean, as postulated by Gordon (1985) for present conditions. Since then, NADW has been flowing steadily across the site, but never stronger than now, as suggested in the separation of Pacific and Atlantic values (Fig. 2).

It should be possible to see the effects of changing bottom water conditions at the site of INMD 115 in the benthic fauna and in physical properties (Fig. 7). Winnowing by bottom currents should affect both sand content and benthic foraminifera composition. Productivity changes may be even more important: supply of food is crucial to foraminiferal abundance and, by fueling bioturbation (resuspension of fines), affects physical properties.

An overall decrease in benthic foraminifera in the Holocene may be ascribed to a general drop in the productivity of overlying waters. A marked decrease in the relative abundance of *Uvigerina peregrina* from Glacial to Holocene supports the inference. This species is abundant in high fertility areas (Lutze and Coulbourn, 1984; Lutze et al., 1986). The less abundant species of *Uvigerina* (*U. senticosa* and *U. proboscidea*) do not follow this trend, however. Other marked faunal changes between Glacial and Holocene include an increase in the relative abundance of *Laticarinina pauperata* as well as porcellaneous species (dominated by *Quinqueloculina* and *Pyrgo* spp.) in the Holocene.

The maximum relative abundance of *N. umbonifera* within the transition emphasizes the fact that the transition is not intermediate between glacial and postglacial conditions. The least sandy (least winnowed?) material occurs within the transition when, according to our hypothesis, NADW shuts down episodically. Similarly, AABW may be much less active as a winnowing agent during such a time, its production being dependent on the supply of NADW to the Southern Ocean (Foster and Carmack, 1976).

Interestingly, sand content is low within the early Holocene as well. Although the NADW was running deep, it might have been running more slowly than now. To keep the NADW-AABW boundary in place, both water masses should have reduced their speed simultaneously. After the climatic optimum (centered on 6000 yr bp), percent sand content increases drastically suggesting an increase in flow of NADW.

We think NADW-production increased in the late Holocene, because we see the effects downstream on the Pacific seafloor, in the form of increased carbonate dissolution (Keir and Berger, 1985). NADW-flow

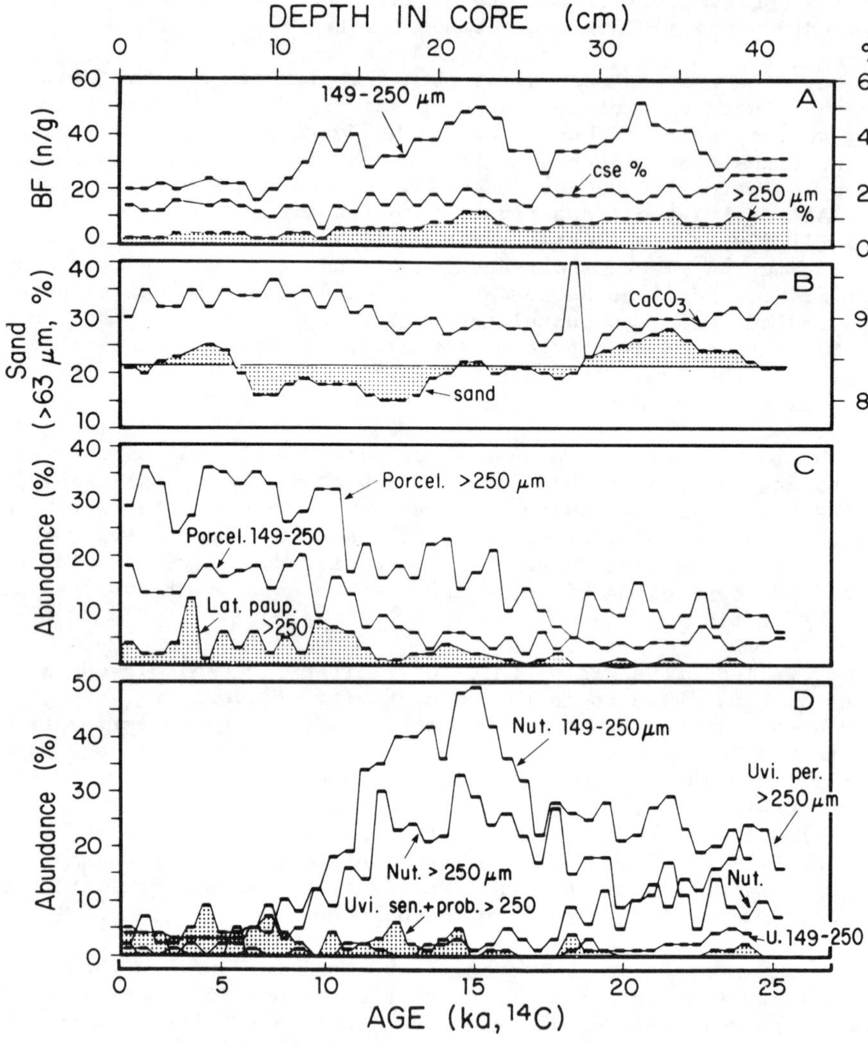

Figure 7. Abundance of benthic foraminifera and physical properties, box core INMD115. A, Number of benthic foraminifera two size fractions per gram of raw dry sediment. CSE%, percent of coarse benthic foraminifera (>250μm) versus total benthic foraminifera >149μm. B, percent $CaCO_3$ and percent sand content. C, relative abundance of porcellaneous species (in two size fractions) and of *Laticarinina pauperata* (in the >250μm size fraction). D, relative abundance of *Nuttallides umbonifera* in two size fractions, *of Uvigerina peregrina, and U. senticosa* + *U. proboscidea* in the size fraction >250μm, and *of U.* sp. in the fraction 149-250μm.

increased because, following the climatic optimum, strengthened winds exported more moisture to the Pacific, increasing the Atlantic-Pacific asymmetry in the density field. The increased demand for heat, for the latent heat export, is being filled by increased cross-equatorial flow and a greater import from the Indian Ocean.

In the carbonate curve of INMD115BX one sees the contrast between glacial and postglacial values, which presumably results from greater dilution with terrigenous material during glacial time. Dust supply suddenly decreased late in the G/H Transition, as is well known from ice cores (Hammer et al., 1985). The system is now on the way toward a reverse transition, into the next ice age (Imbrie, 1981). It appears that the system resists this change by pumping heat into the North Atlantic, which is perhaps the most sensitive part of the climate machine. The compensation mechanism must eventually break down, when the export of latent heat which drives the density-dependent NADW-machine can no longer be made up from the import of warm water from elsewhere. It would appear to be of great interest to model the dynamics of this sequence, and to test for the sense of interference from the present CO_2-induced warming.

REFERENCES

Andree, M., Oeschger, H., Broecker, W.S., Beavan, N., Klas, M., Mix, A., Bonani, G., Hofmann, H.J., Suter, M., Woelfli, W. and Peng, T.H., 1986. Limits on the ventilation rate for the deep ocean over the last 12000 years, *Climate Dynamics*, 1: 53-62.

Baumgartner, A. and Reichel, E., 1975. *The World Water Balance*, Elsevier, Amsterdam, 179pp., maps.

Berger, W.H., 1970. Biogenous deep-sea sediments: fractionation by deep-sea circulation, *Bull. Geol. Soc. Am.*, 81: 1385-1402.

Berger, W.H., 1977. Carbon dioxide excursions and the deep-sea record: aspects of the problem. In: N.R. Andersen and A. Malahoff (eds) *The Fate of Fossil Fuel CO_2 in the Oceans*, Plenum New York, p. 505-542.

Berger, W.H., 1978. Oxygen-18 stratigraphy in deep-sea sediments: additional evidence for the deglacial meltwater effect, *Deep-Sea Res.*, 25: 473-480.

Berger, W.H., 1982a. Climate steps in ocean history - lessons from the Pleistocene. In: W.H. Berger and J.C. Crowell (eds) *Climate in Earth History*, Studies in Geophysics, National Academy Press, Washington D.C., p. 43-54.

Berger, W.H., 1982b. On the definition of the Pleistocene-Holocene boundary in deep-sea sediments, *Sveriges geologiska undersökning, Avhand Uppsat.*, 76 (7), Ser. C (794): 270-280

Berger, W.H., 1985. Carbon dioxide increase and climate prediction: clues from deep-sea carbonates, *Episodes*, 8: 163-168.

Berger, W.H., in press. Ocean ventilation during the last 12,000 years: hypothesis of counterpoint deep water production, *Marine Geology*.

Berger, W.H. and Gardner, J.V., 1975. On the determination of Pleistocene temperatures from planktonic foraminifera, *J. Foram. Res.*, 5 (2): 102-113.

Berger, W.H. and Killingley, J.S., 1982a. The Worthington Effect and the origin of the Younger Dryas, *J. Mar. Res.* (Supplement), 40: 27-38.
Berger, W.H. and Vincent, E., 1986a. Deep-sea carbonates: reading the carbon isotope signal, *Geol. Rundschau*, 75 (1): 249-269.
Berger, W.H. and Vincent, E., 1986b. Sporadic shutdown of North Atlantic deep water production during the Glacial-Holocene transition?, *Nature*, 324: 53-55.
Berger, W.H., Killingley, J.S., Metzler, C.V. and Vincent, E., 1985a. Two-step deglaciation: ^{14}C-dated high resolution $\delta^{18}O$ records from the tropical Atlantic Ocean, *Quat. Res.*, 23: 258-271.
Berger, W.H., Killingley, J.S. and Vincent, E., 1985b. Timing of deglaciation from an oxygen isotope curve for Atlantic deep-sea sediments, *Nature*, 314: 156-158.
Berger, W.H., Killingley, J.S. and Vincent, E., 1987. Timescale of Wisconsin/Holocene transition: oxygen isotope record in the Western Equatorial Pacific. *Quaternary Research*, in press.
Boyle, E.A. and Keigwin, L.D., 1982. Deep circulation of the North Atlantic over the last 200,000 years: geochemical evidence, *Science*, 218: 784-787.
Boyle, E.A. and Keigwin, L.D., 1985. Comparison of Atlantic and Pacific paleochemical records for the last 215,000 years: changes in deep ocean circulation and chemical inventories, *Earth and Planet. Sci. Lett.*, 76: 135-150.
Broecker, W.S. and Van Donk, J., 1970. Isolation changes, ice volumes, and the O^{18} record in deep-sea cores, *Reviews of Geophysics and Space Physics*, 8: 169-198.
Broecker, W.S., Peteet, D.M. and Rind, D., 1985. Does the ocean-atmosphere system have more than one stable mode of operation?, *Nature*, 315: 21-26.
Caralp, M., 1970. Essai de stratigraphie du Pleistocène marin terminal d'après les paléoclimats observés dans des carottes du Golfe de Gascogne, *Bull. Soc. Géol. Fr.*, 712: 403-412.
CLIMAP Project Members, 1976. The surface of the ice-age Earth, *Science*, 191: 1131-1137.
Corliss, B.H., 1982. Linkage of North Atlantic and Southern Ocean deep-water circulation during glacial intervals. *Nature*, 298: 458-460.
Curry, W.B. and Lohmann, G.P., 1982. Carbon isotopic changes in benthic foraminifera from the Western South Atlantic: reconstruction of glacial abyssal circulation patterns, *Quat. Res.*, 18: 218-235.
Curry, W.B. and Lohmann, G.P., 1983. Reduced advection into Atlantic Ocean deep eatern basins during last glaciation maximum, *Nature*, 306: 577-580.
Dansgaard, W., Johnsen, S.J., Clausen, H.B., Dahl-Jensen, D., Gundestrup, N., Hammer, C.V. and Oeschger, H., 1984. North Atlantic climatic oscillations revealed by deep Greenland ice cores. In: J.E. Hansen and T. Takahashi (editors), *Climate Processes and Climate Sensitivity*, Geophys Monogr. 29, Am. Geophys. Union, Washington, D.C., pp. 288-298
Dansgaard, W.,1987. Ice core evidence of abrupt climatic change, this volume
Duplessy, J.-C., Chenouard, L. and Vila, F., 1975. Weyl's theory of glaciation supported by isotopic study of Norwegian Core K 11, *Science*, 188: 1208-1209.

Duplessy, J.-C., Moyes, J. and Pujol, C., 1980. Deep water formation in the North Atlantic Ocean during the last ice age, *Nature*, 286 (5772): 479-482.

Duplessy, J.-C., Delbrias, G., Turon, J.L., Pujol, C. and Duprat, J., 1981a. Deglacial warming of the northeastern Atlantic Ocean: correlation with the paleoclimatic evolution of the European continent, *Palaeogeo., Palaeoclim., Palaeoeco.*, 35 (2-4): 121-144.

Duplessy, J.-C., Bé, A.W.H. and Blanc, P.L., 1981b. Oxygen and carbon isotopic composition and biogeographic distribution of planktonic foraminifera in the Indian Ocean, *Palaeogeo., Palaeoclim., Palaeoeco.*, 33 (1-3): 9-16.

Emiliani, C., 1955. Pleistocene temperatures, *J. Geol.*, 63: 538-578.

Foster, T.D. and Carmack, E.C., 1976. Frontal zone mixing and Antarctic Bottom Water formation in the Southern Weddell Sea. *Deep-Sea Res.* 23: 301-317.

Gordon, A., 1985. Indian-Atlantic transfer of thermocline water at the Agulhas retroflection, *Science*, 227: 1030-1033.

Hammer, C.U., Clausen, H.B., Dansgaard, W., Neftel, A., Kristinsdottir, P. and Johnson, E., 1985. Continuous impurity analysis along the Dye 3 deep core. In: Langway, C.C., Oeschger, H., Dansgaard, W. (eds) *Greenland Ice Core: Geophysics, Geochemistry, and the Environment*, Geophysical Monograph 33: 90-94 American Geophysical Union, Washington, D.C.

Imbrie, J., Abrupt terminations of Late Pleistocene ice ages: a simple Milankovitch explanation, this volume.

Imbrie, J. and Webb, T. II, 1981. Transfer functions: calibrating micropaleontological data in climatic terms. In: A. Berger (editor), *Climatic Variations and Variability: Facts and Theories*, D. Reidel Publ. Co., Holland, 125-134.

Jansen, D. and Erlenkeuser, H., 1985. Ocean circulation in the Norwegian Sea during the last deglaciation: isotopic evidence, *Palaeogeo., Palaeoclim., Palaeoeco.*, 49: 189-206.

Keir, R.S. and Berger, W.H., 1985. Late Holocene carbonate dissolution in the Equatorial Pacific: reef growth or Neoglaciation? In: E.T. Sundquist and W.S. Broecker (eds.) *The Carbon Cycle and Atmospheric* CO_2*: Natural Variations Archean to Present*, Geophysical Monograph, 32: 208-219, American Geophysical Union, Washington, D.C.

Ledbetter, M.T. and Balsam, W.L., 1985. Paleoceanography of the Deep Western Boundary Undercurrent on the North American continental margin for the past 25000 yr. *Geology*, 13: 181-184.

Lorius, C., Merlivat, L., Jouzel, J. and Pourchet, M., 1979. A 30,000-year isotope climatic record from Antarctic ice, *Nature*, 280: 644-648.

Lutze, G.F. and Coulbourn, W.T., 1984. Recent benthic foraminifera from the continental margin of northwest Africa: community structure and distribution, *Marine Micropaleontology*, 8: 361-401.

Lutze, G.F., Pflaumann, U. and Weinholz, P., 1986. Jungquartaere Fluktuationen der benthischen Foraminiferenfaunen in Tiefsee-sedimenten vor NW-Africa: eine Reaktion auf Produktivitaetsaenderungen im Oberflaechenwasser, *Meteor Forsch. Ergebnisse, Reihe C*, 40: 14pp.

Mangerud, J., 1987. The Alleröd/Younger Dryas boundary, this volume.

Mangerud, J., Larsen, E., Longva, O. and Sönstegaard, E., 1979. Glacial history of western Norway 15,000-10,000 B.P., *Boreas*, 8: 179-187.

Mantyla, A.W. and Reid, J.L., 1983. Abyssal characteristics of the World Ocean waters, *Deep-Sea Res.*, 30: 805-833.

Mix, A.C. and Fairbanks, R.G., 1985. North Atlantic surface-ocean control of Pleistocene deep-ocean circulation, *Earth and Planet. Sci. Lett.*, 73: 231-243.

Mix, A.C., Ruddiman, W.F. and McIntyre, A., 1986a. Late Quaternary paleoceanography of the tropical Atlantic, 1: spatial variability of annual mean sea-surface temperatures, 0-20,000 years B.P., *Paleoceanography*, 1 (1): 43-66.

Mix, A.C., Ruddiman, W.F. and McIntyre, A., 1986b. Late Quaternary paleoceanography of the tropical Atlantic, 2: The seasonal cycle of sea surface temperatures, 0-20,000 years B.P., *Paleoceanography*, 1 (3): 339-353.

Newell, R.E., 1974. Changes in the poleward energy flux by the atmosphere and ocean as a possible cause for ice ages, *Quaternary Res.*, 4: 117-127.

Olausson, E., 1965. Evidence of climatic changes in North Atlantic deep-sea cores, with remarks on isotopic paleotemperature analysis, *Progress in Oceanography*, 3: 221-252.

Olausson, E., 1967. Climatological, geoeconomical, and paleooceanographical aspects on carbonate deposition, *Progress in Oceanography*, 4: 245-265.

Redfield, A.C., Ketchum, B.H. and Richards, F.A., 1963. The influence of organisms on the composition of sea-water. In: M.N. Hill (ed.) *The Sea, vol.2*, Interscience, New York, p. 26-77.

Reid, J.L., 1961. On the temperature, salinity, and density differences between the Atlantic and Pacific oceans in the upper kilometre, *Deep-Sea Res.*, 7: 265-275.

Reid, J.L. and Lynn, R.J., 1971. On the influence of the Norwegian-Greenland and Weddell seas upon the bottom waters of the Indian and Pacific oceans, *Deep-Sea Res.*, 18: 1063-1088.

Ruddiman, W.F. and Duplessy, J.-C., 1985, Conference on the last deglaciation: timing and mechanism, *Quat. Res.*, 23: 1-17.

Ruddiman, W.F. and McIntyre, A, 1973. Time-transgressive deglacial retreat of polar waters from the North Atlantic, *Quat. Res.*, 3: 117-130.

Ruddiman, W.F. and McIntyre, A., 1981a. The North Atlantic Ocean during the last deglaciation, *Palaeogeo., Palaeocli., Palaeoec.*, 35: 145-214.

Ruddiman, W.F. and McIntyre, A, 1981b. The mode and mechanism of the last deglaciation: oceanic evidence, *Quat. Res.*, 16: 125-134.

Ruddiman, W.F. and McIntyre, A., 1984. An evaluation of ocean-climate theories on the North Atlantic. In: A.L. Berger et al. (eds.) *Milankovitch and Climate*, Part 2, D.Reidel, Dordrecht, pp. 671-686.

Schnitker, D., 1974. West Atlantic abyssal circulation during the past 120,000 years. *Nature*, 248: 385-387.

Schnitker, D., 1979. The deep waters of the western North Atlantic during the past 24,000 years, and the re-initiation of the Western Boundary Undercurrent, *Marine Micropaleontology*, 4: 265-280.

Shackleton, N., 1977. Carbon-13 in *Uvigerina*: tropical rainforest history and the Equatorial Pacific carbonate dissolution cycles. In:

N.R.Andersen and A.Malahoff (eds.) *The Fate of Fossil Fuel CO_2 in the Oceans*, Plenum, New York, pp. 401-427.

Shackleton, N.J., Imbrie, J. and Hall, M.A., 1983. Oxygen and carbon isotope record of East Pacific core V19-30: implications for the formation of deep water in the late Pleistocene Atlantic, *Earth and Planetary Science Letters*, 65: 233-244.

Stommel, H., 1980. Asymmetry of interoceanic fresh-water and heat fluxes, *Proc Natl. Acad. Sci.* USA, Geophysics, 77 (5): 2377-2381.

Streeter, S.S., 1973. Bottom water and benthonic foraminifera in the North Atlantic - Glacial-Interglacial contrasts, *J. Quat. Res.*, 3: 131-141.

Streeter, S.S. and Shackleton, N.J., 1979. Paleocirculation of the deep North Atlantic: 150,000-year record of benthic foraminifera and oxygen-18, *Science*, 203: 168-171.

Streeter, S.S., Belanger, P.E., Kellogg, T.B. and Duplessy, J.C., 1982. Late Pleistocene paleo-oceanography of the Norwegian-Greenland Sea: benthic foraminiferal evidence. *Quaternary Research*, 18: 72-90.

Weyl, P.K., 1968. The role of the oceans in climate change: a theory of the ice ages, *Met. Monogr.*, 8: 37-62.

Woods, J., 1981. The memory of the Ocean. In: A. Berger (editor), *Climatic Variations and Variability: Facts and Theories*, D. Reidel Publ. Co., Holland, p. 63-83.

RAPID CHANGES IN THE INFLOW OF ATLANTIC WATER INTO THE NORWEGIAN SEA AT THE END OF THE LAST GLACIATION

Eystein Jansen

Dept. of Geology Sect. B, University of Bergen
Allégaten 41
N-5007 Bergen
Norway

ABSTRACT. Both surface ocean temperatures (SST) and the oceanic circulation regime of the eastern Norwegian Sea show evidence of rapid changes during the last deglaciation (approx. 15.000 - 8.000 yr B.P.). SST variations in response to influx of Atlantic surface water operated in phase with the Alleröd-Younger Dryas-Holocene climatic variations. Summer surface temperatures cooled about 5-7°C within less than 500 years at the Alleröd-Younger Dryas transition. Ameliorated SST during the Alleröd was restricted to the easternmost parts of the Norwegian Sea. Deep water formation within the Norwegian Sea commenced at about 11.000 yr B.P. The Alleröd was characterized by high oceanic productivity and a homogenous water column, probably as a result of intrusion of Atlantic waters both at the surface and at intermediate depths.

INTRODUCTION

A number of studies have demonstrated the climatic sensitivity of the Norwegian Sea during the late Quaternary (Kellogg, 1976; Kellogg et al., 1978; Streeter et al., 1982; Jansen and Erlenkeuser, 1985; Jansen and Björklund, 1985). Two major aspects of the paleoceanography of this region are emphasized in this paper: 1. The inflow of temperate Atlantic surface water into the Norwegian Sea, which at present forms a major contribution to the northward heat flux in the Atlantic-Arctic region, and which adds the necessary saltwater component to enhance deep-water formation. 2. The process of deep-water formation which presently occurs as a consequence of winter season cooling and subsequent sinking of surface waters in the areas north of Iceland and in the areas between Jan Mayen and Spitsbergen (Aagaard et al., 1985). Both of these aspects are of more than local significance. The presence of the Atlantic surface inflow has a major climatic impact on northern latitude climates. Deep-water formation within the Norwegian Sea and the overflow of ventilated deep-water into the North Atlantic is a major source of global deep-water renewal and ventilation.

Changes in the circulation and thermal regime of this ocean may also have had an important impact on the adjacent Scandinavian ice-sheet

W. H. Berger and L. D. Labeyrie (eds.), Abrupt Climatic Change, 299–310.

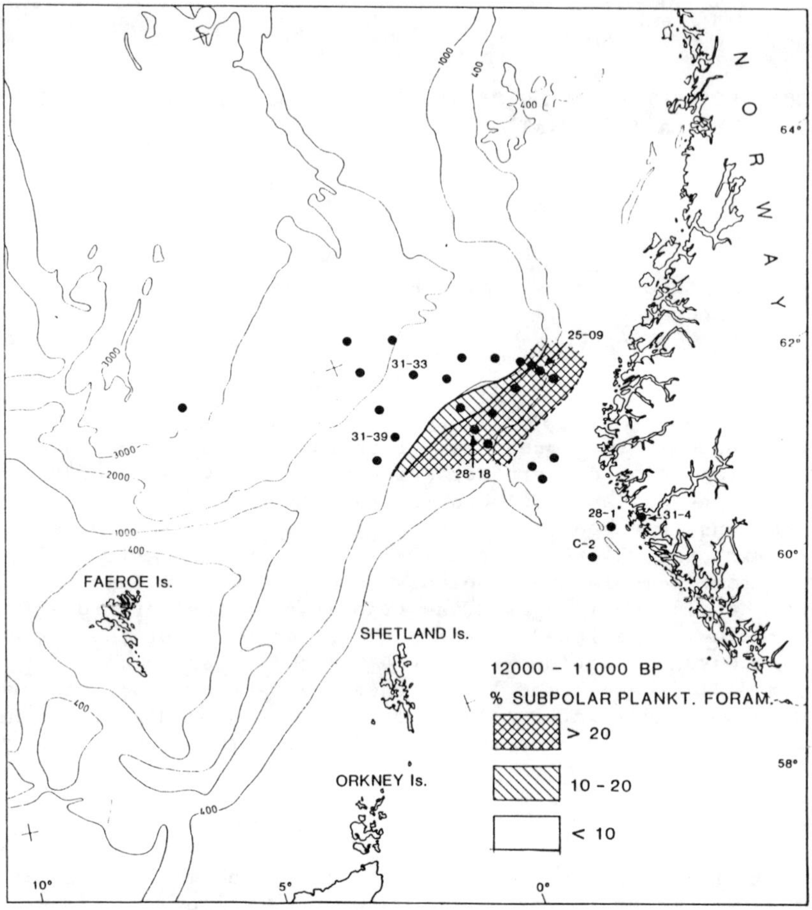

Figure 1. Map showing location of cores presented on Figs. 2-5, and presence of subpolar planktonic foraminifera off Western Norway between 12.000 and 11.000 BP. Each dot on the map represents a core used to define the distribution pattern of subpolar foraminifera, and the hatched area shows the area of influx of North Atlantic water.

during ice ages. Furthermore, recent theories on global climatic change and discussions on the causes of the glacial to interglacial atmospheric CO_2 increase have focused on the role played by changes in high latitude deep-water circulation (Berger and Killingley, 1982; Siegenthaler and Wenk, 1984; Broecker et al., 1985; Berger et al., this volume).

For the purpose of this conference the focus will be placed upon a high resolution record of oceanic change during the rapid transition from full glacial conditions into the present circulation regime. A record of sea-surface temperature changes is obtained from planktonic foraminifera and radiolaria (Jansen and Björklund, 1985). Changes in the vertical water mass structure and circulation regime are investigated by the means of carbon and oxygen isotope analyses of benthic foraminifera. Some of the isotope data were discussed by Jansen and Erlenkeuser (1985) while some are new data not previously published.

STRATIGRAPHY

The location of the cores selected for this presentation is given on Fig. 1. Together they form a transect of cores covering hemipelagic to inshore environments from the continental slope into the outer parts of the fjord areas of Western Norway, crossing the present area of maximum inflow of Atlantic surface water.

The timing of events and correlations between cores are based upon a set of stratigraphic marker horizons defined and discussed by Jansen and Björklund (1985). These are considered isochronous within an accuracy varying between 1000 and 100 years. Shortly summarized these are as follows: 1. Immigration of the benthic foraminifer *Cassidulina teretis* accompanied by an increase in planktonic foraminiferal productivity (13-14.000 yr B.P.). 2. A marked increase in productivity of all planktonic microfossils (about 12.000 yr B.P.). 3. The Vedde Ash-layer (dated by Mangerud et al. (1984) to 10.600 ±60 yr B.P.). 4. The transition into a dominant subpolar environment as defined by foraminifera and radiolaria (9.800 ±200 yr B.P.).

SEA SURFACE TEMPERATURE

Two patterns of surface-water response have been identified. Cores 31-39 (Fig. 2) and 28-18 (Fig. 3) are shown as examples. Since presently available transfer functions do not have any resolution in the temperature range below 5-6°C, no absolute temperature estimates can be assigned to the foraminifer and radiolarian assemblages in the cores. The distribution of planktonic foraminifera in the present Norwegian Sea shows that the border area between cold, polar water masses and warmer, subpolar Atlantic surface water can be defined by the 95% isoline for the polar foraminifer *Neogloboquadrina pachyderma* (sinistrally coiled) in surface sediment assemblages. More than 5% subpolar foraminifera indicates influx of Atlantic surface water (Kellogg, 1976). The content of two radiolarian species with cold water affinities have been used as a radiolarian SST index (*Cycladaphora davisiana* (Morley and Hays, 1983) and *Amphemelissa setosa* (Jansen and Björklund, 1985)).

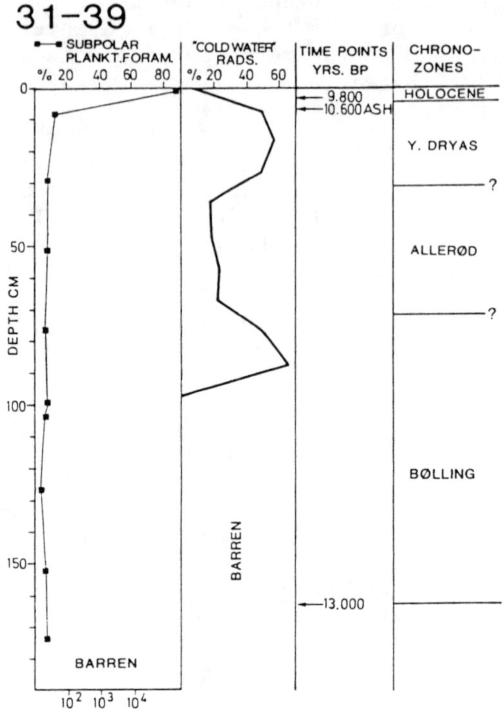

Figure 2. Percentages of subpolar planktonic foraminifera ("warm water" index) and combined percentages of the two radiolarians *C. davisiana* and *A. setosa* ("cold-water" index) in core 31-39 (from Jansen and Björlund, 1985). Time scale according to stratigraphic marker levels as presented in text.

Fig. 2 from core 31-39 shows a record starting at about 13.000 yr BP. Uniform polar foraminiferal fauna until after 10.600 yr BP indicates no temperatures in excess of 5-6°C until the end of the Younger Dryas period.

However, evidence for variation in SST is present in the radiolarian record showing ameliorated temperatures coinciding with the Alleröd chronozone. This difference between radiolarian and foraminiferal response most likely indicates temperature variation in the range between 0 and 5-6°C. However, it may also arise from possible differences caused by foraminifera living in deeper and colder water masses than radiolaria. Yet, the good coherence between radiolarian and foraminiferal data in core 28-18 (Fig. 3) is good evidence against the latter possibility.

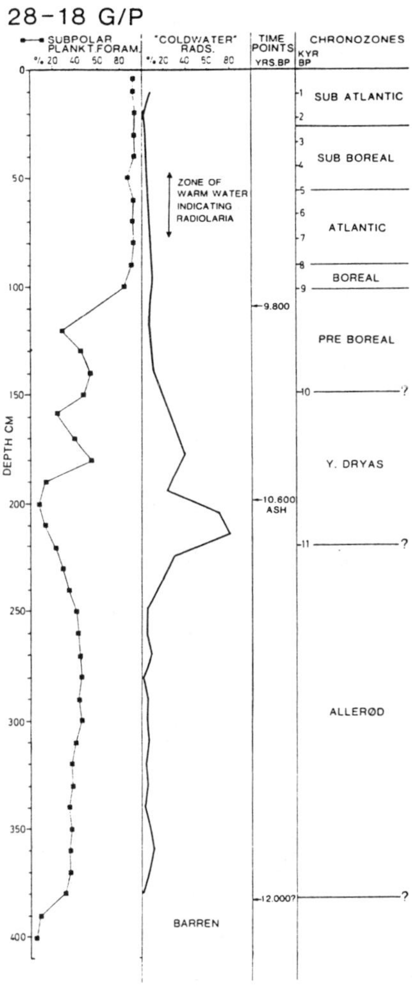

Figure 3. Percentages of subpolar planktonic foraminifera and combined percentages of the two radiolarians *C. davisiana* and *A. setosa* in core 28-18. Time scale according to marker levels as presented in text.

Due to very high sedimentation rates, core 28-18 provides a very detailed record starting at 12.000 yr BP. Both planktonic foraminifera and radiolaria show a period of relatively warm surface waters probably coinciding with the Alleröd chronozone 12.000 to 11.000 yr BP. Based on comparison with planktonic foraminiferal data from the present Norwegian Sea (Kellogg, 1976; T. Johannesssen pers. comm., 1985) a summer SST of about 7°C, similar to those prevailing off the northernmost parts of Norway (at about 80°N), can be assumed for the Alleröd in this area. The much colder temperatures recorded in core 31-33 to the west indicates a strong thermal E-W gradient in the southern Norwegian Sea with a polar front position far to the east, and inflow of warmer Atlantic water entrenched in a narrow wedge crossing over the upper continental slope areas off western Norway. This pattern of inflow is clearly shown by the distribution pattern of subpolar planktonic foraminifera during the Alleröd (Fig. 1).

A sharp decline in SST is seen at the beginning of the Younger Dryas period (Figs. 2 and 3) showing a return to cold temperatures with a summer SST well beneath 5-6°C. By using the assigned time points at 12.000 and 10.600 yr BP in Fig. 3 and assuming a constant sedimentation rate, a rough estimate of the time involved in this change can be made. This indicates that the full change from ameliorated SST to a thermal regime resembling full glacial conditions took place in less than 500 yrs. This time span is in accordance with estimates of the time involved in the deterioriation of climate at the Alleröd/Younger Dryas boundary along the neighboring coastline (Mangerud, this volume). The initial half of the cold Younger Dryas period appears to have been the coldest; SST started to increase at 10.500 yr BP. The last half of the Younger Dryas and the first few hundred years of the Holocene experienced very variable conditions intermittent between cold glacial, and warm Holocene environments indicating a large variability of the inflowing North Atlantic current system. The remainder of the Holocene has been characterized by a stable inflow of North Atlantic water. However, presence of subtropical radiolarians in a zone within the Holocene (Fig. 3) indicate warmer SST at about 7.000 to 4.000 yr BP than at present.

This pattern of SST change during the deglacial period is in close agreement with North Atlantic data, which also show a dramatic drop in SST from the Alleröd into the Younger Dryas at about 11.000 yr BP (Duplessy et al., 1981; Ruddiman and McIntyre, 1981). However, while North Atlantic planktonic foraminiferal data indicate higher temperatures in the time span 13.000 to 12.000 yr BP (correlative with the Bölling period), the Norwegian Sea clearly points to the highest pre-Holocene SST during the Alleröd (12-11.000 yr BP). There is a very good agreement between palynological and glacial geological records from the Scandinavian mainland and the record of SST changes outlined here. Glacial geological evidence show a major retreat of the ice-margin of the Scandinavian Ice Sheet during the Alleröd and a halt or an advance during the Younger Dryas (summarized by Mangerud, 1980). Rapid deglaciation commenced again at 10.000 BP. The same cold->ameliorated->cold->warm sequence is evidenced by vegetational changes (Krzywinski and Stabell, 1984; Larsen et al., 1984) and shallow marine molluscan and foraminiferal assemblages (Mangerud, 1977). At present we are not able to see any major leads or lags by comparing oceanic and terrestrial records. Within the accuracy of our time scale the changes in the ocean, along the ice front, and in the position of the ice margin seem to be occurring at roughly the same time and with roughly the same speed.

OCEANIC CIRCULATION CHANGES

Oxygen and carbon isotope records from benthic foraminifera have been produced from five cores. The oxygen isotope records with

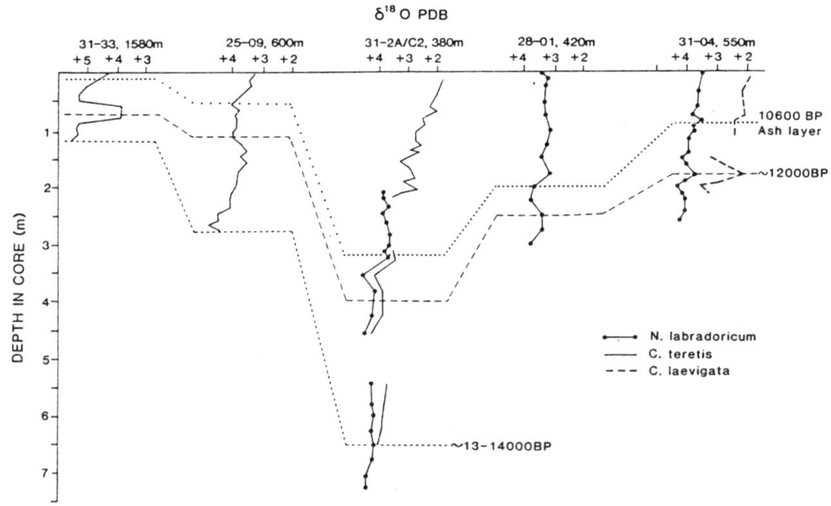

Figure 4. Oxygen isotope records from benthic foraminifera in cores from the Norwegian Sea and nearshore areas off Western Norway. Correlation lines are drawn through marker levels as presented in text. All measurements were made according to standard procedures (Shackleton and Opdyke, 1973). Cores 31-33 and 25-09 were measured at the University of Kiel using a VG MM 602 instrument (Jansen and Erlenkeuser, 1985), parts of core 31-2A/C2 were measured at the University of Cambridge using a MM 903 instrument, while the remainder of the analyses were made at Bergen University using a Finnigan MAT 251 instrument. The inter-laboratory range of error is estimated to be 0.2°/oo.

correlation lines are shown on Fig. 4. Correlation points are made according to the stratigraphic marker horizons listed above (Jansen and Björklund, 1985). The records cover a variable length of time from about 13-14.000 yr BP to the present. The high productivity level at about 12.000 yr BP falls at the end of a peak of low $\delta^{18}O$ values at the end of oxygen isotope Termination 1a, which have been ^{14}C dated to between 13.500 and 12.000 yr B.P. (Duplessy et al., 1981; Mix and Ruddiman, 1985; Berger et al., 1985, Bard et al., this volume).

The Younger Dryas is marked by an increase in $\delta^{18}O$. Such an increase could result both from increased global ice-volume during the Younger Dryas cooling, or from a more local drop in bottom water temperatures along the Norwegian continental margin. Although there is evidence for an advance of the western margins of the Scandinavian Ice Sheet during the Younger Dryas (Mangerud, 1980) as well as the Scottish Ice Sheet (the Loch Lomond Stadial) (Sissons, 1980), no well established data exist from the Laurentide Ice Sheet to substantiate any large

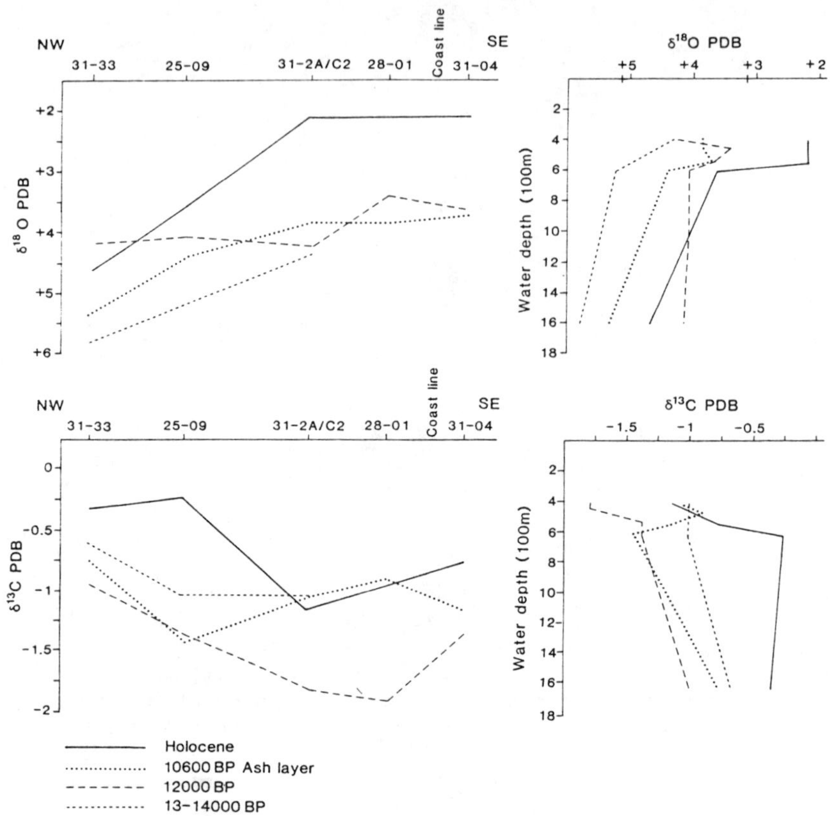

Figure 5. Plots of oxygen and carbon isotopes according to core location on a NW-SE transect and according to water depth at core location. Lines are drawn through correlation points as shown on Fig. 4. Values for *N. labradoricum* are used for cores 31-04 and 28-01. For the rest of the cores values for *C. teretis* have been used.

global Younger Dryas ice-volume increase (see reviews by Denton and Hughes, 1981; Andrews, 1982). However, most oxygen isotope records from high sedimentation rate cores display a plateau in isotope values at about Younger Dryas time (Duplessy et al., 1981; Sarnthein et al., 1982; Berger et al., 1985; Jansen and Erlenkeuser, 1985; Mix and Ruddiman, 1985) indicating at the least a halt in global ice-volume disintegration during the Younger Dryas. Thus the most plausible reason for the Younger Dryas isotope enrichment in Fig. 4 is a drop in bottom water temperature. The largest drop appears in the deepest core at 1600m water depth. This, together with carbon isotopic evidence was taken as a signal of the onset of large scale deepwater formation within the Norwegian Sea by Jansen and Erlenkeuser (1985). Evidence for a concurrent increase in deep-water overflows into the North Atlantic comes from increased sedimentation rates on the Reykjanes Ridge, dated to 11.000 BP (Ruddiman and Bowles, 1976). The evidence for a large increase in deep-water formation in the Younger Dryas is in conflict with the climatic model of Broecker et al. (1985) who suggested that deep-water formation commenced during a climatic warming i.e. the Alleröd.

In order to investigate possible changes in the vertical water mass properties, both oxygen and carbon isotope results for the four stratigraphic marker horizons are plotted on two types of plots: 1. according to their location along a NW-SE transect from the Norwegian Sea and into the outer fjord areas of Western Norway, and 2. according to water depth (Fig. 5).

The present vertical water-mass structure is characterized by presence of warm Atlantic water down to about 600m, with a strong thermocline underneath into the deep-waters, having a temperature close to -1°C. This structure is mimicked by the Holocene oxygen isotope line on Fig. 5 which shows a decline in $\delta^{18}O$ over the shallow water areas, a response to the higher temperatures within the North Atlantic water. Both the 13-14.000 yr BP line and the 10.600 yr BP Younger Dryas oxygen isotope line show a similar pattern as the Holocene. However, the shallow to deep water gradient appears to have been lower during these phases, indicating colder near-surface waters than in the Holocene.

The 12.000 yr BP line is an exception, being a nearly straight line, thus indicating a relatively homogenous water column at least down to 1600m depth during the initial inflow of Atlantic water in Alleröd times. Whether this is a result of dense Atlantic surface waters reaching down to the deep water realm, or a result of a kind of oceanic upwelling as suggested by Jansen et al. (1983), is difficult to ascertain.

The carbon isotope record shows similar evidence. The Holocene line shows more positive $\delta^{13}C$ values for the open ocean than the other records, probably showing a response to better ventilated waters under the Holocene circulation regime. $\delta^{13}C$ in benthic foraminifera may reflect several parameters such as: ventilation of bottom waters (Shackleton et al., 1983; Duplessy et al., 1984), changes in the relative size of the sensitive carbon reservoirs (Shackleton, 1977), productivity in the surface waters and flux of organic carbon to the benthic zone (Zahn et al., 1986; Sarnthein et al., this volume). Thus the variations in carbon isotopes recorded in the pre-Holocene records may be caused by several factors operating together. The 12.000 yr BP level was accompanied by high levels of primary productivity, as reflected by abundant diatom and planktonic foraminiferal assemblages. Thus it is of interest to note that this level, with the exception of core 25-09, shows the lowest carbon istope values, which one would expect under a high productivity area (see further discussion in the paper by Sarnthein et al., this volume). The carbon isotope record for the 13-14.000 yr BP level and the Younger Dryas level at 10.600 are quite similar. They are close to Holocene values in the nearshore areas, and significantly lower than in the Holocene in the Norwegian Sea. This suggests that the sedimentation/productivity regime was quite different from today during the whole pre-Holocene period of the deglaciation.

CONCLUSIONS

1. The southeastern parts of the Norwegian Sea experienced very rapid changes in SST and influx of warm Atlantic water during the period 14.000 to 9.000 yr BP. These changes appear to closely reflect the Bolling-Alleröd-Younger Dryas-Holocene climatic changes as recorded in terrestrial records.

2. Summer SST in the Norwegian Sea changed from about 7°C to temperatures close to full glacial conditions within less than 500 yrs. from the Alleröd to the Younger Dryas.

3. The areas affected by ameliorated SST during the Alleröd were restricted to the extreme eastern parts of the ocean.

4. The deglacial onset of deep-water formation probably took place early in the Younger Dryas period at about 11.000 yr BP.

5. A very homogenous water column during the Alleröd and high productivity at about 12.000 yr BP indicate possible oceanic upwelling caused by the first Atlantic inflow during the last deglaciation.

ACKNOWLEDGEMENTS

J. Ellingsen and E. Irgens made the figures. S. Rolfsen skillfully operated the mass spectrometer in Bergen. Isotope measurements on core 31-2A/C2 were performed in Cambridge with kind guidance from N. Shackleton and M. Hall. P. Ciesielski helped getting the manuscript ready in time. The British Council supported my stay in Cambridge and the Norwegian Research Council for Science and the Humanities supports the National Mass Spectrometry Laboratory in Bergen.

REFERENCES

Aagaard, K., Swift, J.H. and Carmack, E.C., 1985. Thermohaline circulation in the Arctic mediterranean seas, *J. Geophys. Res.*, 90(C3): 4833-4846.

Andrews, J., 1982. On the reconstruction of Pleistocene Ice Sheets: A review, *Quat. Sci. Rev.*, 1: 1-30.

Bard, E., Arnold, M., Duprat, J., Moyes, J., and Duplessy, J.C., 1987. Bioturbation effects on abrupt climatic changes recorded in deep sea sediments. Correlation between $\delta^{18}O$ profiles and accelerator ^{14}C dating, this volume.

Berger, W.H. and Killingley, J.S., 1982. The Worthington effect and the origin of Younger Dryas, *J. Mar. Res.*, 40: 28-37.

Berger, W.H., Burke, S. and Vincent, E., 1987. Glacial-Holocene transition: climate pulsations and sporadic shutdown of NADW produciton, this volume.

Berger, W.H., Killingley, J.S., Metzler, C.V. and Vincent, E., 1985. Two step deglaciation: ^{14}C-dated high resolution $\delta^{18}O$ records from the tropical Atlantic Ocean, *Quat. Res.*, 23: 258-271.

Broecker, W.S., Peteet, D.M. and Rind, D., 1985. Does the ocean-atmosphere system have more than one stable mode of operation? *Nature*, 315: 21-26.

Denton, G.H. and Hughes, T.J., 1981. *The Last Great Ice Sheets*, Wiley, New York, N.Y., 484 pp.
Duplessy, J.-C., Delibrias, G., Turon, J.L., Pujol, C. and Duprat, J., 1981. Deglacial warming of the northeastern Atlantic Ocean: correlation with the paleoclimate evolution of the European Continent, *Paleogeogr. Palaeoclimatol. Palaeoecol.*, 35: 121-144.
Duplessy, J.-C., Shackleton, N., Mathews, R. K., Prell, W., Ruddiman, W. R., Caralp, M. and Hendy, C. H., 1984. Carbon-13 record of benthic foraminifera in the last interglacial ocean: implications for the carbon cycle and the global deep water circulation, *Quat. Res.*, 21: 225-243.
Jansen, E., Sejrup, H.P., Fjaeran, T., Hald, M., Holtedahl, H. and Skarbö, O., 1983. Late Weichselian paleoceanography of the southeastern Norwegian Sea. *Nor. Geol. Tidsskr.*, 63: 117-146.
Jansen, E. and Björklund, K.R., 1985. Surface ocean circulation in the Norwegian Sea 15,000 B.P. to present, *Boreas*, 14: 243-257.
Jansen, E. and Erlenkeuser, H., 1985. Ocean circulation in the Norwegian Sea during the last deglaciation: isotopic evidence, *Palaeogeogr. Palaeoclimatol. Palaeoecol.*, 49: 189-206.
Kellogg, T.B., 1976. Late Quaternary climatic changes: evidence from deep-sea cores of Norwegian and Greenland Seas. In: Cline, R.M. and J.D. Hays (eds.), Investigations of Late Quaternary Paleoceanography and Paleoclimatology, *Geol. Soc. Am. Mem.*, 145: 77-110.
Kellogg, T.B., Duplessy,J.-C. and Shackleton, N.J., 1978. Planktonic foraminiferal and oxygen isotopic stratigraphy of Norwegian Sea deep-sea cores, *Boreas*, 7: 61-73.
Krzywinski, K. and Stabell, B., 1984. Late Weichselian sea-level changes at Sotra, Hordaland, Western Norway, *Boreas*, 13: 159-202.
Larsen, E., Eide, F., Longva, O. and Mangerud, J., 1984. Alleröd-Younger Dryas climatic inferences from cirque glaciers and vegetational development in Western Norway, *Arctic and Alpine Res.*, 16: 137-160.
Mangerud, J., 1977. Late Weichselian sediments containing shells, foraminifera and pollen, at Agotnes, Western Norway, *Nor. Geol. Tidsskr.*, 57: 23-54.
Mangerud, J., 1980. Ice-front variations in different parts of the Scandinavian Ice Sheet 13,000-10,000 years B.P. In: Lowe, J.J., J.M. Gray, and J.M. Robinson, *Studies in the Lateglacial of North-West Europe*, 22-30, Pergamon Press, Oxford.
Mangerud, J., 1987. The Alleröd/Younger Dryas boundary, this volume.
Mangerud, J., Lie, S.E., Furnes, H., Kristiansen, I.L. and Lömo, L., 1984. A Younger Dryas Ash-bed in Western Norway, and its possible correlation with tephra in cores from the Norwegian Sea and the North Atlantic, *Quat. Res.*, 21: 85-104.
Mix, A.C. and Ruddiman, W.F., 1985. Structure and timing of the last deglaciation: oxygen-isotope evidence, *Quat. Sci. Rev.*, 4: 59-108.
Morley, J.J. and Hays, J.D., 1983. Oceanographic conditions associated with high abundances of the radiolarian *Cycladaphora daviasiana*, *Earth Planet. Sci. Lett.*, 66: 63-72.
Ruddiman, W.F. and Bowles, F.A., 1976. Early interglacial bottom current sedimentation on the eastern Reykjanes Ridge, *Mar. Geol.*, 21: 191-210.

Ruddiman, W.F. and McIntyre, A., 1981. The North Atlantic Ocean during the last deglaciation, *Palaeogeogr.*, *Palaeoclimatol. Palaeoecol.*, 35: 145-214.
Sarnthein, M., Erlenkeuser, H. and Zahn, R., 1982. Termination I: the response of continental climate in the tropics as recorded in deep sea sediments, *Bull. Inst. Bassin d'Aquitaine*, 31: 393-407.
Sarnthein, M., Winn, K. and Zahn, R., 1987. Paleoproductivity of oceanic upwelling and the effect on atmospheric CO_2 and climatic change during deglaciation times, this volume.
Shackleton, N.J., 1977. Tropical rain-forest history and the equatorial Pacific carbonate dissolution cycles. In: Anderson, N.R. and A. Malahoff (eds.), *The fate of fossil fuel CO_2 in the oceans*, Plenum Press, New York, 401-428.
Shackleton, N.J. and Opdyke, N.D., 1973. Oxygen isotopic and paleomagnetic stratigraphy of equatorial Pacific core V28-283: oxygen isotope temperatures and ice-volume on a 10^5 year scale, *Quat. Res.*, 3: 39-55.
Shackleton, N.J., Imbrie, J. and Hall, M.A., 1983. Oxygen and carbon isotopic record of core V19-30: implications for the formation of deep water in the Late Pleistocene North Atlantic, *Earth. Planet. Sci. Lett.*, 65: 233-244.
Siegenthaler, U. and Wenk, T., 1984. Rapid atmospheric CO_2 variations and ocean circulation, *Nature*, 308: 624-626.
Sissons, B., 1980. Paleoclimatic inferences from Loch Lomond advance glaciers. In: Lowe, J.J, J.M. Gray, and J.E. Robinson (eds), *Studies in the Lateglacial of North-west Europe*, Oxford, Pergamon Press, 31-43.
Streeter, S.S., Belanger, P.E., Kellogg, T.B. and Duplessy, J.-C., 1982. Late Pleistocene paleo-oceanography of the Norwegian-Greenland Sea: benthic foraminiferal evidence, *Quat. Res.*, 18: 72-90.
Zahn, R., Winn, K. and Sarnthein, M., 1986. Benthic foraminiferal $\delta^{13}C$ and accumulation rates of organic carbon: *Uvigerina peregrina* group and *Cibicidoides wuellerstorfi*, *Paleoceanography*, 1: 27-42.

PALEOPRODUCTIVITY OF OCEANIC UPWELLING AND THE EFFECT ON ATMOSPHERIC CO_2 AND CLIMATIC CHANGE DURING DEGLACIATION TIMES

Michael Sarnthein, Kyaw Winn and Rainer Zahn

Geological-Palaeontological Institute
Christian Albrecht University
Olshausenstr. 40
2300 Kiel
Federal Republic of Germany

ABSTRACT. In addition to variations in the Earth's orbit, changes in atmospheric pCO_2 represent an important factor in creating global climatic and ice volume changes. Atmospheric pCO_2 fluctuations are largely controlled by the exchange rates of CO_2 between the atmosphere and ocean reservoirs, an exchange which greatly depends upon plankton primary productivity in oceanic upwelling regions, where the carbon to carbonate "Rain Ratio" (Berger and Keir, 1984) is high. In order to test this model, a new equation to calculate the local ocean paleoproductivity has been developed. The formula is based on the relationships between carbon accumulation rate, water depth, and carbon-free bulk sedimentation rates (as a "sealing factor") of deep-sea sediments, and is independent of a large range of bottom water O_2 concentration. For comparison with paleoproductivity, $\delta^{13}C$ fluctuations of *Cibicidoides wuellerstorfi* serve to record the total CO_2 dissolved in North Atlantic Deep Water (NADW). In the region of coastal upwelling off northwest Africa, paleoproductivity increased by a factor of three from interglacial to glacial stages depicting a clear 100,000 year cycle. In oceanic "deserts", the productivity varied much less, but approximately with the same cycle during the last 0.5 my. Phase relationships during the last two Terminations show that trade wind strength and related productivity due to upwelling in the east Atlantic started to decrease slightly prior to or simultaneously with global ice melting, synchronously with a drastic increase in atmospheric pCO_2 (after Neftel et al., 1982). On the other hand, CO_2 depletion in the NADW only followed after some 2500 to 4500 years, and thus, cannot have caused the change in atmospheric CO_2. The high latitude insolation balance which causes changes in sea-ice cover and thus, of meridional trade wind intensity, is regarded as the prime factor responsible for this massive feedback mechanism for climatic change.

INTRODUCTION: EXTERNAL CAUSES OF CLIMATIC CHANGE

There is general agreement that a significant share of global climatic variance as monitored by oxygen-isotope curves can be explained by the Earth's orbital forcing according to the Milankovitch theory.

W. H. Berger and L. D. Labeyrie (eds.), Abrupt Climatic Change, 311–337.

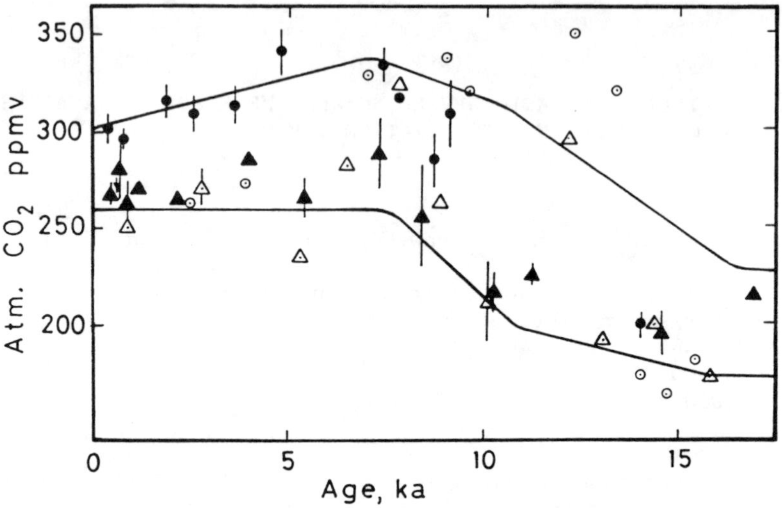

Figure 1. Ice core evidence for a change in atmospheric CO_2 during deglaciation. **Circles**: Greenland cores. **Triangles**: Antarctic cores. Open: Delmas et al. (1980). **Filled with bounding line**: Neftel et al. (1982). After Berger and Keir (1984).

ever, the extent of a linear response to orbital forcing is still under debate. Estimates vary, for example, from 15-25% (Herterich and Sarnthein, 1984) up to 77% (Imbrie et al., 1984). The residual climatic variance, whether small or large, requires consideration of stochastic mechanisms (Hasselmann, 1976) and also, a search for other (non-linear) external forcings unrelated to celestial mechanisms such as the effect of ice isostasy (e.g., Peltier and Hyde, 1984). This is especially true for an explanation of the 100,000-year cycle which is not yet well understood.

Within this context, the ice-core evidence of Delmas et al. (1980) and Neftel et al. (1982) provided a crucial breakthrough. They showed a major change in atmospheric pCO_2 from about 200 to about 300 ppmv during the last deglaciation (Fig. 1). Similar rate changes can be postulated also for earlier Terminations. Accordingly, CO_2 may have been essential in providing the required additional external forcing to accomplish climatic change by means of the "greenhouse effect" (Arrhenius, 1896). The ice core data of global change stimulated a series of studies searching for equivalent CO_2 changes in the ocean as recorded in deep-sea sediments (e.g. Duplessy et al., 1982, 1984; Shackleton et al., 1983; Boyle, 1986). In addition, a number of reservoir models arose, trying to tie this change to large scale processes that might vary the CO_2 exchange rates between the atmosphere and the oceanic deep water. This deep water is the main CO_2 reservoir containing at least 60 times more CO_2 than the atmosphere (Grassl et al., 1984). Broecker (1982; revoked 1984) regarded the nutrient inventory of P and N in the ocean as *the* limiting factor for CO_2 extraction from the atmosphere. More recently, Broecker

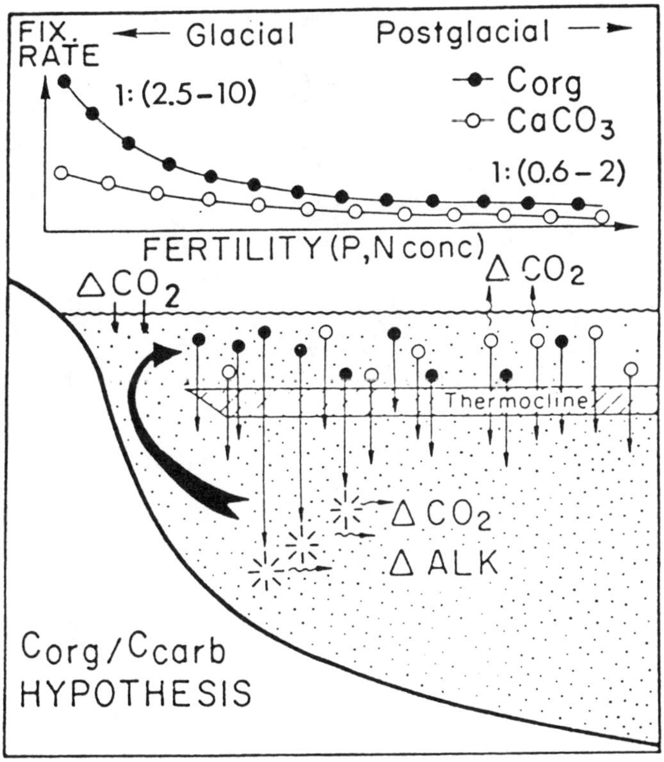

Figure 2. The carbon/carbonate Rain Ratio Model of Berger and Keir (1984).

et al. (1985) and Boyle and Keigwin (1985) considered the intensity of North Atlantic Deep Water (NADW) formation as the main control of differing CO_2 pressure in the atmosphere.

Siegenthaler and Wenk (1984) and Berger and Keir (1984) recognized the possible key role of changes in ocean surface circulation causing ocean productivity changes and thus affecting the atmospheric pCO_2. Berger and Keir (1984) related the influence of productivity on CO_2 to the particle composition ("rain ratio") of plankton fluxes from the surface ocean to the deep sea (Fig. 2). This "biological pump" results in an annual bulk transfer of 23 x 10^9 tons of carbon from the sea surface (and the atmosphere) to greater water depths. The plankton "Rain Ratio Model" of Berger and Keir (1984) is based on data from various sediment traps. These data show that the ratio of organic carbon to carbonate carbon fixed in the particulate plankton flux varies systematically. The ratio increases from 0.6-2.0 below low-fertility areas to 2-10 below high-fertility areas such as zones of oceanic upwelling (Fig. 3a). In the high-fertility regions, the CO_2 extracted from the ocean deep-water by $CaCO_3$ dissolution is by far overcompensated by CO_2 produced by the

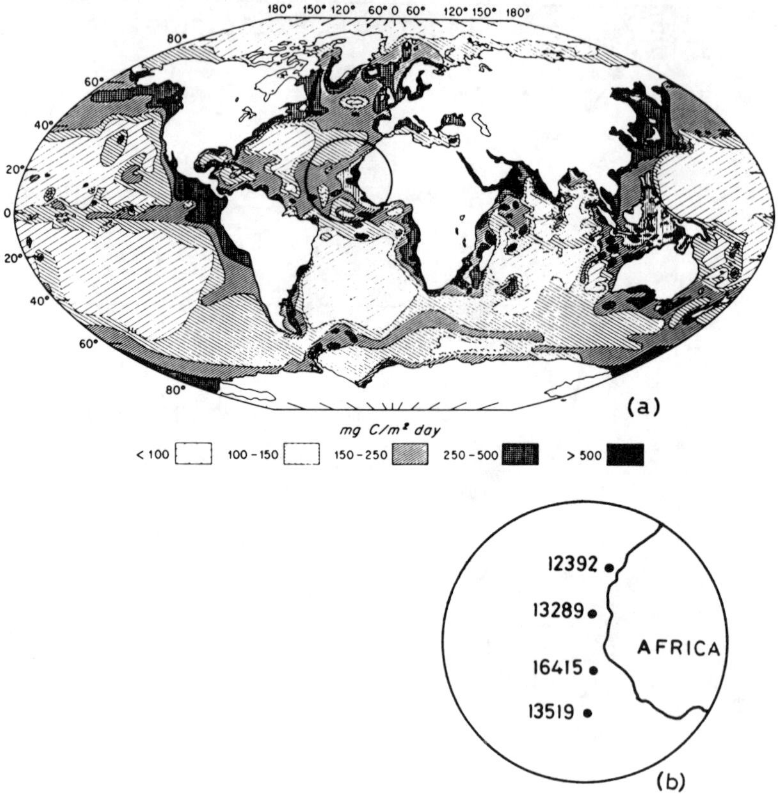

Figure 3. (a). Distribution of primary production in the World Ocean (After Koblentz-Mishke et al. 1970). (b). Core locations offshore from West Africa.

oxidation of carbon particles. On the other hand, CO_2 extraction should equal or exceed CO_2 production below low-fertility regions. In summary, intensity changes of surface-ocean productivity in upwelling zones may rule both the drawdown of total CO_2 by photosynthesis and the amount of CO_2 dissolved or extracted in the deep ocean. This model was also suggested by Flohn (1983) based on modern climate data.

The Rain Ratio Model (P. Weyl in Broecker, 1982) also predicts a change in the proportion of ^{12}C in total CO_2 dissolved in the deep ocean, because organic carbon is isotopically much lighter than dissolved carbon. Until now, however, the Rain Ratio Model has been supported by little hard data; this lack of data also applies to the other CO_2 models. The only direct multiseasonal measurements made in the modern equatorial Atlantic show that the carbon (and CO_2) transfer between the atmosphere and the ocean varies in a small-scale and uneven regional pattern. High carbon fluxes from the atmosphere to the ocean occur in high productivity belts and are largely balanced by a reverse carbon flux outside of these belts (Oudot and Andrie, 1986).

In the present study, we compare partly new data of oceanic paleoproductivity with the benthos ^{13}C record of CO_2 variations in the ocean. Our evidence is obtained from two sediment cores from the East Atlantic, one from a high, the other from a low fertility region (Fig. 3b). In order to test the rain-ratio idea, we compare the phase relationships of paleoproductivity and deep-water oxygenation with that of the benthos oxygen-isotope record of global ice-volume change during the last and penultimate deglaciations (Terminations I and II), and finally, with the aeolo-marine sediment record of paleo-wind strength.

METHODS AND THEORY

The $\delta^{18}O$ record

The ratio of ^{18}O to ^{16}O has been measured in 1 to 20 specimens of the benthic foraminifera species *Cibicidoides wuellerstorfi* (315-400μm) and is reported with respect to the international standard PDB as $\delta^{18}O$ in parts per thousand (^{o}/oo). The data have been measured partly on a VG 602D micromass spectrometer (most values of core 13519; derived from Sarnthein et al., 1984) and partly on a Finnigan MAT 251 mass spectrometer combined with a new, fully automated carbonate preparation line (^{14}C Laboratory, Kiel University, Dr. H. Erlenkeuser and Finnigan Bremen, Dr. K. Habfast). A small part of the data set of core 12392 has been derived from Shackleton (1977) (for further details see Zahn et al., 1986).

The downcore variations in benthos $\delta^{18}O$ are largely related to the changing oceanic isotope composition as caused by the waxing and waning of the great Pleistocene ice sheets (Shackleton and Opdyke, 1973; Duplessy et al., 1985), fluctuations which serve as a standard of global climate history.

The paleoproductivity record

Equatorial and neritic upwelling areas and subpolar regions account for some 45%, and inshore waters for another 20% of the present-day global primary productivity of organic carbon in the world ocean (Koblentz-Mishke et al., 1970) (Fig. 3a). Therefore, a major portion of the CO_2 transfer from the surface to the deep ocean is confined to small parts of the ocean. This regionally-constrained CO_2 transfer was recently confirmed by the direct measurements of Oudot and Andrie (1986). Platt and Harrison (1985) demonstrated that actually only about one third of the total production is new production and the rest is regenerated production. Because of a lack of pertinent global data for new production, this difference has to be kept in mind when paleoproductivity numbers are calibrated in the following text with modern values of total primary production (Table II).

Müller and Suess (1979) were the first to present principles for tracing a productivity distribution pattern back into the past.

Specifically, they described a means to measure ancient values of primary (new!) productivity, i.e. the "paleoproductivity" in (deep-sea) sediments. Their paleoproductivity estimates were based on two variables: (1) the accumulation rates of organic carbon (C_A) as derived from the bulk sedimentation rate (S_B); the percentage ot organic carbon (C), the porosity (Φ); and the density (ρ); and (2) the bulk sedimentation rates (S_B) which represent the burial rate, a sealing effect for organic carbon.

The argument that the two variables C_A and S_B may mathematically be *a priori* correlated (Müller and Suess, 1979, p. 1354) is incorrect because the sedimentation rates of organic carbon (S_C=S_B %C) contained in C_A account for only a negligible proportion (mostly less than 1-3%) of the bulk-sedimentation rates (S_B). However, strictly speaking, the organic carbon free bulk-sedimentation rates (S_{B-C}) should be applied for defining the sealing effect. Pedersen (1983) confirmed by means of the iodine-bromine ratio in organic matter that the estimates of Müller and Suess (1979) are not simply a product of diagenetic effects acting on the organic carbon content, but indeed depict paleoproductivity.

The organic carbon accumulation rate (C_A) in the sediments used for calculating paleoproductivity is a function of the carbon flux near the sea floor (F_C), which in turn is a result of the primary production (P) and the water depth (z). As shown by Suess (1980), the relationship between primary production, carbon flux and water depth is non-linear because carbon consumption occurs at any given depth in the ocean below the euphotic zone, i.e.,

(1) $F_C = kP^a z^b$

and

$$\log F_C = \log k + a \log P + b \log z$$

or

(2) $a \log P = \log F_C - \log k - b \log z$

where k, a and b are constants. Subjecting the original (corrected and supplemented by recent data; Table I) data of Suess (1980) to multivariate analysis and applying a least squares fit for the carbon flux in equation (2), we obtain the following equation:

$$P = 4.90\ F_C^{0.66} z^{0.32}$$

or

(3) $F_C = P^{1.5}/11.11\ z^{0.5}$ (n=38; r=0.96; z≥50)

This water-depth relationship is fairly similar (for realistic water depths) to an equation published by Betzer et al. (1984) who also used a modified data set of Suess (1980).

(4) $F_C = P^{1.41}/2.44\ z^{0.63}$

Equation (3) was used for deducing an improved paleoproductivity equation. Essentially, we based our new equation on the old data set of Müller and Suess (1979) (Table II). However, we recalculated all sedimentation-rate data because new ^{14}C ages are now available for many cores. In addition, new equations which quantify the effects of vertical mixing by bioturbation were incorporated (Berger and Johnson, 1978; Erlenkeuser, 1980). Moreover, a set of recent core data from the equatorial east Atlantic (Tiedemann, 1985) and from the Pacific (Reimers and Suess, 1983a, b; Emerson, 1985) was added. Altogether, data from a total of 39 sites are compiled in Table II. In this set the eastern Atlantic is somewhat overrepresented because only twenty complete core data sets were available from elsewhere.

In summary, organic carbon accumulation rates C_A are a function of paleoproductivity (P) and water depth (equation 3), and of the ("sealing factor") bulk sedimentation rates (S_{B-C}):

(5) $$C_A = kF_C \, S^a_{B-C}$$

Substituting equation (3) for F_C we arrive at:

$$C_A = k \frac{P^{1.5}}{11.11 \, z^{0.5}} S^a_{B-C}$$

or

(6) $$11.11 \, C_A \, z^{0.5} \, P^{-1.5} = k \, S^a_{B-C}$$

where a and k are constants. Linear regression analysis using the data set of Table II then yields a = 1.08 and k = 0.0168 as displayed in Fig. 4 for

(7) $$11.11 \, C_A \, z^{0.5} \, P^{-1.5} = 0.0168 \, S^{1.08}_{B-C} \quad (n=39;\ r=0.98)$$

Solving equation (7) we obtain the paleo-productivity:

$$P^{1.5} = \frac{11.11 \, CA \, z^{0.5} \, S^{-1.08}_{B-C}}{0.0168} = 660 \, C_A \, z^{0.5} \, S^{-1.08}_{B-C}$$

and

(8) $$P = 72.6 \, C^{0.66}_A \, z^{0.32} \, S^{-0.71}_{B-C}$$

By definition, C_A can also be derived from the sediment properties, C, S_B, Φ and ρ:

(9) $$C_A = C \, S_B \rho \, (1-\Phi) \, 10^{-1}$$

where $\rho \, (1-\Phi)$ is the dry bulk density.

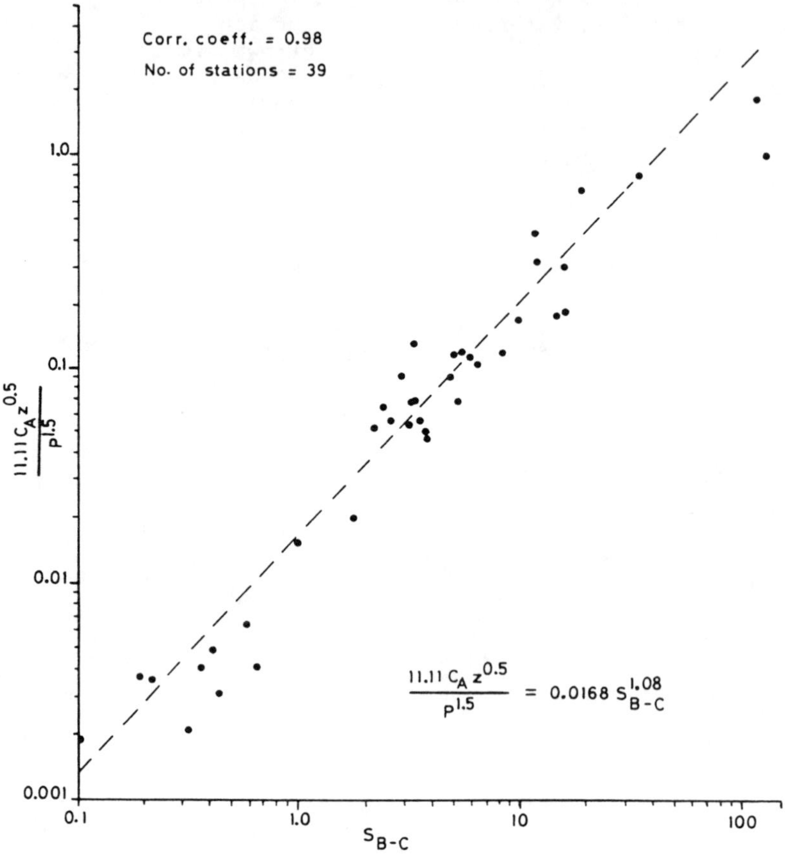

Figure 4. Presentation of the "Sealing Effect": Organic carbon accumulation rates (C_A) and water depths (z) normalized to primary production (P) as a function of the organic carbon free sedimentation rates (S_{B-C}).

Substituting equation (8) for the organic carbon accumulation rate (C_A) in equation (9) we obtain equation (10):

$$(10)\quad P = 15.9\, C^{0.66}\, S_B^{0.66}\, (\rho(1-\Phi))^{0.66}\, S_{B-C}^{-0.71}\, z^{0.32}$$

In this equation, paleoproductivity (P) is measured in $gcm^{-2}y^{-1}$, organic carbon concentration (C) in weight percent, sedimentation rate (S_{B-C}) in cm $1000y^{-1}$, water depth (z) in meters, density of sediment (ρ) in gcm^{-3}, and porosity of sediment (Φ) in percent water/100.

In order to test the validity of this equation, Fig. 5 compares measured primary productivity values (Table II) and modern paleoproductivity values as derived from surface sediment samples using equation (10). The values agree within a factor of 2 with a correlation coefficient of 0.80 (n=39).

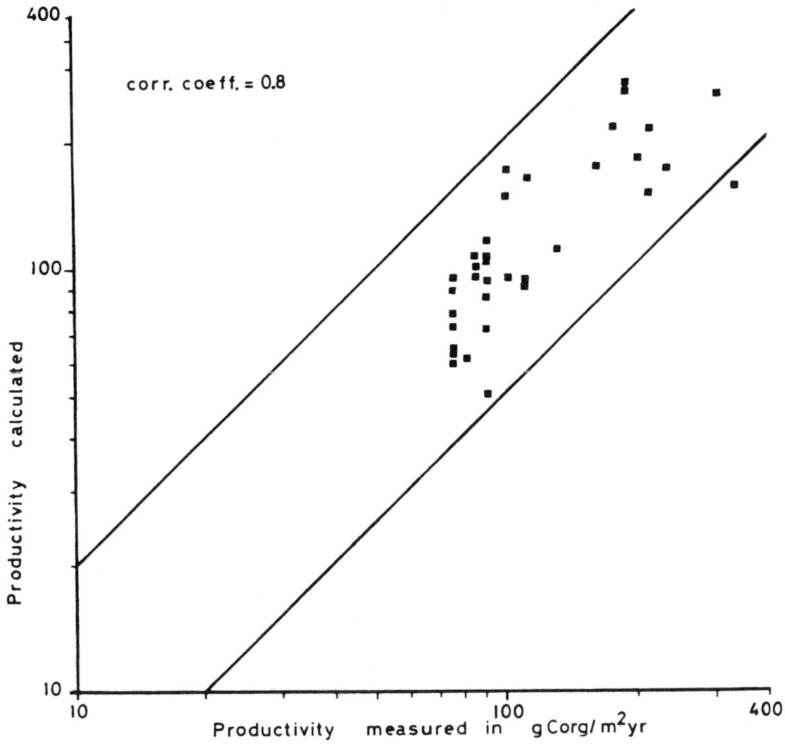

Figure 5. Measured values of surface primary productivity versus estimates calculated from equation (10) based on data in Table II. Most values agree within a factor of 2 or less, as shown by the bounding lines.

For estimating the Cretaceous ocean productivity, Stein (1986) recently used our approach (first presented at the 75th Annual Meeting of the Geologische Vereinigung; Sarnthein et al., 1985) and calculated P from equation (4) of Betzer et al. (1984), the "old" unscreened data set of Müller and Suess (1979) and lumped S_B and S_{B-C}. As a result, he arrived at equation (11), which is remarkably similar to equation (10).

$$(11) \quad P = 5.31\ C^{0.71}\ S_B^{\ 0.07}\ (\rho(1-\Phi))^{0.71}\ z^{0.45}$$

In addition to the variables contained in equation (10), a number of other factors may potentially control the preservation of carbon in the sediment. For example, Emerson (1985) presented a theoretical investigation of the roles of bottom water oxygen concentration, bioturbational mixing, and the carbon degradation rate (i.e., the susceptibility of carbon to degradation). Unfortunately, he did not describe the carbon preservation by means of carbon accumulation rates (C_A) but simply by organic carbon percentages (C). This variable does not compensate for manifold "dilution" effects by the non-carbon sediment supply and thus cannot be used as a measure of carbon flux, primary productivity

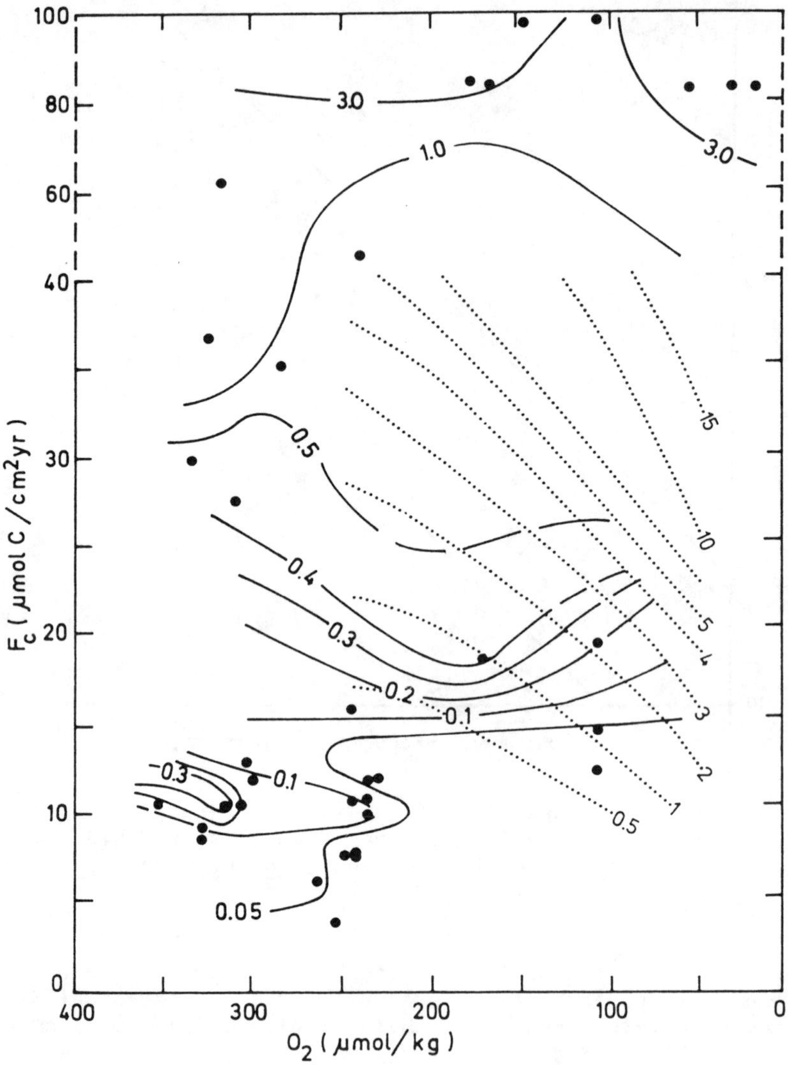

Figure 6. Lines of constant organic carbon accumulation rates as a function of particulate carbon flux C and bottom water oxygen concentrations (O_2). Dotted lines show (expected) content (%) of organic carbon calculated from the model of Emerson (1985).

rates or of carbon preservation (Müller and Suess, 1979). Hence we again tested the role of bottom-water oxygen concentration in the preservation of sedimentary organic carbon by illustrating the interdependence of C_A, F_C, and the bottom-water oxygen concentration (Fig. 6). The results are contrary to those predicted by Emerson (1985). C_A values equating to carbon preservation do not increase with decreasing O_2 concentration (dotted isolines in Fig. 6), because the oxidation of carbon continues as long as any free oxygen is available in bottom water. Only at O_2 concentrations below about 50-100 µmol/kg, accumulation rates of organic carbon tend to increase (Fig. 6, upper right corner).

In a similar manner, the relationship between organic carbon accumulation and bioturbation or the carbon degradation rate do not follow the trends suggested by Emerson (1985), if C_A values are employed instead of %C. The results are mostly contradictory, since they vary by a factor of up to 3 and 4 at a single site (Cochran, 1985), and accordingly, they are discarded.

Based on these findings, we decided to deduce our estimates of ocean paleoproductivity directly from equation (10). The regional applicability of this equation was tested by Tiedemann (1985, modified) who compared productivity data from surface sediments from a large range of different water depths and plankton hauls in the eastern equatorial Atlantic (Figs. 7a and b). The success of this test is largely the result of including the flux-water depth relationship into our paleoproductivity equation.

Finally, the productivity values calculated from equation (10) may be considerably biased because they are based on concentrations of sedimentary total organic carbon (%C), which lump together the proportions of organic matter that are derived from land and from marine plankton. Hydrogen index data from Rock-Eval pyrolysis indicate that terrigenous carbon may contribute as much as 60% of the carbon fraction in deep-sea sediments, especially offshore from river mouths and below dust trajectories (Tissot and Welte, 1984). However, routine quantification of the two different carbon sources is not yet possible by this method because it still contains too many caveats. Another approach to constrain the different sources of carbon is based on the very light $\delta^{13}C$ composition of land derived organic matter (-26°/oo.) as compared to marine organic matter (-18 to -20°/oo.) (Fontugne and Duplessy, 1978). Müller et al. (1983) calculated, as a worst-case estimate, a terrestrial contribution of 40% of the total sedimentary organic matter in modern sediments offshore from the exSpanish western Sahara. This value was considerably decreased during previous phases of enhanced marine plankton productivity, a variation which indicates a general, fairly constant background of dustborne carbon being overlain by a strongly fluctuating input of carbon from marine plankton. Because of a lack of pertinent data, unfortunately, we cannot distinguish between the different sources of carbon supply in this paper.

$\delta^{13}C$ Record of Benthic Foraminifera Tests

Zahn et al. (1986) demonstrated that the $\delta^{13}C$ data from different benthic foraminifera species carry different environmental signals as

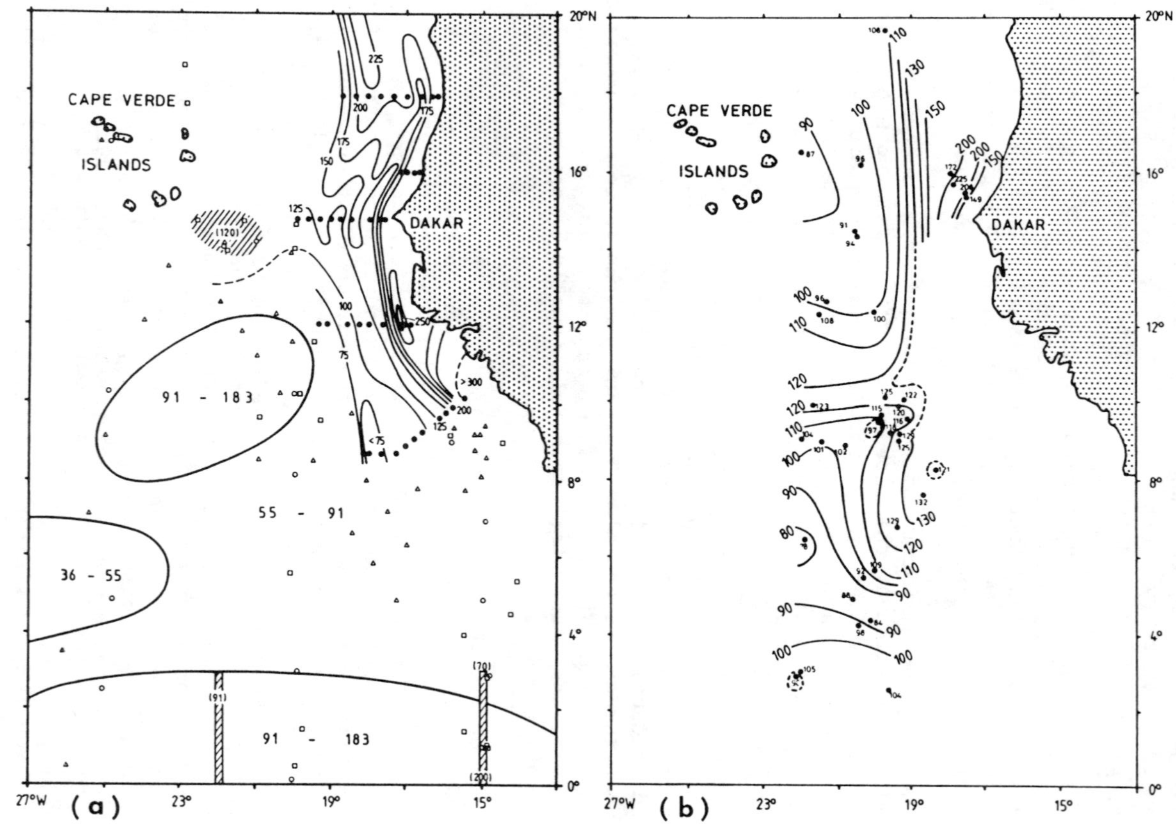

Figure 7. Regional distribution pattern of perennial surface primary productivity measured in the Equatorial East Atlantic (a) the pattern of estimates calculated from surface sediments by means of equation (10). (b) modified from Tiedemann, (1985).

the result of different foraminiferal microhabitats. On the one hand, the $\delta^{13}C$ record of the infaunal *Uvigerina peregrina/U. hollicki* group is strongly, although not linearly influenced by varying local accumulation rates of (isotopically light) carbon and thus, mainly by the local marine plankton production. On the other hand, the $\delta^{13}C$ values of the epifaunal species *C. wuellerstorfi* to a great extent reflect the large scale variations of total dissolved CO_2 in the ambient (regional) bottom water mass such as already postulated by Woodruff et al. (1980), Duplessy (1982), Duplessy et al. (1984) and Duplessy and Shackleton (1985).

Duplessy (op. cit.) mainly related such changes in dissolved CO_2 to fluctuating residence times of the deep water masses, but they can equally reflect fluctuations of ocean productivity (sensu P. Weyl in Broecker, 1982).

In this paper, we compare our paleoproductivity data with $\delta^{13}C$ curves from *C. wuellerstorfi*, which were measured at sediment cores from about 3000m water depth by Sarnthein et al. (1984) and Zahn et al. (1986). Based on the above reasoning, we regard these $\delta^{13}C$ curves as a bulk measure, a "memory" of the total CO_2 gradually accumulated from various high productivity cells in the ambient North Atlantic Deep Water (NADW) while it flows from its formation area in the North Atlantic to the south. Accordingly, we expect that an increase in surface ocean productivity (in addition to the effects of increasing NADW residence time) will result in overall reduced $\delta^{13}C$ values, and vice versa. Our $\delta^{13}C$ data were measured parallel with the $\delta^{18}O$ data.

Paleo-wind Strength

Trade winds are the main driving mechanism of low latitudinal oceanic upwelling. However, they control the plankton productivity and CO_2 transfer within upwelling regions only to a limited degree in a non-linear fashion (Codispoti, 1983). Other major factors are the original nutrient concentration of the upwelled source water and the perturbation of plankton blooms by excessive storminess, which may decrease the degree of carbon utilization.

Despite these complicating factors, it is tempting to evaluate paleo-wind speeds, because the relation between surface wind strength and (merely physical) coastal upwelling intensity is probably linear. This is suggested by the findings of Speth et al. (1978), who observed a linear connection between upwelling controlled zonal sea surface temperature (SST) anomalies and large-scale pressure gradients in the surface atmosphere off northwest Africa.

Paleo-wind speeds can be deduced from the grain-size distribution of aeolo-marine dust sediments based on a formula of Sarnthein et al. (1981). This calculation requires that both the thickness of the turbid atmospheric dust layer and the length of the dust trajectory from the shoreline to the sample location are known. However, the latter variable may be an indeterminant if the (trade) winds blow largely longshore such as offshore from northwest Africa. Here slight variations in wind direction and of sea level may result in large changes of the length of the trajectory, a problem which can be only overcome by careful selection of the location of the sediment core.

RESULTS

Paleoproductivity Variations During the Last 140,000 Years

Fig. 8c presents two records of paleoproductivity change as calculated by equation (10) during the late Quaternary. The record of 'Meteor' core 12392 is derived from the outer influence range of an oceanic upwelling cell offshore from northwest Africa. This high-productivity region is one of the four major trade wind driven upwelling zones of the world ocean attached to an eastern (passive) continental margin (Fig. 3). The record of 'Meteor' core 13519 is obtained from the non-upwelling region of the Sierra Leone Rise, i.e. from an oceanic "desert" (Fig. 3).

In this "desert", the paleoproductivity only varied by 50 to 80% during the last 140,000 years, as compared to more than 300% in the vicinity of nearshore coastal upwelling. In both cores, low and approximately equal productivity levels of about 60-90 gC $m^{-2}y^{-1}$ generally occurred during interglacial stages, whereas high values are characteristic of glacial stages as depicted by the (benthos) $\delta^{18}O$ curve (Fig. 8d). High productivity also occurred in oxygen isotope stage 3, an intermediate state of climate which is poorly understood. In principle, these productivity variations were already recognized by Müller et al. (1983). Their data showed a clear signal of a 100,000 to 110,000 year productivity cycle, which can be traced as far back as approximately 500,000 years, and less distinctly, back to 750,000 years B.P. (Fig. 9).

Unlike the paleoproductivity curves, the $\delta^{13}C$ values of *C. wuellerstorfi* in core 13519 vary almost identically with those in core 12392 through time (Fig. 8a and b). In general, the two $\delta^{13}C$ curves correlate inversely with the strong fluctuations of paleoproductivity found in core 12392, but parallel those of core 13519 only to a limited extent. Likewise, the long-term 100,000 years productivity cycle of core 13519 can hardly be recognized on the $\delta^{13}C$ curve on Fig. 9.

Lowered $\delta^{13}C$ values of *C. wuellerstorfi* generally record a CO_2 enrichment in (here: North Atlantic) deep water (Duplessy, 1982). This enrichment may now be explained by a strongly increased flux of organic carbon from high productivity zones of coastal upwelling, as expected by our model. According to Duplessy (1982), an interglacial to glacial $\delta^{13}C$ decrease of about 0.8% (probably nearer 0.6 $^{o}/oo$) in the NADW should correspond to an additional 1150 GT (or 860 GT) of carbon stored away into the deep ocean (equal to about 150 years of the total new production; combined evidence of Koblentz-Mishke et al., 1970, and Platt and Harrison, 1985). Unfortunately, we cannot verify this number in terms of a paleoproductivity balance for the total north Atlantic or any other ocean, because these data are not yet available through last glacial and interglacial times. Nevertheless, the data obtained from core 12392 and other data sets from neighbouring cores along the east Atlantic continental margin (e.g., Müller et al., 1983, and other

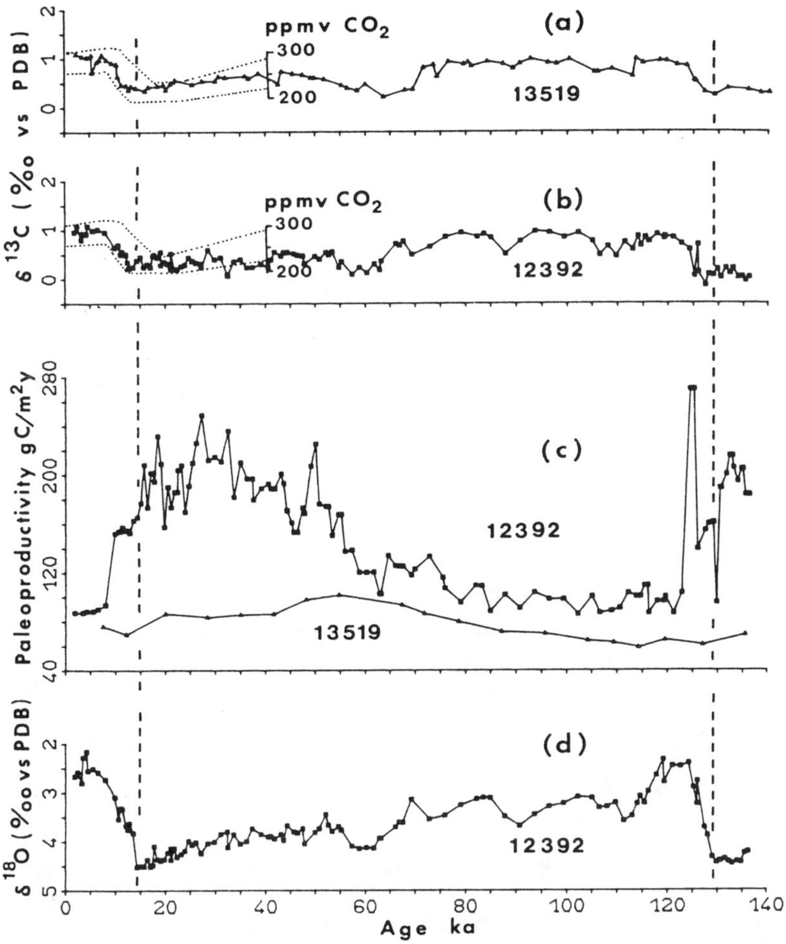

Figure 8. Late Quaternary changes in productivity during glacial and interglacial stages. The $\delta^{18}O$ curve of *Cibicidoides wuellerstorfi* in core 12392 largely reflects changes of global ice volume. The $\delta^{13}C$ curve measured from the same foraminifera specimens documents the total dissolved CO_2 content in ice cores (Neftel et al., 1982). Chronology of core 12392 after Thiede et al. (1982). Time scale of core 13519 after Herterich and Sarnthein (1984; modified).

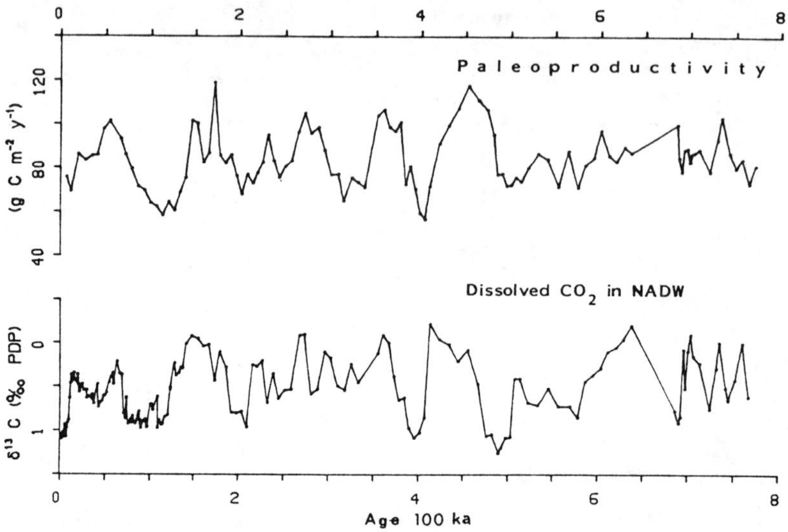

Figure 9. Cyclic paleoproductivity changes based on equation (10) and data of Müller et al. (1983) and $\delta^{13}C$ record (Sarnthein et al., 1984) of dissolved bottom water CO_2 during the last 750,000 years as recorded in core 13519.

unpublished data) make such a linkage highly likely, because coastal upwelling regions may be crucial for inducing major changes of the carbon transfer in the ocean (Flohn, 1983).

In order to test such a scenario, and using the method as Shackleton et al. (1983), the bottom water ^{13}C record of Fig. 8 is juxtaposed with the range of past CO_2 concentrations in the atmosphere as estimated by Neftel et al. (1982) from ice core data (Fig. 1). The scales of atmospheric CO_2 and $\delta^{13}C$ are positioned on the assumption that the ^{13}C difference observed at the top of the core reflects a pre-anthropogenic atmospheric CO_2 level of about 280 ppmv. The scales are adjusted using Broecker's (1982) model to scale the isotopic range for an implied record of surface-water (!) CO_2 content. In Fig. 8, the range of atmospheric atmospheric CO_2 variations indeed appears entirely consistent with the ^{13}C record of bottom water CO_2. The last glacial to interglacial increase of atmospheric CO_2 by about 50% (100 ppmv, equal to 230 GT carbon) parallels the outlined (almost four fold) 860 GT decrease of carbon in the NADW (Duplessy, 1982, modified) and perfectly matches the abrupt reversal of productivity in the coastal upwelling zone. Three quarters of the carbon lost by the deep water were probably added to the concentration in the (1000m thick) ocean surface water, and 1/4 to the atmosphere.

Flohn (1986) estimated a similar order of magnitude for the CO_2 exchange as controlled by oceanic productivity in upwelling regions during modern interannual variations of atmospheric pCO_2.

PHASE RELATIONSHIPS DURING GLACIAL TERMINATIONS I AND II

Temporal leads and lags across climatic reversals provide an important tool for identifying the cause-effect relationships and the external forcing mechanisms controlling (abrupt?) climatic change. Based on its high sedimentation rates (7-10 cm/1000y) and the large number of climatic proxy data already determined, core 12392 gives an excellent opportunity to test, in detail, the phase relationships between changes of paleoproductivity, bottom water and atmospheric pCO_2, and global deglaciation during the last two Terminations.

The onset of Glacial Terminations is defined by a sudden and long term turn in the (benthos) oxygen isotope curve towards lighter values indicative of large-scale glacial melting and global sea level rise (Fig. 8) (Broecker and van Donk, 1970; Ruddiman and McIntyre, 1976). To facilitate discussion, the onset ages of the various steps of Terminations I and II in core 12392 are summarized in Table I and compared with the age of turning points (also stepwise) in the history of productivity and bottom water CO_2 during this time.

Broecker (1982) related the general decrease of ocean paleoproductivity during deglaciation time to limited phosphate and nitrate reservoirs and ultimately to the sea level rise and a flooding of the shelf. However, it is clear from Fig. 8 and Table I that the onset of the massive decrease was either synchronous with the start of both Terminations I and II or occurred slightly earlier. Hence, the decrease in paleoproductivity cannot be the result of glacial melt and sea-level rise. However, the apparent 1500-year lead of the productivity decrease over glacial melting, as suggested at Terminations IA and IIA, may also be incorrect. The deep (early) position of the turning point in paleoproductivity may be caused by post-depositional processes. Possibly, bioturbation lowered the carbon concentration to sediment depths below the actual reversal by the consumption of "fossil" carbon as soon as the contemporaneous flux of carbon to the sea floor was reduced. The 1,500 years lead under discussion (Table III) corresponds to about 15cm sediment thickness, which perhaps would still be accessible to consumption by bioturbational mixing (Rowe, 1985).

Of course, it is legitimate to question the extent to which phase relationships observed in a single core (although obtained from a key location) have general implications, at least for one major upwelling region. For example, the turning point of the paleoproductivity curve at 18,200 y B.P., which is problematic in core 12392, forms the only dominant event in the paleoproductivity record of core 16415 near the southern margin of the same east Atlantic coastal upwelling region, at about 10°N (Fig. 10) and also in core 13519 (Fig. 8). Thus, only a future regional integration of different records will provide a final answer to the exact timing of the paleoproductivity event. For example, Legeckis (1986) depicts contrasting occurrences of El Niño situations in the equatorial Atlantic and the eastern Pacific. However, its full synchroneity or even its lead to global ice melt can already be obtained from the present data.

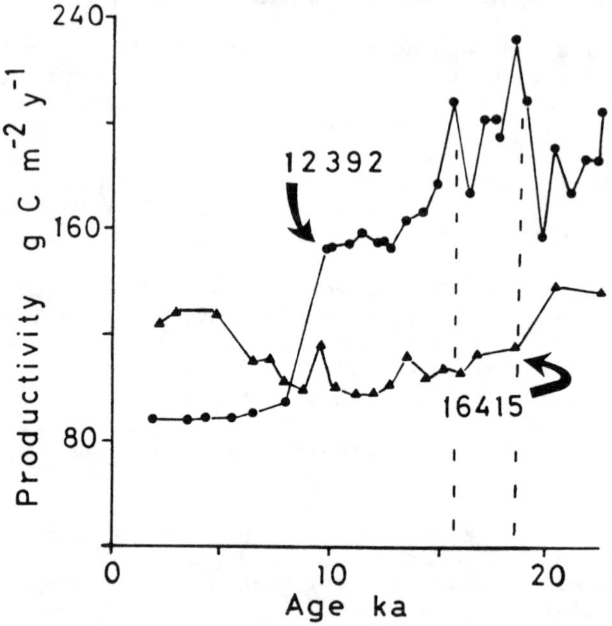

Figure 10. Paleoproductivity record of core 16415. Estimates calculated from equation (10) based on data and stratigraphy of Vogelsang (1985).

During Terminations I and II, the respective reversals in the $\delta^{13}C$ record began (in both cores!) markedly later than those of oxygen isotopes and of paleoproductivity, with a time lag of 2,000 to 4,500 years (Fig. 8; Table I). This result also may strongly affect existing models. Broecker et al. (1985) and Boyle and Keigwin (1985) expected that primarily, an increased formation of NADW gave rise to the large emission of CO_2 from the deep ocean to the atmosphere during deglaciation phases and thereby possibly enhanced the "abrupt" climatic change.

The phase relationships depicted in Fig. 8 however suggest that the CO_2 emission was probably the result of several thousand years of reduced upwelling productivity in low latitudes and diminished CO_2 extraction from the atmosphere. Such a mechanism is in harmony with the ice core record of past atmospheric CO_2 concentrations (Fig. 8). These concentrations had already increased after 16,500 y B.P., which was several thousand years prior to the increase of benthos $\delta^{13}C$, but about synchronous with the reduction of upwelling productivity. Unlike the atmospheric CO_2 reservoir, that of the deep ocean just emerges as a slow system with a delayed response to a quickly changing forcing by ocean productivity at the sea surface.

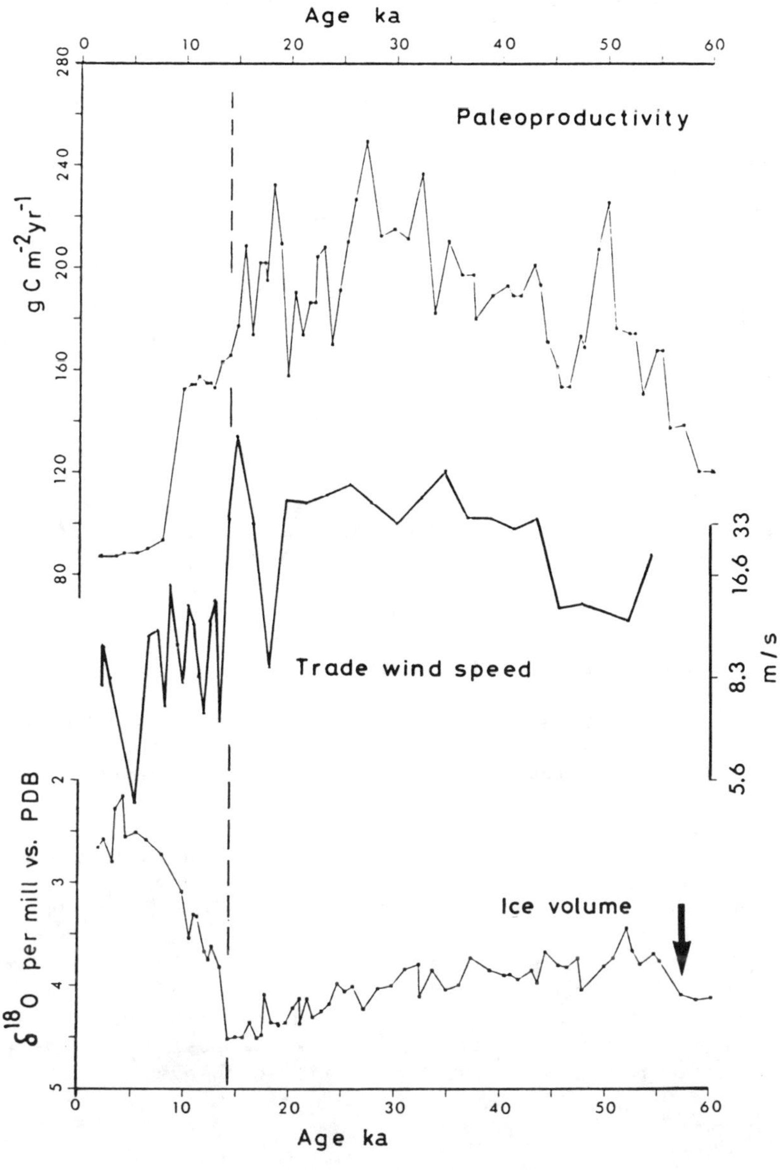

Figure 11. Paleoproductivity and $\delta^{18}O$ record of global ice volume measured in core 12392 verses the history of average trade wind speeds (u). Approximative dust storm wind speeds were calculated from an equation of Sarnthein et al. (1981) and based on grain size data from aeolo-marine dust deposits in core 13289 (Koopmann, 1981) and on the assumption of an almost constant thickness of the turbid atmospheric dust layer (1.2km) and a constant length of dust transport from the shoreline (about 200km).

POSSIBLE ORIGIN OF UPWELLING PRODUCTIVITY CHANGES

Changes of coastal ocean productivity (and the "Rain Ratio Model" of Berger and Keir, 1984) appear established as a prime factor for the massive increase of atmospheric CO_2 synchronous or preceding the global ice melt. Consequently, the focus of our discussion must shift to the question of which factor induced the abrupt paleoproductivity changes.

Labracherie (1980) demonstrated that the upwelling productivity offshore from northwest Africa was increased by an unknown order of magnitude during the last glaciation, because nutrient-rich South Atlantic Central Water (SACW) penetrated farther north as suggested by the frequency of certain radiolarian tracer species. Furthermore, the intensity of the upwelling process itself increased by about 50% during glacial times as deduced from zonal sea surface temperature anomalies (Sarnthein, 1982). This enhanced upwelling appears clearly coupled with enhanced synglacial trade wind speeds (Sarnthein et al., 1982). A continuous speed record across Termination I was tentatively calculated from the well dated core 13289. Its dust deposits are little affected by fluctuations of the length of dust trajectory (rather by a mixture of dust from different sources) (Fig. 11) and reveal a marked, sudden decrease of wind speeds starting near 14,900 y B.P., although one cannot take literally the absolute range of 35-5 m/s because of the many possible pitfalls in the calculation as mentioned above. This event is remarkably synchronous with the onset of global deglaciation and the decrease of productivity found in core 12392.

Possibly, the sensitive response of meridional wind strength to deglaciation was induced by a large-scale break-up of North Atlantic sea ice during the very first phase of deglaciation, a break-up which in turn may have been the most sensitive response to an increase of the insolation budget during this time (Ruddiman and McIntyre, 1981). At least seasonal open-sea conditions in middle high latitudes will effectively reduce the oceanic temperature gradient from high to low latitudes and thus, quickly lead to decreasing meridional winds, which compensate for this gradient. By this mode of interaction, the synchronous change (in geological terms) of upwelling productivity and atmospheric CO_2 concentration with the onset of deglaciation can be easily explained as joint immediate response to changes in the solar insolation regime and may provide an important, highly flexible positive (non-linear) feedback mechanism for accelerating climatic change.

ACKNOWLEDGEMENTS

The authors greatly appreciate the many helpful and stimulating discussions with Erwin Suess, Peter Müller, Wolfgang Berger, Wallace S. Broecker, and Helmut Erlenkeuser and the careful review of this paper by W.H. Berger, H. Flohn, and L. Mayer. This work was supported by the National Program of Climate Research of the Federal Republic of Germany (Grant No. 5138391-KF20041 of the Ministry of Research and Technology)

REFERENCES

Arrhenius, S., 1896. On the influence of carbonic acid in the air upon the temperature of the ground, *Phil. Mag.*, 41: 437.

Berger, W.H. and Johnson, R.F., 1978. On the thickness and the ^{14}C age of the mixed layer in deep-sea carbonates, *Earth Planet. Sci. Lett.*, 41: 223-227.

Berger, W.H. and Keir, R.S., 1984. Glacial-Holocene changes in atmosphere CO_2 and the deep-sea record. In: *Climate Processes and Climate Sensitivity*, edited by J.E. Hansen and T. Takahashi, *Geophys. Monogr.*, 29 (AGU, Washington D.C.), 337-351a.

Betzer, R.R., Showers, W.J., Laws, E.A., Winn, C.D., DiTullio, G.R. and Kroopnick, P.M., 1984. Primary productivity and particle fluxes on a transect of the equator at 153°W in the Pacific Ocean, *Deep-Sea Res.*, 31 (1): 1-11.

Boyle, E.A., 1986. Paired cadmium and carbon isotope data in benthic foraminifera: Implications for changes in oceanic phosphorus, oceanic circulation, and atmospheric carbon dioxide, *Geochim. Cosmochim. Acta*, 50: 265-276.

Boyle, E.A. and Keigwin, L.D., 1985. Comparison of Atlantic and Pacific paleo-chemical records for the last 215,000 years: Changes in deep ocean circulation and chemical inventories, *Earth Planet. Sci. Lett.*, 76 (1/2): 135-150.

Broecker, W.S., 1982. Glacial to interglacial changes in ocean chemistry. In: *Climatic variability in the oceans*, edited by E.B.Kraus and H.P Hansen, *Prog. Oceanogr.*, 11: 151-197.

Broecker, W.S. and van Donk, J., 1970. Insolation changes, ice volumes and the ^{18}O record in deep-sea sediments, *Rev. Geophys. Space Phys.*, 8: 169-198.

Broecker, W.S., 1984. Carbon cycle: Carbon dioxide circulation through ocean and atmosphere, *Nature*, 308: 602.

Broecker, W.S. and Peng, T.-H., 1986. Carbon Cycle: 1985, Glacial to interglacial changes in the operation of the global carbon cycle, *Radiocarbon*, 28 (2A): 309-327.

Broecker, W.S., Peteet, D.M. and Rind, D., 1985. Does the ocean-atmosphere system have more than one stable mode of operation?, *Nature*, 315: 21-26.

Codispoti, L.A., 1983. On nutrient variability and sediments in upwelling regions. In: *Coastal Upwelling. Its Sediment Record*, A, edited by E. Suess and J. Thiede (Plenum Press, New York), 125-146.

Cochran, J.K., 1985. Particle mixing rates in sediments of the eastern equatorial Pacific: Evidence from ^{210}Pb, ^{239}, ^{240}Pu and ^{137}Cs distributions at MANOP sites, *Geochim. Cosmochim. Acta*, 49: 1195-1210.

Delmas, R.J., Ascensio, J.-M. and Legrand, M., 1980. Polar ice evidence that atmospheric CO_2 20,000 yr BP was 50% of present, *Nature*, 284: 155-157.

Duplessy, J.-C., 1982. North Atlantic Deep Water circulation during the last-climatic cycle, *Bull. Inst. Geol. Bassin d'Aquitaine*, Bordeaux, 31: 379-391.

Duplessy, J.C., Shackleton, N.J., Matthews, R.K., Prell, W., Ruddimann, W.F., Caralp, M. and Hendy, C.H., 1984. ^{13}C record of benthic foraminifera in the last Interglacial ocean: Implications for the

carbon cycle and the global deep water circulation, *Quat. Res.*, 21 (2): 244-263.
Duplessy, J.C., Labeyrie, L.L. and Shackleton, N.J., 1985. The oxygen isotope record of benthic foraminifera: Effect of deep water temperature and ice volume changes, *EOS*, 66 (18).
Duplessy, J.-C. and Shackleton, N.J., 1985. Response of global deep-water-circulation to Earth's climatic change 135,000-107,000 years ago, *Nature*, 316: 500-507.
Emerson, S., 1985. Organic carbon preservation in marine sediments. In: *The carbon cycle and atmospheric CO_2: Natural variations Archean to Present*, edited by E.T. Sundquist and W.S. Broecker, *Geophys. Monogr.*, 32 (AGU Washington D.C.), 78-87.
Erlenkeuser, H., 1980. ^{14}C age and vertical mixing of deep-sea sediments. A comment on a paper of Berger and Johnson (1978), *Earth Planet. Sci. Lett.*, 47: 319-126.
Flohn, H., 1983. A climate feedback mechanism involving oceanic upwelling, atmospheric CO_2 and water vapour. In: *Variations in the Global Water Budget*, edited by Street-Perrot et al. (D. Reidel Publ. Co., Dordrecht, Holland), 403-418.
Flohn, H., 1986. Singular events and catastrophes now and in climatic history, *Naturwissenschaften* 73: 136-149.
Fontugne, M. and Duplessy, J.C., 1978. Carbon isotope ratio of marine plankton related to surface water masses, *Earth Planet. Sci. Lett.*, 41 (3): 365-371.
Grassl, H., Maier-Reimer, E., Degens, E.T., Kempe, S. and Spitzy, A., 1984. CO_2, Kohlenstoff-Kreislauf und Klima, I. Globale Kohlenstoffbilanz, *Naturwissenschaften*, 71: 129-136.
Hasselmann, K., 1976. Stochastic climate models, I, *Tellus*, 28: 473.
Herterich, K. and Sarnthein, M., 1984. Brunhes time scale: Tuning by rates of calcium-carbonate dissolution and cross spectral analyses with solar insolation. In: *Milankovitch and Climate*, I, edited by A. Berger et al. (D. Reidel Publ. Co., Dordrecht, Holland), 447-466.
Imbrie, J., Hays, J.D., Martinson, D.G., McIntyre, A., Mix, A.C., Morley, J.J., Pisias, N.G., Prell, W.L. and Shackleton, N.J., 1984. The orbital theory of Pleistocene climate: Support from a revised chronology of the marine $\delta^{18}O$ record. In: *Milankovitch and Climate* (D. Reidel Publ. Co., Dordrecht, Holland), 269-305.
Koblentz-Mishke, O.J., Volkovinsky, V.V. and Kabanova, J.G., 1970. Plankton primary production of the World Ocean. In: *Scientific Exploration of the South Pacific*, edited by W.S. Wooster, 183-193.
Koopmann, B., 1981. Sedimentation von Saharastaub im subtropischen Atlantik während der letzten 25,000 Jahre, *'Meteor' Forsch. Ergeb.* C 35: 23-59.
Labracherie, M., 1980. Les Radiolaires temoins de l'evolution hydrologique depuis le dernier maximum glaciaire au large du Cap Blanc (Afrique du Nord-Ouest), *Palaeogeogr. Palaeoclimatol. Palaeoecol.*, 32 (1/2): 163-184.
Legeckis, R., 1986. Tropical changes at the sea surface, *Nature*, 322: 15.
Müller, P.J., Erlenkeuser, H. and von Grafenstein, R., 1983. Glacial-Interglacial cycles in oceanic productivity inferred from organic carbon contents in eastern North Atlantic sediment cores. In:

Coastal Upwelling: Its Sediment Record, B. edited by J. Thiede and E. Suess (Plenum Press, New York), 365-398.

Müller, P.J. and Suess, E., 1979. Productivity, sedimentation rate and sedimentary organic matter in the oceans - I Organic carbon preservation, *Deep-Sea Res.*, 26: 1347-1362.

Neftel, A., Oeschger, H., Schwander, J., Stauffer B. and Zumbrunn, R., 1982. Ice core sample measurements give atmospheric CO_2 content during the past 40,000 years, *Nature*, 295: 220-223.

Oudot, C. and Andrie, C., 1986. Pressions partielles de CO_2 dans l'Atlantique tropical: variabilité dans les eaux de surface et dans l'atmosphère en janvier et juillet 1983, *Oceano. Acta*, 9 (2), in press.

Pedersen, T.F., 1983. Increased productivity in the eastern equatorial Pacific during the last glacial maximum (19,000 to 14,000 yr B.P.), *Geol.*, 11: 16-19.

Peltier, W.R. and Hyde, W., 1984. A model of the ice age cycle. In: *Milankovitch and Climate*, edited by A. Berger et al. (D. Reidel Publ. Co., Dordrecht, Holland), 565-580.

Platt, T. and Harrison, W.G., 1985. Biogenic fluxes of carbon and oxygen in the ocean, *Nature*, 318: 55-58.

Reimers, C.E. and Suess, E., 1983a. Late Quaternary fluctuations in the cycling of organic matter off central Peru: A proto-kerogen record. In: *Coastal Upwelling. Its Sediment Record. A*, edited by E. Suess and J. Thiede (Plenum Press, New York), 497-526.

Reimers, C.E. and Suess, E., 1983b. Spatial and temporal patterns of organic matter accumulation on the Peru continental margin. In: *Coastal Upwelling Record, B*, edited by E. Suess and J. Thiede (Plenum Press, New York), 311-346.

Rowe, G.T., 1985. The biological fate of organic matter on the NE US continental margin - SEEP Area, 1983/84, *EOS*, 66 (18), 283.

Ruddiman, W.F. and McIntyre, A., 1976, Northeast Atlantic paleoclimatic changes over the past 600,000 years, *Geol. Soc. Am. Mem.*, 145: 11-146.

Ruddiman, W.F. and McIntyre, A., 1981. The North Atlantic Ocean during the last deglaciation, *Palaeogeogr. Palaeoclimatol. Palaeoecol.*, 35: 145-214.

Sarnthein, M., 1982. Zur Fluktuation der subtropischen Wüstengürtel seit dem letzten Hochglazial vor 18,000 Jahren: Klimahinweise und -modelle aus Tiefseesedimenten, *Geomethodica*, 7: 125-161.

Sarnthein, M., Erlenkeuser, H., von Grafenstein, R. and Schröder, C., 1984. Stable isotope stratigraphy for the last 750,000 years; 'Meteor' core 13519 from the eastern equatorial Atlantic', *'Meteor' Forsch. Ergeb. C* 38: 9-24.

Sarnthein, M., Erlenkeuser, H. and Zahn, R., 1982. Termination I: The response of continental climate in the subtropics as recorded in deep-sea sediments, *Bull. Inst. Geol. Bassin d'Aquitaine*, 31: 393-407.

Sarnthein, M., Tetzlaff, G., Koopmann, B., Wolter, K. and Pflaumann, U., 1981. Glacial and interglacial wind regimes over the eastern subtropical Atlantic and NW Africa, *Nature*, 293: 193-196.

Sarnthein, M., Winn, K. and Zahn, R., 1985. Paleoproductivity and the effect of oceanic upwelling on atmospheric CO_2 during Glacial and Interglacial times, *Terra Cognita*, 5 (1): 82.

Shackleton, N.J., 1977. The oxygen isotope stratigraphic record of the Late Pleistocene, *Phil. Trans. Roy. Soc. Lond.*, B 280: 169-182.
Shackleton, N.J., Hall, M.A., Line, J. and Shuxi, C., 1983. Carbon isotope data in core V 19-30 confirm reduced carbon dioxide concentration in the ice age atmosphere, *Nature*, 306: 319-322.
Shackleton, N.J. and Opdyke, N.D., 1973. Oxygen isotope and paleomagnetic stratigraphy of equatorial Pacific core V 28-238: Oxygen isotope temperatures and ice volumes on a 10^5 year scale, *Quat. Res.*, 3: 39-55.
Siegenthaler U. and Wenk, Th., 1984. Rapid atmospheric CO_2 variations and ocean circulation, *Nature*, 308: 624-625.
Speth, P., Detlefsen, H. and Sierts, H.-W., 1978. Meteorological influence on upwelling off northwest Africa, *Dt. hydrogr. Zt.*, 31 (3): 95-104.
Stein, R., 1986. Surface-water productivity as inferred from sediments deposited in oxic and anoxic deep-water environments of the Mesozoic Atlantic. In: *Proc. of Blackshale Conference*, Hamburg, 1985, edited by E. Degens, 20 ms-p, in press.
Suess, E., 1980. Particulate organic carbon flux in the oceans - Surface productivity and oxygen utilization, *Nature*, 288: 260-263.
Thiede, J., Suess, E. and Müller, P.J., 1982. Late Quaternary fluxes of major sediment components to the sea floor at the northwest African continental slope. In: *Geology of the northwest African continental margin*, edited by U. v. Rad et al. (Springer, Heidelberg), 605-631.
Tiedemann, R., 1985. Verteilung von organischem Kohlenstoff in Oberflächen-sedimenten und die örtliche Primärproduktion im äquatorialen Ostatlantik, 0-20°N, 15-25°W, *Unpubl. M.Sc. Thesis*, *Kiel Univ.*, 74 pp.
Tissot, B.P. and Welte, D.H., 1984. *Petroleum Formation and Occurrence*, Springer, Berlin, 2nd edition.
Vogelsang, E., 1985. Hochauflösende Zeitreihen der Termination I aus Sedimenten des äquatorialen Ostatlantiks, *Unpubl. M.Sc. Thesis, Kiel Univ.*, 85 pp.
Woodruff, F., Savin, S.M. and Douglas, R.G., 1980. Biological fractionation of oxygen and carbon isotopes by recent benthic foraminifera, *Mar. Micropaleont.*, 5: 3-11.
Zahn, R., Winn, K. and Sarnthein, M., 1986. Benthic foraminiferal $\delta^{13}C$ and accumulation rates of organic carbon: *Uvigerina peregrina* group and *Cibicidoides wuellerstorfi*, *Paleoceanogr*, 1 (1): 27-42.

TABLE I

Organic carbon flux rates in the water column and mean "surface" primary production rates compiled from data sets of Suess (1980), Smith and Hinga (1983) and Emerson (1985).

Latitude (°)	Longitude (°)	Water Depth (m)	Carbon flux (gm^{-2} yr^{-1})	Primary production (gm^{-2} yr^{-1})
31.5 N	55.9 W	976	0.89	40
31.5 N	55.9 W	3694	0.32	40
31.5 N	55.0 W	5367	0.45	45
13.5 N	54.0 W	389	2.46	50
13.5 N	54.0 W	988	1.44	50
13.5 N	54.0 W	3755	0.63	50
13.5 N	54.0 W	5068	0.62	50
15.3 N	151.5 W	378	1.30	40
15.3 N	151.5 W	2778	0.40	40
15.3 N	151.5 W	4280	0.32	40
15.3 N	151.5 W	5582	0.24	40
32 nm SE	Bermuda	3200	0.77	40
0.6 N	86.1 W	2650	2.28	100
27.7 N	78.9 W	675	2.60	72
33.5 N	76.2 W	1345	5.40	85
38.3 N	69.6 W	3650	4.20	160
24.9 N	77.7 W	2000	2.10	72
36.7 N	122.2 W	50	158.00	500
36.7 N	122.2 W	250	92.00	500
36.7 N	122.2 W	700	42.00	500
36.7 N	122.2 W	50	33.00	150
36.7 N	122.2 W	250	19.00	150
36.7 N	122.2 W	700	17.00	150
32.8 N	144.4 W	75	25.00	100
32.8 N	144.4 W	575	5.30	100
32.8 N	144.4 W	1050	4.40	100
38.8 N	72.5 W	2200	6.30	100
38.5 N	72.0 W	2750	2.30	160
55.6 N	15.4 E	55	65.00	305
15.1 S	75.4 W	50	240.00	1200
15.1 S	75.5 W	70	130.00	600
15.1 S	75.5 W	100	110.00	600
6.6 N	92.8 W	3500	1.44	91
8.8 N	104.0 W	3100	1.68	108
34.0 N	65.8 W	5140	0.72	72
32.5 N	117.8 W	1193	9.80	273
32.3 N	117.5 W	1230	9.80	273
32.5 N	120.6 W	3815	9.80	307

TABLE II

Core data used in the derivation of the productivity formula

Core No./ Station	Water depth	C	ρ (1-Φ)	S_B	S_{B-C}	C_A	Productivity measured	Productivity estimated
	(m)	%	g/cm³	cm/ka	cm/ka	g/m²yr	g/m²yr	g/m²yr
12392-1	2575	0.35	0.9780	5.05	5.02	0.17286	75	87
12310-3	3076	0.34	0.6504	3.80	3.78	0.08403	90	70
12327-4	2037	1.34	0.4065	8.33	8.14	0.45374	130	108
12328-4	2798	1.66	0.5691	15.08	14.65	1.42462	230	167
12329-4	3315	0.61	0.6233	2.56	2.54	0.09733	90	105
12336-1	3645	0.44	0.6775	2.20	2.18	0.06558	75	93
12337-4	3085	1.46	0.5691	4.90	4.78	0.40713	160	168
12344-3	711	2.59	0.5691	16.45	15.73	2.42468	210	146
12345-4	966	3.67	0.5691	16.45	15.43	3.43574	210	206
12347-1	2710	2.34	0.5691	12.48	11.98	1.66195	175	212
13209-2	4713	0.51	0.5691	3.20	3.17	0.09288	85	98
13521-1	4504	0.37	0.6450	3.50	3.48	0.08353	90	84
16017-2	812	1.81	0.9879	6.80	6.68	1.21591	200	179
16402-1	4203	0.46	0.6037	5.24	5.20	0.14552	110	89
16403-2	4234	0.45	0.6310	3.66	3.63	0.10393	110	92
16404-1	4787	0.42	0.6460	3.17	3.15	0.08601	95	94
16405-1	4870	0.43	0.7501	3.40	3.37	0.10966	95	105
16415-1	3841	0.61	0.6500	2.70	2.67	0.10705	90	114
16432-2	4515	0.52	0.6447	5.40	5.35	0.18103	90	103
V15-141	5934	0.72	0.6500	2.90	2.86	0.13572	100	145
V15-142	5885	0.92	0.6500	3.30	3.25	0.19734	100	169
10127-2	5686	0.28	0.3710	0.19	0.19	0.00197	50	61
10132-1	5004	0.22	0.5300	0.58	0.58	0.00676	75	60
10140-1	5144	0.33	0.3735	0.41	0.41	0.00505	75	64
10141-1	5189	0.31	0.3735	0.36	0.36	0.00417	75	62
10145-1	4599	0.21	0.5300	0.32	0.32	0.00356	100	58
10147-1	4619	0.23	0.5300	0.43	0.43	0.00524	100	61
10175-1	5164	0.40	0.3968	0.23	0.23	0.00365	75	78
7706-37	370	13.20	0.2702	50.00	34.10	17.83320	185	255
7706-39	186	13.20	0.2268	160.00	126.62	47.90016	330	154
7706-41	411	19.60	0.1910	160.00	112.47	59.89760	330	250
7706-44	580	7.83	0.4880	13.00	11.33	4.96735	185	278
7706-49	3970	3.23	0.2300	20.00	18.91	1.48580	110	160
7706-50	4902	0.78	0.3500	6.00	5.92	0.16380	90	92
7610-08	2060	1.50	0.2904	10.00	9.74	0.43560	100	93
MANOP H	3500	0.90	0.1130	0.66	0.65	0.00671	90	49
MANOP M	3100	1.20	0.1700	1.00	0.98	0.02040	75	73
MANOP C	4450	0.36	0.3820	1.80	1.79	0.02475	80	60
MANOP S	4915	0.90	0.2240	0.10	0.10	0.00202	75	94

TABLE III

Interpolated ages (y B.P.) of major turning points in the $\delta^{18}O$ deglaciation record, the productivity history of coastal upwelling, and in the $\delta^{13}C$ record of deep-water CO_2 concentration as measured in Meteor core 12392. Age estimates of the last 30,000 years are largely based on radiocarbon ages of planktonic foraminifera larger than 125m.

	$\delta^{18}O$ (*C.wuellerstorfi*)	Onset of Paleoproductivity decrease	Onset of $\delta^{13}C$ increase (*C.wuellerstorfi*)
Onset Termination I			
Step B:	10,500	9,800	
Step A:	14,250[1]	15,700	12,700[5]
		?18,400[3]	
Onset Termination II		124,500[4]	
Step B:	125,000[2]		
Step A:	130,000[2]	131,700	127,400[5]

1) This turning point is about 1,500 years younger than that found in a well dated, high-resolution sediment core 700km further further south (Sarnthein et al., 1982), but similar to that of core 13519.
2) By analogy with Ruddiman & McIntyre (1976) and Imbrie et al. (1984).
3) Identification ambiguous, because it is based only on a single measurement.
4) A preceding intra-Termination productivity culmination near 125,000 y B.P. is regarded as real because it is supported by independent tantamount evidence from a synchronous, extreme $\delta^{13}C$ minimum of *Uvigerina peregrina* (Zahn et al., 1986).
5) Only a single turning point occurs in each Termination.

Part VII.

Modeling Climate and its Record

MODELING FUTURE CLIMATE CHANGE AFFECTING NUCLEAR WASTE DISPOSAL: AN OUTLINE

Richard G. Craig

Department of Geology
Kent State University
Kent, Ohio 44242

ABSTRACT. The Department of Energy must select and operate a permanent repository for high-level nuclear waste in a geologic medium which must remain stable for 100,000 years or more. To assess site stability, a hierarchy of models will be used for forecasts of climate changes that could affect nuclide transport. Regional scale statistical models will be solved with boundary conditions defined by General Circulation Models (GCMs). Boundary conditions for GCMs will be set by a dynamic model of global ice sheet growth driven by Milankovitch parameters. Uncertainty will be computed with a Monte Carlo approach. The forecasts are to be validated against the modern climate and reconstructions of the geologic record.

INTRODUCTION

Nuclear waste is produced from approximately 100 commercial nuclear reactors in the United States. In the typical reactor, about 100 fuel rods must be replaced each year. These are still at a very high temperature and will emit radioactivity above environmentally safe levels for hundreds of thousands or millions of years (NAS, 1983). No permanent safe storage has yet been constructed. These wastes continue to accumulate at the reactor sites; usually they are kept in pools of water.

The Nuclear Waste Policy Act of 1982 mandated the Department of Energy (DOE) to identify a permanent nuclear waste repository in a geologic medium. The site cannot accept nuclear wastes until it is licensed by the Nuclear Regulatory Commission (NRC). Criteria the site must satisfy are based on: engineering aspects of site design and operation, transportation considerations and geologic issues regarding site stability, among others.

The period over which the stability of the site must be assured is set by the Environmental Protection Agency as 10,000 years. In published regulations (10CFR960), the DOE has chosen a "conservative" approach so that conditions considered favorable for the selection of a site must remain favorable for 100,000 years. This is the time span for

W. H. Berger and L. D. Labeyrie (eds.), Abrupt Climatic Change, 341–356.

which forecasts of gross climatic change have already been published (Berger, 1978; Imbrie and Imbrie, 1980; Kukla et al., 1981; Imbrie, 1985; Kanari, Fuji and Horie, 1984).

The long-term processes considered must include all factors which could significantly disrupt the wastes during that time. Therefore, site characterization will consider the impacts of a full glacial and a full interglacial climate upon site stability. Two alternative approaches exist to predict the effect of future climates: modeling or statistical analysis of previous climate states. In this manuscript I describe plans for modeling that have been developed.

The modeling has two major foci. Climatic change of direct interest to DOE and NRC must be site-specific and detailed; this is to be produced with local models. Climatic change that can affect the sites -- and which is most likely to occur over the time span of concern -- is global glaciation of the scale recorded in the Quaternary. All of the sites will be affected to some extent by this global change. Thus, separately from the local modeling, there is need for a global model of climate change to provide consistent boundary conditions for all of the sites.

This paper will describe the research strategy for modeling global climate change and the specific plans for modeling local climate change at the Hanford Nuclear Reservation in Washington. Characteristics of a model already developed for the site at Yucca Mtn., Nevada will be described to illustrate the conceptual approach being followed. I do not intend to describe a finished project here. By providing a detailed specification of the method, I hope to encourage constructive criticism which will lead to further improvements.

Three sites have been chosen for investigation for placement of the first nuclear repository: Hanford (Washington), the Nevada Test Site (NTS) and, the Palo Duro Basin (Texas) (Figure 1). These areas have survived an extensive and lengthy selection process by DOE. An important criterion for selection was the perceived stability of the groundwater system. Because of the depth of placement of the repository, direct exposure by erosion is unlikely. At the NTS and at the Hanford site, volcanic activity is possible but has a low probability. Transport of some nuclides within the groundwater system is nearly certain. The major question is the rate of transport. If the rates are now sufficiently high, or if the rates were to increase in the future, the stability of the repository could be compromised.

POTENTIAL IMPACT OF CLIMATE CHANGE

Because the planned depth of storage is approximately 1km, the current state of the groundwater flux is not likely to change quickly. It is assumed that only high amplitude (on the scale of glacial-interglacial cycles) or long period (many thousands of years) variations in the boundary conditions will produce significant changes; sensitivity analyses of this point will be done. Variations in recharge conditions

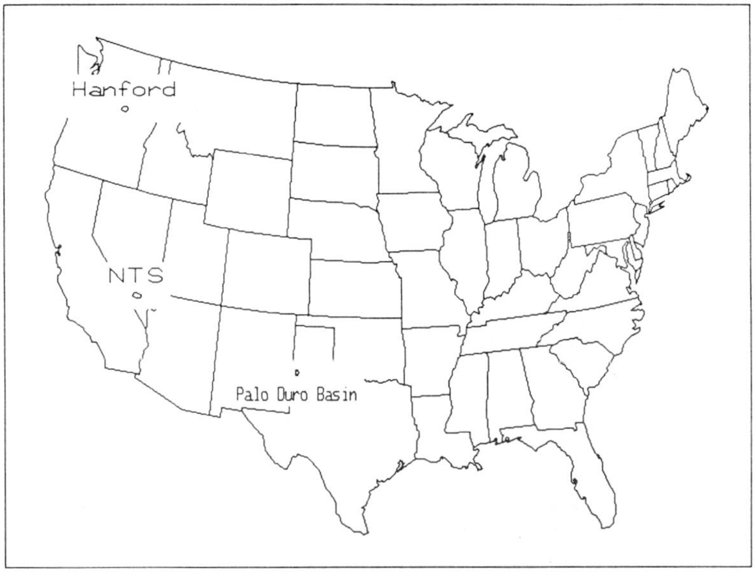

Figure 1. Sites being considered for nuclear waste storage.

due to climatic change are of special interest. Climate change in the Quaternary Period provides a guide to the magnitude of change that is likely in the next 100,000 years.

The history of groundwater change during the Quaternary Period is poorly known, so it is necessary to estimate recharge to the groundwater system from a model of local climate parameters (including temperature and precipitation). This local model must remain consistent with the hemispheric and global scale processes that accompany major climate changes which, in turn, are computed from global scale models. Consistency in the estimates of the three sites can be improved since each of the local estimates will be based upon boundary conditions specified from the same global scale model.

Intermediate between climate change and its effects upon groundwater flux is the response of the rivers and lakes. At each of these sites effluent streams provide a large percentage of the recharge to the groundwater system. The geometry of the surface water system can control the discharge areas of the groundwater system. An example is given by the former occurrence of a 200m deep lake in Death Valley, the discharge area for the groundwater system draining Yucca Mt. Such lakes did form at various times during the Quaternary Period in the southwestern U.S. (Smith and Street-Perrot, 1983).

At Hanford, lobes of ice at the edge of the Cordilleran ice sheet invaded river valleys, creating ice dams that backed up lakes behind them. The largest of these, Lake Missoula, repeatedly drained cataclysmically in some of the largest floods known in the earth's history

(Baker and Bunker, 1985). The Pasco Basin, site of the Hanford Nuclear Reservation, lies in the path these flood waters took. Future glaciations would likely be accompanied by such floods and these may increase groundwater flux through the repository.

MODELING PROCEDURES

Scenarios Being Considered

Four scenarios are being given special consideration in evaluating these repositories:

(1) no change in climate
(2) a new glacial epoch
(3) an interglacial
(4) a "super interglacial epoch".

The first of these is a benchmark case. It will be used to evaluate the capability of the model to reconstruct the best-known of all climate states -- today's. Quantitative measures of the uncertainty in model forecasts will be obtained from its ability to reproduce modern temperature, precipitation, runoff, and vegetation patterns. Specific statistical tests of this question are being constructed.

Glacial conditions will probably return during the next 100,000 years and possibly within the next 10,000 years (Berger, 1981). Because of the expected concurrent changes in the groundwater system, the climate of such periods must be accurately modeled. This exercise will form the central feature of the studies to be done. It is considered the climatic state most likely to result in disruption of the groundwater system of the repository.

The third focus of attention will be the next interglacial. Although we might consider the present interglacial sufficient experience to allow estimation of the behavior of the groundwater and climate systems during an interglacial, there are enough climatic excursions known in the Holocene, and the detailed history of the climatic system in that time is well enough documented, to warrant a modeling exercise. We can determine the magnitude of changes in the climate and groundwater systems (compared to today's) to be expected even without a major climatic shift.

Another scenario considered increasingly likely is a "super interglacial." Increased amounts of carbon dioxide and other trace gases and aerosols may lead to such a climate state (Broecker and Peng, 1984). This would create a hydrologic system unlike any known in the Quaternary Period. For this reason, only modeling can suggest the actual characteristics of such a climate state.

Climatic variables of concern

For the purposes of application to a groundwater model, the variables of greatest concern are precipitation and temperature. It is insufficient to consider only annual averages because of the highly seasonal nature of climate in these areas. Only certain months provide positive balances of available moisture over large parts of these areas. The remaining months show deficits of moisture. If only considered on an annual basis, the summed evapotranspiration would yield unrealistically low estimates of available moisture. This would produce a bias in the estimate of infiltration (and groundwater recharge) and make the groundwater flux appear lower than appropriate. Therefore, both precipitation and temperature are estimated on a monthly basis (Figure 2a-b; Craig and Roberts, 1983).

From these, effective moisture (recharge and runoff) are computed by subtracting an estimate of evapotranspiration from precipitation (Figure 2c; Craig and Roberts, 1984). Evapotranspiration estimates take into account temperature, insolation, and vegetation type. Evapotranspiration of soil moisture and the vegetation intercept term are separate from the estimate of evaporation from the surface of water bodies. For this, a more physically-based equation following the ideas of Benson (1981) is applied. The equation of Benson will be expanded to include advection terms. The availability of estimates of effective moisture forms the basis of a model of the surface hydrologic system (Figure 2d; Singer and Craig, 1983).

To estimate precipitation and temperature, regional controls must be considered. These include sea surface temperatures (SSTs) and atmospheric circulation on the one hand, and the orographic influence of local topography on the other. Since the climates at the NTS and at Hanford are controlled by the temperatures of the Pacific Ocean and the dominantly zonal westerlies, we can compute the local precipitation and temperature regimes by accounting for the variability of the source of moisture (the Pacific Ocean) and the direction of movement of that moisture. Local orographic effects are estimated from a digital elevation model. Large changes in sea level influence absolute elevation and so the magnitude of adiabatic heating and cooling. Sea level changes also affect the distance an air mass may travel over the continent before reaching the area of concern. This will affect the "continentality" of the site. Variables that must be known to estimate local climate are summarized in Table 1.

TABLE 1. Variables needed to estimate changes in climate in the western U.S. over 100,000 years.

Variable	Application
Topography	Orographic effects
	Configuration of coast
	Drainage systems
	Local exposure and shading
SSTs	Precipitable water
Wind direction	Orographic effects
	Source of moisture

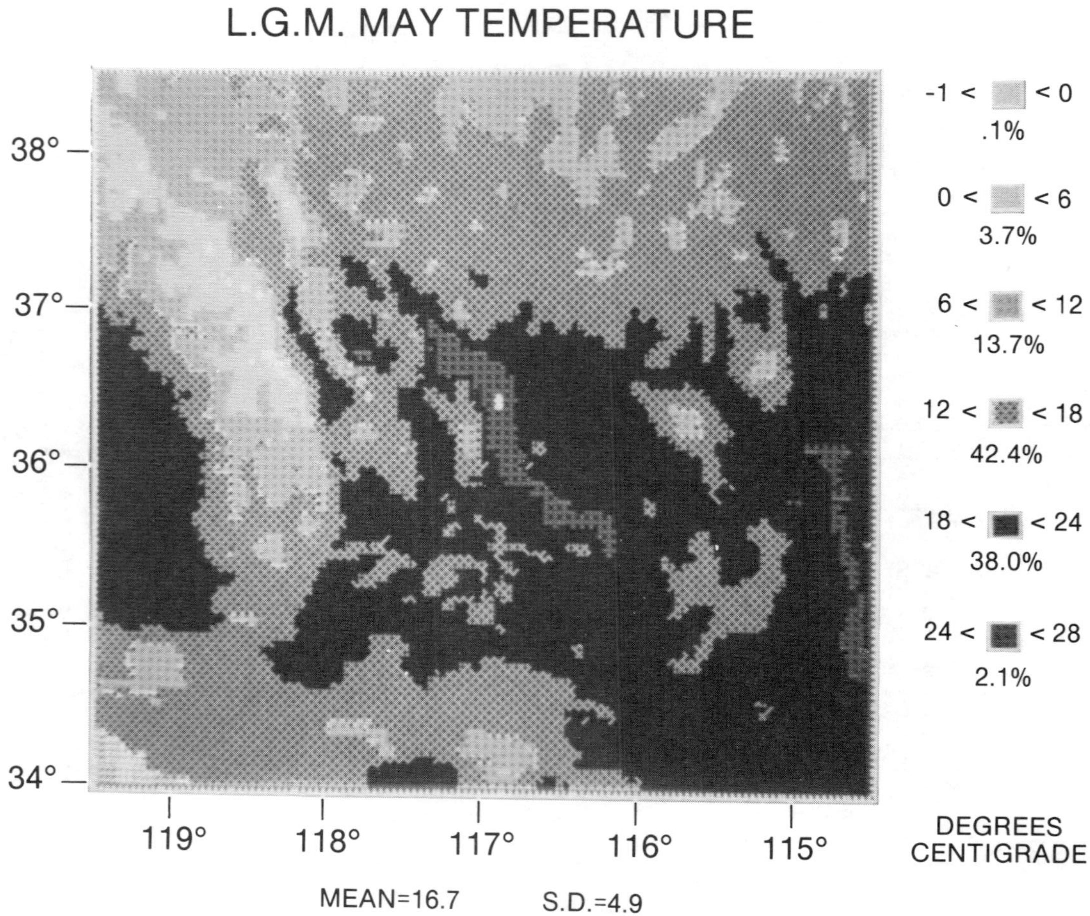

Figure 2. Examples of the simulation of modern and last glacial maximum (LGM) climate in the southwestern United States. (a) LGM temperature for the month of May

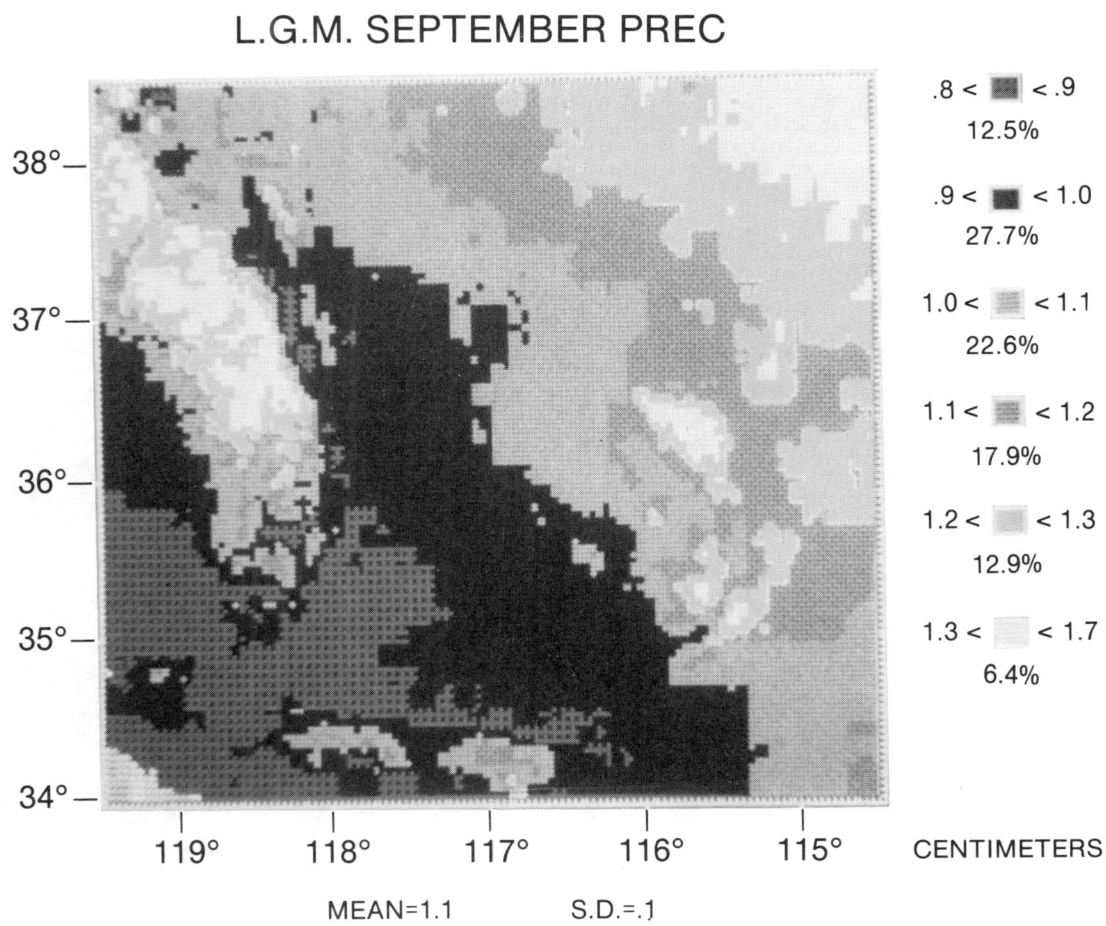

Figure 2. (b) LGM precipitation for the month of September

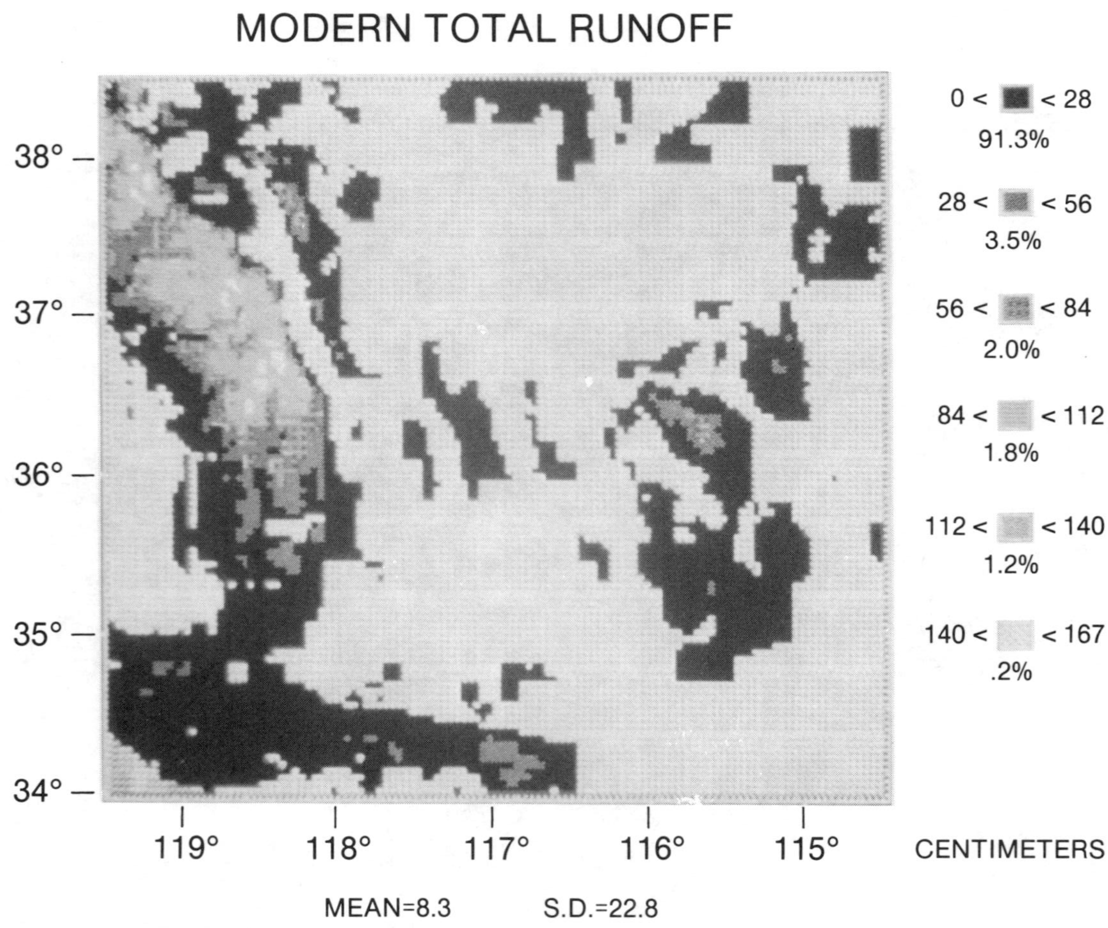

Figure 2. (c) modern runoff summed over the whole year

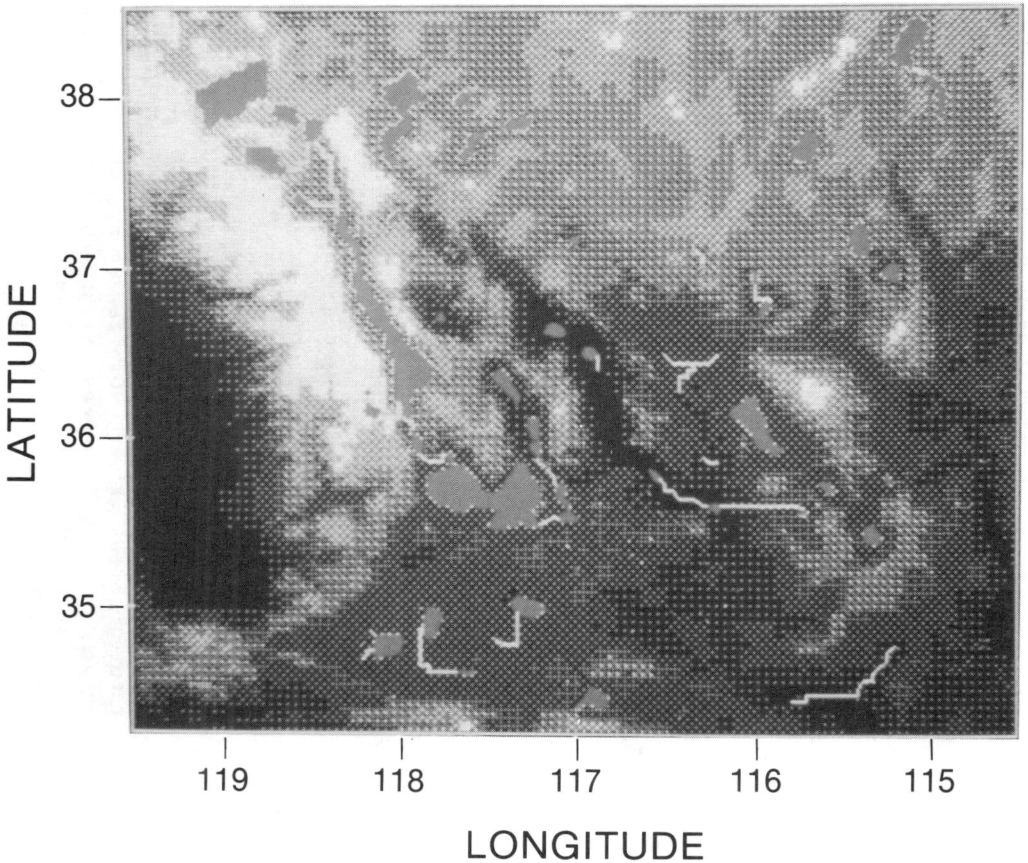

Figure 2. (d) the lake system in equilibrium with simulated LGM runoff.

Difficulties at the Global Scale

At the global scale, our ability to estimate long-term climatic changes is based primarily upon the ice volume record from oxygen isotope studies. The oxygen isotope records are never exactly duplicated by "global driver" models (Imbrie and Imbrie, 1980; Imbrie 1985; Kukla et al., 1981). This lack of fit is a measure of our skill in reproducing (and, by inference, in forecasting) the climates of the earth by each given model. Although models may differ in their nature, complexity and sophistication, all defensible models should fit the record equally well.

The oxygen isotope record is not a perfect documentation of the volume of ice on the earth (Mix and Ruddiman, 1984). Since the future forecasts are driven by models whose parameters were estimated in a biased fashion the forecasts themselves can be expected to be suboptimal.

Each model may have been tuned to a different data set, as illustrated by the two models offered by Imbrie. The original model (Imbrie and Imbrie, 1980) was tuned to five oxygen isotope records from actual cores. The more recent, and presumably better model (Imbrie, 1985) was tuned to the SPECMAP curve (Imbrie et al., 1984). Since the data sets used for tuning differ, a direct comparison of the percent variance explained would be an inappropriate method of choosing between the two. For inter-model comparisons and for estimating uncertainty in our forecasts a "Monte Carlo" procedure is being applied.

COMPONENTS OF THE MODEL

The global driver

Much progress has been made in understanding the history of, and controls on, major climate change throughout the Quaternary Period. A consensus has been reached that the important climate changes arise from changes in the insolation regime of the earth due to variations in its orbit. Several workers have succeeded in formulating quantitative models that can reproduce the gross fluctuations in global ice volume as recorded in oxygen isotope records from deep sea cores (see, for example: Imbrie, 1985). Those equations can be solved for the next 100,000 years knowing the future orbital values. Orbital parameters can be solved with equal accuracy in the future as in the past (Berger, 1984). Alternative models of climatic variability that must be considered ignore the Milankovitch effects (Hasselmann, 1976).

Solution of the GCMs requires information about the geographic distribution of ice throughout the world (Gates, 1976). Once the total *volume* of ice is established for a given time period, that *distribution* can be estimated using a two-dimensional dynamic ice model (Mahaffy, 1976). Several such models exist that estimate the mass balance of the ice sheet using empirical terms dependent upon insolation under a computable orbital configuration (Budd and Smith, 1979; Oerlemans and Van der Veen, 1984). The more sophisticated of these incorporate feedback effects due to isostatic adjustments within the crust and upper mantle (Peltier, 1980). Some also account for sea level changes and iceberg calving at ocean margins (Pollard, 1982; 1983). The thermodynamics of the ice sheet must also be considered. The empirical basis of the estimate of mass balance terms introduces a major difficulty in use of these models for future conditions. These applications will require careful calibration of the relation using glacial geomorphology.

General Circulation Models

Regional patterns of atmospheric circulation are solvable using general circulation models of the atmosphere (AGCMs) and these are routinely used for many research problems including carbon dioxide and nuclear winter studies as well as paleoclimatic reconstructions (Kutzbach and Otto-Bliesner, 1982). For this project, the GCM to be used is the Community Climate Model (version 6) of the National Center for Atmospheric Research (Washington et al., 1980). Although that model is capable of seasonal simulations with a mixed-layer ocean, the initial sensitivity experiments are being done with fixed insolation and ocean temperatures. The first questions to be studied are the effects of the most dramatic changes anticipated upon the climatic variables in the western U.S. Boundary conditions that will be changed include sea surface temperatures (SSTs), ice extent and elevation and, carbon dioxide concentration in the atmosphere. Sea surface temperatures will be obtained from the CLIMAP (1981) data set and from published studies with mixed layer models using increased carbon dioxide levels.

Solutions obtained from the GCMs will provide estimates of the regional atmospheric circulation patterns at seasonal extremes. The solutions for the grid cells immediately surrounding the repository sites will establish boundary conditions for the local climate models. These compute local climatic response to the given boundary conditions. Precipitation and temperature computed from the GCM will help determine the hydrologic balance of the Cordilleran ice sheet in the modeling effort for the Hanford area.

Local Response: Climate

Before the local climate model can compute the climate state corresponding to the given boundary conditions, it is necessary (in the case of Hanford) to determine the current configuration of the Cordilleran ice sheet. This information is used to adjust the estimates of local climate for effects due to the proximity of that ice sheet, including anticyclonic circulation induced by decreased atmospheric pressure over the ice sheet and katabatic winds descending from the surface. The model of the Cordilleran ice sheet will be a dynamic one patterned after the work of Mahaffy (1976) and of Budd, Jenssen and Smith (1984).

Estimates of the local climatic response to the boundary conditions of the given scenario will be based upon methods already developed for the NTS (Craig and Roberts, 1983, 1984; Craig, 1984). More deterministic mesoscale models of the climate system are now under development in an experimental mode for modern climate. They have a number of limitations in their current configuration that preclude their use for this application. Difficulties with the use of mesoscale models include the very long integration times required for the level of detail and spatial coverage required. Current mesoscale models are designed to forecast short-term weather systems rather than long-term climate. They are usually integrated over a period of several days to a month. For forecasts of the monthly average climate for a one month period, integration over perhaps a year for perpetual monthly boundary conditions or many years of seasonal climates would be required. This is neither computationally feasible nor is there promise that the resulting values would improve upon those of a more stochastically-oriented approach. Indeed, parameterization of these smaller-scale phenomena appears to be the best available method at this time.

The model to be used is a statistical one which is based upon the modern relation between sea surface temperature, sea level, primary atmospheric circulation in the area of interest, topography and climate variables as recorded at modern climate stations. A grid of topographic values is established for the area of concern. The topographic data allows estimation of the orographic effects upon temperature and precipitation. It is also used to interpolate the climate estimates to finer grid spacing as needed. Sea surface temperature information is interpolated, using a cubic polynomial, from the August and February estimates of CLIMAP (CLIMAP Project Members, 1981). Wind directions can be obtained directly from the AGCMs. For the time of the last glacial maximum, sea levels estimated by Denton and Hughes (1981) are used.

Equations are developed using step-wise regressions to estimate mean monthly temperature and precipitation for each month. Since the SSTs and winds are more global in nature and are estimable from GCMs, they form boundary conditions that can reflect the varying global climate. Improvements in the model to allow computations of the adjustment of the local winds to the influence of topography will be made for the model at Hanford. The Nevada model has been tested by its ability to estimate modern climates and those of the last glacial maximum.

The level of spatial resolution required is fixed by the requirement that the major drainage connections within the area must be accurately computed. The steep relief and the narrow drainage ways of the Basin and Range near Death Valley mandated a grid spacing of 2.5′ of latitude and longitude. The grid used involved over 8000 points. To reduce computational demands, the climatic variables are computed at a spacing of 7.5′ and interpolated to the finer spacing using the topographic data as weighting factors. Very little experimentation with other grid spacings has yet been done.

Impact: Hydrology and Vegetation

The final component of the system model provides estimates of the response to this local climate change. First, response of the vegetation system is modeled. This then allows estimates of evapotranspiration. The relative amounts of runoff and infiltration are computed next. With this information, the flow of rivers and the size of lakes is computed.

Vegetation change is estimated by a statistical model of the relation between vegetation and climate. A contingency table is computed from modern observations and maps. Vegetation type and climate at each point within the area of concern are tabulated. A conditional probability model of the vegetation/climate relation is developed from the contingency table to forecast the dynamic response of vegetation to climate change. Such methods have been applied to the Nevada Test Site but are still in an early stage of development (Hose and Craig, 1985). Once the vegetation types are estimated, evapotranspiration rates are computed with the Blaney-Criddle equation (Blaney and Criddle, 1950). Evapotranspiration is subtracted from precipitation to give available moisture for runoff and infiltration.

The second local response to climate change occurs in the local surface water system. The topographic grid is used to compute drainage directions for each grid cell. The entire drainage system is computed and each closed basin, drainage divide and outlet is found (Singer and Craig, 1983; see also Figure 2d).

This information has two applications. First, recharge rates form boundary conditions to groundwater models that compute the flux through the repository. Second, and equally important is the ability to directly compare the estimated patterns to that observed in the geologic record under analogous conditions. This forms one of the most important tests of the entire modeling procedure that is available.

CONCLUSIONS

The components for a useful model of long-term climatic change are now available. Although each of these separate pieces will continue to advance in sophistication, it now appears possible to synthesize a holistic prognostication of the major climatic events in specific regions for the next 100,000 years. Current limitations of these models do not preclude the synthesis of these components into a more comprehensive and testable model.

There is now much alarm about the short-term implications of the rise in carbon dioxide in the atmosphere. Just as troublesome are the long-term results. Heating of the earth by several degrees centigrade could, within a few hundreds of years, melt significant portions of the ice caps that have remained relatively stable for thousands of years. The configuration of these ice caps is a basic control upon the current climatic system. Release of water from these ice caps would raise sea level, modify oceanic circulation and the energy balance of the earth. When combined with the other natural changes tending to lead us into another glacial stage, the complex interactions can be forecast with the procedures described here.

The global modeling effort is designed to provide a consistent estimate of future global climate change to ensure that the evaluation of individual sites is based upon the same assumptions about the controlling factors. Other countries are also faced with nuclear waste disposal; it would be rational to share the results obtained by modeling global climate change. This could set the stage for dealing with these changes.

ACKNOWLEDGEMENTS

Work has been supported by contracts from the Department of Energy through its Pacific Northwest Laboratories operated by Battelle Memorial Institute, and through the Basalt Waste Isolation Project at Hanford Washington operated by Rockwell International. At several times in the development of these plans they have benefited from the contributions of a group of consultants and an advisory panel. Consultants on the project have included: E. Barron, J. Gardiner, and N. Pisias. The advisory panel includes: J. Imbrie, T. Hughes, W. Washington, and L. Gates. Reviews of plans have also been received by S. Manabe and J. Kutzbach. I thank the editors for their helpful suggestions. Any errors in concept or detail remain the responsibility of the author, of course.

REFERENCES

Baker, V.R. and Bunker, R.C., 1985. Cataclysmic Late Pleistocene Flooding from Lake Missoula: A Review, *Quaternary Science Reviews*, 4: 1-41.

Berger, A.L., 1978. Long-Term Variations of Daily Insolation and Quaternary Climatic Changes, *Journal of Atmospheric Sciences*, 35: 2362-2367.

Berger, A.L., 1981. Climatic Variations and Variability: Facts and Theories, D. Reidel Pub. Co., Holland, 795 pp.

Berger, A.L., 1984. Accuracy and frequency stability of the earth's orbital elements during the Quaternary. In: A. Berger, J. Imbrie, J. Hays, G. Kukla and B. Saltzman (eds.), *Milankovitch and Climate*, Part 1, D. Reidel Pub. Co., pp. 3-10.

Benson, L.V., 1981. Paleoclimatic Significance of Lake-Level Fluctuations in the Lahontan Basin, *Quaternary Research*, 16: 390-403.

Blaney, H.F. and Criddle, W.D., 1950. Determining Water Requirements in Irrigated Areas from Climatological and Irrigation Data, *U.S. Dept. Agric. Soil Conserv. Service Tech. Paper*, 96: 1-48.

Broecker, W.S. and Peng, T.H., 1984. The Climate-Chemistry Connection. In: J.E. Hansen and T. Tokahashi, eds., *Climate Processes and Climate Sensitivity*, American Geophysical Union, Geophysical Monograph 29, 327-336.

Budd, W.F. and Smith, I.N., 1979. The Growth and Retreat of Ice Sheets in Response to Orbital Radiation Changes, *International Association of Hydrological Sciences*, 131: 369-409.

Budd, W.F., Jenssen, D. and Smith, I.N., 1984. A Three-Dimensional Time-Dependent Model of the West Antarctic Ice Sheet, *Annals of Glaciology*, 5: 29-36.

CLIMAP Project Members, 1981. Seasonal Reconstruction of the Earth's Surface at the Last Glacial Maximum, *Geological Society of America Map and Chart Series* MC-36, 18 pp.

Craig, R.G., 1984. Climate and Lakes of the Death Valley Drainage System During the Last Glacial Maximum, *Report* to: Department of Energy, Pacific Northwest Laboratories, 157 pp.

Craig, R.G. and Roberts, B.L., 1983. Estimating Orographically-Influenced Climate in the Southwest U.S. Using Regression-based Equations, *Geological Society of America Abstracts with Programs*, 15: 549.

Craig, R.G. and Roberts, B.L., 1984. Seasonal Patterns of Precipitation and Temperature in the Southwest U.S. at the Time of the Last Glacial Maximum, *American Quaternary Association, Abstracts with Programs*, 8th Biennial Meeting.

Denton, J.H. and Hughes, T., 1981. *The Last Great Ice Sheets*, A Wiley-Interscience Publication, John Wiley & Sons, New York, 484 pp.

Gates, W.L., 1976. Numerical Simulation of an Ice-Age Climate With a Global General Circulation Model, *Journal of Atmospheric Science*, 33: 1844-1873.

Hasselmann, K., 1976. Stochastic Climate Models, Part I, Theory, *Tellus*, 28: 473-484.

Hose, D.R. and Craig, R.G., 1985. A Conditional Probability Model of Vegetation/Climate Relations in the Southwestern U.S., *Geological Society of America Abstracts with Programs*, 17: 614.

Imbrie, J., 1985. A Theoretical Framework for the Pleistocene Ice Ages, *Journal of the Geological Society of London*, 142: 417-432,

Imbrie, J.H., Hays, J.D., Martinson, D.G., McIntyre, A., Mix, A.C., Morely, J.J., Pisias, N., Prell, W.L. and Shackleton, N.J., 1984. The Orbital Theory of Pleistocene Climate: Support from a Revised Chronology of the Marine ^{18}O Record. In: A. Berger, J. Imbrie, J. Hays, G. Kukla, B. Saltzman, eds., *Milankovitch and Climates*, Part 1, D. Reidel Pub. Co., pp. 269-306.

Imbrie, J. and Imbrie, J.Z., 1980. Modelling the Climatic Response to Orbital Variations, *Science*, 207: 943-953.

Kanari, S., Fuji, N. and Horie, S., 1984. The Paleoclimatological Constituents of Paleotemperature in Lake Biwa. In: A.L. Berger, J. Imbrie, J. Hays, G. Kukla, and B. Saltzman, eds., *Milankovitch and Climate*, Part 1, D. Reidel Pub. Co., pp. 405-414.

Kukla, G., A. Berger, R. Lotti, and J. Brown, 1981. Orbital Signature of Interglacials, *Nature*, 290: 295-300.

Kutzbach, J.E. and Otto-Bliesner, B.L., 1982. The Sensitivity of the African-Asian Monsoonal Climate to Orbital Parameter Changes for 9000 Years BP in a Low-Resolution General Circulation Model, *Journal Atmospheric Sciences*, 39: 1177-1188.

Mahaffy, M.W., 1976. A Three-Dimensional Numerical Model of Ice Sheets: Tests on the Barnes Ice Cap, Northwest Territories, *Journal Geophysical Research*, 81: 1059-1066.

Mix, A.C. and Ruddiman, W.F., 1984. Oxygen-isotope analyses and Pleistocene ice volumes, *Quaternary Research*, 21: 1-20.

NAS, 1983. *A Study of the Isolation System for Geologic Disposal of Radioactive Waste*, National Academy of Sciences Press, Washington, D.C.

Oerlemans, J. and van der Veen, C.J., 1984. *Ice Sheets and Climate*, D. Reidel Publishing Co., 217 pp.

Peltier, W.R., 1980. Models of Glacial Isostasy and Relative Sea Level. In: Bally, A.W., P.L. Bender, T.R. McGetchin and R.I. Walcott, (eds.), *Dynamics of Plate Interiors*, Washington, D.C., Geodynamics Series, Volume 1, American Geophysical Union, pp. 111-128.

Pollard, D., 1982. A Simple Ice-Sheet Model Yields Realistic 100kyr Glacial Cycles, *Nature*, 296: 334-338.

Pollard, D., 1983. Ice-Age Simulations with a Calving Ice-Sheet Model, *Quaternary Research*, 20: 30-48.

Singer, M.P. and Craig, R.G., 1983. A Computer Simulation Model of the Growth and Desiccation of Pluvial Lakes in the Southwestern United States, *Geological Society of America Abstracts With Programs*, 15: 689.

Smith, G.I. and Street-Perrot, F.A., 1983. Pluvial Lakes of the Western United States. In: H.E. Wright, Jr. and S.C. Porter, (eds.), *Late-Quaternary Environments of the United States*, Volume 1: *The Late Pleistocene*, 190-212.

Washington, W.M., Semtner, A.J., Jr., Meehl, G.A., Knight, D.J. and Mayer, T.A., 1980. A General Circulation Experiment with a Coupled Atmosphere, Ocean and Sea Ice Model, *Journal of Physical Oceanography*, 10: 1887-1908.

DETECTION OF ABRUPT CHANGES IN DEEP-SEA SEDIMENT CORES: THE FORWARD PROBLEM

H.N. Dalfes*

National Center for Atmospheric Research
Boulder, CO 80307-3000
U.S.A.

ABSTRACT. Post-depositional processes alter climate proxy signals recorded in deep-sea cores. These processes along with timescale problems and measurement noise make the detection of abrupt events a difficult problem. The need for a dual approach which combines *forward* and *inverse* approaches is discussed. The forward approach is illustrated with examples of numerical solutions of an advection-diffusion equation representing bioturbation processes in a variable sedimentation rate environment. Requirements for a comprehensive inverse approach are also discussed.

INTRODUCTION

Understanding abrupt events in climate history requires the estimation of at least the amplitude and the timing of the change in relevant climate variables. At geologic time scales, analysis and interpretation of various climate proxies from deep sea cores constitute the principal tool for approaching this problem. Main sources of uncertainty encountered in such interpretation efforts can be classified in three groups:

-Uncertainties due to the conceptual (and quantitative modeling) of the actual relations between climate variables and proxy quantities;

-Chronostratigraphic problems due to processes disturbing the proxy record during and after its sedimentation;

---

*Author's present address:

TÜBITAK/TBAE
Yer Bilimleri Bölümü
P.O. Box 74
41410 Gebze, Kocaeli
Turkey

W. H. Berger and L. D. Labeyrie (eds.), Abrupt Climatic Change, 357–363.

-Uncertainties and limitations due to field and laboratory sampling procedures.

In this paper, we are concerned with the second and third groups of problems in the context of *abrupt climatic change*. The study of such isolated climate events is to be distinguished from studies of a more general nature of the climate variability over several time scales which emphasize spectral characterizations and where "quasi-stationarity" assumptions are very often made to permit Fourier analyses. Abrupt changes on the other hand are singular events best analyzed in the time domain.

Here we are proposing a dual approach to the problems of information recovery from deep sea cores. This approach is based on the premise that the best way to test the properties of any information recovery scheme is to know the correct answer in advance. We will present in the next section the *forward* problem of simulating the distribution of a tracer in a sampled deep sea core, given tracer and sedimentation rates histories. We will conclude by discussing avenues of future research where forward and inverse solutions will be considered together to allow development of reliable information recovery schemes.

THE FORWARD PROBLEM

Numerous physical and biogenic processes take place between the time of deposition of a sediment layer containing the tracer(s) of interest and the time the sediment is sampled in a laboratory. Here we will concern ourselves only with the bioturbation phenomenon in a variable sedimentation rate environment, neglecting for the moment many other important processes such as compaction, dissolution, etc. This stage will be followed by a sampling procedure which mimicks sampling and measurement noise.

Bioturbation: a Diffusion-Advection Model

The problem of tracer distribution in bioturbated ocean cores have already received considerable attention (Goldberg and Koide, 1962; Guinasso and Schink, 1975; Officer and Lynch, 1982, to name a few). Most of these authors used simulated bioturbation processes as a constant coefficient advection-diffusion problem in a diffusive slab on a non-diffusive medium. Here we will consider the same equation, but allow the advection speed, i.e. the sedimentation rate to depend on time. The advection-diffusion equation is:

$$\frac{\partial c}{\partial t} = \frac{\partial}{\partial z}\left(D\frac{\partial c}{\partial z}\right) - v\frac{\partial c}{\partial z} \qquad (1)$$

where c is the tracer concentration per unit mass of sediment, z depth measured in a downcore coordinate system, v time dependent sedimentation rate and D diffusion coefficient.

This equation is considered along with the following boundary conditions:

$$vc_1 - D\frac{\partial c_1}{\partial z} = vc_0 \qquad z = 0$$

$$D\frac{\partial c_1}{\partial z} = 0 \qquad z = d \qquad (2)$$

$$c_1 = c_2 \qquad z = d$$

where c_1 and c_2 are the concentrations in the mixed layer of depth d and in the underlying non-diffusive medium. Fig. 1 describes the geometry of the problem.

A diffusion-advection modeling of bioturbation may not be totally adequate due to the finite size of the burrowing species. On the other hand, by introducing a diffusion time scale to this problem, it allows one to distinguish between cases where advection (i.e. transition through the bioturbation zone) is important and cases where mixing dominates. One should also mention that for high D diffusion coefficient values, the model tends asymptotically toward the well-mixed layer approximation of Berger and Heath (1968).

In the following treatment we will assume that the diffusion D is constant, i.e. independent of depth in the bioturbation layer and time). This constraint is relatively easy to eliminate, but can only be done at the expense of adding other adjustable parameters.

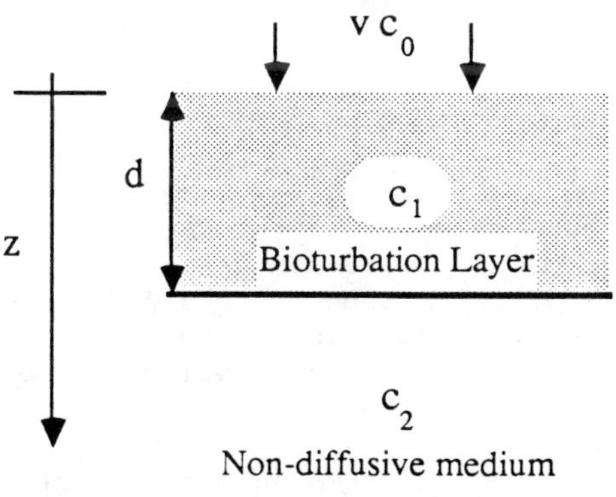

Figure 1: Schematic description of the bioturbation phenomenon formulated as an advection-diffusion problem.

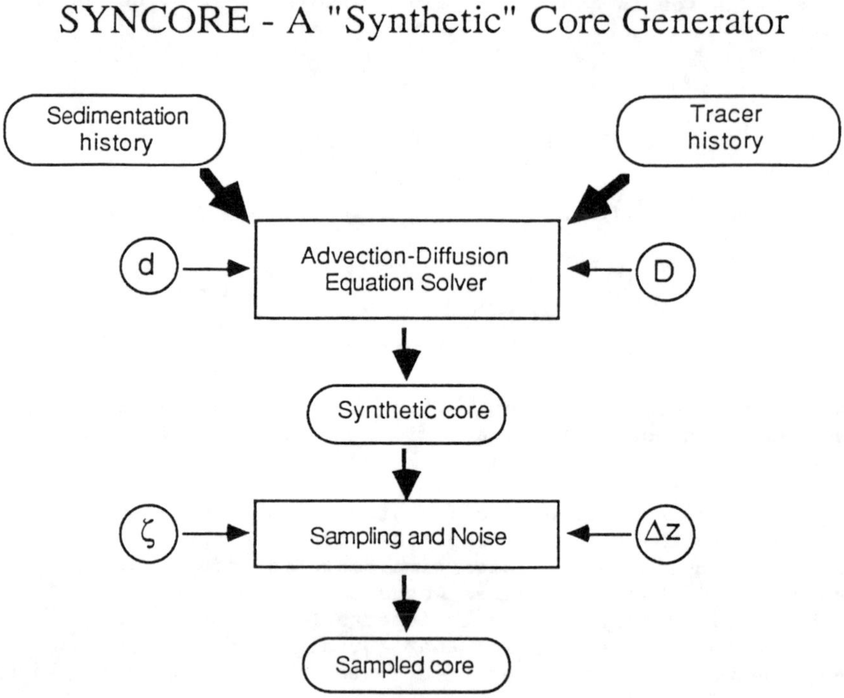

Figure 2: Schematic description of SYNCORE

SYNCORE: A Synthetic Core Generator

The software package SYNCORE is a system still under development to simulate biogenic and physical process in deep sea sediments to produce *synthetic core* data. It is the *forward* component of the dual approach proposed in this paper.

The basic structure of the present version of SYNCORE is displayed on Fig. 2. The approach consists of solving the advection-diffusion equation given histories of tracer concentration ($c_o(t)$) and the sedimentation rate ($v(t)$). This is done using a *fully implicit* second order numerical scheme. The time step size is on the order of 50 years. We call this fine time resolution representation of the tracer concentration as a function of depth in the *synthetic core*, $c(z_i)$.

At the second stage which tries to mimick laboratory core sampling and measurement noise, this synthetic core is sampled at equal depth intervals by piecewise linear interpolation of $c(z_i)$ and the resulting number is multiplied by a pseudo-random number:

$$s(i\Delta z) = c(i\Delta z)\ (1 + \zeta) \qquad (3)$$

where ζ is a Gaussian random number with a zero mean and a standard deviation equal to the relative measurement error. The function $s(i)$ is the resulting sampled core.

Examples

Here we present two example outputs from SYNCORE. Though tracer magnitude and histories are chosen to give a "realistic" look to the simulations, they do not represent attempts to simulate particular observed data.

The first example (Fig. 3) is a step function change in the tracer concentration integrated along with a constant sedimentation rate history. Values for the mixed layer parameters are chosen within ranges mentioned in the literature. The standard deviation of the relative sampling/measurement error is 0.1, a value fairly realistic for isotopic measurements. One observes the expected spreading of the high values upward in the core; i.e. a "smoothing" effect.

The second example (Fig. 4) is chosen to demonstrate the flexibility of our numerical approach to the forward problem: a fairly complex tracer and sedimentation history can be handled. Such input functions would require painfully complex algebra for analytical approaches. A variable sedimentation rate along with experimental measurement noise renders the visual detection of the initiation and termination times of tracer events very uncertain.

CONCLUSIONS: TOWARD A DUAL APPROACH

We described a numerical model of the bioturbation process in a variable sedimentation rate environment. Examples show the flexibility of such an approach compared to previous analytical solution of the same equation. We were also able to include realistic sampling and measurement noise effects to the resulting solution to produce what we would like to call a "synthetic" core.

As mentioned earlier, we are proposing a research program that will combine the forward and the inverse technique for a better information recovery from deep-sea cores. The forward problem as described here is an initial effort in this direction. In the future, the numerical model can be extended to include other post-depositional effects such as porosity change with compaction, small time-scale fluctuations in the sedimentation rate etc. The numerical approach to the solution make it very flexible.

Once a "realistic" enough signal is generated, it can provide a testbed for a variety of inversion techniques. The knowledge of the original signal and of the magnitude of the disturbing processes will permit one to estimate the skill of a given technique, or to compare various approaches among themselves.

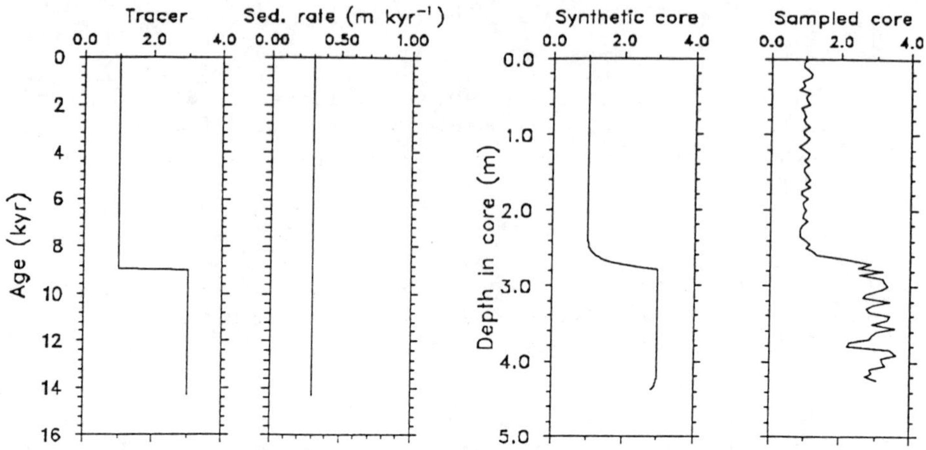

Figure 3: The output of SYNCORE for a step function tracer history and constant sedimentation rate. For this case, the mixed layer depth d = 0.20m and the diffusion coefficient D = $10m^2kyr^{-1}$. For such a high D value, the model is almost equivalent to a *well-mixed layer* model.

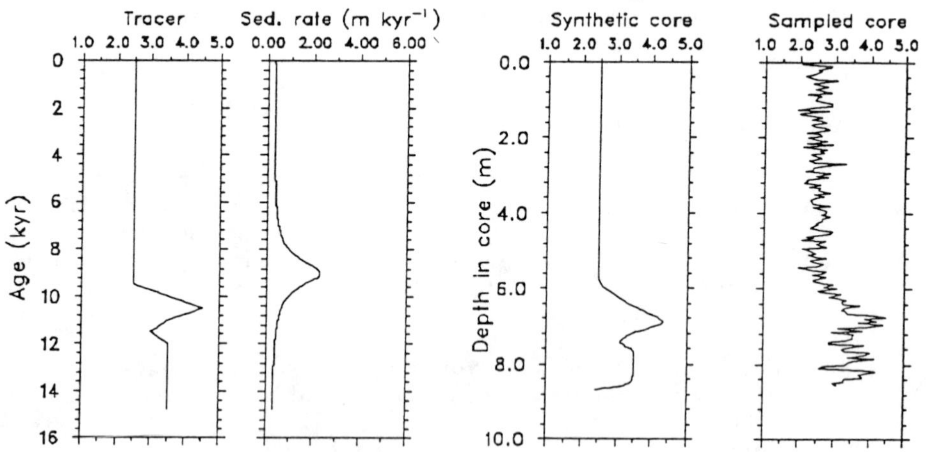

Figure 4: The output of SYNCORE for a complex tracer and sedimentation rate history. For this case, the mixed layer depth d = 0.08 and the diffusion coefficient D = $0.1m^2kyr^{-1}$.

REFERENCES

Berger, W.H. and Heath, G.R., 1968. Vertical mixing in pelagic sediments, *J. Mar. Res.*, 26: 134-143.

Goldberg, E.D. and Koide, M., 1962. Geochronological studies of deep sea sediments by the ionium/thorium method, *Geochim. Cosmochim Acta*, 26: 417-450.

Guinasso, N.L., Jr. and Schink, D.R., 1975. Quantitative estimates of biological mixing rates in abyssal sediments, *J. Geophys. Res.*, 80: 3032-3043.

Officer, C.B. and Lynch, D. R., 1982. Interpretation procedures for the determination of sediment parameters from time-dependent flux inputs, *Earth Planet. Sci Lett.*, 61: 55-62.

ABRUPT TERMINATIONS OF LATE PLEISTOCENE ICE AGES: A SIMPLE MILANKOVITCH EXPLANATION

John Imbrie

Department of Geological Sciences
Brown University
Providence, R. I. 02912-1846

ABSTRACT. Abrupt changes of climate known as Terminations mark the end of a number of late Pleistocene ice ages. This paper argues that Terminations were a normal part of the climate system's response to orbital forcing during the late Pleistocene, and that it is unnecessary to invoke special mechanisms which operate only during these episodes of rapid change.

Two types of abrupt climatic change can usefully be distinguished. *Type 1* is a simple state change, i.e., a rapid and statistically significant change in some observed measure of the climate system's state that occurs in the absence of any fundamental change in the system's structure, or any long-term change in its external boundary conditions. Familiar examples of this type occur as part of the annual cycle: the first killing frost, and the onset of the monsoon. *Type 2* is a rapid change of climate that reflects a long-term change in the system's boundary conditions. Such events must occur occasionally, for example, as slow, tectonically driven changes in geography permit new modes of atmospheric or oceanic circulation.

This paper considers abrupt changes in climate that occurred at the end of several late Pleistocene ice ages. Kukla (1970) found evidence for a number of these abrupt warming events in the soil sequence of central Europe, where the stratigraphic boundary between an ice-age loess and the overlying interglacial soil is sharply marked by a boundary line (a "Markline"). Broecker and van Donk (1970) found isotopic evidence for the same sequence of climatic events in deep-sea cores. Each event in this sequence was called a *Termination*, and was numbered in (reverse) order beginning with Termination I, approximately 12,000 years ago.

The main point of this paper is that Terminations are abrupt climate changes of Type 1. They are a normal part of the climate system's response to Milankovitch forcing in the same way that an abrupt onset of a monsoon is part of the system's linear and non-linear response to the annual march of the sun. There is, however, this difference: that whereas the seasonal cycle can be satisfactorily modeled as a response to forcing at one frequency (the annual cycle) and its harmonics, the

W. H. Berger and L. D. Labeyrie (eds.), Abrupt Climatic Change, 365–367.

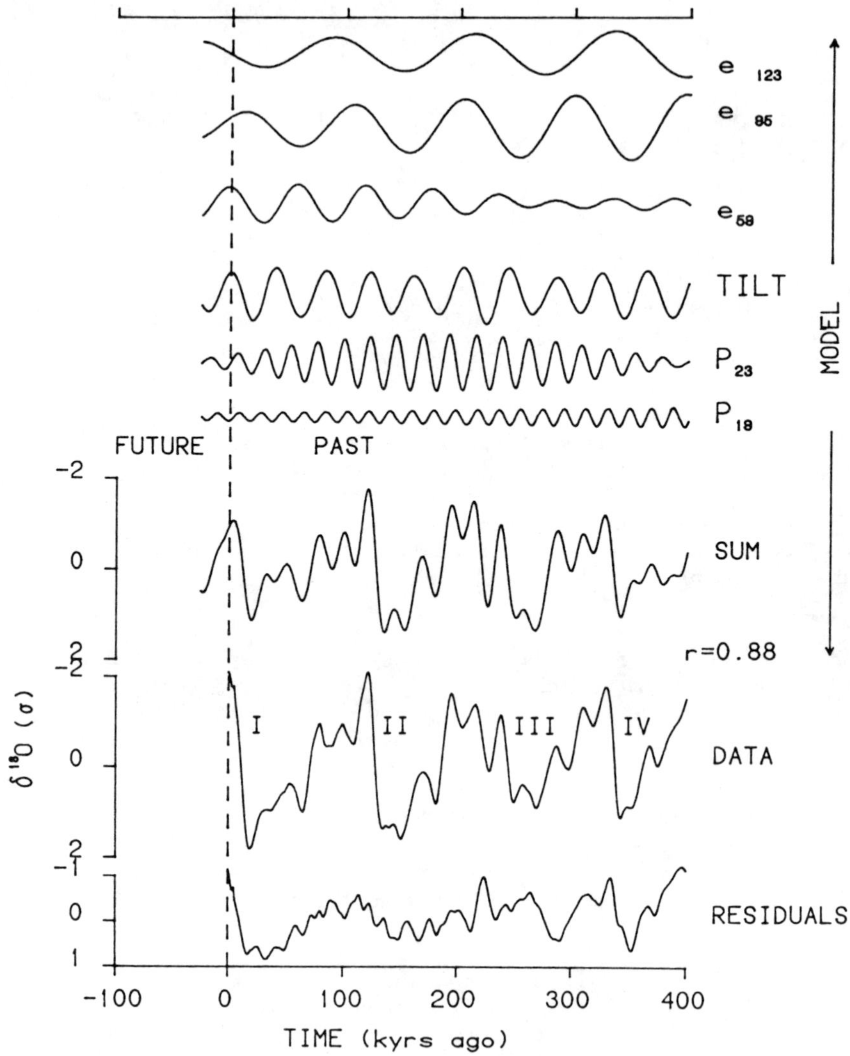

Figure 1. Model of $\delta^{18}O$ variations over the past 400,000 and next 25,000 years. The model curve is the sum of six components, each of which is an orbital curve multiplied by an appropriate gain and shifted by an appropriate phase (see Imbrie, 1985, for details). Roman numerals indicate Terminations I - IV, which occur about 12, 128, 245 and 339 KY B.P. in the data (Imbrie et al., 1984) and in the model.

ice-age cycle must be modeled as the sum of responses to Milankovitch forcing at a number of frequencies (the main frequencies of variation in precession, obliquity, and eccentricity) and their harmonics (Imbrie, 1985).

When such a model is made for the $\delta^{18}O$ record of climate for the past 400,000 years, it becomes clear that at least the major features of this record -- including the particular features known as Terminations I - IV -- are satisfactorily accounted for (Fig. 1). Both the timing and the approximate amplitude of all of the major changes are reproduced with surprising fidelity. As the plot of residuals demonstrates, however, the model $\delta^{18}O$ levels during the interglacial intervals which follow hard upon Terminations I, II, and IV are systemically somewhat low (isotopically heavy). This suggests some non-linear mechanism which is not accounted for by this low-order Fourier model -- a situation quite similar to that encountered in modeling those abrupt changes of climate which occur as a regular part of the annual cycle.

This result suggests that a number of different causal mechanisms must be involved in producing Terminations. Each mechanism acts continuously over long intervals of time with a sinusoidal pattern of rate changes that is determined by the frequency of the corresponding orbital variation. A Termination results when the (positive) sum of the contributions from the different mechanisms is large. Although non-linear mechanisms must act to enhance the linear response during interglacial regimes, there seems to be nothing special about the dynamics of climate during Terminations -- or at least nothing that is not explained by considering the operation of individual mechanisms acting over longer intervals. In fact, it would be difficult to sort out these different mechanisms by confining one's attention exclusively to the termination events themselves.

REFERENCES

Broecker, W.S. and van Donk, J., 1970. Insolation changes, ice volumes, and the O^{18} record in deep-sea cores,. *Reviews of Geophysics and Space Physics*, 8: 169-197.

Imbrie, J., 1985. A theoretical framework for the Pleistocene ice ages, *J. Geol. Soc. London*, 142: 417-432.

Imbrie, J., Hays, J.D., Martinson, D.G., McIntyre, A., Mix, A.C., Morley, J.J., Pisias, N.G., Prell, W.L. and Shackleton, N.J., 1984. The orbital theory of Pleistocene climate: support from a revised chronology of the marine $\delta^{18}O$ record. In: A.L. Gerger et al. (eds.) *Milankovitch and Climate*, Part I, D. Reidel Publ. Co., pp. 269-305.

Kukla, G.J., 1970. Correlation between loesses and deep-sea sediments, *Geol. Fören. Srockholm Förh.*, 92: 148-180.

SIMULATION OF PALEOCLIMATIC TRACERS USING ATMOSPHERIC GENERAL CIRCULATION MODELS

S. Joussaume[1], J. Jouzel[2]

[1]Laboratoire de Météorologie Dynamique
24. rue Lhomond 75231 Paris Cedex 5 - France 2

[2]Laboratoire de Géochimie Isotopique-LODYC
(UA.CNRS 1206) CEA/IRDI/DESICP, DPC.
91191 Gif sur Yvette Cedex -France

ABSTRACT. In this article we review the main aspects of the modeling of atmospheric tracers using General Circulation Models currently under development in the LMD (Laboratoire de Météorologie Dynamique) and the GISS (Goddard Institute for Space Studies) Atmospheric General Circulation Models (AGCM's). This original approach is illustrated by examples concerning the water isotope species (HDO and $H_2{}^{18}O$) and the desert dust particles, which are important climatic tracers of past climates.

INTRODUCTION

Atmospheric general circulation models (AGCMs) can make essential contributions in the study of past climates and abrupt climatic changes; they can simulate atmospheric circulation for different snapshots in the past and they allow investigation of mechanisms of climatic change (see Schneider, Peteet and North, this volume). It is important to compare model results with observations in order to check the validity of the various assumptions underlying the simulations (model parameterizations or boundary conditions). However, this comparison is not always easy since it requires the translation of paleodata for sea, ice or land cores into climatic parameters, like temperature, precipitation or wind field. This is not always straightforward, for example for water isotopes or desert dust contents of deep ice cores.

To facilitate the interpretation of tracers within paleoclimatic records we have included in AGCM simulations of past climates the atmospheric cycles of important climatic tracers (Joussaume et al., 1984a,b; Joussaume, 1985; Jouzel et al., 1987). On one hand this approach allows us to obtain a more direct comparison with observations. On the other hand, the knowledge of both the climatic response of tracers and of atmospheric parameters should help us in translating paleodata into such parameters.

W. H. Berger and L. D. Labeyrie (eds.), Abrupt Climatic Change, 369–381.

This approach has been applied to water isotope species (HDO and $H_2{}^{18}O$) in the LMD (Laboratoire de Météorologie Dynamique) AGCM and the GISS (Goddard Institute for Space Studies) AGCM, and for desert dust particles in the LMD AGCM. The choice of these tracers was motivated by their important changes with climate, especially during the last ice age maximum (18 ka BP) as seen from deep ice cores, and by the problems arising during interpretation (Thompson, 1977; Lorius et al., 1979; Petit et al., 1981).

Concerning water isotopes, present day data display a good correlation between isotope ratios of precipitation and surface air temperature in high latitudes (Dansgaard, 1964; Lorius and Merlivat, 1977) which is well understood by simple models of isotope cycles (Jouzel and Merlivat, 1984). However, the isotope atmospheric cycle is quite complex and is sensitive to the entire atmospheric circulation (sources, transport, precipitation), not just to temperature. Large changes might result therefore during a glacial/interglacial oscillation. A change is indeed observed in the relation between HDO and $H_2{}^{18}O$ ratios at the last glacial maximum with a decrease of the deuterium excess $\delta D-8\delta^{18}O$ (with $\delta^{18}O$ and δD isotopic ratios $H_2{}^{18}O/H_2{}^{16}O$ and $HDO/H_2{}^{16}O$ versus the standard mean ocean water reference) interpreted as an increase in the relative humidity over moisture source regions (Jouzel et al., 1982). To investigate the impact of the whole atmospheric circulation on the relations isotope-temperature and HDO-$H_2{}^{18}O$, AGCMs, which explicitly simulate the global and regional features of atmospheric dynamics and thermodynamics with fully detailed hydrological cycles, are the only appropriate tools.

Concerning desert dust particles, a large increase in the amount of dust in ice cores has been observed during the last glacial maximum. The dust, in the 1 to 5 μm range comes from continents, but its geographical origin is not known precisely. According to Petit et al. (1981), this great increase could be a global signal linked with enhanced aridity and stronger atmospheric circulation. It is interesting to test this hypothesis by simulating desert dust cycles in an AGCM.

In the following, we will present a brief overview of the modeling of water isotope and desert dust cycles. Then some experiments will be presented: first, experiments for January present day conditions performed to test the modeling with the LMD and GISS AGCMs; second, preliminary results from an experiment for the 18 ka BP climate with the GISS AGCM.

MODELING

The evolution in time of a tracer is governed by source and removal processes, large scale transport and diffusion processes. Source and removal processes are highly dependent on the nature and type of particles and will differ for water isotopes and desert dust particles. Some modeling processes will depend on parameterizations used in the AGCM itself, like removal or diffusion processes. The AGCM modeling will not be described here, but can be found in Sadourny and Laval (1984) for the LMD AGCM, and in Hansen et al. (1983) for the GISS AGCM. More details on the water isotope and desert dust modeling in the LMD AGCM can be found in Joussaume (1983), Joussaume et al. (1984a,b) and Joussaume (1985). The isotopic simulations recently obtained using the GISS model were, partly, presented at the conference (Jouzel et al., 1986). A detailed description of the isotopic model and interpretation of the isotopic results for modern simulations will be published soon by Jouzel et al. (1987).

Isotopes

Isotopic differentiation occurs during the water atmospheric cycle because of the slight differences in the physical properties (vapour pressure and molecular diffusivities) between $H_2{}^{16}O$, the main component of water, and the isotopic species $H_2{}^{18}O$ and HDO.

The modeling of the water isotope cycles in AGCMs follows the basic water cycle: surface evaporation and condensation, water vapour transport, atmospheric condensation processes, subsequent evaporation of precipitation and ground hydrology, with differentiation at each phase change. In both models, water vapour fluxes at the surface are treated basically in the same way: a drag law is used over ocean and no fractionation assumed over other surface types. Advection is a critical process in isotope modeling because it is necessary to avoid the production of negative values for water vapour and to transport isotopic ratios. Different solutions have been found for both models and lead to satisfying results within the limits of the accuracy of the schemes used. For precipitation, isotope modeling follows the distinction introduced in AGCMs between convective and non-convective systems.

As a whole, the philosophy of isotope modeling is similar in both models, the main differences being due to the characteristic dynamics and physics of the two models.

Desert Dust

Dust is raised by winds in desert zone areas. The model uses parameterization depending on the surface wind speed. The flux formulation is defined in arbitrary units, assuming all processes are linear. Desert zone areas where dust is raised must also be defined. Two approaches are possible: either prescribe the source regions as known from observations, or generate source regions by the model itself as regions of dry soil. The second approach has to be used for paleoclimate simulations where source regions are unknown. The first approach is ideally the best one to compare with observations, but is somewhat limited for now as we do not know the source regions for fine dust well; however, we have used it for the experiment results presented here, with desert regions based on the Unesco map (1979). It is important to point out that the definition of source areas is indeed the most critical point in the modeling.

The main removal process for particles in the 1 μm range is the rainout process, in which the dust particles act mainly as condensation nuclei. The corresponding residence time as simulated by the model is about 10 days. Sedimentation velocity is weak. Transport and diffusion parameterization are very similar to water isotope species.

NUMERICAL EXPERIMENTS

Water Isotopes

The following simulations have been performed to date:

- January present day climate experiments using a low resolution version of the LMD AGCM (32 points in longitude, 24 points in latitude, 11 levels) fully described in Joussaume (1983) and Joussaume et al. (1984a). A July present day climate experiment with a higher resolution of the LMD AGCM (64x50x11) is now being analyzed.

- A 3-year simulation for present day climate and a 3-year simulation for 18 ka BP conditions using also a low resolution version of the Model II GISS AGCM (36 points in longitude, 24 points in latitude, 9 levels) to be described in Jouzel et al. (1987).

Here, we will briefly discuss January simulations for both models and compare them with available data from the IAEA (International Atomic Energy Agency) network. Preliminary results for the annual cycle and 18 ka BP experiments will then be presented.

The geographical charts of simulated mean oxygen 18 content of precipitation are given on Figures 1a and 1b along with mean January observations (Figure 1c). The general $\delta^{18}O$ decrease from the equatorial to the polar regions seen in data is well shown by both models with for example, the -8°/oo isocontours in quite comparable latitudinal bands for the three charts. The contrasts between oceanic and continental areas are also well marked; in particular there is a good representation of continental gradients. A noticeable difference between the two models is seen over Antarctica. Available data give January values between -40 and -50°/oo over the large part of central Antarctica in agreement with GISS results. Minimum values are significantly higher in the LMD simulation; this tendency might be due to an overestimation of simulated temperatures as confirmed when comparing the simulated and observed $\delta^{18}O$ relationships.

The relationships between the isotopic content of precipitation and surface air temperature are shown on Figures 2a to 2c along with the least square straight lines for temperatures above and below 15°C. The

Figure 1. Mean oxygen 18 content of precipitation for January a) LMD AGCM, b) GISS AGCM, c) observations (J.R. Lawrence in preparation). Values in ^{o}/oo versus the standard mean ocean water reference. Isolines every 2°/oo (models), and every 4°/oo, below -8°/oo (observations). Values above -2°/oo are dotted and values between -8 and -4 ^{o}/oo are shaded.

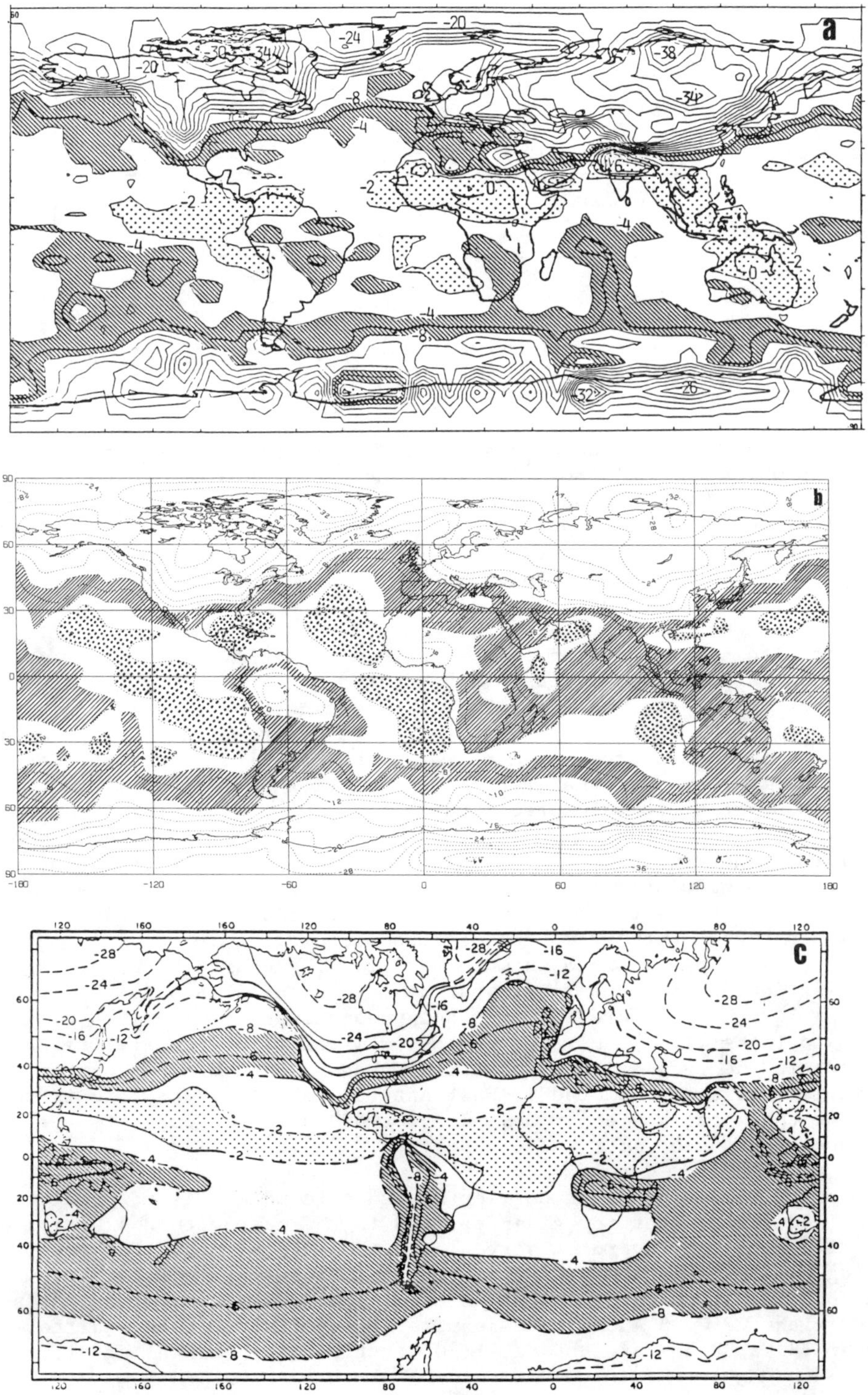
a
b
c

$\delta^{18}O$/Temperature T>15°C s slope in °/oo°C^{-1}	LMD GISS IAEA	s = 0.13 s = - 0.15 s = - 0.04	r = 0.22 r = - 0.21 r = - 0.05
$\delta^{18}O$:Temperature T<15°C s slope in °/oo°C^{-1}	LMD GISS IAEA	s = 0.60 s = 0.55 s = 0.59	r = 0.95 r = 0.90 r = 0.86
$\delta D/\delta^{18}O$ regression line r correlation coeff.	LMD GISS IAEA	$\delta D = 8.0\ \delta^{18}O + 8.3$ $\delta D = 8.1\ \delta^{18}O + 11.6$ $\delta D = 7.8\ \delta^{18}O + 8.5$	r = 0.99 r = 1.00 r = 0.98

Table 1. Regression lines for the statistical relationships $\delta^{18}O$-temperature and δD-$\delta^{18}O$ for January as simulated by the LMD and GISS AGCMs compared to IAEA data compiled over 20 years.

higher range (T > 15°C) encompasses all the tropical and equatorial sites for which there is no $\delta^{18}O$-dependency as previously noted by Dansgaard (1964) and clearly observed in Figure 2c. In these regions, the $\delta^{18}O$ distribution in precipitation is dominated by the amount of precipitation (the so-called amount effect - Dansgaard, 1964), which is well reproduced in both models. Below 15°C, the linear relation between $\delta^{18}O$ and temperature existing in the data is simulated, with an excellent agreement between observed and predicted $\delta^{18}O$ temperature gradients. There is also a very good agreement between the IAEA observations and model results for the relationship between deuterium and oxygen 18 ratios very closed to Craig's (1961) observations for annual means. The slopes and correlation coefficients of these relationships are given in Table 1. This convincingly demonstrates the capability of both AGCMs to realistically reproduce the main characteristics of the deuterium and oxygen 18 distributions in precipitation.

Turning now to the yearly time-scale, the δD and $\delta^{18}O$ seasonal cycles are well simulated in the 3-years GISS simulation (Jouzel et al., 1986, 1987). However, in some areas like Greenland this cycle is not reproduced although the temperature seasonal cycle exists. This deficiency needs to be examined in more detail. The chart of the average $\delta^{18}O$ distribution for the first year of the 3-year GISS simulation is shown on Figure 3a. Here again, model predictions compare well with IAEA data (IAEA, 1981), however the maximum values are underestimated by about 2% in the tropical and equatorial bands. At the other extreme, the minimum values predicted in East Antarctica (-54°/oo) are in the range of those registered in Dome C (= -50°/oo) and Vostok (= -57°/oo) ice cores (Lorius et al., 1979. 1985).

One of the purposes of developing isotopic models is to predict isotopic distributions for given periods in the past with the ultimate objective that this approach serves as a theoretical basis for interpreting paleodata in terms of climatic variations. As a first step, a 18 ka BP simulation using the CLIMAP (1981) boundary conditions was performed with the GISS AGCM. The results are on Figure 3b as a difference between 18 ka BP and present day $\delta^{18}O$ distributions (Jouzel et al.,

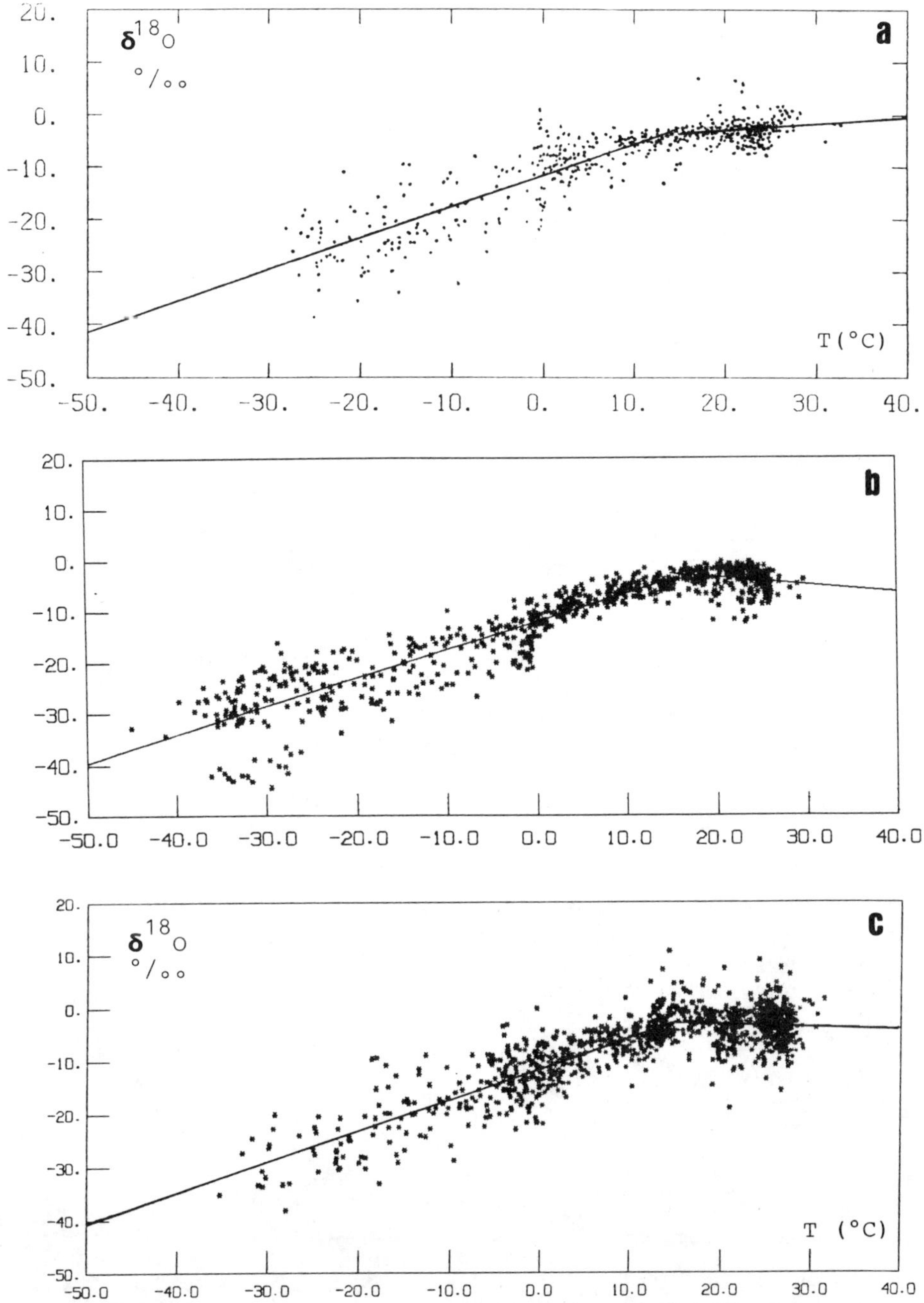

Figure 2. Mean oxygen 18 content of precipitation as a function of mean surface air temperature for January a) LMD AGCM, b) GISS AGCM, c) observations from the IAEA network. Continuous lines correspond to regression lines (Table 1).

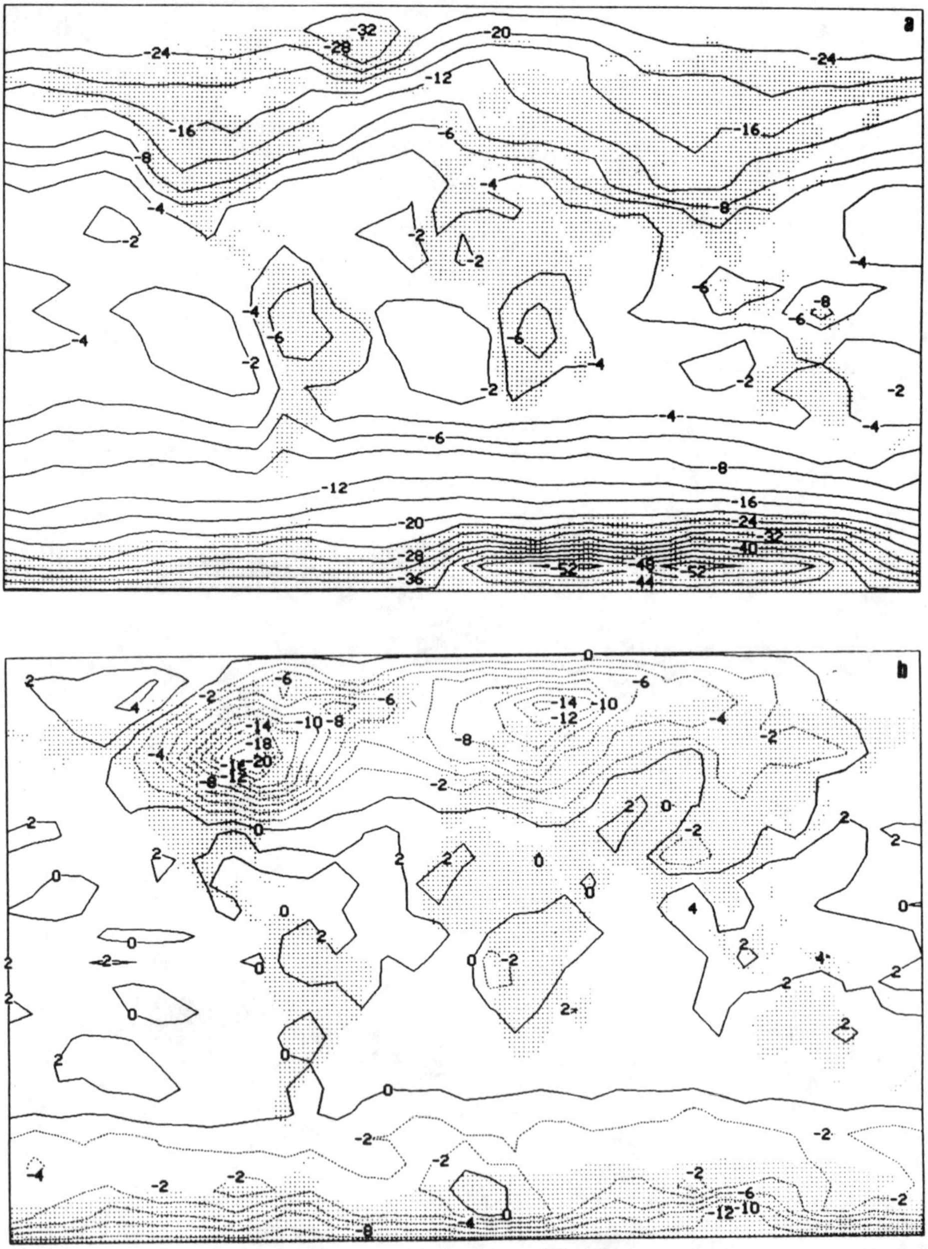

Figure 3. Distribution of the annual mean oxygen 18 content of precipitation simulated with the GISS AGCM a) modern climatic conditions, isolines every 2°/oo down to -8°/oo and every 4°/oo below, b) 18 ka BP - modern difference, isoline every 2°/oo.

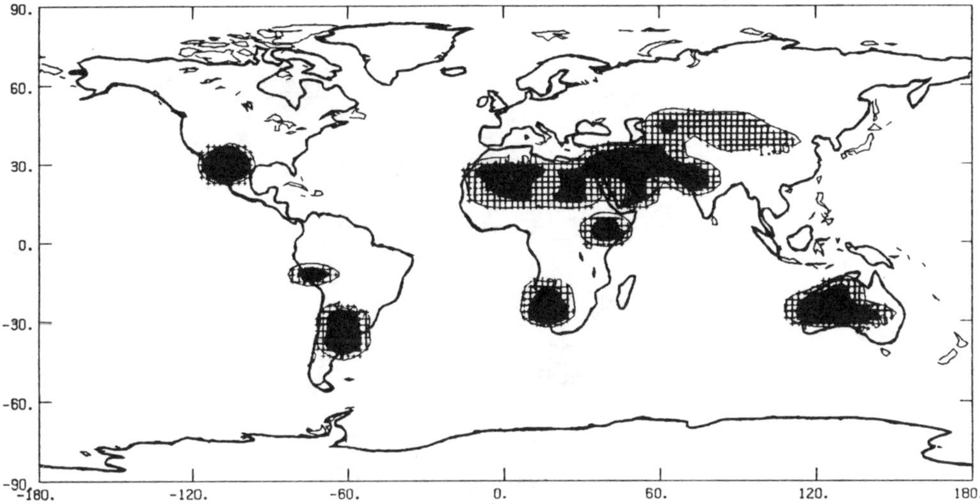

Figure 4. Mean simulated input flux of dust towards the atmosphere in arbitrary units - isolines 100., and 400. (maximum 1700.).

1986). This preliminary chart must be considered only as an illustration of a work in progress, however, some interesting features can be noticed: there is no clear signal in the tropical and equatorial regions where, as expected, the predicted $\delta^{18}O$ are lower for 18 ka BP than for present day conditions over polar regions.

Desert Dust

Simulations for January present day climate have been performed with the low resolution version of the LMD AGCM (32x24x11). Averages over the last 20 days of a 80 day run are presented here.

The prescribed extent of source regions, following the Unesco (1979) desert map, is displayed on Figure 4 which shows the simulated mean daily input flux of dust towards the atmosphere. The flux intensity, calculated by the model as a function of the surface wind speed, varies with time and space with a maximum for the Mexican desert. Globally, the Saharan and Arabian sources prevail over the other sources.

The atmospheric content of dust is displayed on Figure 5 with the vertically integrated mass of dust. Higher values are found over source regions, with however a space shift of maxima compared to the input flux, resulting from transport, convective diffusion and removal processes. In the tropics, dust from Sahara, Atacama and Namibia is partly transported towards the west by easterlies. This is well confirmed by observations for Saharan+ dust (Prospero et al., 1981).

At higher latitudes, transport is dominated by westerlies, especially in the roaring forties. Transport of Asian dust over the Pacific Ocean has effectively been observed (Duce, 1980). The extent of dust plumes is increased with height where stronger winds prevail, however

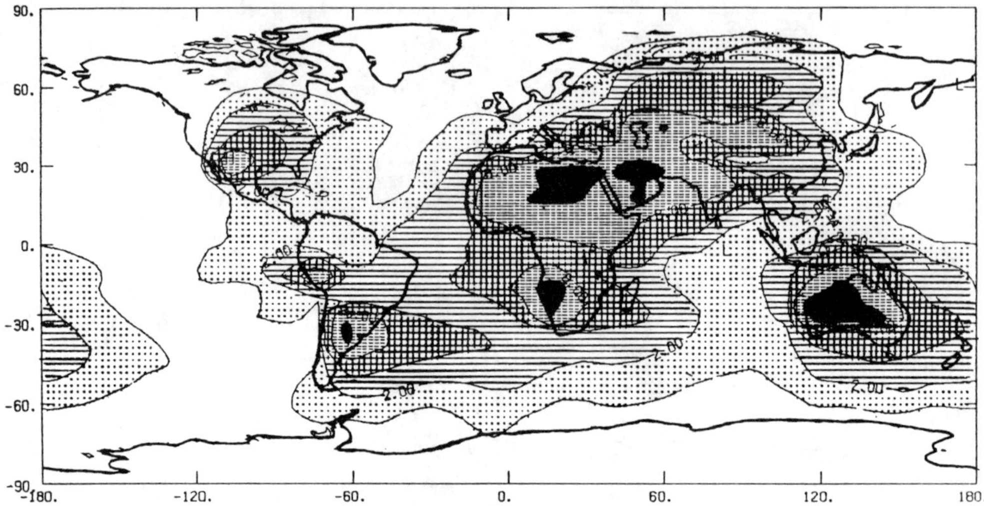

Figure 5. Mean vertically integrated mass of dust in arbitrary units. Isolines and increasing shading : 100., 200., 400., 800. and 1600. (maximum 2540.).

the concentration is strongly decreased by a factor 20 between 900 and 500mbar and a factor 1000 between 900 and 200mbar.

To obtain more information about dust transport, we have included a differentiation of dust according to its geographical origin. We have separated 6 source regions: Sahara, South Africa, Arabian and Thar deserts, Central Asia, Australia and America. We display in Figure 6 the vertically integrated mass of Saharan dust which represents 26% of the total amount of dust from all the sources. The transport over the Atlantic towards the Inter-Tropical Convergence Zone is well outlined and is in good in agreement with observations (Prospero et al., 1981); South America is reached in 10-13 days by the Saharan dust plume. Model results also display an important transport towards the east, north and south. The simulated mean percentage of the mass of dust on the East side over the source region is about 30% whereas the West component is about 10%. Compared to D'Almeida (1985), this eastward transport seems to be overestimated while the westward one is underestimated. This might be due to an underestimation of the surface easterlies, the Harmattan, simulated by the model.

The use of different dust types allows us to estimate the relative contribution from different continents to dust deposits. This type of information is of obvious interest over ice sheets for paleoclimatic studies, however, it has to be interpreted cautiously since it may be particularly sensitive to the definition of source regions or to the diffusion in the model. For Antarctica, South American dust dominates by a factor 3 to 4 in Central and Western Antarctica, whereas Australian dust prevails in Eastern Antarctica by a factor 4. In Greenland, Central American dust prevails largely, except on the western side where

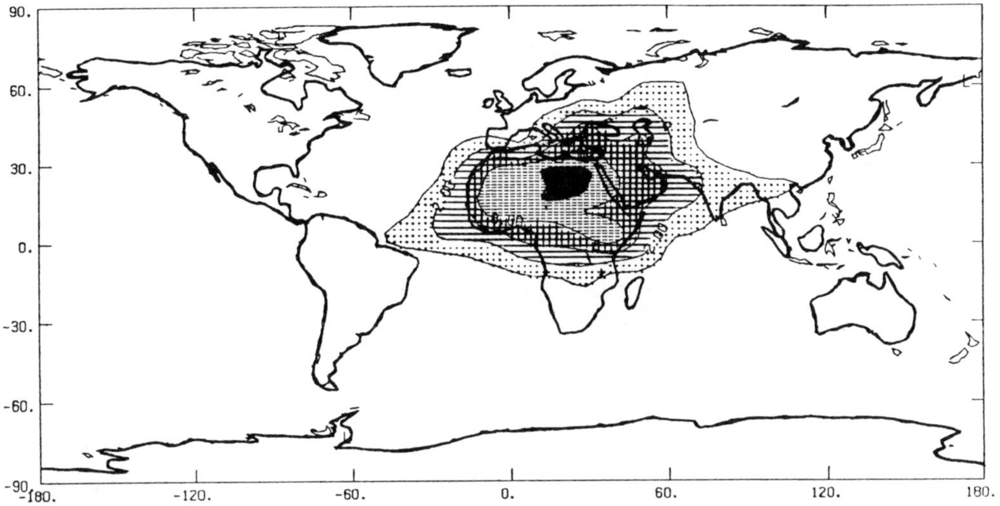

Figure 6. Mean vertically integrated mass of dust from Sahara. Same units, isolines and shading as Figure 5 (Maximum 1730.).

Saharan dust is equally represented. As a concluding remark, it is important to point out that even if our dust results are qualitatively reasonable, a more quantitative validation is difficult as few data are available. To improve the comparison with data, we are now performing forecasting experiments for special events, well monitored by satellite.

CONCLUSION

Although the new modeling approach presented in this article is still at an early stage of its development, the brief overview of available results given in this article, clearly shows the capability of AGCM's to reproduce desert dust and water isotope atmospheric cycles. This confidence is supported by comparison of model predictions with data but also by the comparison of distributions obtained from two AGCM's with very different characteristics for their dynamics, microphysics and hydrological cycle. A number of critical modeling problems such as numerical advection schemes and convective parameterizations need further improvements, sensitivity of the model results to different parameters has to be examined and further comparison between 18 ka BP and present day simulations to be performed, but there is no doubt that AGCM's are adequate and unique tools to treat the very complex problem of global scale atmospheric tracers. When fully developed, it can be reasonably hoped that this approach will contribute to a better interpretation of paleodata in terms of climatic changes.

ACKNOWLEDGEMENTS

This work has been realized with R. Sadourny (LMD), G. Russell (GISS), R. Suozzo (SIGMA-DATA), R. Koster (MIT), W.S. Broecker and J. White (Columbia University) and we particularly thank them for their essential contribution. Computer time for the LMD model was contributed by the Centre de Calcul Vectoriel pour la Recherche.

REFERENCES

CLIMAP Project Members, 1981. Seasonal Reconstructions of the Earth's Surface at the last Glacial Maximum, *Geological Society of America*, Map and Chart Series MC-36.

Craig H., 1961. Isotopic variations in meteoric waters, *Science*, 133: 1702-1703.

Dansgaard, W., 1964. Stable isotopes in precipitation, *Tellus*, 16: 436-468.

D'Almeida G., 1985. Report on the study of transport of Sahelian particulate matter using sulphotometer observations, WMO/TD n° 45.

Duce B.A., Unni, C.K. and Ray, B.J., 1980. Long-range atmospheric transport of soil dust from Asia to the Tropical North Pacific, *Science*, 209: 1522-1524.

Hansen J., Russell, G., Rind, D., Stone, P., Lacis, A., Lebedeff, S., Ruedy, P. and Travis, L., 1983. Efficient three-dimensional global models for climate studies: models I and II, *Monthly Weather Review*, 111 (4): 609-662.

IAEA, 1981. Statistical treatment of environmental isotope data in precipitation, *Technical report series 206*. IAEA Vienne.

Joussaume S., 1983. Modélisation des cycles des espèces isotopiques de l'eau et des aérosols d'origine désertique dans un modèle de circulation générale de l'atmosphère, Thèse de 3ème cycle, Université de Paris VI.

Joussaume S., Jouzel, J. and Sadourny, R., 1984a. A general circulation model of water isotope cycles in the atmosphere, *Nature*, 311: 24-29.

Joussaume S., Rasool, I., Sadourny, R. and Petit, J-R., 1984b. Simulation of desert dust cycles in an atmospheric general circulation model, *Annals of Glaciology*, 5: 204-207.

Joussaume S., 1985. Simulation of airborne impurity cycles using atmospheric general circulation models, *Annals of Glaciology*, 7: 131-137.

Jouzel J., Merlivat, L. and Lorius, C., 1982. Deuterium excess in an East Antarctic ice core suggests higher relative humidity at the oceanic surface during the last glacial maximum, *Nature*, 299: 688-691.

Jouzel J. and Merlivat, L., 1984. Deuterium and oxygen 18 in precipitation: modelling of the isotopic effect during snow formation, *J. Geophys. Res.*, 89: 11749-11758.

Jouzel J., Broecker, W.S., Koster, R., Russell, G., Suozzo, R and White, J., 1986. Modern and 18 ka BP simulations of the HDO and $H_2{}^{18}O$ atmospheric cycles using the NASA/GISS general circulation model: *Conference on abrupt climatic changes S10*, Ref. 86-8: 149-152.

Jouzel, J., Russell, G., Suozzo, R., Koster, R., Broecker, W.S. and White, J., 1987. Simulations of the HDO and $H_2{}^{18}O$ atmospheric cycles using the NASA/GISS general circulation model: the seasonal cycle for present day conditions, submitted to Journal of Geophysical Research.

Lorius C. and Merlivat, L., 1977. Distribution of mean surface stable isotope value in East Antarctica: Observed changes with depth in the coastal area - *IAHS-AISH* (International Association of Hydrological Sciences) 118: 127-137.

Lorius C., Merlivat, L., Jouzel, J. and Pourchet, M., 1979. A 30,000 yr isotope climatic record from Antarctic ice, *Nature*, 280: 644-648.

Lorius C., Jouzel, J., Ritz, C., Merlivat, L., Barkov, N.I., Korotkevich, Y.S. and Kotlyakov, V.M., 1985. A 150,000-year climatic record from Antarctic ice, *Nature*, 316: 591-596.

Petit J-R., Briat, M. and Royer, A., 1981. An ice age aerosol content from East Antarctic ice core samples in past wind strength, *Nature*, 293: 391-394.

Prospero J.M., Glaccum, R.A. and Nees, R.T., 1981. Atmospheric transport of soil dust from Africa to South America, *Nature*, 289: 570-572.

Sadourny R. and Laval, K., 1984. January and July performance of the LMD general circulation model. In Berger, A. and Nicolis, C. (eds), *New Perspectives in Climate Modelling, Developments in Atmospheric Science*, 16, Elsevier, 173-198.

Schneider, S.H., Peteet, D.M., and North, G.R., this volume. A climate model intercomparison for the Younger Dryas and its implications for paleoclimate data collection.

Thompson L.G., 1977. Microparticles, ice sheets and climate: *OSU Institute of Polar Studies*, Report n° 64, 148pp.

Unesco, 1979. Carte de la répartition mondiale des régions arides, *Technical Notes*, MAB n° 7.

CLIMATE SENSITIVITY AND PAST CLIMATES: EVIDENCE FROM NUMERICAL STUDIES

J.F.B. Mitchell

Meteorological Office
Bracknell
Berkshire RG12 2SZ
England

ABSTRACT. The use of three dimensional general circulation models to study past climates is considered. The importance of internal feedbacks (due to changes in water vapour, sea-ice and clouds) is reviewed, and the physical basis of the relevant parametrizations is assessed. At present, models can provide qualitative guidance on large-scale features of past climates for periods when boundary conditions can be specified accurately and are changing sufficiently slowly.

INTRODUCTION

The causes of climate change are very much a matter for speculation. There is evidence that variations in the earth's orbit may be responsible for oscillations between glacial and interglacial periods. Major volcanic eruptions such as that of Krakatoa in 1883 deposit large quantities of aerosol in the stratosphere where they may remain in sufficient quantities to reduce the solar radiation reaching the surface for several years. Oscillations may occur due to internal variations generated within the climate system. Anomalies in deep ocean temperature may persist for a century or more; variations in the major ice sheets will occur over much longer timescales.

One recurring problem in the search for explanations of past climate changes is that the mechanisms proposed do not provide a large enough forcing on their own to explain the observed changes in climate. Thus it seems necessary that large perturbations stem from amplification of the response within the climate system. In this paper, various mechanisms which amplify or damp the response of climate to imposed perturbations are reviewed and assessed, using material from numerical studies of CO_2 and climate. Clearly it is essential to establish that the sensitivity of a model to perturbations is correct if that model is to be used to assess the factors contributing to past climate changes. Recent work has stressed the contribution of feedbacks to globally averaged changes (for example, Dickinson, 1985). Here we review internal feedbacks due to changes in water vapour, sea-ice and cloud as deduced from numerical studies of the effect of CO_2, assess their physical significance (that is, are the model feedbacks realistic?) and consider some

W. H. Berger and L. D. Labeyrie (eds.), Abrupt Climatic Change, 383–397.

consequences of the existence of regional variations in climate change. For those not familiar with general circulation models, an account of the construction and use of general circulation models (GCMs) and their use is given by Mitchell (1983a). There have been several "snap-shot" studies of the last glacial maximum (Manabe and Hahn, 1977; Williams et al, 1974; Gates, 1976; Hansen et al, 1984) in which the effect of prescribing sea surface temperatures and continental ice sheets for 18,000 years before present was investigated. Manabe and Broccoli (1985) using a model with a simple representation of the ocean and sea-ice have assessed the impact of changing only the continental ice sheets from their present day configuration to that 18,000 years before present. Mitchell (1976, 1977) and more recently Kutzbach and his co-workers (for example, Kutzbach and Street-Perrott, 1985) have studied the effect of changes in the earth's orbital parameters on climate. This is of particular interest during the Holocene, and especially around 9000 years ago when the seasonal variation of insolation was much more pronounced than at present. Applications of numerical models to the Younger Dryas are discussed by Schneider, Peteet and North (this volume). In these studies some or all of the feedback mechanisms to be discussed play a role. In fact, the sensitivities found within these models is largely a consequence of the weighting of positive and negative feedbacks.

GLOBALLY AVERAGED EFFECTS

The relative contribution of individual feedback processes has traditionally been analysed using a simple globally averaged energy balance model (for example Hansen et al, 1984; Schlesinger and Mitchell, 1986; Dickinson, 1985). A variable, ΔT, the departure of globally averaged temperature from equilibrium, is given by

$$C\, d\Delta T/dt + \lambda \Delta T = \Delta Q \qquad (1)$$

where C is the heat capacity of the system, t is time, ΔQ is the imposed external perturbation and λ is the feedback parameter. At equilibrium, $\Delta T = \Delta Q/\lambda$; the smaller λ, the weaker the restoring term on the left hand side of the equation, and the greater the sensitivity of the model (see Appendix). We define the 'no feedback' case as that in which the unperturbed earth atmosphere system behaves like a black body with an effective emitting temperature of 255K and has a short wave reflectivity of 30%, corresponding to $= 3.7\ Wm^{-2}\ {}^\circ C^{-1}$.

We now consider the principal physical mechanisms known to modify λ, and estimate their magnitude from studies of the effect of increased atmospheric CO_2. Where necessary, we have assumed that doubling atmospheric CO_2 concentrations warms the troposphere/surface system by 4 Wm^{-2} (for example Lal and Ramanathan, 1984), which from the equation (1) gives an equilibrium warming of 1.1°C in the absence of feedbacks (Table 1).

TABLE 1.
Typical Values of Internal Feedbacks

Feedback	Contribution to feedback factor (Wm^{-2} $°C^{-1}$)	Feedback factor λ (Cumulative, Wm^{-2} $°C^{-1}$)	Change in surface temperature, °C, on doubling ($\Delta Q = 4$ Wm^{-2})
None		3.7	1.1
Water Vapour	-1.5	2.2	1.7
Sea ice/snow	-0.4	1.8	2.2
Cloud cover	-0.9	0.9	4.4

Water Vapour

Increasing temperature by 1K from 288K increases the saturation vapour pressure of water by about 7%. Thus, it is to be expected that the atmospheric water vapour content will increase with temperature. Manabe and Wetherald (1967), on the basis of the observed seasonal variation of atmospheric humidity, concluded that relative humidity, (the percentage of saturation) was *approximately* constant. This conclusion is supported by numerical studies (as shown in Fig. 1, from Mitchell et al., 1987). This being the case, an increase in atmospheric temperature will be accompanied by an enhancement of atmospheric water vapour, increasing the opacity of the atmosphere to long wave radiation and thus enhancing the warming of the atmosphere and surface.

Estimates of the change in λ due to water vapour feedback based on published work are in fair agreement whether from radiative convective models assuming fixed relative humidity (-1.8 Wm^{-2} $°C^{-1}$, Manabe and Wetherald, 1967; -1.4 Wm^{-2} $°C^{-1}$, Hansen et al., 1984) or from three dimensional models which generate their own humidity structure (-1.6 Wm^{-2} $°C^{-1}$, Mitchell, 1983b; -1.5 Wm^{-2} $°C^{-1}$, Hansen et al., 1984). All the figures quoted are for conditions of average cloudiness and the results from general circulation models take into account the geographical and seasonal variations in atmospheric lapse rate.

Snow and Sea Ice

An increase in atmospheric temperature will tend to reduce highly reflective snow and sea-ice cover, leading to increased absorption of insolation and further warming. Clearly, the strength of this feedback will depend on the areal cover of snow and sea-ice, and hence the temperatures of the unperturbed simulation. In general, the colder the control simulation, the greater the sensitivity of the model, as illustrated in Fig. 2 (from Spelman and Manabe, 1984). Removal of sea-ice

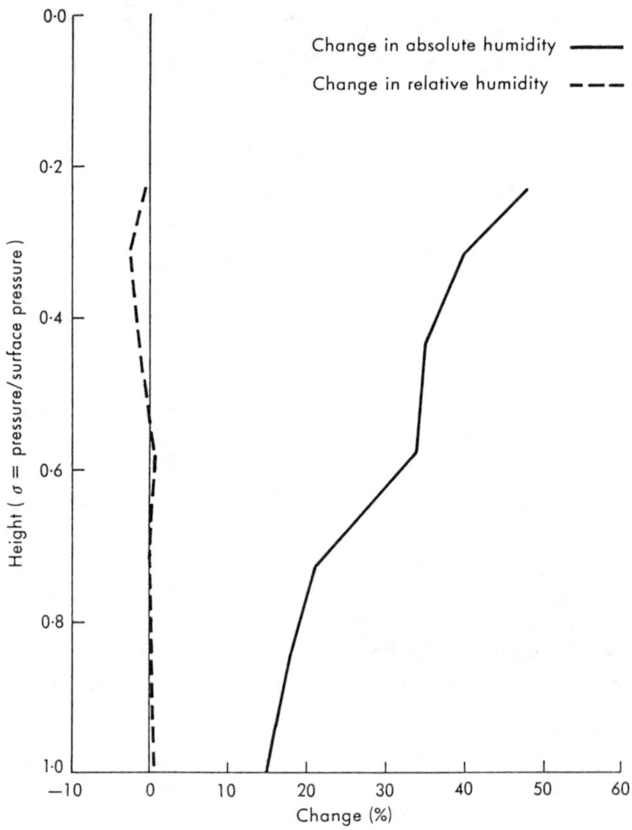

Figure 1. Globally averaged annual mean changes in specific humidity (solid line, percentage of unperturbed value) and relative humidity (dashed line, percentage of saturation) simulated in a GCM experiment in which CO_2 was doubled and sea surface temperatures were increased by 2°C (see Mitchell et al., 1987). The vertical coordinate is the ratio of in situ pressure over surface pressure, denoted by σ, $\sigma = 0.5$ corresponds to about 5500m on average.

greatly increases the thermal inertia of the surface and reduces the amplitude of the annual cycle - this will enhance a warming in winter and reduce it in summer. Robock (1983), using observational data, estimated the potential effects of changes in snow and ice area, in albedo (due to surface melting) and, in the case of sea-ice, in thermal inertia on sensitivity. He deduced a net contribution to λ of -0.6 Wm^{-2} $°C^{-1}$. This is somewhat higher than the -0.3 and -0.4 Wm^{-2} $°C^{-1}$ found by Hansen et al. (1984) using two versions of a three dimensional climate model which for present day conditions produced respectively 15% less and 23% more ice cover than observed.

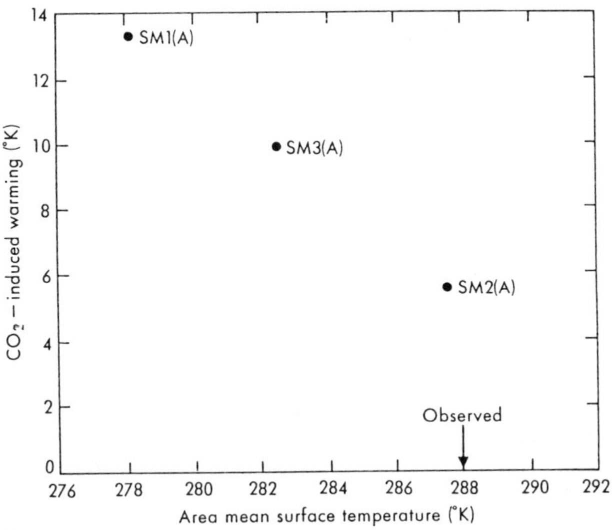

Figure 2. Relationship between an idealized 3-dimensional climate model's sensitivity to a quadrupling of CO_2 and the simulated control temperature (from Spelman and Manabe, 1984).

It must be recognized that in all these studies, simple parametrizations of the thermodynamics of sea-ice had been used, and in almost all cases, the advection of sea-ice, and possible changes in ocean salinity and the advection of heat by the ocean have all been neglected. Thus, the above estimates of the strength of sea-ice feedback must be regarded as tentative. Perhaps the most important points to emerge are that such a feedback will be stronger during glacial rather than interglacial periods, as snow and ice cover are more extensive, and that the associated temperature changes are likely to be greater in winter than in summer.

Cloud Cover Feedback

Clouds reflect solar radiation and absorb and emit infra-red radiation. An increase in cloud cover enhances both the reflection of solar radiation, producing a surface cooling, and the emission of long wave radiation, producing a warming. Current evidence suggests that the former effect is dominant at most latitudes where insolation is significant. The long wave effects may be more important for high, thin cirrus. In addition, an increase in cloud height will also tend to warm the surface, as radiation from the cloud top is emitted to space at a lower temperature.

Simulations with increased CO_2 using three dimensional climate models show a consistent decrease in middle and upper tropospheric cloud at most latitudes, with increases in cloud in the lower stratosphere, particularly in middle and high latitudes (Schlesinger and Mitchell, 1986).

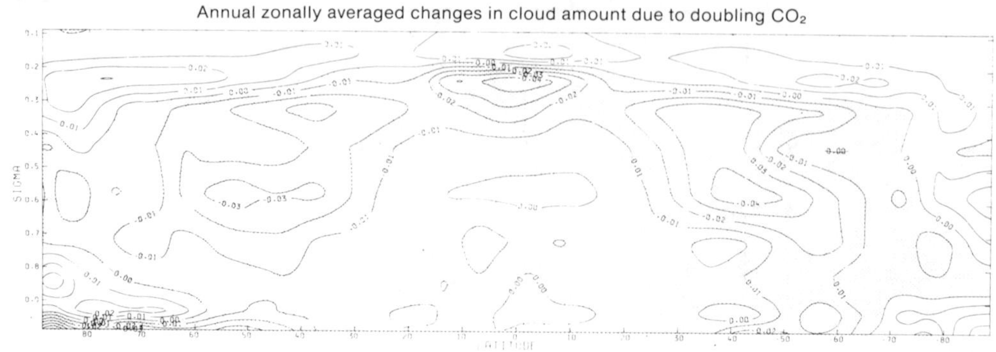

Figure 3. Zonally averaged changes in cloud (annual mean) due to doubling atmospheric CO_2 (see also Fig. 4). Contours every 2% of total cover, areas of decrease are shaded. The vertical co-ordinate is the σ (=pressure/surface pressure).

Similar changes have been found in an experiment (Mitchell et al., 1986) using a version of the Meteorological Office 11-layer model in which CO_2 amounts were doubled (Fig. 3). Estimates of contribution of the cloud changes include -0.9 Wm^{-2} $°C^{-1}$, (from the ex periments of Mitchell et al., 1986) and -1.1 to -0.8 Wm^{-2} $°C^{-1}$ found in the experiments of Hansen et al. (1984). (In the latter experiments the cloud optical thickness was prescribed to decrease with height; since cloud height increased there was an additional positive feedback through the change in optical properties).

It is important that the mechanisms producing the changes in cloudiness in the model simulations are fully understood. The vertical profiles of globally averaged changes in radiative heating, frequency distribution of the level at which moist convection terminates (convective cloud top) and relative humidity from the Meteorological Office experiment are shown in Figs. 3 and 4. On introducing the globally averaged changes in CO_2, water vapour, temperature and cloud in a one-dimensional version of the radiation scheme used in the experiment, it is found that the large cooling 'spike' near 230mbs is due mainly to the increase in water vapour, and a similar though slightly less pronounced cooling is found at the same level in an earlier experiment with prescribed cloud. The simulated changes in radiative heating rates make the atmosphere more unstable below 230mbs, reducing the frequency with which moist air from convective parcels mixes with environmental air below that level and increasing the frequency of mixing above. Thus relative humidity decreases through most of the troposphere below 230mbs, and increases above, with associated changes in cloudiness. These changes are consistent with an increase in the height of the tropopause. The observed tropopause is highest in low latitudes and in summer when insolation is strongest; thus there is at least circumstantial observational evidence that the models' response to tropospheric warming is physically correct.

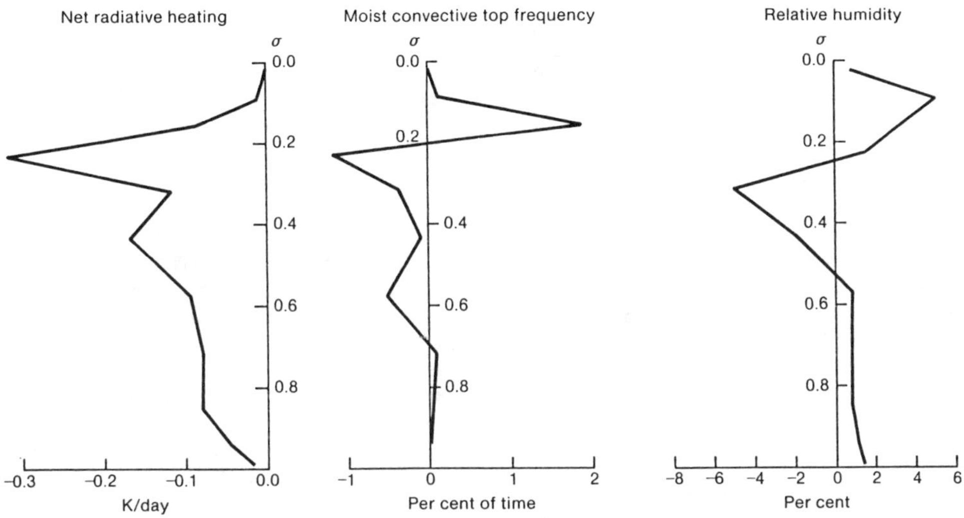

Figure 4. Vertical distribution of globally averaged changes due to doubling atmospheric CO_2 concentrations in a version of the Meteorological Office 11-layer model with interactive cloud coupled to an oceanic mixed layer (Mitchell et al., 1986). Results are averaged over three summers (June to August) from a seasonal integration. The vertical coordinate is σ (= pressure/surface pressure).

a. Changes in net radiative heating of the atmosphere (degrees Kelvin/day).

b. Changes in frequency distribution of level of termination of moist convection (cumulus top) (percentage of total time).

c. Changes in relative humidity (percentage of saturation).

The details of the above changes are obviously sensitive to the parametrization of radiation, particularly with respect to water vapour, and the vertical transport of water. The vertical resolution of most models is poor near the tropopause (generally 3 layers above 250mb and 4 in the case of the Meteorological Office model) and the vertical extent of cloud is usually limited; for example, in the version of the Meteorological Office model used here it is confined below 200mb. The parametrization of cloud has to date been based on relative humidity, although attempts have been made to devise a parameterization of cloud liquid water (Charlock and Ramanathan, 1985). This approach has the advantage that cloud optical properties can be related in a consistent manner to the liquid water content (see for examples, Stephens, 1978), but such liquid water content parametrizations are fairly arbitrary as yet.

Cloud Optical Properties

The results considered up to this point have assumed that cloud optical properties are invariant with climate. A general increase in atmospheric specific humidity might be expected to lead to an increase in cloud liquid water content and hence the reflectivity of cloud. Stephens (1978) has derived simple parametrizations of the relation between cloud optical properties and cloud liquid water content, based on observational data. He found a gradual increase in cloud reflectivity (albedo) with cloud liquid water content, although the long wave emissivity of thin cloud (for example, cirrus) may also increase significantly. An increase in albedo would cool the surface; an increase in the long wave emissivity of high cloud would have the opposite effect. It is by no means clear how one can relate cloud liquid water content to large scale model variables (for example temperature, humidity, etc.). Nevertheless, Somerville and Remer (1984) have attempted to relate cloud liquid water content to temperature, and find that the resulting negative feedback is sufficient to counteract all the positive feedbacks in Table 1. In view of the uncertainties in their parametrization, the magnitude of cloud optical feedback is extremely uncertain, and is a priority for future research.

REGIONAL CHANGES

Since good palaeoclimatic data are generally available over only limited regions, information concerning global mean sensitivity is not particularly useful for testing against the record. Here, we consider three factors which must be borne in mind when comparing model and palaeoclimatic data.

Scale

Current general circulation models use a horizontal grid of 250 to 800km. In practice, both because of the numerical smoothing necessary in all models to ensure computational stability, and the year-to-year variability inherent in non-linear models (and the real world), it is advisable to consider regions rather than individual grid points. Even then, the regional response of models to a given perturbation is likely to vary considerably from model to model. For example, one can conclude from numerical studies that increased atmospheric CO_2 may produce a drying of the mid-latitude continents in summer (Schlesinger and Mitchell, 1986). Not all models produce such a drying, and amongst those that do, the exact latitudes at which the drying occurs varies. Consequently, it would be unwise to compare observational data from a small region with results from a single numerical study.

Regional Anomalies

Feedbacks act locally as well as globally. Thus, in the experiment using the Meteorological Office model described above, changes in soil moisture and in cloud cover reinforce one another as can be seen over the northern mid-latitude continents in summer (Fig. 5). The reduction in soil moisture reduces evaporation and hence the moisture necessary to form cloud; the reduction in cloud increases the energy available for evaporation and enhances the surface drying prevalent at that time of year. Manabe and Wetherald (1986) also find that the simulated drying

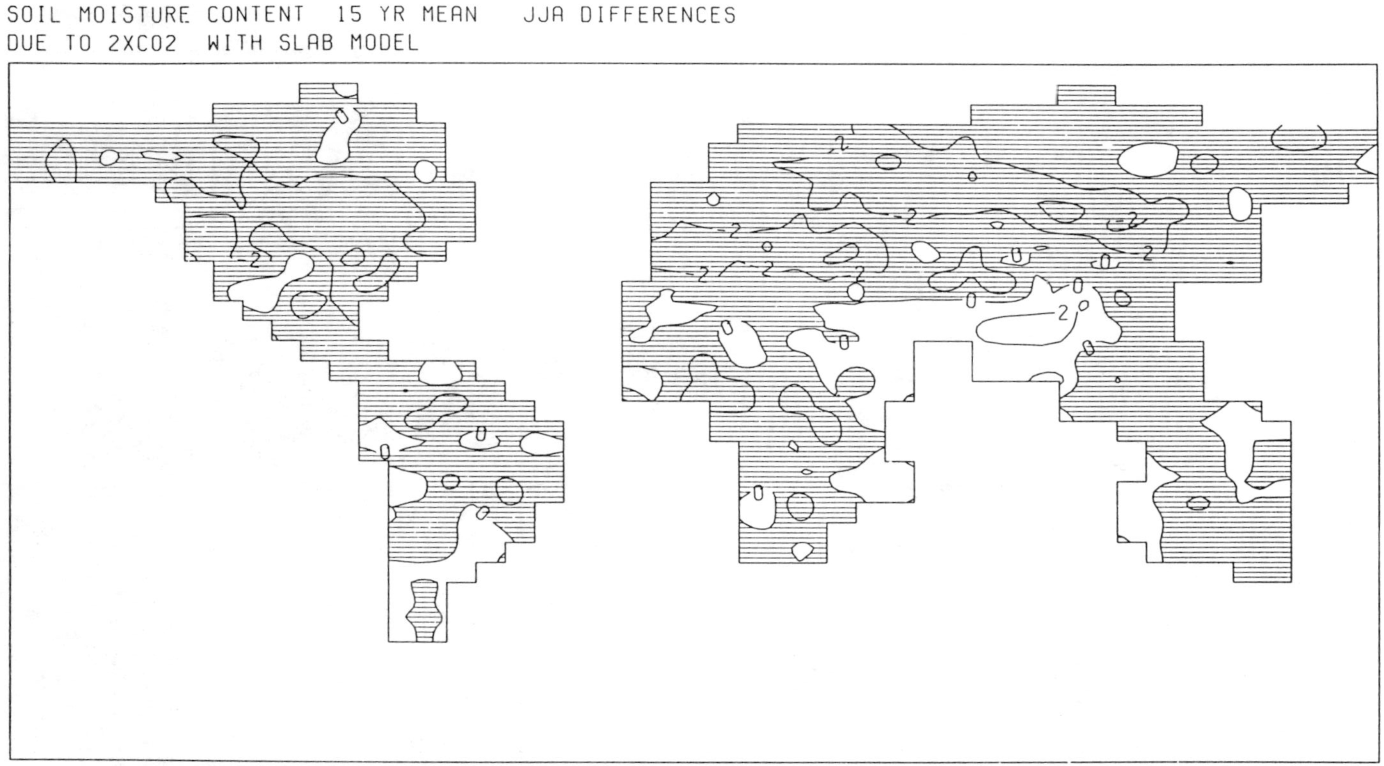

Figure 5. Regional changes in summer (June to August) due to doubling atmospheric CO_2 in a version of the Meteorological Office 11-layer model (see also Fig. 3).

a. Soil moisture (depth of water accumulated in soil). Contours every 2cm, regions of decrease are shaded.

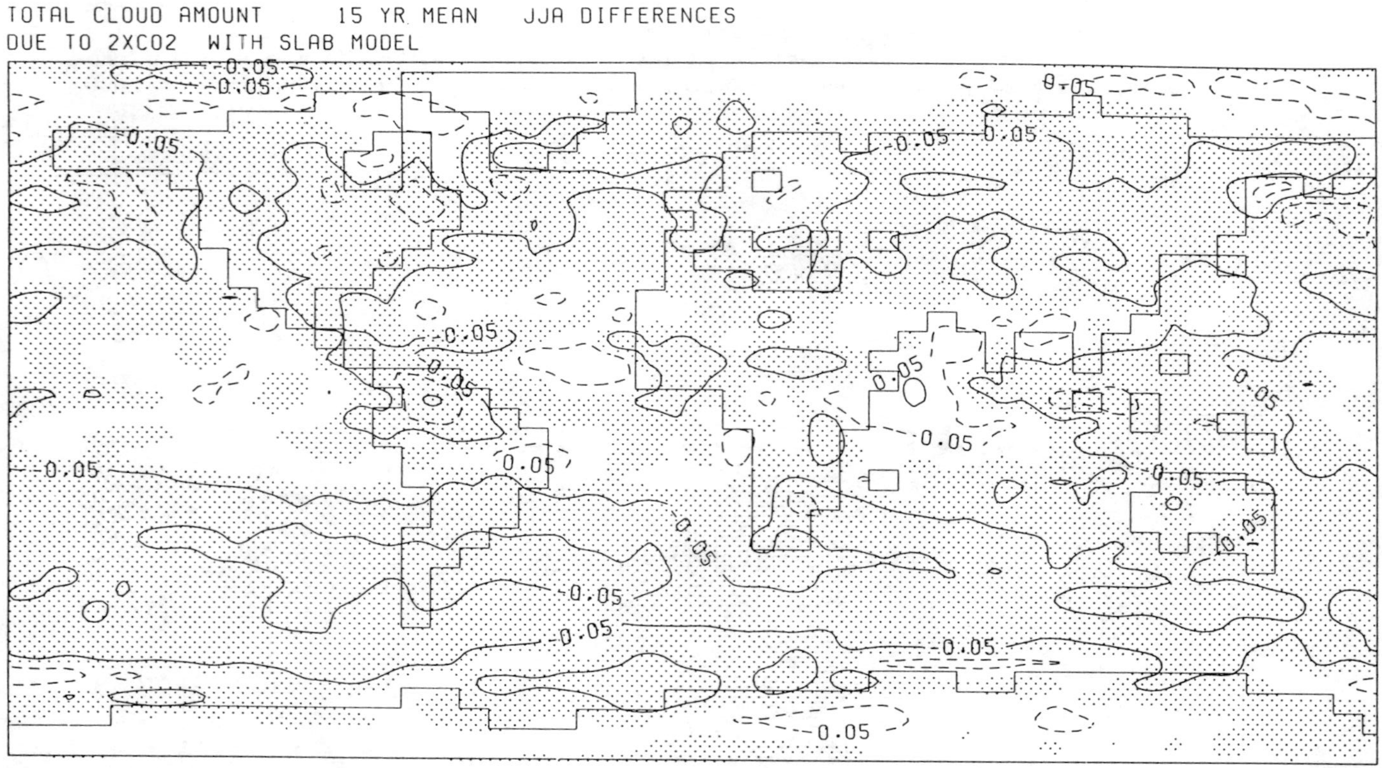

Figure 5 b. Total cloud. Contours every 5% of total cover, areas of decrease are shaded.

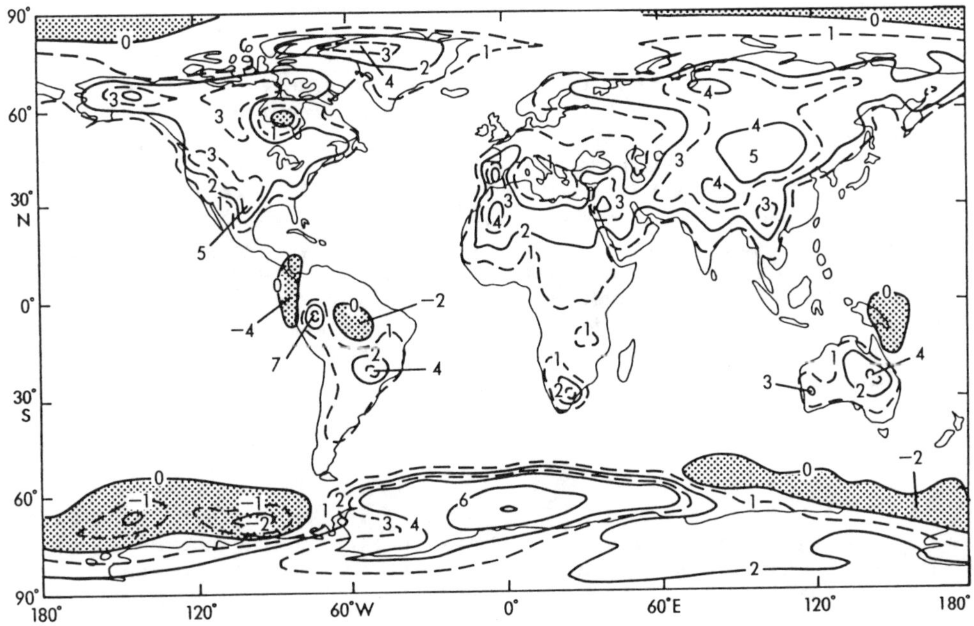

Figure 6. Changes in continental surface temperature due to increasing atmospheric CO_2 ten-fold, holding sea surface temperatures unchanged (June to August). Contours every 2°C, areas of decrease are shaded. Additional contours at -1, 1, 3.

of the northern mid-latitude continents in summer due to increased CO_2 is enhanced by cloud feedback.

Transient Effects

Because of its smaller thermal inertia, the continental surface will respond more quickly than the ocean to changes in external forcing. As an extreme example, consider the response of a model in which sea surface temperatures are prescribed, that is, the ocean has an infinite heat capacity. On increasing CO_2 by a factor of 10, temperature increases of up to 5°C occur in the center of continents (Fig. 6), demonstrating that large changes in land-sea temperature contrast can occur if the ocean is not in thermal equilibrium. Thus it cannot necessarily be assumed that the equilibrium and transient response of climate to changes in boundary conditions will be the same. The importance of the transient response of climate to increased CO_2 has been considered further by Schneider and Thompson (1981) and Bryan et al, (1982).

CONCLUDING REMARKS

General circulation models can be used to assess the influence of specified boundary conditions on equilibrium climate. Thus, they are a powerful tool for studying periods when the relevant boundary conditions are well known and stationary or, at most, changing on a time scale larger than the response time of the model. They may provide both insight concerning the physical mechanisms leading to climate change and guidance on past climate in regions where palaeo-climatic data are poor or sparse. Even so, it must be recognized that there is some uncertainty in the sensitivity of models to a given perturbation (i.e. doubling atmospheric CO_2 concentrations) so results are qualitative rather than quantitative. The parameterization of sea-ice and especially cloud must be more physically based if these uncertainties are to be reduced. Given good widespread palaeo-data from a period in which the boundary conditions are well established, one could validate the sensitivity of climate models directly from observational data. However, until there is both a significant improvement in the quality and coverage of past data and a better agreement in the regional changes produced by models in response to the same perturbation, one cannot have much confidence in such validation.

ACKNOWLEDGEMENTS

I would like to thank C.A. Wilson, W.J. Ingram and W.M. Cunnington for their help in the studies of changes in cloud using the Meteorological Office 11-layer model, and P.R. Rowntree for comments on the manuscript.

APPENDIX - STABILITY, SENSITIVITY AND FEEDBACKS

In a recent paper, Hansen et al. (1984) made an analysis of feedbacks in a three dimensional climate model, based on certain simplifying assumptions and procedures derived from studies of feedbacks. In this note, we derive some of the results used by Hansen et al. using elementary calculus, and clarify the underlying assumptions.

Suppose the rate of change with time dT/dt of a variable T in the climate system (for example, temperature) can be expressed in the form

$$dT/dt = F(T, \{X_i\}, \{Y_j(T)\}) \qquad (1)$$

where F (for example, heat flux per unit heat capacity of the system) can be expressed as a function of T, parameters X_i assumed to be independent of T (for example, the solar constant, atmospheric CO_2 concentration) and parameters Y_j which are known to be dependent on T (for example, planetary albedo, atmospheric emissivity). Equilibrium values of (1) are given by dT/dt = 0, so (1) can be expanded (to first order) about an equilibrium value T_e by

$$dT/dt = - \lambda \; (T - T_e) \qquad (2)$$

where $\lambda = dF/dT \mid T{=}T_e$. If $\lambda > 0$ perturbations on T decay exponentially to zero with e-folding time $1/\lambda$, but if $\lambda < 0$, they grow exponentially and the equilibrium is unstable (see for example, Fraedrich 1978, 1979).

From (1)

$$\lambda = -\left(\frac{\partial F}{\partial T} + \Sigma_J \frac{\partial F}{\partial Y_j} \frac{dY}{dT_j}\right) \qquad (3)$$

assuming that the Y_j are independent and ∂ denotes a partial derivia-tive.

Consider small changes ΔT_e in equilibrium T_e due to perturbations ΔX_o in X_o. Since $F\ (T_e,\ \{X_i\},\ \{Y_j\ (T_e)\}\ ) = 0$

$$\frac{dF}{dT} \Delta T_e + \frac{\partial F}{\partial X_o} \Delta X_o = 0 \qquad (4)$$

where dF/dT, $\partial F/\partial X_o$ are evaluated at $T=T_e$, whence the sensitivity of T_e changes in X_o is given by

$$\frac{dT_e}{dX_o} = -\left(\frac{dF}{dT}\right)^{-1} \frac{\partial F}{\partial X_o} = 1/\lambda \quad \frac{\partial F}{\partial X_o} \qquad (5)$$

which, with (3) may be rearranged to give

$$\frac{dT_e}{dX_o} = -\frac{\partial F/\partial X_o}{\partial F/\partial T} \quad \frac{1}{1+\Sigma \frac{\partial F}{\partial Y_j} \frac{dY_j}{dT} \left(\frac{\partial F}{\partial T}\right)^{-1}} \qquad (6)$$

In Hansen et al's paper, T is the effective radiating temperature of the earth atmosphere system, and F is the net heat flux "into the planet", $\partial F/\partial X_o$ is the rate of change in F due to changing an independent parameter (in their example, the solar constant), $\partial F/\partial T$ is the rate of change in net heating of the planet due to changes in long wave cooling to space ($\partial F/\partial T < 0$), and $-\frac{\partial F/\partial X_o}{\partial F/\partial T}$ gives sensitivity of T_e to changes in X_o in the absence of changes (feedbacks) in the $Y_j(T)$. The amplification due to feedbacks is

$$\frac{1}{1 + \Sigma_j \frac{\partial F}{\partial Y_j} \frac{dY_j}{dT} \left(\frac{\partial F}{\partial T}\right)^{-1}} = \frac{1}{1-g} \qquad (7)$$

where $g = \Sigma_J\, g_j$, with $g_j = -\frac{\partial F}{\partial Y_j} \frac{dYj}{dT} \left(\frac{\partial F}{\partial T}\right)^{-1}$

is the "gain" defined in Hansen et al. The expression (6) is valid provided F can be linearized about T_e; and the feedbacks implicit in the $Y_j(T)$ are independent. Further estimates of λ and feedback factors are given by Watts (1980) and Dickinson (1985).

REFERENCES

Bryan, K., Komro, F.G., Manabe, S. and Spelman, M.J., 1982. Transient climate response to increasing atmospheric carbon dioxide, *Science*, 215: 56-58.

Charlock, T.P. and Ramanathan, V., 1985. The albedo field and cloud radiative forcing produced by a General Circulation Model with internally generated cloud optics, *J. Atmos. Sci.*, 43: 1408-1429.

Dickinson, R.E., 1985. Climate Sensitivity. In: *Advances in Geophysics*, Vol. 28, (ed. S. Manabe), Academic Press, pp. 251.

Fraedrich, K., 1978. Structural and stochastic analysis of a zero-dimensional climate system, *Quart. J. R. Met. Soc.* 104: 461-474.

Fraedrich, K., 1979. Catastrophes and resilience of zero dimensional climate system with the albedo and greenhouse feedbacks, *Quart. J. R. Met. Soc.*, 105: 147-167.

Gates, W.L., 1976. Modelling the Ice-Age Climate, *Science*, 191: 1138-1144.

Hansen, J., Lacis, A., Rind, D., Russell, L., Stone, P., Fung, I., Ruedy, R. and Lerner, J., 1984. Climate sensitivity and analysis of feedback mechanisms. In: *Climate Sensitivity and Analysis of Feedback Mechanisms*, J. Hansen and T. Takahashi (eds.), Geophysical Monograph 29, 130-163, American Geophysical Union, Washington DC.

Kutzbach, J.E. and Street-Perrott, F.A., 1985. Milankovitch forcing of fluctuations in the level of tropical lakes from 18 to 0 kyr BP, *Nature*, 317: 130-134.

Lal, M. and Ramanathan, V. 1984. The effects of moist convection and water vapour radiative processes on climate sensitivity, *J. Atmos. Sci.*, 31: 118-133.

Manabe, S. and Broccoli, A.J. 1985. The influence of Continental Ice Sheets on the Climate and Ice Age, *J. Geophys. Res.*, 90: 2167-2190.

Manabe, S. and Hahn, D.G., 1977. Simulation of the tropical climate of an Ice Age, *J. Geophys. Res.*, 82: 3889-3911.

Manabe, S. and Wetherald, R.T., 1967. Thermal equilibrium of the atmosphere with a given distribution of relative humidity, *J. Atmos. Sci.*, 24: 241-259.

Manabe, S. and Wetherald, R.T., 1986. Reduction in summer soil wetness induced by an increase in atmospheric carbon dioxide, *Science*, 232: 626-628.

Mitchell, J.F.B., 1976. The effect on climate of changing the parameters of the earth's orbit: two summer integrations incorporating a fully interactive surface, *Met 0 20 Tech. Note II/72*, Meteorological Office, Bracknell, Berkshire RG12 2SZ..

Mitchell, J.F.B., 1977. The effect on climate of changing the earth's orbital parameters: two summer integrations with fixed sea surface temperatures, *Met 0 20 Tech. Note II/100*, Meteorological Office, Bracknell, Berkshire RG12 2SZ..

Mitchell, J.F.B., 1983a. The physical basis of climate modelling. In: *Palaeoclimate Research and Research Models*, A. Ghazi, (ed.), D. Reidel Publishing Company, Dordecht/Boston/Lancaster, 34-51.

Mitchell, J.F.B., 1983b. The seasonal response of a general circulation model to changes in CO_2 and sea surface temperature, *Quart. J.R. Met. Soc.*, 109: 113-152.

Mitchell, J.F.B., Wilson, C.A. and Cunnington, W.M., 1987. On CO_2 climate sensitivity and model dependence of results, *Quart. J.R. Met. Soc.*, 113: 293-322.

Mitchell, J.F.B., Wilson, C.A., Ingram, W.J. and Cunnington, W.M., 1986. Numerical studies of the effect of increased atmospheric carbon dioxide on climate, *Final report of CEC Contract CL-114-UK(H)*.

Robock, A., 1983. Ice and snow feedbacks and the latitudinal and seasonal distribution of climate sensitivity, *J. Atmos. Sci.*, 40: 986-997.

Schlesinger, M.E. and Mitchell, J.F.B., 1986. Model projections of equilibrium climatic response to increased CO_2 concentration. In: *Projecting the Climatic Effect of Increasing Carbon Dioxide*, US Dept of Energy report DOE/ER-0237, MacCracken and Luther, (eds.), Climatic US Dept. of Energy, Washington DC.

Schneider, S.H. and Thompson, S.L., 1981. Atmospheric CO_2 and climate: Importance of the transient response, *J. Geophys. Res.* 86: 3135-3147.

Somerville, R.C.J. and Remer, L.A., 1984. Cloud optical thickness feedbacks in the CO_2 climate problem, *J. Geophys. Res.* 89: 9668-9672.

Spelman, M.J. and Manabe, S., 1984. Influence of cceanic heat transport upon the sensitivity of a model climate, *J. Geophys. Res.*, 89: 571-586.

Stephens, G.L., 1978. Radiation profiles in extended water clouds II: Parametrization schemes, *J. Atmos. Sci.*, 35: 2123-2132.

US National Academy of Sciences, 1983. *Changing Climate*, Carbon Dioxide Assessment Committee, National Academy Press, Washington DC, 496 pp.

Watts, R.G., 1980. Climate models and CO_2 induced climate changes, *Climate Change*, 2: 387-408.

Williams, J., Barry, R.G. and Washington, W.M., 1974. Simulation of the atmospheric circulation using the NCAR Global Circulation Model with Ice Age boundary conditions, *J. Appl. Meteorology*, 13: 305-317.

A CLIMATE MODEL INTERCOMPARISON FOR THE YOUNGER DRYAS AND ITS IMPLICATIONS FOR PALEOCLIMATIC DATA COLLECTION

Stephen H. Schneider[1], Dorothy M. Peteet[2] and Gerald R. North[3]*

[1]National Center for Atmospheric Research**
P.O. Box 3000
Boulder, Colorado 80307

[2]NASA/Goddard Space Flight Center
Goddard Institute for Space Studies
2880 Broadway
New York, New York 10025

[3]Department of Meteorology
Texas A & M University
College Station, Texas 77843

ABSTRACT. It has been suggested by a number of paleoclimatologists that formation of a low salinity meltwater layer in the North Atlantic ocean during the period(s) of most rapid volumetric glacial retreat following the Wisconsin glacial maximum could have led to an extension of North Atlantic sea ice sufficient to cause a temporary, regional (at least) reversal of the longer term deglacial warming trend. In particular, it is speculated that such an extension of North Atlantic sea ice may have been responsible for the Younger Dryas (10,800-10,000 yr. BP) cold interval, evident in European proxy climatic records. Others have suggested (in the context of ice age initiation) that one feedback mechanism whereby equatorward extension of North Atlantic sea ice could amplify the regional high latitude temperature response is through a reduction in the poleward transport of heat by ocean currents as sea ice extends equatorward. At the same time, there is evidence that the Younger Dryas interval was not merely a European event, but may have been worldwide in scope. Some of the above-mentioned hypotheses and feedback mechanisms are quantitatively investigated using several energy balance climate models as well as general circulation model experiments in which (NCAR model) all oceanic grid points in the North Atlantic Ocean north of 45° latitude will be considered ice covered and sea ice;

*The author's contribution was conducted while an employee of Goddard Space Flight Center, NASA, Greenbelt, Maryland.
**The National Center for Atmospheric Research is sponsored by the National Science Foundation.

W. H. Berger and L. D. Labeyrie (eds.), Abrupt Climatic Change, 399–417.

or land and sea ice as well as orbital elements are imposed appropriate to 10,500 yr. BP (GISS GCM). All of these results are intercompared. All sets of results show that there are substantial seasonal and regional differences in the climatic response for these particular forcings. Because the climatic signals are so different from season to season and from place to place, these model results help to suggest a priority list indicating where paleoclimatologists might dig through the geologic record in order to uncover remains which might help to tell us the seasonal and regional distribution of past climates. As that field evidence becomes available, then not only will we have an opportunity for an explanation of a very intriguing, rapid and dramatic past climatic change, but we will have another opportunity to independently verify the fidelity of our climatic models.

INTRODUCTION

It has been suggested by a number of paleoclimatologists that formation of a low salinity meltwater layer in the North Atlantic Ocean following the Wisconsin glacial maximum could have caused a temporary, regional (at least) reversal of the longer term deglacial warming trend. For example, Mercer (1969), Johnson and McClure (1976) and Ruddiman and McIntyre (1981) suggested that such an influx of ice and meltwater may have been responsible for the Younger Dryas (11,000-10,000 yr. BP) cold interval, most evident in European proxy climatic records. Others (i.e. Johnson and Andrews, 1979) have suggested (in the context of ice age initiation) that one feedback mechanism whereby equatorward extension of the North Atlantic polar front could amplify the regional high latitude temperature response is through a reduction in the poleward transport of heat by ocean currents. At the same time, there is some evidence that the Younger Dryas interval was not merely a European event, but may have been more worldwide in scope (e.g., see Moore, 1981; Harvey, 1980; Rind et al., 1986). Broecker et al. (1985) suggest that such intervals may have been marked by fluctuations in the formation rate of North Atlantic Deep Water, which would affect thermohaline circulation worldwide. Mathematical climatic models applied to the Younger Dryas are one method of quantitatively investigating the kinds of suggestions mentioned above.

MODEL VERIFICATION: CLIMATIC OPTIMUM

One of the problems facing climatologists when evaluating climate model predictions on controversial issues such as CO_2 or other trace gas increases and their effects on climate is model verification. Since *direct* validation is only possible by monitoring the earth over time, most attempts to validate climate model sensitivity are based on finding "surrogate" forcing and response analogs such as the seasonal cycle. The ability of a model to produce the 15°C Northern Hemisphere or 5°C Southern Hemisphere seasonal cycle of surface temperature in response to seasonal solar forcing is indeed a powerful test of model validity, but the seasonal test certainly is not sufficient to validate a model's sensitivity to all conceivable radiative forcings, including greenhouse gas increases (e.g., see the discussion in Schneider and Londer, 1984). Clearly, other independent surrogate verification tests are needed. Another source of validation is paleoclimatology, because geological evidence indicates that the earth's climate history has included conditions very different from those at present.

For example, one of the most successful simulations to date was performed by Kutzbach and Guetter (1984), in which they used the NCAR Community Climate Model in an attempt to explain the warmest period in recent climatic history, the so-called "climatic optimum" that occurred between about 9 and 5 thousand years ago. It was a time when summertime northern continental temperatures were probably several degrees warmer than at present and monsoon rainfall was more intense in parts of Africa and Asia. Kutzbach and colleagues found that the "optimum" could be explained simply by the fact that the tilt of the earth's axis (its obliquity) was slightly greater than that at present, and, at the same time, the orbit was such that the earth was closer to the sun (i.e., perihelion) in June rather than in January as it is now. These variations do not make substantial changes in the annual amount of solar radiation received on earth, but do change substantially the difference between winter and summer heating periods. About 5-10% more solar heat in the Northern Hemisphere summer and a comparable amount less in winter occurred 9,000 years ago compared to today (Berger, 1978). Another way to view the forcing at 9000 yr. BP is to consider it a test of a model's sensitivity to periodic solar forcing at the annual frequency. The reason for this analogy is that as the earth goes around the sun in its elliptical orbit it sees an apparent periodic solar constant (to first order in eccentricity). The relative phase of this periodic forcing and that of the seasonal cycle can lead to constructive or destructive interference in the annual harmonic of the temperature field. The ability of a model to reproduce this effect quantitatively as a function of position is a test of its sensitivity to forcing at the annual frequency. Note that the latitudinal distribution of this periodic forcing is the same as the mean annual latitude distribution in strong contrast to the latitudinal distribution of the seasonal heating (North, Mengel and Short, 1983a, p. 246). This forcing at 9000 yr. BP was sufficient in Kutzbach's simulations to substantially alter mid-continental warming in summer months, which led to the enhanced monsoon rainfall and river runoff in the model. Approximately 2-4°C warming was obtained throughout most of central Eurasia and part of west central North America in Kutzbach's simulation. These results agree qualitatively with paleoclimatic evidence for these regions and support the explanation of enhanced seasonality for changes in the fossil record.

YOUNGER DRYAS: GISS AND NCAR GENERAL CIRCULATION MODELS

Another appropriate interval for investigation with climate models is the Younger Dryas. During the deglaciation following the last glacial maximum, conditions around the globe were rapidly approaching the warmer interglacial values that we have been experiencing for the last 10,000 years. However, approximately 11,000 years ago, a dramatic cooling occurred rapidly, and lasted for 500 to 1,000 years before the uninterrupted present interglacial finally set in. This large climatic

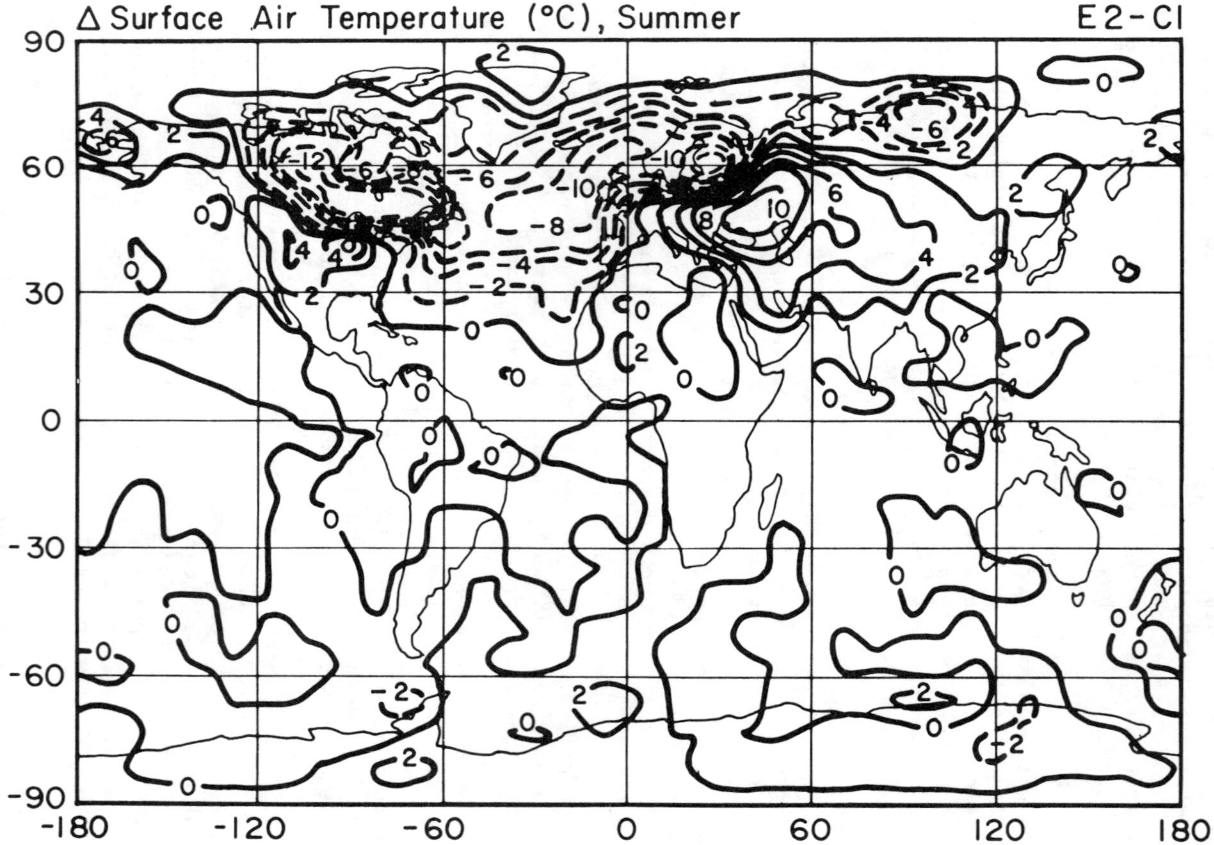

Figure 1. Changes in summer surface air temperature produced by the GISS GCM (Rind et al., 1986) when 11,000 yr BP orbital values and ice extent along with ice-age N. Atlantic sea surface temperatures were compared with modern conditions. Note large warmings in eastern Europe and western North America despite the presence of large ice sheets and cold North Atlantic temperatures.

excursion, the Younger Dryas, was most intense in Europe, especially northwest Europe and the British Isles. As mentioned earlier, various reasons for this oscillation have been proposed, primarily dealing with a fresh water influx into the North Atlantic. However, insolation 11,000 yr. BP was similar to values at 9,000 yr. BP and would, in the absence of other forcings, have suggested northern latitude summer warmth rather than cooling between 11,000-10,000 yr. BP.

To test the relative importance of various physical forcing mechanisms, a group at Goddard Institute for Space Studies and Lamont Doherty Geological Observatory (Rind et al., 1986) performed several experiments with their 8° x 10° resolution 3-dimensional General Circulation Model in which 11,000 yr. BP insolation, colder North Atlantic surface temperatures, and ice sheets were included. Compared with the present summer climate, their results (Fig. 1) produced a cooling in western Europe of some 10°C, a qualitatively expected result given the cold North Atlantic they imposed as boundary conditions in their model. In central Eurasia, a warming of some 4-10°C resulted, partially a result of enhanced seasonal insolation increase and partially ascribed to atmospheric subsidence at the edge of the ice sheet. The additional warming due to this subsidence, associated with the jet stream exit region, is explored in more detail in Rind (1986) and is debated by Schneider (1986) and Rind and Peteet (1986). Fig. 2 shows the GISS results for winter, in which the decreased solar radiation and the downstream cooling effects of a colder North Atlantic and glacial ice combine to result in very cold mid-continental temperatures.

Independently, Schneider and S. L. Thompson at NCAR performed general circulation model experiments (using the Chervin, 1986 seasonal version of the NCAR CCM) inspired by the Younger Dryas (Figs. 3 and 4). In these model experiments solar radiation was left at present values, but sea ice was imposed as a fixed boundary condition down to about 45° N latitude over the whole North Atlantic--not because Schneider and Thompson believed that to have been precisely a Younger Dryas reconstruction, but simply as a climate sensitivity experiment to show how a regional change in the character of the North Atlantic could cause a climatic response that could be studied differentially from place to place and from winter to summer. Note in these results that during the summer (Fig. 3), the climatic effects of this noticeably altered North Atlantic are primarily felt near the coastline, hardly spreading inland. In the summer case the solar heating of the land is, of course, relatively intense and even though the oceans are extremely cold the primary mechanism controlling the mid-continental summertime temperatures is summertime levels of solar radiation. Thus the oceans summertime influence is concentrated primarily in coastal areas. In contrast, for present-day conditions in winter when the seasonal solar input is much less, the normally moderate winter ocean surface temperatures maintain a relatively equable downstream climate, particularly in the western parts of mid-latitude continents. Thus, if the Younger Dryas really were characterized by a substantial increase of sea ice, one would expect downstream winter temperatures to be very severely depressed, as indeed they are in Fig. 4 for Europe and even northeastern United States and Canada.

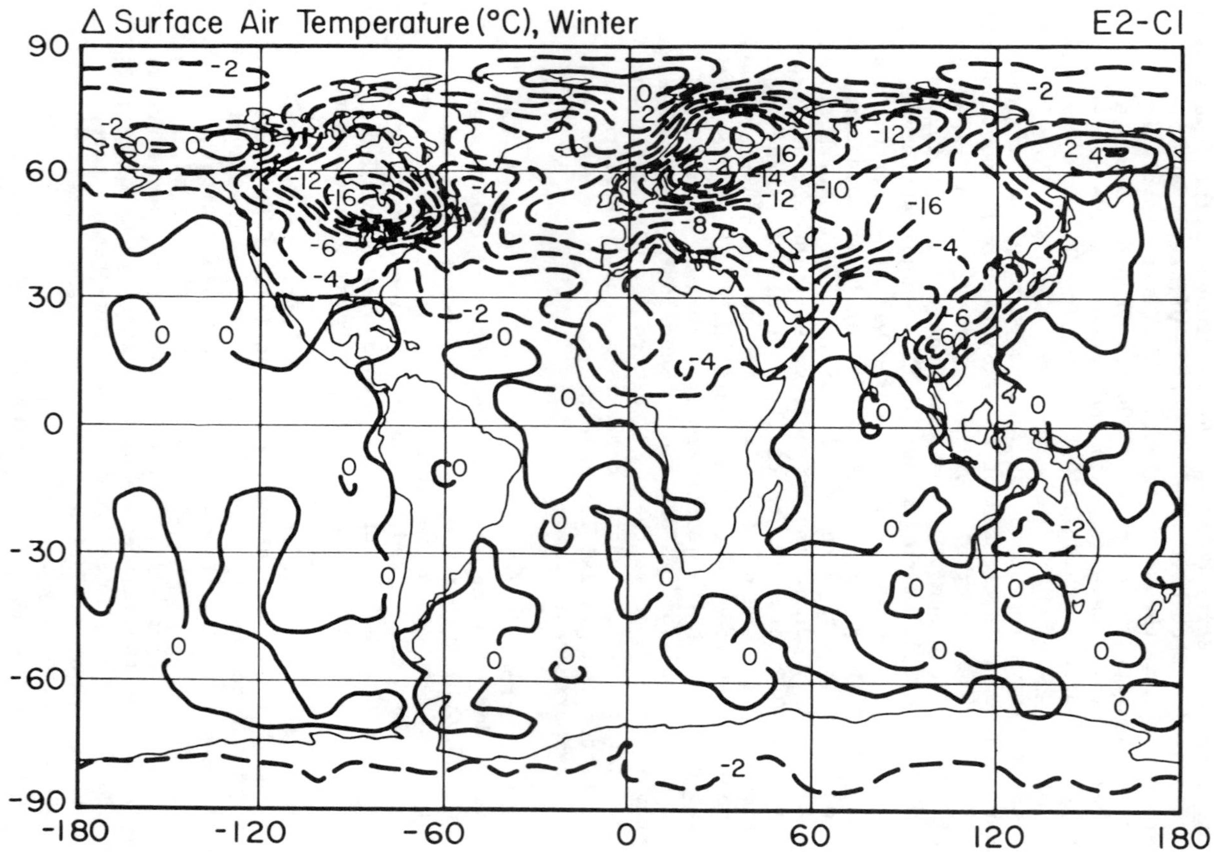

Figure 2. Changes in winter surface air temperatures produced by the GISS GCM when 11,000 yr. BP orbital values and ice extent estimates along with ice-age North Atlantic sea surface temperatures were compared with modern conditions. Note extreme cooling in northeastern Europe.

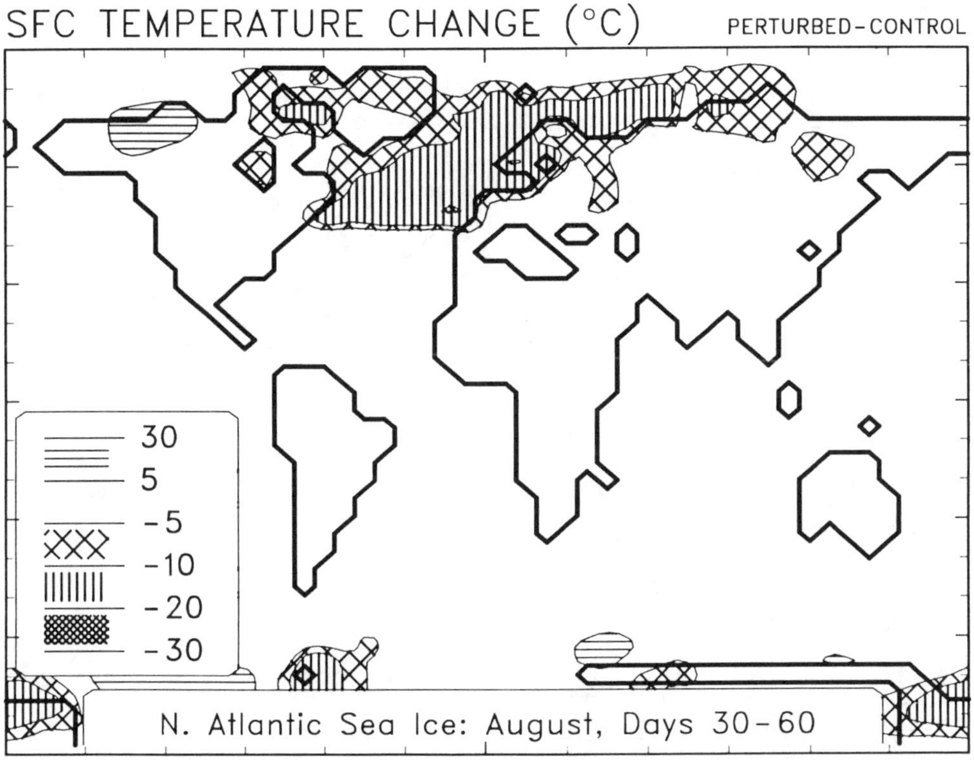

Figure 3. Surface temperature change in the NCAR Community Climate Model for an imposed sea ice cover in the North Atlantic Ocean down to about 45°N for August (Schneider and Thompson, 1986, unpublished).

To help identify the relative importance of various physical forcing factors, the GISS-Lamont group also performed sensitivity studies in which they used colder North Atlantic sea surface temperatures along with today's boundary conditions. Fig. 5, though qualitatively similar to the *average* of the NCAR extreme season results on Figs. 3 and 4, is not quantitatively comparable for several reasons. First, the GISS North Atlantic input surface temperatures are different from the NCAR experiments. Second, the GISS results are for surface *air* temperatures whereas the NCAR results are for surface land temperatures. Clearly, further experimentation and comparisons are needed. Knowing the sensitivity of surface temperatures to external forcings like increased solar radiation is, of course, a long-standing, primary goal of climatic modeling. Therefore, it is important to try to extract and delineate any differences that might exist among models. To aid in this process, it is useful to turn to other models, including some of differing structure that are found in the hierarchy of climatic models available in the literature.

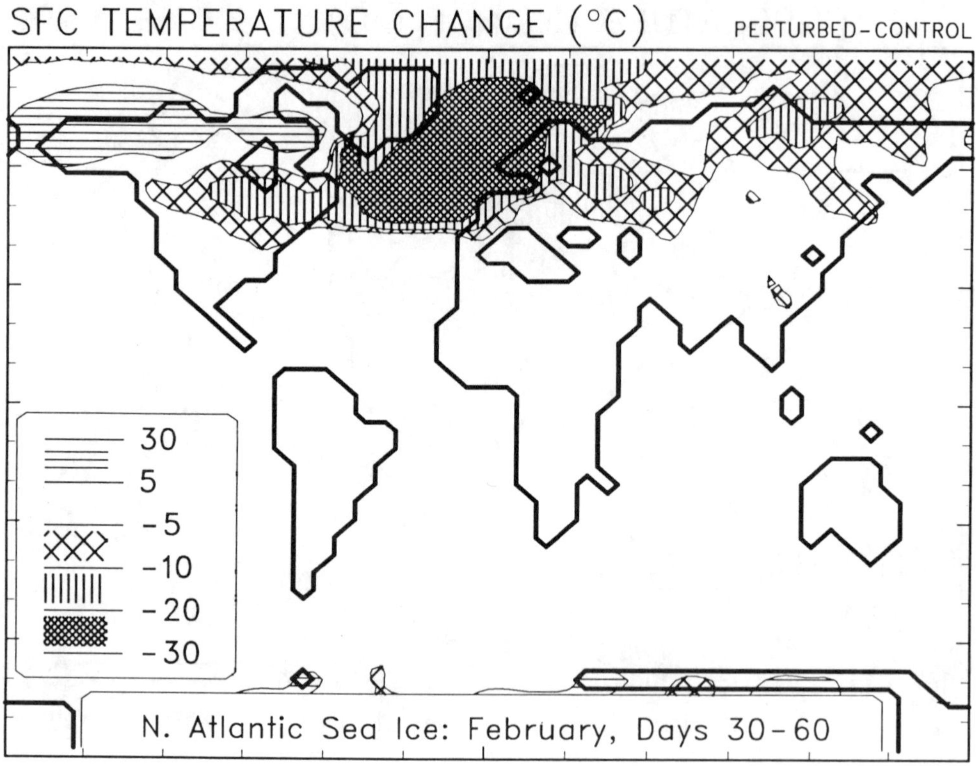

Figure 4. Same as Fig. 3 except for February

PROXY DATA, ENERGY BALANCE AND GENERAL CIRCULATION MODEL INTERCOMPARISONS

One example is the two-dimensional energy balance model of North, Mengel and Short (1983b). Although it does not explicitly compute windfields, it parameterizes their effects on horizontal heat flow in the atmosphere through two-dimensional eddy diffusion. Despite the simplicity, the model does remarkably well in separating the amplitude and phase of the first two harmonics of the seasonal cycle of surface temperature using a spectral model whose horizontal resolution is equivalent to 1500km. To see the effects of the geographically constrained thermodynamic response to the changes in sea ice and perihelion phase, the Younger Dryas case is a good test case. Fig. 6 shows the summer warming from a single change: 11,000 yr. BP insolation values. In Eurasia, the +4°C warming is close to the NCAR CCM results of Kutzbach and Guetter (see Fig. 7). Fig. 8 shows the marginally reduced temperatures for the January, 11,000 yr. BP insolation values. However, when sea ice is imposed in the North Atlantic down to 45°N and 11,000 yr. BP insolation is used, the results also qualitatively agree with the GCMs: cooling over the oceans and western Europe in summer (Fig. 9) with +3°C warming in Eurasia, and with intense cooling over the entire North

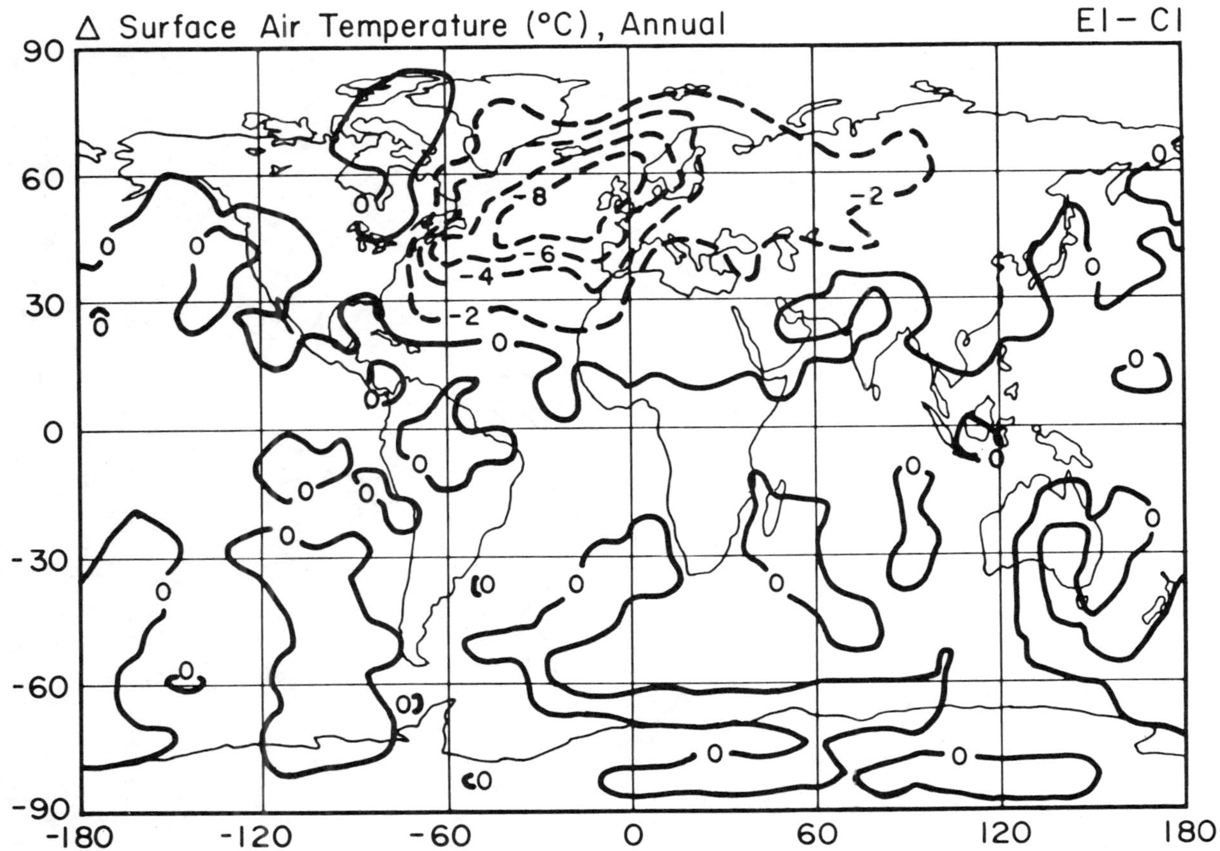

Figure 5. Changes in annual surface air temperature produced by the GISS GCM when ice-age North Atlantic sea surface temperatures replaced modern values.

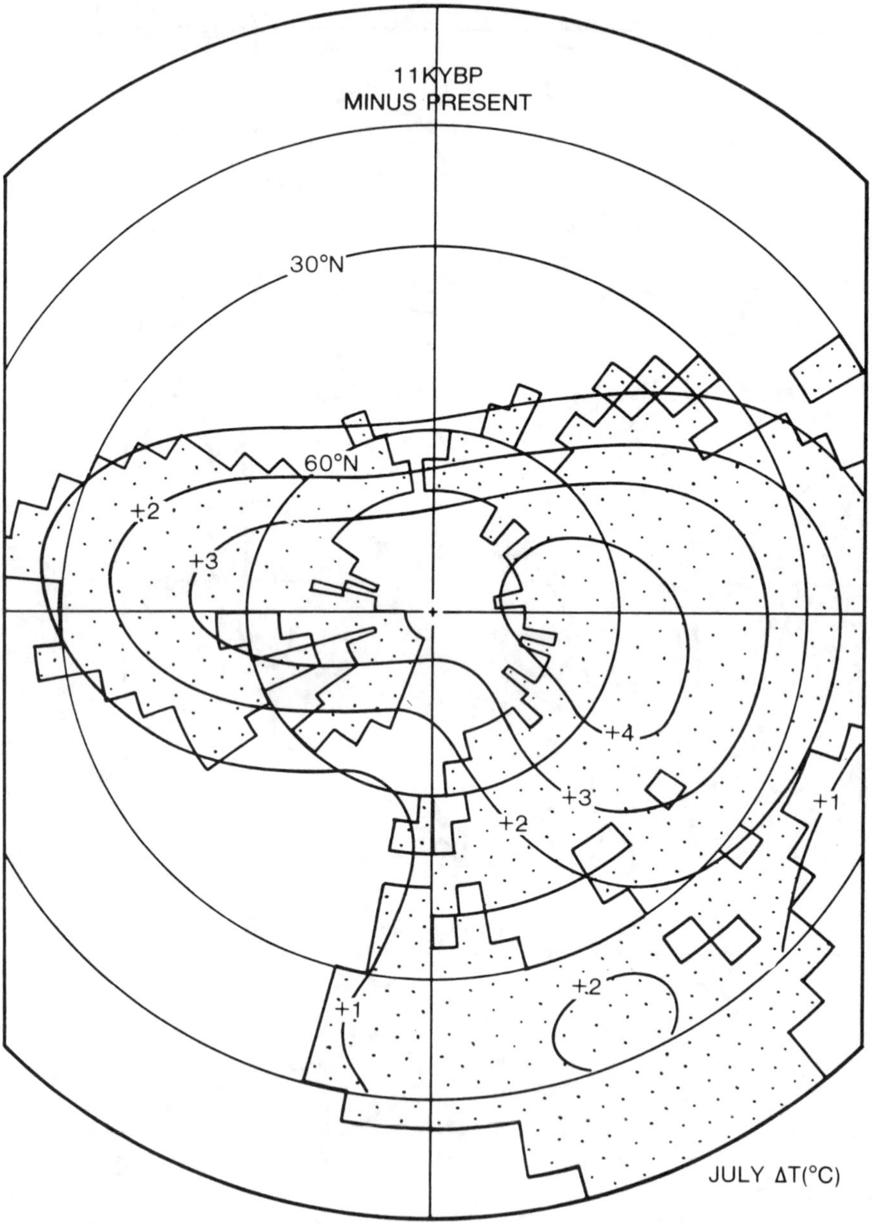

Figure 6. Simulation of the change-from-present July average climate (month following Northern summer solstice) with the linear two-dimensional energy balance model of North, Mengel and Short (1983b) with orbital elements as they were 11 ky BP. In this simulation no changes were made in the sea surface properties of the North Atlantic.

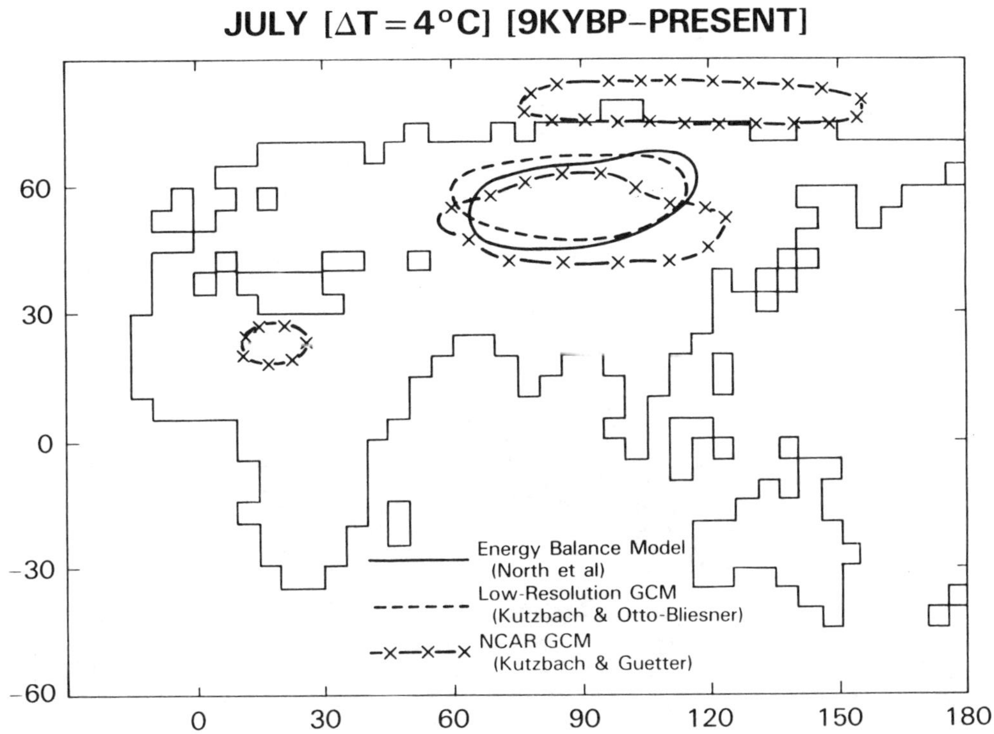

Figure 7. A comparison of energy balance and general circulation model (GCM) responses to increased summer insolation (see Crowley et al., 1986). Time interval is about 10,000 years ago. The 4°C contours indicate the difference between the model response for 10,000 years ago and the present (earlier time period warmer). The energy balance model is from North et al. (1983b); the low-resolution model and the NCAR models are from Kutzbach and Otto-Bliesner (1982) and Kutzbach and Guetter (1984), respectively.

Atlantic sector in January (Fig. 10). The results, especially for January, appear more symmetric (Fig. 10) around the North Atlantic basin compared, for example, to the Schneider/Thompson results (Fig. 4) for a number of possible reasons. The eddy diffusion parameterization in the 2-D EBM tends to smooth out temperature fields without the explicit advective effects of westerly winds (the latter are simulated only in the GCMs, of course). Second, in the NCAR results, for example, in addition to the advective effects of the westerlies there may be an apparent atmospheric long-wave response to the imposed North Atlantic sea surface forcing in which a ridge of high pressure builds over Alaska and a trough develops over north-eastern North America. This actually causes warming in the Arctic sector of Alaska. It is simply too soon to determine if this shift in the long waves seen on Fig. 4 is statistically significant in the model (let alone a realistic feature). More runs are needed to obtain adequate model-generated data necessary for statistical

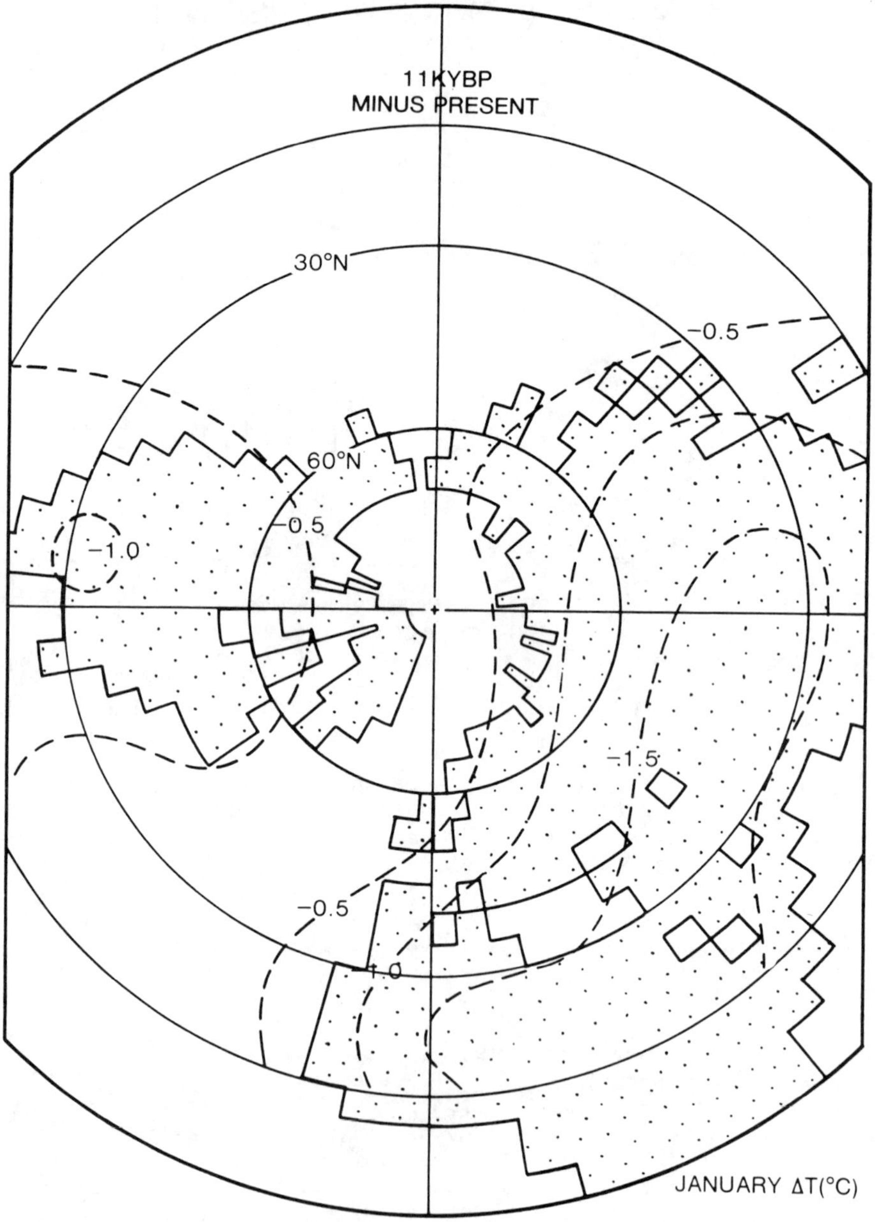

Figure 8. Same as Fig. 6 except that the month is January.

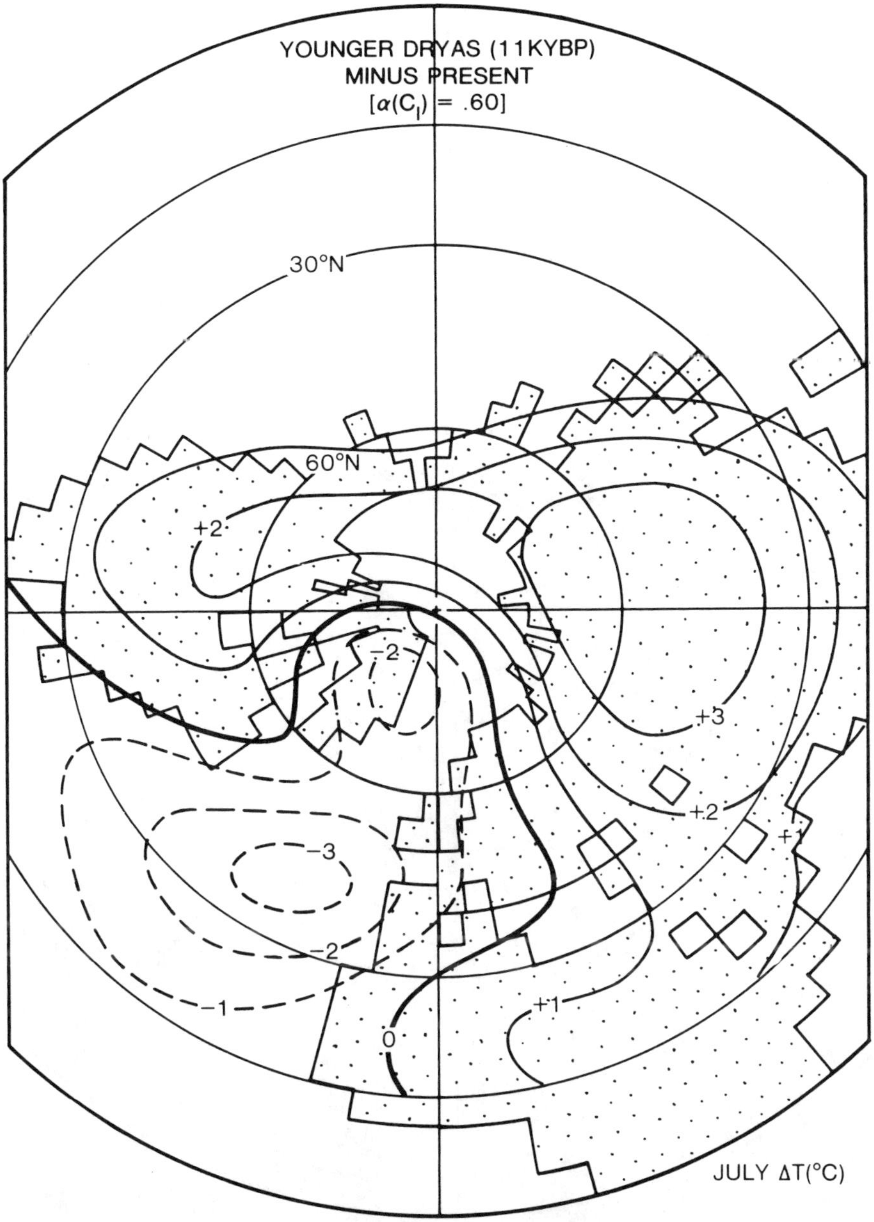

Figure 9. Same as Fig. 6 except that perennial sea ice is assumed to cover the North Atlantic poleward of 45° latitude. The sea ice has albedo 0.60 and it also has the effective heat capacity characteristic of perennial sea ice (about 1/7 that of the oceanic mixed layer). Note that in this figure and the next the changes in mid-Asia are hardly affected by the conditions in the North Atlantic.

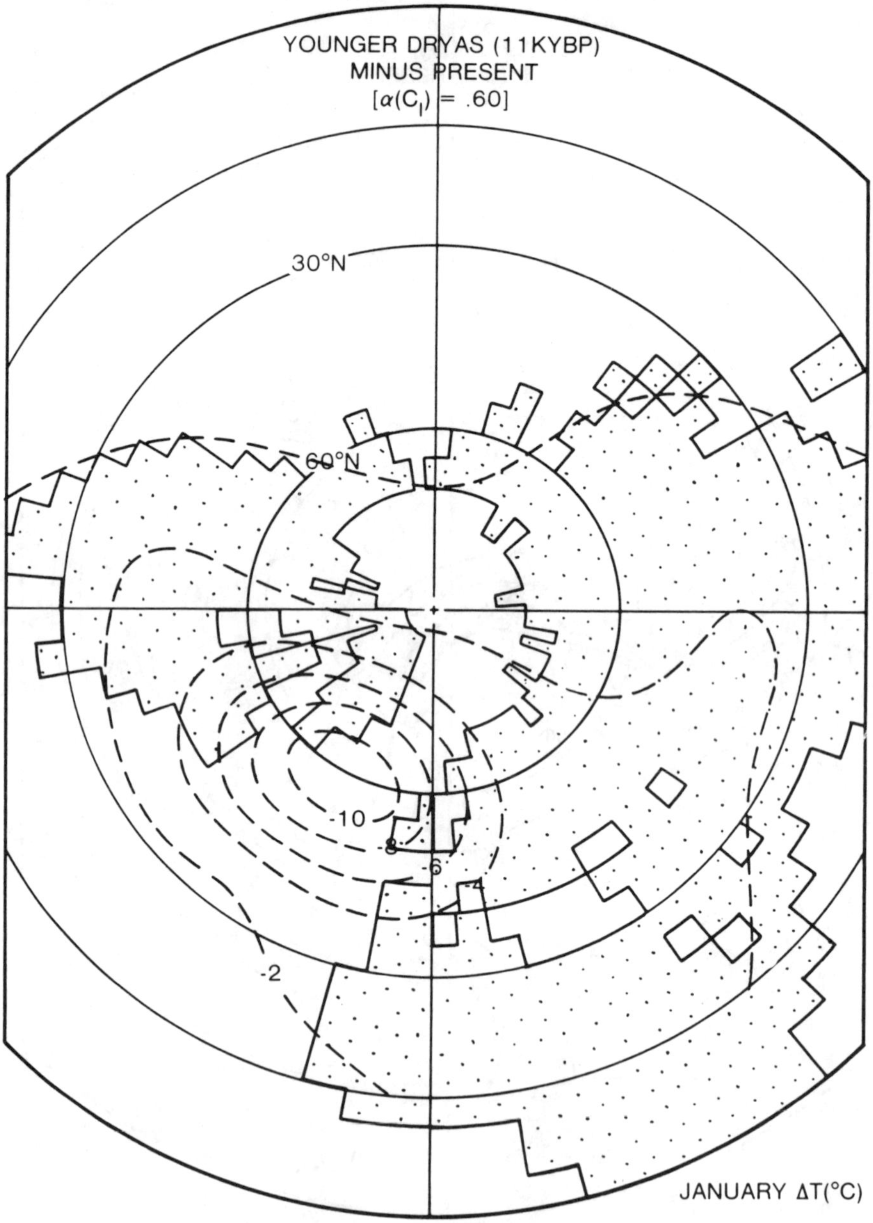

Figure 10. Same as Fig. 9 except for January.

significance tests. In any case, the GCM results do suggest that world-wide "teleconnections" could be associated with cold North Atlantic waters in the Younger Dryas. Independently, Adem (1985) performed an experiment with a geographically resolved energy balance model putting in the solar insolation values of 10,000 years ago, and obtained the Eurasian summer warming of about 4°C, comparable to North's results on Fig. 6.

Various model calculations show that these mid-continental summer temperatures should have been some 2-10°C warmer 9,000 to 11,000 years ago compared to today. As models differ in their quantitative predictions of temperature change, paleoclimatic data may be useful for the first time to help discriminate between models and model experiments. An extensive data set, including annual mean and seasonal amplitudes e.g., from pollen (Pons et al., 1985), beetles (Coope, 1985), and glacial data (Sissons and Sutherland, 1976) over Eurasia and central North America are needed to help resolve such questions of model sensitivity. Moreover, since the seasonal and regional distribution of temperature response is somewhat different in all the models, having seasonally and geographically resolved empirical and model-generated data may also help both to explain the Younger Dryas climatically and to aid substantially in our confidence in climate model sensitivity in general. One more result from yet another kind of model can help to elaborate further on the latter point.

OCEAN SECTOR CALCULATIONS

The possibility that meltwater could either increase the amount of sea ice or decrease the thickness of the North Atlantic mixed layer has been investigated using an energy balance climate model which is zonally averaged except for retention of land sea resolution, and in which sea ice and land snow cover fractions and thicknesses are computed in terms of explicit model physics (model adapted from Harvey, 1986). In particular, zonally averaged sea ice fractions are computed using the zero layer model of Semtner (1976) and the ice fraction and thickness evolution equations of Hibler and Ackley (1983) with prescribed, equatorward ice advection velocities. Lateral melting and ice free oceanic mixed layer temperatures are computed in a manner similar to that of Parkinson and Washington (1979).

The model is first run to equilibrium for present day conditions. The effect of a deglacial meltwater lens is then represented in two ways: (1) Sea ice is extended southward in the Northern Hemisphere into mid-latitudes; and (2) both the mean and the amplitude of the prescribed seasonal mixed layer depth variation are reduced poleward of 40°N. To investigate the relative importance of these changes, the model is run to a new equilibrium with these changes. Also, the dependence of the model sensitivity to these changes, both in the zones where they occur and globally, is investigated as various sea ice and model processes (including oceanic heat transport) are suppressed or modified. In particular, model sensitivity is compared for cases in which the oceanic horizontal heat diffusion coefficient is independent of the zonal ice fraction, and scaled by the zonal ice-free fraction. These results will

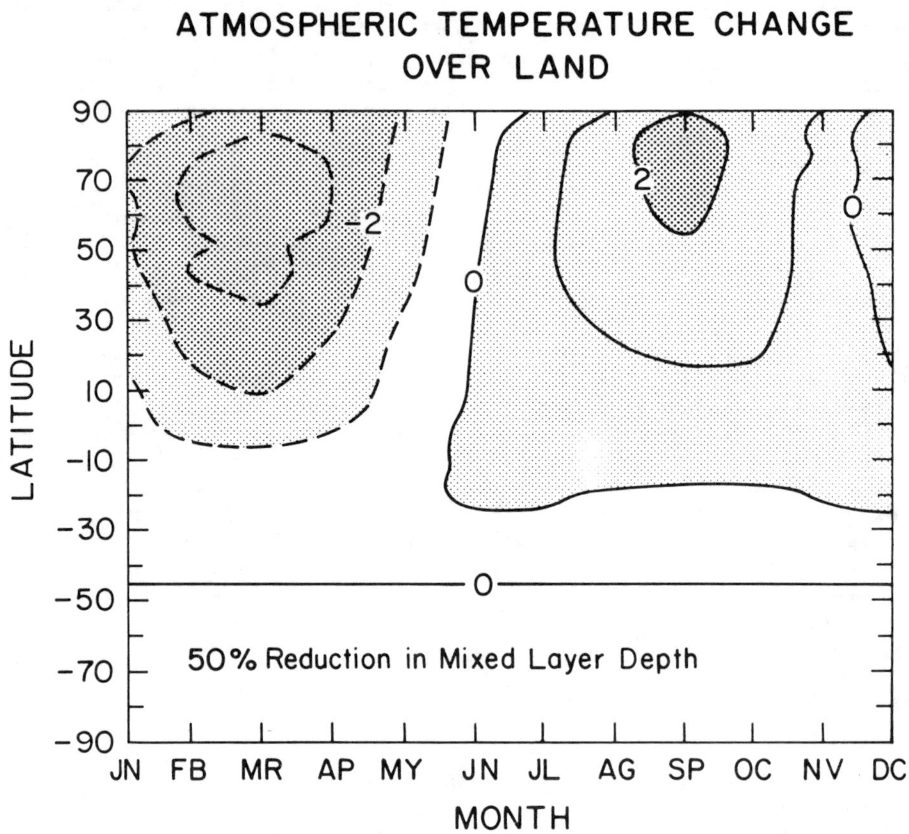

Figure 11. Latitude and seasonal difference between perturbed and control runs for atmospheric temperature over land in the energy balance model of Harvey (1986) from an imposed reduction in mixed layer depth by 50%, poleward of 40°N. (From Harvey and Schneider 1987.)

be reported elsewhere (Harvey and Schneider, 1987). Some preliminary results are reported here. For example, the reduction of the oceanic mixed layer depths by half leads to a few degrees C reduction in atmospheric surface temperatures over land in winter and spring, and a slightly smaller temperature *increase* in summer and fall (Fig. 11). This occurs because a thinner mixed layer has less thermal inertia and thus the seasonal temperature range increases. However, an increase in sea ice in the 45-65°N zone (70% cover) leads to a very different climatic signal than that from the reduced mixed layer case. Fig. 12 shows that increased mid-latitude sea ice leads to very substantial cooling over land in winter and a much lesser cooling in summer - but cooling, nonetheless. (There is, of course, an even more substantial cooling over the ice-covered oceans in both seasons - not shown here.) These results

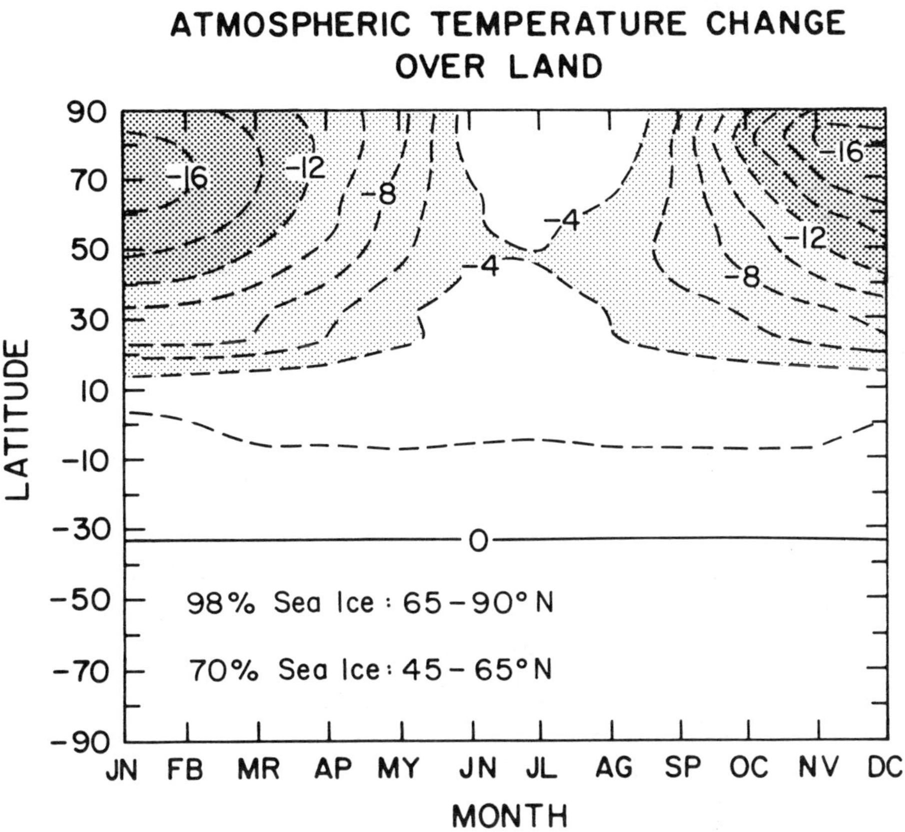

Figure 12. Same as for Fig. 11 except for a case in which sea ice is held fixed at 98 percent poleward of 65°N and at 70% between 45°N and 65°N. Note that the character of the downstream (i.e., over land) seasonal signals is qualitatively different for the two forcings on Figs. 11 and 12.

are zonally averaged, and proportionately larger regional responses, especially in coastal Western Europe, would be anticipated to occur in the three-dimensional GCMs which indeed were seen in Figs. 1-4.

OUTLOOK FOR MODELER - FIELD WORKER COLLABORATIONS

The Younger Dryas appears to be one of the most promising areas for future research in the combined fields of climate modeling and paleoclimatic data analysis. Indeed, we believe that the time is ripe for considerably greater interactions among modelers and those who reconstruct the climatic records. Participants from both subdisciplines would benefit greatly from advanced communication with the other groups. For example, modelers will be aided in model verification and data collectors might be advised by modelers where to look for certain kinds of proxy data. It is our hope that the examples described here will help to stimulate the kinds of interdisciplinary cooperation required not only to solve problems common to various disciplines, but also to help society by providing surrogate verification of future climate predictions. Such verification efforts will probably prove necessary before any forthright policy response to present predictions is likely.

ACKNOWLEDGMENTS

The authors wish to thank L.D. Danny Harvey, John G. Mengel, David Rind, David A. Short, and Starley L. Thompson, for their helpful comments on early drafts of the manuscript.

The National Center for Atmospheric Research is sponsored by the National Science Foundation.

REFERENCES

Adem, J., 1985. The climate of ten thousand years ago: A numerical simulation, *Geof. Int.*, 24: 383-407.

Berger, A., 1978. Long-term variations of daily insolation and Quaternary climate changes, *J. Atmos. Sci.*, 35: 2362.

Broecker, W.S., Peteet, D.M. and Rind, D., 1985. Does the ocean-atmosphere system have more than one stable mode of operation?, *Nature*, 315: 21-25.

Chervin, R.M., 1986. Interannual Variability and Seasonal Climate Predictability, *J. Atmos. Sci.*, 43: 233-251.

Coope, G.R., 1985. Evidence from insect fossils for sudden and intense climatic changes during the last 45,000 years. Paper presented at Conference on Abrupt Climatic Change, Biviers, October 16-22, 1985.

Crowley, T.J., Short, D.A., Mengel, J.G. and North, G.R., 1986. Role of seasonality in the evolution of climate during the last 100 million years, *Science*, 231: 579-584.

Harvey, L.D.D., 1980. Solar variability as a contributing factor to Holocene climatic change, *Prog. Phys. Geog.*, 4: 487-530.

Harvey, L.D.D., 1986. A three-level energy balance climate model with topography, explicit radiative transfer, and explicit sea ice and snow mass budgets, NCAR Cooperative Thesis No. 97, University of Toronto and the National Center for Atmospheric Research.

Harvey, L D.D. and Schneider, S.H., 1987. Climate modeling sensitivity studies applied to the Younger Dryas: Energy balance model results (to be submitted to *Quat. Res.*).

Hibler, W.D. III and Ackley, S.F., 1983. Numerical simulation of the Wedell Sea pack ice, *J. Geophys. Res.*, 88: 2873-2887.

Johnson, R.G. and Andrews, J.T., 1979. Rapid ice sheet growth and initiation of last glaciation, *Quat. Res.*, 12: 119-124.

Johnson, R.G. and McClure, B.T., 1976. A model for Northern Hemisphere continental ice sheet variation, *Quat. Res.*, 10: 319-353.

Kutzbach, J.E. and Otto-Bliesner, B.L., 1982. The sensitivity of the African-Asian monsoonal climate to orbital parameter changes for 9000-yr. BP in a low-resolution general circulation model, *J. Atmos. Sci.*, 39: 1177-1188.

Kutzbach, J.E. and Guetter, P.J., 1984. The sensitivity of monsoon climates to orbital parameterization changes for 9000 years BP: Experiments with the NCAR General Circulation Model. In: *Milankovitch and Climate, Part 2*, A.L. Berger et al. (eds.), D. Reidel, Dordrecht, 801-820.

Mercer, J.H., 1969. The Alleröd oscillation: a European anomaly?, *Arct. Alp. Res.*, 1: 227-234.

Moore, P.D., 1981. Late glacial climatic changes, *Nature*, 291: 380.

North, G.R., Mengel, J.G. and Short, D.A., 1983a. Climatic response to a time-varying solar constant. In: *Weather and Climate Responses to Solar Variation*, B.M. McCormac (ed.), Colorado Associated University Press, Boulder, 243-255.

North, G.R., Mengel, J.G. and Short, D.A., 1983b. Simple energy balance model resolving the seasons and the continents: Application to the astronomical theory of the Ice Ages, *J. Geophys. Res.*, 88: 6576-6586.

Parkinson, C. and Washington, W.M., 1979. A large-scale numerical model of sea ice, *J. Geophys. Res.*, 84: 311-337.

Pons, A., de Beaulieu, J.-L., Guiot, J. and Reille, M., 1985. The Younger Dryas as a model of abrupt climatic change. Paper presented at Conference on Abrupt Climatic Change, Biviers, October 16-22, 1985.

Rind, D., 1987. Components of the Ice Age Circulation. *J. Geophys. Res.*, in press.

Rind, D. and Peteet, D., 1986. Comments on S. H. Schneider's Editorial "Can Modeling of the Ancient Past Verify Prediction of Future Climates?", *Climatic Change*, 9: 357-360.

Rind, D., Peteet, D., Broecker, W.S., McIntyre, A. and Ruddiman, W.F., 1986. Impact of cold North Atlantic sea surface temperatures on climate: Implications for the Younger Dryas Cooling (11-10k). *Climate Dynamics*, 1: 3-33.

Ruddiman, W.F. and McIntyre, A., 1981. The North Atlantic ocean during the last deglaciation, *Paleogeogr., Paleoclim., Paleocol.*, 35: 145-214.

Schneider, S.H., 1986. Response to Rind and Peteet Comment, *Climatic Change*, 9: 361-362.

Schneider, S.H. and Londer, R., 1984. *The Coevolution of Climate and Life*, Sierra Club Books, San Francisco, 563 pp.

Semtner, A., 1976. A model for the thermodynamic growth of sea ice in numerical investigations of climate, *J. Phys. Oceanogr.*, 6: 379-398.

Sissons, J.B. and Sutherland, D.G., 1976. Climatic inferences from former glaciers in the southeast Grampian Highlands, Scotland, *J. Glaciol.*, 17: 325-346.

INDEX